GALILEO

GALILEO

J.L. HEILBRON

OXFORD
UNIVERSITY PRESS

OXFORD
UNIVERSITY PRESS

Great Clarendon Street, Oxford OX2 6DP

Oxford University Press is a department of the University of Oxford.
It furthers the University's objective of excellence in research, scholarship,
and education by publishing worldwide in

Oxford New York

Auckland Cape Town Dar es Salaam Hong Kong Karachi
Kuala Lumpur Madrid Melbourne Mexico City Nairobi
New Delhi Shanghai Taipei Toronto

With offices in

Argentina Austria Brazil Chile Czech Republic France Greece
Guatemala Hungary Italy Japan Poland Portugal Singapore
South Korea Switzerland Thailand Turkey Ukraine Vietnam

Oxford is a registered trade mark of Oxford University Press
in the UK and in certain other countries

Published in the United States
by Oxford University Press Inc., New York

British Library Cataloguing in Publication Data
Data available

Library of Congress Cataloging in Publication Data
Library of Congress Control Number: 2010933141

Typeset by SPI Publisher Services, Pondicherry, India
Printed in Great Britain
on acid-free paper by
Clays Ltd, St Ives plc

ISBN 978–0–19–958352–2

PREFACE

It will be useful and perhaps reassuring to state that this is not the biography of a mathematician. To be sure, Galileo enjoyed such epithets as "divine mathematician" and "Tuscan Archimedes," and he spent the first half of his career, from 1589 to 1610, as a professor of mathematics. Moreover, historians rank him as the first to introduce effective quantification into physics. For all that, he was no more (or less!) a mathematician than he was a musician, artist, writer, philosopher, or gadgeteer. His last disciple and first biographer, Vincenzo Viviani, boasted that his master could compete with the best lutanists in Tuscany, advise painters and poets on matters of artistic taste, and recite vast stretches of Petrarch, Dante, and Ariosto by heart. But his great strength, Galileo said when negotiating for a post at the Medici court in 1610, was philosophy, on which he had spent more years of study than he had months on mathematics.[1]

Galileo's intense study of the moon, sun, and planets in 1609/10, when he was the only man on earth anatomizing the man in the moon and the satellites of Jupiter, called upon his skills as an observer and draughtsman, his dexterity as a craftsman, and his knowledge of perspective and foreshadowing, and not at all on his ability as a mathematician. The hurried little masterpiece, *Sidereus nuncius* (1610), which described his unprecedented observations and astonishing deductions, quickly compelled assent. Years of reading the poets and experimenting with literary forms enabled him to write clearly and plausibly about the most implausible things. If ever a discoverer was perfectly prepared to make and exploit his discovery, it was the dexterous humanist Galileo aiming his first telescope at the sky.

Perhaps the best single-word descriptor for Galileo is "critic." He was a true connoisseur of the arts and sciences, able, says Viviani, to talk intelligently, and with apt quotations, on virtually every respectable subject with all sorts of people. As a connoisseur, he argued the excellence of painting over sculpture, monody over counterpoint, one version of Dante's Hell over another, and, in mathematics, Archimedes over everybody. Galileo was the embodiment of baroque *buon gusto* in matters of art and science. That, of course, did not mean that he was judicious or well mannered. He often deviated from good taste in criticizing others. This indulgence, coupled with an inventive wit and the adolescent pleasure, which he never outgrew, of scoring off people, made him powerful enemies even among those who respected his gifts.

Galileo would have done well in any of several professions. He might have chosen as his brother Michelangelo did, and followed their father's path as a musician; or do as he said he would have preferred had he had a free choice, and become a painter.[2] He could have been a man of letters, the confidential secretary to a duke or cardinal, or better, a grand duke or pope. About the only profession for which he was unsuited was the one for which his father intended him: medicine. Galileo pitched on mathematics, for which he had a knack, as an escape from doctoring. The remedy worked. He had the good fortune to invent a clever proposition that enabled him, at the age of 20, to demonstrate several theorems in the style of Archimedes that impressed a few mathematicians in Italy. These theorems represented the high point of Galileo's mathematics. He did not publish them in their time, but fifty years later, as an appendix to his last and most technical book, *Discourses on two new sciences* (1637). He never made much use of algebra, disliked complicated calculations of the sort that delighted Kepler, and avoided geometrical questions more difficult than those with which he debuted.

As a young man, carefree despite the family's straitened finances, esteemed by friends from among the best families in Florence, clever, witty, sociable, versed in literature and music, and with a gift for geometry and a taste for gambling, Galileo did not resemble much the troubled inventor of modern science familiar from the usual histories. His friends would not have expected him to become the sworn enemy of Aristotle, the champion of Copernicus, the standard-bearer of mathematics, the *bête noire* of the Jesuits, or the best-known of all martyrs to academic freedom. Galileo would have become none of these things had he not had to work for a living.

This book appears on the 400th anniversary of Galileo's announcement of the riches his telescope revealed in the heavens. This was an event of world-historical importance. Another, coincident anniversary, the centennial of the completion of the national edition of Galileo's correspondence, manuscripts, and printed works, marks an accomplishment of equal importance in the smaller world of Galileo scholars. For a hundred years this inspired edition by Antonio Favaro and the specialized studies he spun from it have guided the investigations of biographers of Galileo and historians of the Scientific Revolution. The number of directly relevant useful publications now extends to many thousands. The last decade alone has been enlivened by the appearance of many serious book-length biographies and monographs. Why, apart from centennial obeisance, burden the world with another? Is there anything fresh to say?

Yes. Galileo's biographers tend to rush their gladiator prematurely into an imaginary arena filled with pig-headed philosophers and fire-spitting priests. He did spend time arguing with and suffering from such people. But Galileo the gladiator and martyr of science began as Galileo the patrician humanist. I hope to have introduced something fresh by locating Galileo more firmly among Florentine cultural institutions than others have done. That makes space for character development. Galileo underwent a sort of epiphany under the impetus of the telescopic discoveries he made at the age of 45. He had published very little, and nothing of importance, up to that time. He had many good ideas, but held them back, partly from a feeling of financial insecurity and partly from a cultivated circumspection. He saw no reason to burden the world with scattered results, half-treatises, imperfect theories, or unproved assertions.

When he had armed himself with the telescope, however, he declared all he knew and more. To the surprise of his colleagues and against their advice, he attacked philosophers, theologians, and mathematicians, taunted the Jesuits, jousted with everyone who contested his priority or his opinions. He became a knight errant, quixotic and fearless, like one of the paladins in his favorite poem, Ariosto's *Orlando furioso*. This change in behavior, which won him a continually lengthening list of enemies, made his disastrous collision with a pope who for many years had been his friend and admirer intelligible and even inevitable. Restoring to Galileo his youth in Pisa and Florence, his maturation in Padua, and his megalomaniacal middle age in the Medici court not only gives his history what he might have called *momento* but also helps to fix his role as hero of the Scientific Revolution. He was a great man in the sense that he changed the world in a way others could not, not by inventing telescopic astronomy or finding a few principles of motion, but by bringing in his special idiom some fundamental problems in the culture of his time so crisply into conflict that they could not be avoided or resolved.

Galileo lived for 78 years, many of them in the eye of a storm. He had friends, enemies, and correspondents of all sorts: mathematicians, philosophers, literary people, bureaucrats, princes, cardinals, and characters from the heroic poems he knew almost by heart. The minor actors easily slip from the memory. A genealogical table of the Galilei and a glossary of most people mentioned in the text apart from such household names as Einstein and God, but including characters from Ariosto and Tasso, follow Chapter VIII.

Galileo's mathematics seldom goes beyond plane geometry and the rules of proportion, and most of his published writings on physics and astronomy remained at the level of popularized science in his own day. Neither his geometry nor his science should present a technical challenge to today's educated lay person. On the contrary: the main obstacle to overcome in understanding Galileo's work is thinking that he was engaged in the same enterprise as modern physicists. As will appear, many of his apparent errors are mathematical jokes, rhetorical exaggerations, or wishful thinking. Failure to appreciate that his genre was not ours has prompted many unnecessary arguments and misunderstandings, which it will be a pleasure not to review here. What criticism is offered of his work stays within Galileo's own terms of reference.

It is a great pleasure to thank colleagues for their help and support, not only in customary acknowledgment of debt, but also in gratitude that, despite fashionable widespread cynicism about the academy, generosity and encouragement abound in some corners of the Republic of Letters. Among these generous Republicans are Paolo Galluzzi, Massimo Bucciantini, Ernan McMullin, Nick Jardine, Maurice Finocchiaro, Jim Bennett, Sven Dupré, Peter Watson, Louise Clubb, Mario Biagioli, Mike Shank, José Ferreirós, and Jed Buchwald. Sunspots provided occasion for supererogatory expressions of solidarity: Franz Daxecker very kindly supplied his books and expertise about Christoph Scheiner, and Eileen Reeves and Albert Van Helden with equal kindness sent the proofs of their new and definitive translations of the main documents in the squabble between Scheiner and Galileo. My thanks also go to keepers of the treasure houses of the Republic, especially the staffs of the Upper Reading Room of the Bodleian Library, the Taylorian Library, and the Museum of the History of Science, all in Oxford, and of the Museo Galileo (Istituto e Museo di Storia della Scienza) in Florence.

Publishing books is as essential to the Republic of Letters as writing them. Latha Menon, who commissioned this book, and her colleagues at Oxford University Press UK, Emma Marchant and Claire Thompson, are exemplary intermediaries between script and print. Correcting errors before printing them is another necessary operation. Alison Browning, Marita Hübner, and Cameron Laux policed the text vigilantly. The Republic is not self-sustaining and would be a poorer place if it were. For the photographs reproduced in Plate 16 I am indebted to Stephen Markeson. For help in keeping in touch with the wider society I thank, and apologize to, everyone who has listened to my nascent notions about Galileo, especially my wife and sounding board, Alison Browning, my advisor on Venetian courtesans, Wanda Case-Goody, and the Friday night regulars at the Rose & Crown in the Oxfordshire village of Shilton.

Galileo by Francesco Villamena, first published in Galileo, *Istoria e dimostrazioni intorno alle macchie solari* (1613), then in the *Assayer* (1623), and again, in the form used here, in Lorenzo Crasso, *Elogii* (1666).

CONTENTS

PICTURE ACKNOWLEDGEMENTS

Plates

11. Federico Cesi by Pietro Facchetti, 1612 © Accademia Nazionale dei Lincei, Rome/Istituto e Museo di Storia della Scienza, Florence; Virginio Cesarini by Anthony van Dyck, early 1620s © The State Hermitage Museum, St. Petersburg/RIA Novosti/TopFoto

12. "Allegory of the Active and Triumphal Church of the Dominican Order" by Andrea Bonaiuti (Andrea da Firenze), 1365–67, detail © Basilica of Santa Maria Novella, Florence/2010 Scala, Florence/Fondo Edifici di Culto, Min. dell'Interno

13. "Divine Providence" by Pietro da Cortona, 1630s, detail © Galleria Nazionale d'Arte Antica, Palazzo Barberini, Rome/Araldo de Luca/Corbis

14. Thomas Campanella by Francesco Cozza, date unknown © Bibliothèque Municipale de Lille (cod. 690, c. 349r); Christoph Scheiner by Christopher Thomas Scheffler, 18th century © Stadtmuseum, Ingolstadt

15. "Glorification of illustrious Tuscans" by Cecco Bravo, 1636, detail © Casa Buonarroti, Florence/2010 Scala, Florence

16. Façade of the Palazzo dei Cartelloni, Via Sant' Antonio 11, Florence, detail, and statue of Galileo in the Uffizi Loggia, Florence, by Aristodemo Costolli; both photos © Stephen Markeson, 2009

I

⸻ ∞ ⸻

A Florentine Education

1.1 UPBRINGING

On the façade of a house near the central train station in Florence there is an immense inscription in stone setting forth Galileo's feats. Commissioned by his hagiographer Viviani, it includes among the encomia the information that Galileo was born on the day, almost to the hour, of the death of Michelangelo, "the divinely endowed spirit" whom God had appointed to instruct the Florentines in art, poetry, and architecture. "God himself compensated you," so Viviani's stony billboard addresses Michelangelo's posterity, "and enhanced your glorious annals with the birth of your patrician Galileo, most blissful initiator, father, prince, and guide of philosophy, geometry, and astronomy."[1]

The story lacks a week of the truth. Michelangelo died on 9 February 1564. Galileo's mother, Giulia Ammannati, uncooperative as usual, delayed his birth to 16 February. We know the day and time of the event from two birth charts that Galileo later drew up for himself.[2] Although Galileo did not receive Michelangelo's soul fresh from its owner, he inherited a genius of michelangelic proportions and the taste of Michelangelo's generation. Galileo was a humanist of the old school. He much preferred Ariosto, the darling poet of the sixteenth century, to Tasso, who would be a favorite of the seventeenth. He rated writers like Machiavelli higher than the prosateurs of his own time.[3] He did not like mannerism in art, distortions, extravagances, anamorphoses. He was a stickler for decorum. He stayed with the geometry of the Greeks rather than employ the algebras of his contemporaries. He had little interest in the advanced planetary astronomy of Tycho and Kepler. He always began with Aristotle. He was not an innovator by temperament. And, we are told, he liked to wear clothes that were fifty years out of date.[4]

Although Galileo was born in Pisa, the hometown of his recalcitrant mother, he prided himself on being a noble of Florence through his father, Vincenzo Galilei, a musician and musical theorist. Vincenzo's nobility did not imply wealth but the right to hold civic office and he lived in the straitened circumstances usual in his profession. His marriage to Giulia, whose family dealt in cloth, was a union of art and trade. When they married in 1562, he received two bolts of cloth from his bride's brother, and, what went further, a hundred ducats and a year's rent on a house.[5] For a time the new husband worked in Pisa at the textile trade but soon moved to Florence where he set up as a musician and came to the attention of Giovanni de' Bardi, a man of the world who had fought the Turks and returned to Florence to patronize the arts. Bardi became a true friend and reliable patron. He sent Vincenzo to Venice to study with the top musical theorist of the time, Gioseffè Zarlino. That was in the 1560s. Vincenzo also spent time in Rome collecting madrigals. He returned to Pisa or Florence and Giulia often enough to father six or seven children, three of whom played parts in Galileo's life—his younger sisters Virginia and Livia, and his younger brother Michelangelo. During the early 1570s, Vincenzo lived in Florence, building up his reputation as a theorist in Bardi's circle, which called itself the Camerata after the small room in which it met, and devising an attack on the polyphony and tuning he had learned from Zarlino.[6]

Vincenzo was addicted to his music. He played the lute whenever and wherever possible, "walking in town, riding a horse, standing at the window, lying in bed." We know this odd fact from his most important book, *Fronimo* (1568, 1584), which contains many compositions for two lutes, one accompanying the other.[7] Galileo grew up playing second lute to his compulsive father. He could have learned freethinking also from Vincenzo, who liked to say, as Galileo did too, that people who invoke authority to win arguments are fools.[8] Galileo's mother Giulia knew the value of authority. It was said that she brought Galileo before the Holy Office (the Inquisition) in Florence for calling her names—*puttana, gabrina* (harlot, crone)—and that the obliging inquisitor issued an admonition to the exasperated son. The choice of compliment, "gabrina," adds verisimilitude to the story; the nasty ugly old witch Gabrina is a character in Ariosto's *Orlando furioso*, which Vincenzo admired as much as Galileo. Like Gabrina, Giulia did not improve with age. A year

before she died in 1620, Galileo's brother Michelangelo, then a court musician in Bavaria, expressed his surprise at learning that their mother was "still
so terrible." Her more impressive performances included coming to blows
with Galileo's mistress, suborning his servant to spy on him, and trying to
steal spectacle lenses he kept in his shop.[9] Perhaps Galileo worried about
congenital madness. It is suggestive that the main question he put to the
horoscopes he cast for his daughters was how rational they would be.[10]

While Vincenzo traveled or sojourned in Florence, Galileo lived in Pisa,
sometimes with Giulia, at the house of a relative, Muzio Tedaldi. In these
circumstances he began Latin and other necessary subjects under a teacher
of whom it is known only that his name was Jacopo Borghini and that he
charged 5 lire a month.[11] At the age of 11 the boy began a more regular course
of study at the mother monastery of the Vallombrosan order (a variety of
Benedictines) near Florence. Vallombrosa was a beautiful place certain to
appeal to a romantic teenager awakening to art, literature, and spiritual
life. During his stay the monks entertained the artist Federico Zuccari, then
working on sketches of the seven deadly sins for his frescoes in the cupola of
the Florentine cathedral, and a little later Giovanni Stradano came to paint
the portraits of the monastery and its enchanted forest (Plate 1).[12] Vallombrosa's monks were distinguished for their learning, notably the abbot, "a
man of rare and acute mind," at home, we are assured, in theology, astrology, mathematics, rhetoric, cosmology, and "all the other sound arts and sciences." Galileo's lifetime flirtation with the art of astrology may well have
begun at Vallombrosa. A younger boy, his ill-fated friend Orazio Morandi,
was to join the Vallombrosans and rise to abbot of its seat in Rome, which he
turned into an astrological research institute. Its archives held a birth chart
for Galileo.[13]

The attractions of Vallombrosa, its setting, arts, learned monks, and
stable life, persuaded Galileo that he had a religious calling. Very likely also
the mixed impulses of early adolescence gave him twinges of conscience
and piety that he thought to assuage by joining his teachers. He may have
worn a monk's garb for a while. He did not complete his novitiate, however, or perhaps enter fully into it because his father abruptly removed him
from the temptation. Although Vincenzo had praised the monkish life of
withdrawal—"truly blessed is he who has fled the irritations and tiresome
vanities of the world, the intemperance, ambition, pride, adulation, anger,
deceptions, and so on of the court, and goes to live a solitary and peaceful

life"—he had no wish to see his heir buried in a monastery.[14] Galileo's first steps back into our vale of tears took him to Pisa. "I am glad that you have Galileo back and that you want to send him here to study," Tedaldi wrote in July 1578, though, with grain at 15 lire a sack, a growing boy would cost something to maintain.[15] That the price of bread for Galileo should have been a consideration suggests that playing the lute in all possible places had not made Vincenzo wealthy.

After the escape from Vallombrosa, Vincenzo tried to place Galileo in Pisa's Collegio di Sapienza, set up by Cosimo I de' Medici in 1543 for smart kids from impoverished families. Galileo qualified for admission in every particular except that, at 14, he was four years under age. He remained at home until, in September 1580, at 16, he could enroll at the University of Pisa, an old foundation given new life by Cosimo I, to begin the study of medicine.[16] The course required some knowledge of Aristotelian physics. Pisa had two able expositors of the subject, Francesco Buonamici and Girolamo Borro (or Borri). Each taught a rigid Aristotelianism without agreeing on its principles, although both had a deep respect for the integrity of the philosophy they interpreted and a deep disdain for the compromises required to subject Greek philosophy to Christian theology.[17] Accordingly, they had no patience with the system of Thomas Aquinas, then recently recognized by the Council of Trent as official Catholic teaching and enthusiastically so received by the Dominicans and the Jesuits. Without the prophylaxis of Thomism, Aristotle's philosophy contained teachings toxic to Christians. Among the most frightful of these were the eternity of the universe, the mortality of the soul, and a deity incapable of knowing individuals: no creation, no last or intermediate judgment, no afterlife, no Providence, no Christianity. It took courage to be a strict Aristotelian in a Catholic university in Galileo's time.

The bravos Borro and Buonamici set the agenda for Galileo's physics. Their diverting lectures on motion and gravity did not lie on his immediate path, however, since physicians did not need to know why bodies fall in order to put their patients underground. Nor did the medical course lie on his direct path, except as an obstacle. An honorable way to finish with it unexpectedly opened through the Medici court, which customarily moved to Pisa for a period around Easter. Among its hangers-on was a mathematician, Ostilio Ricci, who would end his career as mathematician to the Grand Duke Ferdinando I. In 1583, however, when he met the disaffected medical student, Ricci was merely the instructor of the grand-ducal pages. Galileo heard a lecture or two of Ricci's on Euclid. With that instruction he mastered Euclid's *Elements* almost on his own and showed such promise that his teacher advocated his release from medicine. Vincenzo agreed most reluctantly. He too liked mathematics, or anyway arithmetic, which he employed in his musical theorizing, but he knew that mathematicians were no more prosperous than lutanists.[18]

Galileo continued at Pisa until 1585, when, like many a noble youth, he left university without a degree. Meanwhile he had continued with his reading. Ricci introduced him to the work of Archimedes, who would be to Galileo what Virgil was to Dante: an ancient shade sure of the path, the one through Hell, the other through mathematics. To Galileo Archimedes would always be "superhuman," the exemplar of the mathematician pure and applied: the theorist of cones and balances, of quadratures and barycenters, but also the inventor of machines to catch counterfeiters, destroy navies, and do the work of giants.

1.2 GAP YEARS

Mathematics

During the 1580s Ricci began to teach at yet another institution established under the civic-minded despot Cosimo I: the Accademia del Disegno set up in 1569 for the instruction of artists, sculptors, and architects. The first directors provided for instruction in mathematics and money to pay for one lecture a week, on Sundays, open to anyone who preferred science to sermons. The stipend, one ducat a month, was a fourth or a fifth of the salary of a beginning university professor of mathematics, not enough to keep the post filled until the engagement of Ricci in 1589. His lecturing covered Euclidean geometry and its application to architecture and perspective, which he also taught privately. One of his venues was the home of Bernardo Buontalenti, a friend and collaborator of Vincenzo Galilei's patron Bardi and a favorite architect of the Medici. Among those who heard Ricci lecture at Buontalenti's was Lodovico Cardi, who, under the nickname of Cigoli, occupied the van of Florentine painters of his time. Galileo may have been present at Buontalenti's and possibly attended the Disegno, where Cigoli taught.[19] In any case, the two became fast friends and Galileo learned how to draw.[20]

With Ricci's encouragement and Vincenzo's agreement, Galileo spent his first post-nongraduate years preparing himself desultorily for the insignificant post of a mathematics professor. According to the traditional ranking of the sciences, philosophers and physicians occupied the top of the academic ladder, mathematicians and grammarians the bottom. Galileo was not the sort to sit contentedly on lower rungs. Nor did he expect to. Some well-placed mathematicians were arguing, occasionally successfully, that their work deserved greater respect than philosophers accorded it. Three of these mathematicians were important for Galileo's career. The eldest and most aggressive, Giovanni Battista Benedetti (1530–90), served as court mathematician to the Duke of Savoy; his most important work for Galileo was a book of *Physical and mathematical speculations* that replaced several important Aristotelian positions about motion with propositions identical to ones Galileo would adopt.[21] The second in age and first in institutional power was Christoph Clavius (1537–1612), professor of mathematics at the central Jesuit university, the Collegio Romano, from 1565 until his death (Plate 6). He fought energetically and effectively for a conspicuous place for mathematics in the Jesuit curriculum and established a quasi-research group at the college, but he did not try to subvert standard physics with its results.[22] He would support Galileo's career and the mathematicians at the Roman College would help advance it—up to a point.

The third of Galileo's early mathematical mentors, the Marchese Guidobaldo del Monte (1545–1607), was the most conventional and the most useful. A man of the world as well as of books, del Monte had fought against the Turks (we shall meet several more such veterans) and served briefly as inspector of the fortresses of Tuscany before retiring to his castle near Urbino to work at mathematics. Galileo adopted del Monte's approach to simple machines, the *Liber mechanicorum* (1577), then considered the best modern book on the subject, and del Monte adopted Galileo as his scientific son.[23] He and his brother, Francesco Maria del Monte, no mathematician but, as a cardinal, a powerful patron, helped to impose Galileo on the University of Pisa as its lecturer in mathematics in 1589. Benedetti, Clavius, and del Monte were role models. Galileo began, as Clavius ended, a professor; ended, as Benedetti began, a court mathematician; and, though far from a hereditary marquis, had the same sense of the dignity of his profession as del Monte.

Mathematics had crouched low among Aristotelian sciences for two reasons. First, it dealt with abstractions like lines without thickness and planes without body, and so operated in a world of make-believe. When its conclusions did pertain to physical things, they tended to be the least interesting things knowable. Who cares how tall Socrates was? Quantitative accidents—dates, scores, balance sheets—annoy and bore most people. Still, although the height of Socrates does not signify, that of the pole star does; and since knowledge of it allows us to navigate, it must relate to some sort of truth. Another unexceptionable example of the sometime truth of mathematics frequently invoked at the time argued from the spherical shape of the earth to the conclusion that the sun cannot shine on its entire surface at once.[24] It was no less a truth if argued the other way around, from the limited illumination to the spherical shape. This brings us to the second reason for depreciating mathematics: the nature of its proofs.

In Aristotelian logic, the strongest demonstration (*demonstratio potissima*) is the perfect syllogism, of which the form "all B are A, all B are C, therefore all A are C" is the exemplar. In practice, the premises of a physical proposition (all B are A, all B are C) were agreements among philosophers based on the repeated and confirmed experiences of rational animals.[25] The question naturally arose whether mathematics made use of the *demonstratio potissima*. Perhaps yes, because Aristotle recommended geometrical proofs; but perhaps no, because we lack direct experience of the abstract entities of the geometers. Furthermore, the major premise of geometrical propositions is often far from any axiom or first principle and may not involve directly or at all the defining property or "essence" of geometrical figures. Euclid argued from an axiom about parallel lines that the sum of the angles in a triangle is two right angles. How could that axiom be a proper definition or representation of the essence of triangles, which cannot be constructed from parallel lines?[26]

Most people interested in such questions in the sixteenth century denied that mathematics met the standard of the *demonstratio potissima* because it could not establish its premises by the intuitive certainty of sense experience.[27] Or, if its premises be allowed, its demonstrations might be very powerful but the knowledge so gained relatively little, as inapplicable to the sensory world; or, if applicable, in "mixed sciences" like astronomy, no longer certain, since astronomy, like all physical sciences, had to take its principles from observation and experience. Assertive mathematicians replied that their methods differed from, but were no less powerful than, the best syllogistic demonstration. Clavius took this position, as did his student Giuseppe Biancani, who would become an authority among the Jesuits.[28] Others argued that mathematics had the same logical structure as physics and stronger demonstrations when applied to mixed sciences like astronomy.[29] Consideration of these life-and-death questions added excitement to the existence of a mathematician. If granted equality in certainty and relevance with physics, mathematics might be the Archimedean lever with which, if he could find a place to put it, a mathematician might move the world. Galileo would expend great effort and incur much ill will in seeking the place and working the lever.

Galileo did not have the means of a marquis or the reputation to qualify as a court mathematician. That left teaching. He recruited private pupils, perhaps among his friends in Florence, certainly among seekers of science in and around Siena. In 1588 he found a job for three months with his friends the Vallombrosans tutoring a brother in perspective at their monastery in Passignano; he received a total of 58 lire (around 8 scudi), which, if continued for a year, would have brought him half the salary of a poorly paid professor.[30] The same year, 1588, he tried for the junior chair of mathematics at the University of Bologna. His application opens with an arithmetical error: it gives his age as "around 26," though he was only 23. Apparently he thought his youth incongruous with his mathematical pedigree, which consisted of his time with Ricci and his experience of teaching in Florence and Siena. It continued with his attainments. "He has the most excellent judgment [in mathematics] and in many other things he has studied, notably in the humanities and in philosophy." The electors preferred Giovanni Antonio Magini, Galileo's senior by nine years, a practiced astrologer and astronomer, diligent calculator, and, later, Galileo's occasional friend. No doubt they took the safer choice. Magini was a graduate of their university. Also, Bologna's senior professor of mathematics, Pietro Antonio Cataldi, doubted the correctness of a premise that underpinned the mathematical proofs Galileo submitted as his strongest credential. The professor of mathematics at the University of Padua, Giuseppe Moletti, agreed with Cataldi that Galileo was a good geometer with a defective premise. Del Monte and Clavius also had trouble with it. As will appear, it was unexceptionable, although (and this may have been the trouble) it did not deliver what Galileo's proofs needed.[31]

Music

Magini's victory extended Galileo's opportunities to cultivate literature and music in Florence. Among the many cultural gurus who helped develop his taste and character during the four years he lived at home between leaving

Pisa without a degree and returning as a professor, the most influential was his father. Vincenzo Galilei was then engaged in defending the modernizing program he set out in 1581 in his *Dialogo della musica antica e della moderna* (1581). Musical theory needed overhauling, according to Galilei, because the current orthodoxy, as represented by Zarlino, did not correspond to practice. Zarlino accepted Ptolemy—the same deserving Ptolemy whose cosmology Galileo would savage—and cleaved to him although the simplest experiment would convince an educated ear that modern music did not use the Ptolemaic diatonic scale. Galilei's spokesman in his *Dialogue on ancient and modern music* is his patron Bardi, who explains the matter to an astute and inquisitive mutual friend, the practical musician Piero Strozzi. They agree as a condition of the discussion that "we always set aside (as Aristotle says in the eighth book of the Physics) not only authority but seemingly plausible reasoning that may be contrary to any perception of truth." Vincenzo Galilei was a theorist of integrity. He did not just reject authority but showed by "sensory experience and necessary demonstration," as Galileo would say, just where the authority went astray. Ptolemy had succumbed to a numerology that appeared to secure a tuning based on the characterization of musical intervals by the different lengths of otherwise identical vibrating strings. A string half the length of another under the same tension sounded the octave; with lengths in the ratio 2:3, the fifth; 3:4, the fourth; 4:5 and 5:6, the major and minor third. In recommending Ptolemy's scheme, Zarlino liked to dwell on the perfection of the number six, from which all the ratios descend, as well as their simplicity, all being of the form $n/(n + 1)$.[32]

Galilei dismissed the numerology as nonsense and Zarlino's claim that singers naturally adopted Ptolemy's diatonic tuning in polyphonic music as false. As Galilei discovered under the guidance of a Florentine living in Rome, Girolamo Mei, who had made a thorough study of ancient Greek music, Ptolemy's authority was an artifact. Moderns had decided to prefer him to his ancient competitors just as, with equal arbitrariness, they had chosen Aristotle over Plato in philosophy. Mei very generously shared his knowledge of the alternatives; all he required to unlock his generosity was assurance that Galilei too was a Florentine patrician.[33] That being given, Galilei learned that the Greeks had used many modes and monodies, and knew nothing of the four-part harmony that Zarlino taught and ranked as the most perfect of musical forms. A romantic as well as a scholar, Mei understood that the main purpose of music was not rational entertainment but appeal to the emotions, and that Zarlino's play with the number six was mumbo jumbo. Galilei's *Dialogue* employs Mei's principles, and sometimes his very words, which came to this: Greek music had more emotional power than modern counterpoint because it was monodic and polymodal.[34] Music that accompanied singing had to be matched to words even at the cost of violating rules of progression or harmony.[35]

Adopting Galilei's battle cry, "counterpoint...is an enemy of music," Bardi incorporated some of his protégé's ideas in the musical interludes he arranged for three successive Medici weddings in the 1580s. These inter-mezzi, which framed stage plays, were showpieces incorporating compli-cated stage machinery, singers, and musicians. Galilei contributed at least one piece to these extravaganzas, a lament based on the speech the miser-able Ugolino muttered at the very pit of Dante's Inferno. The obvious inver-sion of the performance, in which the play framed and incorporated the intermezzi, developed rapidly in Florence in the 1590s and eventuated in 1600 in Jacopo Peri's *Erudice*, the first opera whose score has survived. Cigoli provided the scenery. The occasion was still another Medici wedding. The movement to which Galilei and the Camerata belonged was not an inconse-quential mobilization of argumentative musical theorists, but an important component of an artistic and aristocratic culture. Galileo had deep roots in this culture and tendrils to Peri and the Camerata.[36]

Zarlino complained of his former student's betrayal and said so in print, in 1588. In reply, Galilei described additional experiments to show the misfit between the Italian ear and the Pythagorean and Ptolemaic tunings; issued the so-called "laws of Galilei" (two successive consonances of the same size do not produce a consonance, and the difference between an octave and a consonance is a consonance); and aimed a very neat blow against the stand-ard numerology. Why take the lengths of vibrating strings as the definition of pitch? If the tensions in chords of equal length are preferred, the fifth cor-responds to the ratio 9:4, the octave to 4:1, and so on.[37] Vincenzo probably had Galileo's help in the refinement of the experiments that underlay these ratios. The experiments included tests of similar strings stretched by une-qual weights and strings of unequal cross-sections bearing similar weights. Father and son studied deviations from unison caused by the nature, struc-ture, and form of the strings, and the manner of plucking. A particularly interesting investigation showed that a steel string and one of gut tuned to unison in the octave would not agree perfectly when stopped at a fret. The best that can be done when instruments play together or accompany singers is to forget the perfect ratios of the various modes and tune everything to equal temperament.[38]

Two propositions in Vincenzo's final investigations sound very much like Galileo. Referring to a "simple-minded assertion of Zarlino," father or son points out that to get a tone and its octave simultaneously from a single string, it must be stopped at a third, not a half, of its length. The other proposition in Galileo's style states that a sliding transition from one note to another on a fretless string instrument does not take place continuously, but through a series of discrete steps so small, "almost like atoms," that the ear cannot distinguish them. But mathematics can. "There are few things that cannot be weighed, numbered, and measured."[39] One of these few things is poetry.

Poetry

Galileo's social life during his gap years did not center on mathematicians, of which there were few in town besides the grand duke's lecturer Ricci, his astrologer Raffaelo Gualterotti, and his globe maker Antonio Santucci.[40] Galileo preferred literary men, Florentine nobles like him though wealthier, of which there were many. Two of them played important parts in Galileo's story. One, Giovanni Battista Ricasoli Baroni, a studious and disturbed young man, died early under circumstances soon to be related. The other, Giovanni Battista Strozzi (il giovane), 17 years Galileo's senior, had distinguished himself early among the cultivated aristocratic young men of Florence. He took even longer than Galileo not to graduate from the University of Pisa, some seven years spent with Borro, Buonamici, and literary dilettantes. If Galileo had had any money, he would have faced the same career choice as Strozzi did: marriage, church, or court. None appealed to Strozzi. His inamorata wed another; for the church he had no calling; and he was far too rich to serve the Medici. In 1590 he went to live with the Oratorians in Rome and soon became intimate with the grandees and prelates there. He would be helpful when Galileo undertook to instruct the papal court on astronomy.

Strozzi and Ricasoli were leading lights of a serious literary club, the Accademia degli Alterati, composed, etymologically, of altered, twisted, false, angry, and befuddled poets. There is good reason to believe that Galileo was a member. Several other friends belonged, including Bardi, and, although the Academy's incomplete records do not mention Galileo, he was present to hear Ricasoli deliver a lament over the death of Grand Duke Francesco I.[41] During his gap years, Galileo cultivated exactly the same subjects that then occupied the Alterati. His pertinent extant compositions include two handfuls of poems, extensive "Notes" on Petrarca's *Rime* and Ariosto's *Furioso*, "Considerations" on the *Gerusalemme liberata* of Torquato Tasso, outlines for a play, and, though neither poetry nor criticism, two lectures on Dante's *Inferno*, almost none of them published in their time.[42] Except for the lectures and a satirical poem, they cannot be dated precisely. Rather than agonize inconclusively over their chronology, we will consider his "literary" writings when they best fit his story.[43] On this principle, the criticism of Ariosto and Tasso began during the late 1580s when a controversy over their relative merits engaged all the literary types in Florence.[44] The controversy had its lighter side as academicians struggled to decide whether Homer or Virgil, Ariosto or Tasso was the greater poet, whether Tuscan was the best of all languages, and whether Aristotle's principles of poetry should regulate Italian romances. Galileo's notes and considerations supplied weapons for himself and his friends for deployment in the Italian theatre of the warfare between ancients and moderns.[45]

The Alterati changed its director or regent every six months. During each term, members placed compositions in a large vase excavated now and then by a "censor" and a "defender." These officials argued the merits and faults of the various pieces before the academy, which chose a few that it wanted to hear at length.[46] A fragment by Galileo about good taste represented the tone and type desired. It explains how a judicious critic differs from a pedant. Both can tell that a bald toothless woman without a nose is not a perfect incarnation of feminine beauty. "But it does not follow that she would be very beautiful if she had teeth, nose, and hair, but only if in these and every other part there was an excellence difficult to describe and represent...The intelligence of the pedant extends only to the total of missing parts...To him all eyes, mouths, and bodies are equally beautiful: and he would unhesitatingly prefer a woman who has a beauty spot to one who does not although the second has all her parts most beautifully proportioned and the first is without any grace or symmetry."[47] Beauty is not a matter of addition but of harmony and fitness. Galileo never could bring himself to accept distorted-circle, that is, elliptical astronomy.

After much debate, the Alterati decided that sonnets were the highest form of poetry. Galileo tried his hand at them, and mayhap buried some in the capacious urn. Here is a specimen, in which a couplet of Galileian inventiveness follows a Petrarchan opening, and the rest proceeds on standard lines:

> Now that the sun has plunged, its golden curls ablaze,
> Into Ocean's waves against Iberia's shores,
> The alleyways discharge a multitude of whores
> In beautiful platoons attracting every gaze.
> The woolen mills close up within the city's maze
> Silk weavers stop their work behind their shuttered doors
> Birds wheel overhead as a thund'ring tower pours
> Forth noise of clanging bells that puts them in a daze.
> Townsfolk also flock, to their ancient bridge to tell
> The gossip of the day and take some harmless play
> Florentines of old thus came to buy and sell.
> So I leave you my sweet hope, my dear, fare-thee-well
> I'll go to the piazza, where I'll end the day.[48]

The platoon of harlots is well observed as their rush to vespers to atone for past, and to attract new, business was as sure a sign of evening as the suspension of other trading.[49]

By and large critics have not appreciated Galileo's poetry. One of his modern admirers, who ranked him the best master of Italian prose after Machiavelli, remarks of his sonnets that we are lucky there are not more of them, and offers the example of the change from Petrarchan sentiment to social studies in the poem just quoted.[50] On the positive side, Galileo's literary criticism is still essential reading for expositors of Ariosto and Tasso.[51] The anonymous editor of the "Considerations on Tasso" who wrote at the end of the eighteenth century allowed that most of Galileo's criticisms were sound and that, if reduced to system, "they would constitute a science of action in poetry."[52] Several selections of Galileo's writings compiled to help Italian children learn to write and think contain ample excerpts from his literary criticism. Like his science, they illustrate how to combat pedantry and error, reject authority, and build on reason and experience. Favaro lost no time after finishing his exhaustive edition of Galileo's works in scavenging them for texts suitable for use in general instruction. It cannot be said that their study has imbued Italian academic writing with the "simple, geometrical, positive, objective" style that made Galileo the greatest master of Italian prose between Machiavelli and Manzoni, nay, the best Italian writer of all time.[53]

The literary sport in Florence during the 1580s incorporated and culminated decades of disputes over the nature and merit of Dante's tri-part poem, the *Divine Comedy*. Do not think it poetry merely because it rhymes and is sometimes sublime. According to the criteria deducible from Horace's *Ars poetica*, the main guide to literary theory of the early cinquecento, a good poem should edify as well as delight.[54] No doubt, Dante's work gave pleasure; but did it also promote civility, virtue, morality? Yes, because it taught the ways and wages of sin; no, because it did so through stories that aroused disgust rather than pity, and has as its hero not a great personage but a peevish poet. Granting that it is a poem, is it a tragedy, comedy, or epic? Dante called it a comedy, and so it is, because it contains low characters and ends happily. Yet does its narrative and invention, together with the loftiness of its theme, not make it an epic?[55]

To unravel this conundrum, and to ravel many more, critics writing after 1550 had Aristotle's *Poetics* in good editions with useful commentaries. According to this authority, an epic, "poetry that mainly narrates," must meet the tests of unity of action, coherence, and plausibility as well as furnish pleasure and utility with appropriate decorum, imitation, and invention. Critics who founded their *buon gusto* on the literal word of Aristotle complained that Dante's *Comedy* has no unity. Each of its three parts comes to its own dénouement through a vast quantity of disparate discursive episodes of doubtful taste. As for invention, where are the peripaties (reversals of fortune), discoveries, and sufferings? Dante's journey runs in a straight line towards an end too easily glimpsed at the beginning. Worse yet, a well-behaved poet should say little for and about himself; Dante is never off the stage. Worst of all, the story violates the essential rule of verisimilitude: "It is fundamentally wrong to make up plots…compounded of improbable incidents."[56] And what does Dante offer? A week's journey through the earth, up a gigantic mountain, across the planetary spheres, all the way to God, chaperoned by the shade of an ancient poet and the ghost of a departed love; interviews with sinners who though dead feel the torments of the flesh and though suffering unspeakably speak, well and clearly, about the errors of their ways; and conversations with the saved and the saints, arrayed eternally without jealousy in order of merit. And there is another violation, more monstrous, perhaps, than all the rest: learnedness. Dante strews bits of astronomy, physics, cosmology, theology, unsuitable to poetry and destructive of pleasure, throughout his comedy.

The great Dante debate reached its height in 1587. In that year, the Accademia Fiorentina, founded in 1540 and promoted by the Medici to defend and propagate Tuscan language and literature, established two lectureships on Italian poetry and heard praises of the divine comedian from, among others, Strozzi, Don Giovanni de' Medici, and Jacopo Mazzoni (Plate 6).[57] Strozzi's friend Don Giovanni, an illegitimate son of Cosimo I, was a poetical engineer, musician, alchemist, and astrologer, a man of immense reach, tolerant enough to befriend his Jewish librarian and brave enough to challenge Galileo over some problems in applied mathematics.[58] Mazzoni, who was to be Galileo's special friend and colleague at Pisa, was 40 in 1588, "a man of the highest level in science, at home in all languages, a most perfect master in all faculties." Mazzoni had studied at Padua, worked on calendar reform, edited the Index of Prohibited Books, and begun his defense of Dante 15 years before his performance in Florence.[59] That proved a great success. Strozzi, also a lifetime defender of Dante, praised Mazzoni's lectures as "full of profound and marvellous ideas;" the Alterati elected him a member; and Grand Duke Ferdinando demanded that he accept a professorship at Pisa. Mazzoni's teaching met every expectation. He was surrounded by students and professors except when he dined with the grand ducal family, "seasoning the meal with his arguments."[60] Called to Rome to teach by Pope Clement VIII, who had a fondness for philosophy, Mazzoni was about to be made a bishop when death, having cut down the last of his brothers without issue, obliged him to marry instead.[61]

In his famous lecture to the Accademia Fiorentina, Mazzoni located the core of poetry in credible imitation. The poet is not restricted to the false. He can imitate the truth, even in matters scientific, philosophic, and theological, provided that he does it poetically, that is, with accessible and credible images. Mere versification of a true history or a world picture would not qualify, however; for although the result might be credible, it would not be imitation. "Concerning this I have written at length in my new book [*Della difesa della Commedia di Dante*, available at all good book stores], where I also show with what tact Dante has at times introduced either a philosopher or a theologian to discuss matters pertinent to the contemplative sciences in an understandable fashion, never deviating from the credible."[62]

The loyal Florentine academicians and the independent Alterati had been at pains to demonstrate the propriety and elevation of Dante's use of science against those who, like the influential critic Ridolfo Castravilla, objected to poetry's condescending to "scholastic matters." "Truly, when I consider Dante's *Comedy*, I see nothing but a medley, miscellany, and muddle of learning that he could have had from any old monk."[63] The muddle began in Hell with a complicated staging for which the Florentines had devised a suitable landscape. It met with objections. The Accademia Fiorentina deemed it necessary to reply. It was a job soon discharged by the Academy's young practical mathematician, Galileo.[64]

1.3 CHARACTER ANALYSIS

Critical insights

As the Dante dispute culminated, the practiced Florentine academicians undertook to judge the relative merits of Ariosto and Tasso. *Orlando furioso*, published first in 1516 and definitively in 1532, was a product of the high Renaissance, of the time of Raphael, Michelangelo, and Machiavelli. It violates almost every one of Aristotle's rules except the pleasure principle.[65] The *Furioso* is a mixture of Greek myth and medieval romance, and, by operation of its magic, of *Harry Potter* and *The Lord of the Rings*, and fuller than any of them with episodes that do not advance its scarcely discernible plot. For invention it has few peers: magic, antimagic, spells and counterspells, brave knights and beautiful maidens in profusion, courtly and carnal love, dragons, duels and battles, jousts, Amazons. Through it runs an irony that reduces the too-virtuous knights, the too-beautiful maidens, and the two-timing magicians to their human equivalents. A master of soap opera and cliff-hangers, Ariosto breaks off a fight to the death, a seduction, shipwreck, or hair-breadth escape to return to an equally fraught situation interrupted a thousand lines earlier. "I must stop here as there is an English knight demanding to have his turn." "If my story pleases you, you can find it again later."[66]

Two of these stories will indicate the pace and playfulness of the whole. While the lovesick ninny Orlando disappears for several thousand lines, his beloved Angelica has many adventures with men and monsters. In one of these, which later inspired a famous painting by Ingres, a community of terrified Scots ties her naked to a rock to appease the hunger of an orc. As dinnertime approaches so does the knight Ruggiero on a flying horse. He rescues the luscious maiden and lands on a soft meadow. Too eager for his reward, he gets entangled in his armor while Angelica, capitalizing on his fumbling, deploys the magic ring she wears—the only thing she wears—to make herself invisible. Poor Ruggiero, at last stripped for action, gropes about like a blind man. "Many a time he hugged the empty air, hoping to clasp the damsel in the same embrace."[67]

Thus saved from himself, Ruggiero was able to concentrate his attention on his true love, the lady warrior Bradamante. The tale required that they get together as the Adam and Eve of the House of Este, under whose patronage Ariosto composed his poem. Bradamante spends most of her time looking for Ruggiero. One evening she is surprised by darkness in a place with only one castle where she might find shelter. The castle had unusual house rules. The first knight to arrive was welcomed (or two or three, if they came together); a subsequent applicant would be housed only if he could defeat the incumbent(s) in a duel. A similar rule held for women with beauty rather than valor as the criterion. Three touring Swedish champions were settling down in the castle when Bradamante demanded accommodation. She easily unseated the Swedes and took her place at the fireside wearing her helmet. When she removed it she presented her host with a puzzle. There already was a woman guest in the castle. Though exquisitely beautiful, she was a hag in comparison with Bradamante. Could a person who had obtained entry on one ground also evict on another? The rules did not cover the case. Bradamante improved the paradox: if the earlier guest had been more beautiful than she, would she have had to leave? The answer can be found at the end of canto 32.[68]

The inventiveness, irony, and liveliness of Ariosto's beautifully told tall tales made them immensely popular. Some 25,000 copies of the *Furioso* were printed in the sixteenth century. Verses set to music, by Byrd and Bardi among others, were sung by admirers high and low, literate and not. Men of such different good taste as the Galilei, Montaigne, Cervantes, La Fontaine, Voltaire, Lessing, Goethe, Hegel, Foscolo, and Croce prized the adventures of Ruggiero, Bradamante, Astolfo, Angelica, and the rest.[69] Who cared whether the *Furioso*, the high entertainment of the age, sinned against the conventions? Tasso's admirers. Many of them hailed *Gerusalemme liberata* (1581), curiously also a product of Este patronage, as more profound and better behaved if less entertaining than Ariosto's masterpiece.

During the half-century that separated the poems, much had happened to endorse the sombre, even melancholy, tone of the *Liberata*. The Council of Trent had tightened doctrine and discipline to meet the Protestant threat. The popes had set up the Index of Prohibited Books, resurrected the Roman Inquisition, and nourished the Society of Jesus. The Turks, though defeated at the naval battle of Lepanto in 1571, still menaced the seas as pirates and the land as conquerors. The age required a new heroic poem. Tasso's theme, the great Christian epic of the establishment of the Kingdom of Jerusalem in 1099 under Godfrey of Bouillon, hit the mark. It became a favorite among the Jesuits and almost an official poem of the Counter Reformation.[70] None-theless it has its budget of magic spells and chivalrous adventures. Not only were they expected in a heroic poem, but without them Tasso could not have accounted for his actors, whose heads he supposed to be filled with the old romances.[71] Of irony, however, there is not a hint. Tasso's idea of fun was a river whose water causes imbibers to laugh themselves to death. "This river also makes me laugh my head off [Galileo scoffed] since no one in the entire work enjoys, has enjoyed, or is willing to enjoy, an iota of laughter."[72]

What the *Liberata* lacks in rollicking good fun and ironic distance it almost makes up for in seriousness of purpose, unity of plot, and depth of charac-ter. Tasso's crusaders are not perfect; their heroism stutters, they suffer from uncertainty, indecision, the human condition. Godfrey sometimes behaves stiffly or foolishly. Despite his age and wisdom, he swallows the sorceress Armida's autobiography of virtue and chastity, although her purpose in vis-iting his camp is to seduce as many of his men as she can. His blindness and her slowness irritated Galileo. "Madonna Armida, knock it off [*stare i madri-galetti*], otherwise Godfrey if he has any brains will figure out that you are a cheat and send you to a bordello..." While Godfrey should have been worry-ing about supplies, troops, the approaching Muslims, and so on, "you [Tasso] consume 100 stanzas and more retailing four sluttish tricks of Armida and describing the poltroonery of fifty champions who abandon the army and their honor to follow her."[73] Later critics tend to agree with Galileo's censure of Tasso's prolixity and misplaced grandiloquence, although they do not enter into conversation with him and his characters as Galileo did.[74]

The first serious assertion that the wordy, elevated, melancholy, psycho-logically penetrating Latinate *Liberata* was superior to the light-hearted, superficial, ironic, playful, popular *Furioso* reached Florence in 1584. Among the first to the portcullis was Ricasoli, who, giving direction to the Alterati, offered the general thesis that "Ariosto merits greater praise than Tasso."[75] Galileo's "Considerations on Tasso" often parallel the discussions at the academies, even in their more whimsical parts, although with a bite that has caused his later admirers puzzlement and even embarrassment.[76] Galileo's cuts and thrusts should be regarded as material for sprightly debates among brash young literary men, private notes for knockdown confrontations with members of the Alterati, who tended to favor Tasso, not formal literary criticism for later depth analysis.[77] He could play very well—against Ariosto as well as Tasso—the Florentine game of culling barbarisms, Latinisms, inept words, pedanticisms, stuffing, lard, veneer.[78]

Galileo's substantive criticism of Tasso may provide deeper insights into his than into the poet's character. Perhaps his most frequent serious charge is poverty of invention. "I've always thought that this poet is, in his inventions, above all mean, poor, and miserable, whereas Ariosto is magnificent, rich, and marvelous." Leafing through the *Liberata* is like wandering through a collection of objects "unusual for their antiquity or rarity, but in fact trivial things, a petrified crab, a dried chameleon, a fly and a spider in amber, some of those dolls said to come from Egyptian tombs, and, as for pictures, something sketched by Baccio Bardinelli or Parmigianino, and a thousand other small things; but as soon as I enter the *Furioso*, I see opening up a wardroom, a tribune, a royal gallery with a hundred ancient statues by the most celebrated sculptors...and a great many vases of crystal, agate, lapis lazuli, and other gems, full to overflowing with precious, marvelous things, all excellent."[79] From which we may infer that Galileo prized inventiveness when it produced great and beautiful things. Pedestrian wonders or out-of-the-way oddities meant nothing. He would observe this principle in holding back his small or partial discoveries until the telescope revealed to him the greatest wonders of the age, and in recommending some of his nicer geometry and cosmological speculations as marvels.

Love both carnal and committed is also often on Galileo's mind. He objects to most of Tasso's love scenes that the parties have no clue what to do. The great champion Tancredi, called out to do battle with the ferocious Muslim hero Argante, stops en route smitten by the beauty of the warrior maid Clorinda. Galileo objects. "God give me patience with this man! Ah, Tancredi, you coward, so there are your heroic acts! Ah, to be chosen above all others to chastise Argante, and in exchange you stop to make love! What a hero! And what a nice place you have found for wooing, at least half a mile

from the lady…Ah, Dio, Sig. Tasso, are these really your heroes…? While
you, messer Ariosto, have Mandricardo leap out of the bed where he is lying
naked with Doralice at the sound of a call to arms."[80] Tancredi's subsequent
lovemaking consists of a few grunts and the unwitting killing of Clorinda,
and of surrender when grievously wounded to the chaste Erminia, who is
better at pharmacy than at flirting.[81]

Tasso could write love scenes, from experience it was said, one of which,
the cooing of Armida and her love-captive Rinaldo, has inspired many
paintings and at least one opera. The scene in Armida's pleasure garden just
before the arrival of two knights sent by Godfrey to remind Rinaldo of his
duty contains the line, "She swoons in his caress / cheeks flushed and bare /
While silver beads of sweat their charms enhance." Galileo: "I've never seen
sweat go white except around a horse's testicles."[82] He objected to explicit
descriptions even in Ariosto. He did not scruple to rewrite couplets in the
Furioso to remove references to a girl's thighs and French kissing.[83]

A usually sound critic has deduced that "Galileo never knew the dreams of
love." The same authority found evidence that Galileo showed a deep insight
into female psychology in one of his notes on Ariosto. Angelica, needing
a manly escort, must choose between Orlando and the King of the Circas-
sians. She chooses the king, not because of his status but because she thinks
that she can govern him more easily than she could the superhero. Galileo:
"A marvelous description of female behavior: and this is one of the reasons
that most women prefer people of lower condition to men held in high
esteem."[84] It seems rather the jejune generalization of a jilted young man.
That Galileo had been rejected in this way may explain his admiration for the
behavior of the ferocious Rodomonte when Doralice jilted him for Mandricardo. Rodomonte thereupon prayed that his friend King Agramante, who
had endorsed and enforced her choice, would lose his kingdom so that he,
Rodomonte, could restore him to it, "and make him see that a true friend
ought to be favored, right or wrong, even were the whole world against
him."[85] Perhaps it is far-fetched to imagine Galileo as Rodomonte and the
father or brothers of the girl who may have jilted him as Agramante.

Besides love and invention, Galileo's criticism shows a constant concern
for verisimilitude. The line is difficult to define where the miraculous appears
at every turn. Ariosto's wonders seem realistic, because coherent and kept
at an ironic distance; Tasso's often fail, because unnecessarily implausible
and presented as true. Where Tasso gratuitously violates mathematics and

physics Galileo pounces. From a position on the walls of Jerusalem Erminia identifies individual crusaders and defenders skirmishing on the plain. "Now if we calculate that the action takes place over a mile away…Erminia could see things that in our time cannot be distinguished even at a furlong." She must have had an early telescope. Armida's palace offends against all Galileo's rules for verisimilitude. It has a garden inside and, within the garden, hills, valleys, woods, caves, rivers, and ponds. It sits on the top of a mountain on an islet in the Canaries. "The palace must have a circumference of hundreds of miles, though it is built on a mountain peak; the base of the mountain must be thousands of miles around; the islet in the Canaries on which it rests must be the largest island in the world; which runs against the truth, since all the Canaries are very small."[86] The realistic treatment of the marvelous in Ariosto's style became a frequent and powerful literary technique with Galileo. His appreciation of tall tales told realistically would help him to slide easily from the hypothetical and probable to the true and necessary, as in his eventual rendition of the Copernican system as something akin to revealed truth.

Delving deeper, we may find that Tasso was out of Galileo's psychological depth. Tancredi is not a coward but a Hamlet character. Godfrey is a good leader because he listens and deliberates. Galileo misses the emotion in the affair of Tancredi and Clorinda, Oscar Wilde's notorious proposition that "each man kills the thing he loves." He does not like "effusive sentiment or delicate melancholy."[87] Critics have slated his criticism as too strict and mathematical, as in his concern over the size of Armida's estate.[88] Was it his classicism or his conservatism that made him "deaf to the more modern in Tasso's pathetic treatments"?[89] The cause lies deeper: Galileo could not tolerate ambiguity in character any more than in geometry. He came to judge contemporaries, his friends and opponents, in the same black-and-white terms he applied to fiction. So, although he could praise Tasso occasionally for a well-turned phrase, as in his description of the façade of Armida's palace, and even for a good speech, as in Argante's retort to Tancredi at the commencement of their final duel (I weep not for my fate but for Jerusalem's), he begins and ends his "Considerations" without a nuance.[90] At the beginning: "Tasso performs his work roughly, jerkily, and crudely for want of everything needed for it." At the end: "The rest of the stanza is feeble as usual, meaningless, unexpressive, portraying nothing with his usual generalities."[91]

Galileo's aversion to complexity of character, which is of a piece with his depreciation of history, is the obverse of his compulsive attention to linguistic details.[92] He preferred perfecting Ariosto's language to sounding Tasso's psychology. "The persuasive force of Galileo's analysis diminishes as the inspiration and poetic return of the pages examined increase."[93] As in literature, so in physics Galileo dealt more comfortably with the accidents than with the essences of things. Again and again we shall find him avoiding causal accounts in favor of mathematical descriptions. The avoidance produced his best work and some of his worst "marvels," just as it resulted in apt but also in excessive criticism. Galileo's ducking of causal connections in physics and discomfort with depth of character in literature, and his reliance on mathematics in offense and defense, had the same psychological roots. Initially the ducking, the discomfort, and the reliance protected him from risk-taking; but in time they blinded him to the risks he ran. Galileo's particular genius, his literalness, black-and-white judgments, hypochondria, and shallow psychological depth perception made a Manichean personality.

A shrewd critic has described Galileo's argumentative style as a chivalric duel to the death. "He never takes his adversary by abrupt frontal attack, but after a courteous greeting stands back to await the first blow. Going on the defense, he entices his opponent to advance. Suddenly he strikes where least expected, and, profiting from the surprise, presses in, pushes back, knocks out his adversary, and withdraws without taking any further notice of the combat."[94] An apt image. "*Orlando furioso* was always present in [Galileo's] mind."[95] Galileo was an Orlando in argument, and, like Ariosto's paladins, developed the skill of slicing his adversaries in half without their noticing it.

We need not limit the consequences for science of Galileo's addictive reading to his style of argument or to the clean, crisp, limpid, precise, assured, ironical, natural, direct style that, he told Viviani, he acquired from frequent readings of Ariosto.[96] There may be much more. "Ariosto, the true painter of the beauty of nature, helped Galileo in some way in reading the heavens and revealing the earth, for pictures in the imagination contribute not a little to inform and direct the meditations of the philosopher."[97] We know the effect of reading and rereading stories of derring-do, of chivalrous knights, of maidens in distress, of a single hero victorious against an army. We know it from the behavior of another reader of the *Furioso*, the melancholic knight of the rueful countenance, Don Quijote de la Mancha. Was Erminia's telescopic vision present in Galileo's mind when he perfected his spyglass? Did he recall the competent anonymous nymph of the *Furioso*, who, "with simple words moved the earth and stopped the sun," when he meditated about the

Copernican theory?[98] Did he remember Astolfo's voyage to the moon when he began to explore the heavens?

Madness

Duke Astolfo, Orlando's caring cousin and fellow paladin, flew to the moon in a chariot pulled by winged horses to look for the great champion's wits. That was perfectly reasonable since all things lost here end up there. The moon turned out to be a second earth, with rivers, lakes, plains, valleys, mountains, and spacious forests filled with accessible nymphs.[99] In his lunar travels, Astolfo came across lost reputations, abandoned loves, discarded treaties, forgotten charities, faded beauties, everything, indeed, except folly, "which abides with us on earth," before coming to a mountain of brains kept in bottles to avoid evaporation. "The one containing the mighty brain of Orlando was the biggest of them all." Astolfo's amazement at these wonders was nothing compared to his shock on finding a small vial containing some of his own wits, which he had not missed, and other vessels filled with the brains of people he had deemed perfectly reasonable. There were wits lost in loving, in putting faith in princes, in seeking honors, wealth, or some other imagined good. Philosophers, mathematicians, astronomers, and poets were particularly well represented: "Di sofisti e d'astrologhi raccolto / e di poeti ancor ve n'eran molto." Astolfo inhaled his missing wits and returned to earth with Orlando's and the conviction that everyone is more or less crazy.[100]

The earliest documents in which Galileo gives an account of himself concern madness. The account is a byproduct of Galileo's testimony in a court of law about Ricasoli's mental state during the spring and fall of 1589.[101] The question on which the case turned was whether Ricasoli was in his right mind when he made his will late in 1589 in favor of a cousin to the exclusion of his sister. The defense (which sought to uphold the will) tried to show that Ricasoli had normal dealings with bankers and tradesmen during the time he supposedly was mad. Would it be right for a merchant to take advantage of a deranged gentleman? Following this line of inquiry, the court elicited information that may explain how Galileo, who had no fortune or regular income, could afford to run with the Ricasolis. Would it be right to play cards for money with a mad man? And did you, Galileo, not gamble at cards with the unfortunate Giambattista, and win considerable sums? Galileo admitted to gambling with Ricasoli and others but claimed to have forgotten the stakes and the winnings. Another gambler, Jacopo de' Medici, testified that on one occasion he had seen "one of the Galilei" at cards, which suggests that Galileo's teenaged brother Michelangelo might have been involved in the play too. Galileo recommended himself to Ricasoli and the others by his general ability and well-stocked mind, and almost certainly also by his dexterity at casting horoscopes; but he also owed his easy familiarity with them and his pocket money to his ability at games of chance.[102]

A calculation by Galileo of the relative frequencies of specific throws with three dice has survived. Its main operative result is that 10 turns up more frequently than 9 once in 108 throws. Only a frequent player could hope to use this information to advantage.[103] As a good gambler Galileo occasionally bluffed by raising the stakes on a losing hand—a technique he later identified with the propensity of his philosophical opponents to add reckless worthless arguments to bad ones. This criticism applied better to him. His later claims about experimental results and theoretical insights contained a quantity of bluff.

Galileo's account of his fellow gambler's madness is a well-observed clinical history of progressive "melancholy." One evening while the two were lying in bed together (nothing need be read into this) Ricasoli suddenly announced that he had to leave Florence immediately because he had committed great crimes for which the Medici had condemned him to death. For much of his flight he had the company of his cousin, Giovanni Ricasoli, and of Galileo, who took care to keep Ricasoli's relatives informed about their travels. Toward the end of the walkabout, after Galileo had left the company, Ricasoli wrote the disputed will giving his estate to Giovanni. Galileo testified to Ricasoli's progress from fear, hypochondria, and sleeplessness to neglect of person, clothes, and health—a sequence similar to Ariosto's description of Orlando's descent into insanity, which Galileo thought particularly persuasive.[104] Symptoms included fleeing shelter in the middle of the night, talking of the dead as if they lived, going about in mourning, praying endlessly, and, in one classic instance, running away from a priest whom he suspected of coming to administer the last rites before the Medici murdered him. After a hard and dangerous trip made more perilous by the need to dodge the priest, Ricasoli's party reached Genoa. He immediately announced that the following day they would flee to Turin.

That decided Galileo to try to place his troubled friend under medical care. A stratagem was necessary since Ricasoli supposed that the Medici employed doctors as well as priests in their assassinations. At that time a young nun in Florence aptly named Maria Maddalena de' Pazzi was having widely publicized visions that revealed among much else that melancholic fits came directly from God. Perhaps with her revelations in mind Galileo found a monk in Genoa willing to tell Ricasoli that a nun there had had a vision that a Florentine nobleman afflicted with melancholy would seek refuge in Genoa. According to the fake vision, God had decided that Ricasoli had suffered enough. "And because the Almighty operates most often by natural means, he wanted the gentleman to seek relief in medicines for the causes that, with God's approval, have caused these humors in the gentleman's body." Ricasoli accepted this charade, compounded of true concern, invention, and good humor, and put himself in the care of a doctor. Galileo went back to Florence intending to return to Genoa with fresh clothes for the *furioso*. But in a few days Ricasoli was on the move again, this time to Milan, where he made his will. Galileo saw him only once again, "more melancholy than ever."[105]

In alarm Galileo advised Ricasoli's uncle Lorenzo Giacomini that Ricasoli "needed very great and urgent care." Giacomini knew something about madness. In a famous lecture to the Accademia Fiorentina in 1587, he had agreed with Aristotle that the *furor poetico* was an affliction of the black bile, that is, a form of melancholy; when enhanced by astrological influences and wine, it stimulated "potent phantasms," powerful imaginings, compulsive behavior, "arising not from one's own discourse and judgment, but from nature."[106] Alas! doctors neither of the body nor of the soul could cure Ricasoli's *furor*, and his expectation of an early death was soon fulfilled.

The court wanted to know how Galileo had arrived at his diagnosis of melancholy:

> Q. How do you know when someone is out of his mind?
>
> Q. By many signs, particularly believing in things entirely false and impossible.
>
> Q. How do people deranged by melancholy humors act?
>
> A. Some think themselves wild beasts, others that they have monstrous limbs, others that they are dead, and most, "according to the doctors," fear that they will meet a violent death.
>
> Q. Is a gentleman who prays all the time of sound mind and is prayer a form of madness?

A. Praying continually is not a symptom of madness since prayer is not a form of madness.

Q. Is it reasonable or not to think about death and is it a good idea for a reasonable and prudent man to prepare for it through prayer or otherwise?

A. Thinking about a natural or violent death may be reasonable as is preparing for it through prayer and other means.

Q. Is everyone who suffers from melancholy humors mad?

A. Those who suffer much from melancholy humors can be supposed mad, since melancholy humors are among the sorts of infirmity that according to doctors attack the mind; and people who suffer from such humors cannot be deemed very reasonable for only those so qualify who in all their actions make perfect use of memory, speech, and imagination.

Q. Have you seen people in Florence with melancholy humors who are not thought crazy?

A. No.

Q. Can people who take medicine continually have problems and become delirious about their disease?

A. There are diseases in which the victim usually is delirious, full of imaginings, and upset by medication.

Galileo had learned something about the symptoms and treatment of disorders caused by melancholic humors in his medical course. As he pointed out to the court, almost everyone given to study was melancholic. That was an old association, taught by astrology and confirmed by the pasty faces of bookworms. A disease caused by a surfeit of melancholic humors that impeded operations of the mind was something altogether different.[107]

Galileo was to have bouts with melancholy all his life. He was often ill and in debt, and when solvent and well, worried and hypochondriac. Like his poor friend Ricasoli, he did many self-destructive things and imagined himself pursued by a legion of enemies. The last and greatest of his melancholic acts was his late-in-life challenge to the Roman Catholic Church. That raised an individual's melancholia to world importance. We know from the great anatomist of melancholy, Galileo's younger contemporary Robert Burton, that the condition could produce a conviction of knowing the truth and a missionary zeal to impose it on others; in its severest form it brought

the megalomania that causes its sufferers to found religions.[108] Before he
contracted this advanced form of melancholy around 1610, Galileo exhibited only the mild melancholic symptoms of uncertainty, protectiveness, circumspection, ironic humor, and scholarly arrogance. As a young man and
even well into middle age, he did not appear likely to suffer from paranoia or
hypochondria. He was strong and robust despite recurrent illness, tall and
stocky, with a light complexion, slightly reddish hair, and a look between a
grin and a leer (Plate 3).[109]

Several copies of the transcript of Galileo's testimony about Ricasoli
exist with vigorous unfriendly marginal comments ("lies...untruths...sto-
ries"). The annotator, presumably a lawyer for the defense, supported these
charges by portraying Galileo as irresponsible and unstable. The game with
the priest and the nun was a sacrilege if true and a perjury if false. The supposed sale of his testimony for the sum required to establish his sister in a
convent, a miserable 150 ducats, was doubly contemptible. He was a scoun-
drel, a beggar, the "sfratato figlio d'un maestro di suonare," the unfrocked
son of a music teacher.[110] Galileo's extravagant anger at any accusation of
prevarication ("the greatest abomination") and his tendency to overkill
challenges to his authority may have roots in these baseless attacks on his
character.[111]

2

─⊶⊷⊷─

A Tuscan Archimedes

2.1 HELL AND MATHEMATICS

Galileo's first known public lectures defended the traditional geography
of the Inferno before its guardian, the Accademia Fiorentina. The tradition began with the circle of architects and mathematicians around Filippo
Brunelleschi, creator of the magnificent dome of the Florentine cathedral.
Perhaps, as some think, the design of Hell, for which Brunelleschi's biographer Antonio Manetti has the credit, is just Brunelleschi's dome reversed;
and, indeed, with Zuccari's frescoes of the damned around its walls and the
protruding lantern at its top, the cupola does look like Manetti's Hell upside
down (Plate 15).[1] The first version of the *Comedy* to illustrate his vision, Girolamo Benivieni's edition of 1506, disclosed the plan in a half-dozen maps and
a dialogue between himself and Manetti that later often appeared as a pro-
logue to the great poem.[2]

Manetti's Inferno as interpreted by Benivieni reached perfection in a lecture to the Accademia Fiorentina in 1541 by Pier Francesco Giambullari.
After consulting a mathematician he drew out his discourse to fill a small
octavo of 150 pages, which he dedicated to Cosimo I for "having given
shelter in the Honored Womb [so it is, *Onorato Grembo*] of your Florentine
Academy to every muse wanting to develop her most beautiful ideas in [the
Tuscan] language."[3] Just as the academicians relaxed in their womb with the
latest news from Hell, a rival interpreter from Lucca, Alessandro Vellutello,
declared Manetti's construction ridiculous and impossible. It occupied, he
thought, an open space equal to one-sixth the volume of the earth, which
he ruled out on the impertinent ground that it could not be mechanically

stable. Vellutello crammed the whole of Hell within a hundred miles of the earth's center.[4]

The Florentines threw back the charge of violating gravity and one of them made a model, to exact measure, to show the stability of the standard model, "for love of truth…and a desire to defend Manetti."[5] Dante had known what he was about. The geography of Hell is a reflection of Aristotle's classification of knowledge. Physics, that is, natural philosophy, dominates down through Circle VII, where the Violent reside, as indicated by the naturalistic landscapes. In Circle VIII, that of the Deceivers, mathematics rules: "and thus we will see the poet dwelling on the 'how much' and not the 'why'." At the pit of Hell, with Satan and the traitors, there is only metaphysics, which considers pure form, pure evil, without any sensible matter.[6] All of which is perfectly credible.

Dante did not give enough information to draw a map to scale and Manetti's plan as presented by Benivieni and Giambullari was not easy to grasp. The key datum is the circumference of the ninth of ten concentric ditches making up the eighth circle. The fact that Dante supplied this datum to characterize a region reserved for Fraudsters should have been a warning to mathematicians who thought to exploit the information that the ninth ditch has a circumference of 22 miles.[7] The number implies a diameter of 7. With the further information that the diameter of the tenth or innermost ditch is 3.5 miles, and assuming equal spaces between the ditches, Manetti had worked out that the diameter of the outermost ditch of the eighth circle is 35 miles.[8] Galileo's task was to derive the infernal plan from this meager specification and to show "how wrongly the virtuoso Manetti and the most learned and noble Accademia Fiorentina have been slandered by Vellutello." The way would be painful, Galileo told the academicians; he would have to use mathematical terms, Greek and Latin terms, far from the pure Tuscan in which they delighted.

In Manetti's plan, the Inferno occupies a spherical sector made by rotating an equilateral triangle, with vertices on the earth's surface and center, around its height as axis. Galileo displayed his authority and competence by declaring that this sector does not amount to one-sixth the volume of the sphere, as the corresponding figure of a hexagon in a circle might suggest, but, as he knew from Archimedes, to one-fourteenth. Hell is a hollow cone that cuts the earth's surface in a circle whose diameter equals the earth's radius r (Figure 2.1). Dante supplied measures, though not in the *Divine comedy*: arc AJB, centered on Jerusalem, is 3,400 miles (r = 3,245 miles).[9]

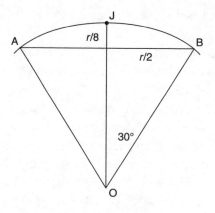

FIG. 2.1 The cone of Hell. O is the earth's center, *r* its radius, J Jerusalem; arc AJB = 3245 miles.

The Inferno comprises nine "circles" arranged in eight levels, like tiers in an amphitheater. The first four circles have no subdivisions; the fifth and sixth, which occupy the same level, have three between them. The sixth level or seventh circle also has three subdivisions, making ten distinct arenas from the first circle (Limbo) down to the seventh (the Violent). A great pit below the Violent, which Dante and Virgil negotiate on dragon-back, ends at the Deceitful, the eighth circle with its ten ditches. Within the innermost ditch is a deep well guarded by giants, at the bottom of which is the frozen ground of the Treacherous, distributed into four regions tight around the earth's center. There stands Satan, or rather his navel; from the waist down he is inverted and stiff in ice; from the waist up, he dines without digesting, chewing simultaneously the worst of traitors, Brutus, Cassius, and Judas, in his three mouths. To find the distances from the devil's belly button to the centers of all the circles based solely on the diameters of the ninth and tenth ditches of Malebolge (the eighth circle), and also to calculate their widths and circumferences, demanded a grand architectural vision.

Galileo illustrated his vision with two maps, of which one or more versions may survive in a portfolio of Hell pictures dating from around 1590. It contains illustrations by Stradano and by Galileo's friends Cigoli and Luigi di Piero Alamanni (Plate 2).[10] While his auditors kept their eyes on some such maps, Galileo continued with his mathematics. He divided the radius JO into eighths and the arc BJ, which is 1,700 miles long, into intervals of a hundred miles (Figure 2.2). To find Limbo, the first tier, he drew the lines O1 and κK. The piece I, cut off from κK by OB and O1, is the floor of Limbo in the plane of the drawing; the ring described by rotating I around the axis OJ is the tier

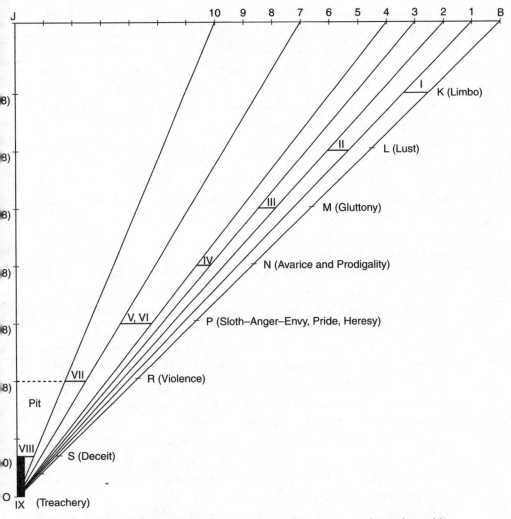

FIG. 2.2 The mathematics of Hell. The horizontal lines I, II…indicate the widths of the levels; the divisions along JB, intervals of 100 miles; σO is the Well of the Giants, not drawn to scale.

or circle of Limbo. The veriest tyro in geometry then could work out that the width of I is 87.5 miles. Galileo made a fuss of it and attributed the principle of the calculation to Archimedes. The width of the second circle, home of the Lustful, is the piece of λL cut off between O1 and O2; it comes to 75 miles. The third circle, of the Greedy, and the fourth, of the Avaricious, are made in the same way: III = 62.5 miles, IV = 50 miles.

To make room for the three divisions of level five, that is, Circle V (the Slothful–Angry–Envious and the Proud) and Circle VI (the Heretical), 300 miles on earth must be set aside, giving 37.5 miles for the width of each division. Circle VII (the Violent), with its three divisions, takes up another 300 surface miles, for a total width of 75 miles, 25 per division. This brings the leading edge of the tier to 175 miles from the axis OJ and leaves a distance of $r/4$ to be apportioned between the eighth and ninth circles. We know that the width of the widest ditch in VIII is 17.5 miles. Let it correspond to the remaining 700 miles on the earth's surface. Then it lies at a distance of $17.5r/700 = r/40$ from the earth's center, and the great pit between VII and VIII has a depth of $r/10$. The height of the Giant's Well and the Siberia of the traitors amounts to $r/40 = 81$ and $3/22$ miles. Galileo and Manetti were very precise. Greater precision can be obtained from the information that the torso of a giant is about 30 spans or 3 frieslanders or three times the height of the great bronze pinecone at St Peters; that a giant frozen waist deep set Dante and Virgil at the bottom of the well; and that Lucifer is of such a size that his arm is to a giant what a giant is to Dante. Galileo follows up these data, but we will not follow him.

The challenger Vellutello proceeded oppositely to Manetti, from the center upward, using the same dimensions for the devil and the ditches but otherwise building his stingy Stygia from the neat but arbitrary rule that the rise between consecutive levels equals the radius of the lower one. The pit between the seventh and eighth circles has a height of 140 miles and a constant diameter; elsewhere the descent slopes, but not along lines directed toward the center. These circumstances—the cylindrical pits and the non-radial inclines—gave Galileo an opening through which to introduce the heavy if irrelevant artillery of physics.[11]

Giambullari had allowed that the canopy of Manetti's amphitheater—an unsupported chunk of land 3,246 miles in diameter and, at its greatest depth, 408 miles thick—would fall into the empty space below it. For this insight he claimed the authority of Aristotle. In the text in question, Aristotle taught that a body possessing absolute heaviness (earth) moves in a right line toward the center of the earth if nothing blocks the fulfillment of its destiny.[12] A stone lintel can sustain itself though not lying on a line through the earth's center because "cohesiveness" holds it together as gluons do quarks; but, as Giambullari deduced from Aristotle and Benivieni inferred from common sense, cohesiveness will not span 3,200 miles.

Galileo disagreed. After criticizing Vellutello for building structures subject to rock slides, he attempted to save Manetti's conical Hell, where, apart from the canopy, all the tiers are supported by solid earth. And the canopy? It too is stable, according to Galileo, who here, perhaps for the first time, opposed Aristotle with a pseudo-experiment. "It is easily shown that its thickness is sufficient; for a [model] with an arch of 30 braccia needs only a thickness of four braccia, or even one of a single braccio, or perhaps a half, to sustain it." Manetti's construction easily provides this security. "Whence we can persuade ourselves that in its overall description his Inferno is very much more credible than Vellutello's."[13]

In this typically clever argument, Galileo replaced physics by mathematics, scaling up his model without regard to the strength of his material. Later he gave prolonged attention to the problem of the cohesion of solids, and, as will appear, arrived at the non-Aristotelian conclusion that literally nothing holds them together.[14] The form of Galileo's argument for the stability of Manetti's vault often recurs in his later polemical writings. The rhetorical move, "not only is it stable under a scale model, but also under one a quarter or even an eighth as thick," became more fateful, though not always more reliable, when transferred from the imaginary world of souls and devils to the supposedly real one of philosophers and priests.

Many parallels in order and content suggest that Galileo studied Manetti's system in Benivieni's dialogue. Benivieni has his interlocutors praise the invention of an off-stage inventor, a technique Galileo would employ with himself as beneficiary; and, again like Benivieni, Galileo would leave his readers no alternative but acceptance of the invention. Benivieni: "If it is not as you say, it is almost impossible that anything else could agree so well." Manetti: it is all demonstrated, "by commutated proportion." And then, with Manetti off stage: "what more is there to say?"[15] Perhaps this: when asked to assign exact places to certain sinners, Manetti backed off. "My scythe does not extend to such crops." Benivieni acknowledged that the task is not appropriate to a mathematician.[16] Mathematical argument pursued resolutely in its own right can preempt other considerations, for example, prudence and forbearance. Dante bothers with exact numbers only in describing the deceitful and the treacherous. Galileo did not take the lesson. He would be brought down by pushing the rhetoric and application of mathematics beyond prudence, and in a manner not free from deceit.[17]

2.2 BARYCENTRIC EXERCISES

Galileo launched his mathematical career on a few propositions in the Archimedean idiom and an instrument to implement the line of thought that Archimedes had started in the bathtub. As Galileo knew from Vitruvius via Buonamici, Archimedes solved theoretically and without invasive procedures a royal problem with plebian applications: did a crown offered as pure gold contain a large amount of silver? Archimedes took advantage of the difference in density of the two metals. Since density (δ) is weight (W) divided by volume (V), he needed a way to measure the volume of the crown and of samples of the pure metals. His method, inspired by the overflowing bath, was to drop the objects into a full basin and measure the amount of water each displaced. Then by weighing them in air he had the information to deduce their densities. The crown's fell out between those of gold and silver. Thus did the clean mathematician detect the foul counterfeiter.[18]

Galileo judged this method ill-suited to the dignity of Greek geometry. According to our revisionist historian of science, Archimedes used a special balance, in which he weighed the samples first in air and then in water. There was no need for nudity or mopping up. Galileo designed the necessary apparatus (a little balance, a *bilancetta*), "which I believe to be the same as Archimedes used since besides being very accurate it depends on propositions found in his works."[19] In Figure 2.3, the object *h* under investigation hangs from B and a counterweight *d* from A; when weighed in air, they are in equilibrium around the center C of the balance arm AB. Now plunge *h* in water: according to Archimedes' theorem, it loses weight by an amount equal to the weight of water it displaces. To recover equilibrium, *d* must be moved toward C, say to E; then $W{:}w = CA{:}AE$,

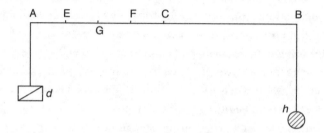

FIG. 2.3 Schematic of Galileo's *bilancetta*.

where W is the weight of h and of d in air and w is the weight of water displaced.[20] This solves the problem since, taking δ(water) = 1, δ(h) = W:w = CA:AE. Let weights of silver and of gold equal to the weight of the crown in air successfully occupy the place of h, and let the equilibrium positions of d be E for gold, F for silver, and G for the crown; then its proportion of gold to silver by weight is as GF to GE.[21]

Galileo did not quit here, as Archimedes had, but, with an inspiration both practical and musical, showed how to measure the required distances accurately. Wrap a couple of windings of a very fine steel wire around E, F, and G, and fill the spaces between them with tightly wound coils of a very fine brass wire. The distances GF and GE can be determined by counting the turns of wire, best done by running a sharp stiletto down them so that both by ear and feel you can keep the count accurately. But be careful (Galileo advised) when interpreting the results to note that the order reverses expectation: the distance of G to the silver mark F indicates the gold content, that from G to the gold mark E the silver.[22] A similar reversal underpinned the theorems on which Galileo's early reputation as a mathematician rested.

In his treatise on floating bodies, Archimedes stated that the center of gravity of the figure generated by rotating a parabolic section around its axis lies a third of the way from base to apex.[23] He offered no proof, perhaps because he thought it too easy to bother with. That is not how Del Monte's teacher and Archimedes' editor Federico Commandino of Urbino saw it. "[It is] a most difficult and most obscure matter." No one had proved anything about centers of gravity of solid bodies, as far as he knew. "Unless therefore I love my own productions excessively, I believe that my treatment will be of considerable use and a great pleasure to scholars." Galileo did not like it, nor have later connoisseurs.[24]

Commandino used Archimedes' standard procedure, aptly named exhaustion, which approximates the area (or volume) of an unknown figure by inscribing and circumscribing figures with known area (or volume) in or around it. In Figure 2.4, successive disks of height $a/4$ and ordinates p, q, r begin the process of exhausting the paraboloid. The volume of the disks is $(\pi a/4)(p^2 + q^2 + r^2)$. Their individual centers of gravity fall at distances $a/8$, $3a/8$, $5a/8$ above O, and their combined center of gravity or barycenter at some point X near Q. If AO were a horizontal balance and OX = x, the situation of Figure 2.5 with $k = \pi a/4$ would result. For equilibrium, $x(p^2 + q^2 + r^2) = (a/8)(p^2 + 3q^2 + 5r^2)$. Since we deal with a parabola for which the ordinate along the

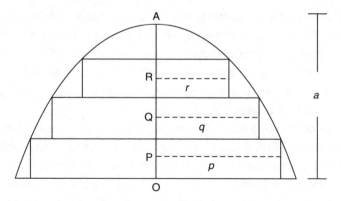

FIG. 2.4 Archimedes' method of exhaustion of a paraboloid.

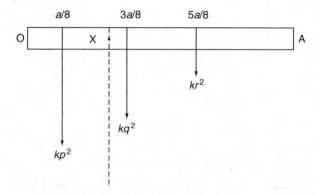

FIG. 2.5 The center of gravity X of the disks of Fig. 2.4.

axis AO is proportional to the square of the abcissa, $r \propto 3a/8$, $q \propto 5a/8$, $p \propto 7a/8$, if, with Commandino, we substitute the ordinates for the disks' radii. The solution to this bit of algebra is $x < a/3$. With more, smaller disks x would rise toward and perhaps beyond $a/3$. Commandino removed this last possibility in the classical manner by considering disks circumscribed around the paraboloid. This time the center of gravity of the disks occupies a point Y > $a/3$ above O and will decline with more, smaller disks. The center of gravity of a paraboloid of revolution therefore lies a third of the way along its axis from its base to its vertex.[25]

Galileo preferred the time-honored technique of solving a different problem from the one proposed. He offered the situation depicted in Figure 2.6a, in which a balance arm suspended from its center of gravity at X supports a series of weights at equal intervals, the second being twice, the third three times, and the fourth four times as heavy as the first. Suppose D to be at the center of the arm and let all the weights *a, b,...* have the same value; they have different labels in the figure to make the analysis easier to follow. The center of gravity of the five *a*'s is at D; of the four *b*'s, half way between C and D; of the *c*'s, at C; of the *d*'s, half way between A and C. If they are all moved into these positions, the balance arm would remain in place: X does not move. The situation is that of Figure 2.6b: the same distribution of weights obtains, but in reverse order and with half the distance between successive points of suspension. (Here is where the *bilancetta* reversal comes in.) X must therefore divide the two balances AB, AD in the same proportion: AX:BX = AD:AB

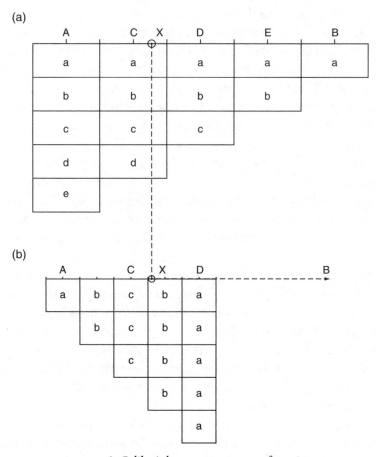

FIG. 2.6 Galileo's lemma on centers of gravity.

= 1:2. Hence the point X lies one-third of the way from A to B. The center of gravity of the paraboloid of revolution is an anticlimax. Let it be divided into disks of equal height as in Commandino's scheme. Since the weight of each disk is proportional to the area of its base, and since, by the "symptom" of the parabola, this area is proportional to the distance of the center of the base from the vertex, Galileo recovered the situation of the balance with an arithmetically increasing load. The center of gravity of a parabolic section therefore lies along the axis one-third of the way from the base to the vertex.[26]

The goal of Galileo's play on the balance was to determine the center of gravity of solid figures. He began with the progression of weights just analyzed and added to it four others, starting the second under E, the third under D, and so on. That places a weight of 1 under B, 3 under E, 6 under D, etc., or, in an obvious shorthand $\{1,3,6,10,15\}$. The sequence $\{1,2,3,4,5\}$ has its center of gravity X_1 at AB/3, counting from A. X_2, the corresponding point for $\{0,1,2,3,4\}$, lies at AE/3, that is, at C; similarly X_3 lies at $(2/3)$AC, X_4 at $(1/3)$ AC, and X_5 (for $\{0,0,0,0,1\}$) at A. Reversing as before, the weights would be in equilibrium on a lever arm AX_1 = AB/3 with center of gravity at point Y_1, where $AY_1:BY_1 = AX_1:AB = 1:3$. The barycenter of $\{1,3,6,10,15\}$ therefore lies at C, one fourth of the way from A to B. It would be a good guess that the barycenter Z_1 of staggered sequences $\{1,3,6,10,15\}$, which produce the distribution $\{1,4,10,20,35\}$, lies at AZ_1 = AB/5. Galileo stopped with two sequences, $\{1,3,6,10,15\}$ + $\{0,1,3,6,10\}$ = $\{1,4,9,16,25\}$, which gives loading proportional to the squares of the distances. He deduced that its barycenter Y must lie between those of the constituent sequences, Y_1 and Y_2, that is, between AC and 3AC/4.[27]

The same can be said for the position of the center of gravity G of a right cone whose axis equals the lever AB, and whose cross-sections are approximated by a set of inscribed disks of the same height, as in Figure 2.7, and by a similar set of circumscribed ones. The barycenter G_i of the inscribed system lies below, that of the circumscribed system G_e above, G. This is all that Galileo could get from his technique of reversing weights around their center of gravity. To show that G divides the axis in the ratio 1:3, he had to employ the method of exhaustion. He had not developed the appropriate analogy with the balance. That would have required dividing the line AB ever more finely and charging each division according to the law of squares. For five divisions, $\{1,4,9,16,25\}$, the barycenter falls at $(10/11)$AC;

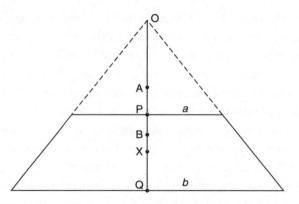

FIG. 2.7 The center of gravity of a truncated cone.

for 9 divisions,{1,4,9...81}, at (54/57)AC. With increasingly finer divisions, it approaches C as closely as desired. This would have corresponded to diminishing the heights of the cone's inner and outer cylinders *ad libitum*, driving G_i and G_e toward coincidence at a point one-fourth of the way from the cone's base to its vertex. Instead, Galileo abandoned the procedure of his lemma and applied the method of exhaustion directly to the cone in a standard *reductio ad absurdum*.[28]

This result sets up the final proposition, which Galileo regarded as his special "invention." It states, in the jabberwocky formulations of the old geometry, that the frustum of a cone or pyramid has its center of gravity at a point that divides the distance between the smaller and larger bases as "three times the greater base plus twice the mean proportional between the greater and the smaller bases plus the smaller base is to triple the smaller base plus the said double of the mean proportional distance plus the greater base." In plain algebra, if *a* and *b* are the radii (or sides) of the bases, then, if X in Figure 2.7 is the center of gravity of the frustrum, PX:QX = $(3b^2 + 2ab + a^2):(3a^2 + 2ab + b^2)$. Galileo demonstrated this equivalence through a welter of opaque proportions based on the clear background idea of the law of the lever—but not employing his ingenious lemma. Taking the known centers of gravity of the small and large cones (at A and B, respectively) and the unknown center of the frustum at X, he had AB:XB = ([B]−[A]):[A], where the bracketed quantities indicate the volumes of the two cones. From this equation and the known relationships PA = OP/4, OP:OQ = *a:b*, [A]:[B] = $a^3:b^3$, the desired ratio, PX:QX = (PB + BX):(QB−BX), can be obtained with the expenditure of a little algebra.[29]

Galileo's theorems showed his mastery of the ancient methods but did not command immediate consent. Florentine adepts to whom he showed them spied a *petitio principii*, or vicious circle, in his trick of reversing weights without moving their center of gravity. Had he proved or just assumed that the center of gravity would remain in place during the mental manipulation of the weights hanging from it? Galileo appealed to outside experts. He first approached the dean of Jesuit mathematicians, Clavius, to whom he delivered his propositions in person in Rome toward the end of 1586. Clavius encouraged him, not only because of the theorems but also for the sake of their common discipline. Still, mathematics must be correct as well as respected, and Clavius suspected that his new protégé had indeed committed the *petitio* of which he stood accused. As for the golden proposition, the center of gravity of a truncated cone, it was too much for Clavius. "I've not yet had time to review this demonstration. I'm waiting for an opportunity to refresh my mind about the subject..."[30] A similar answer came from Guidobaldo del Monte, who happened to have in press a paraphrase of Archimedes on equilibria and centers of gravity, "a refined and profound science," which, he thought, still lacked a good demonstration of the barycenters of paraboloids and conoids. Galileo's allegedly circular proofs would not do.[31] In fact, Galileo's reversing trick is sound but irrelevant.

Galileo's answers to these objections were modest but firm: he wrote with due deference to social position and seniority but perfect equality in mathematical matters. He sent a more refined version of his argument; Clavius was not convinced, and hedged. "I am not an oracle." The Archimedean martinet del Monte soon came around: he had not understood at first, owing (he said) to the conciseness of the argument. Everything now agreed perfectly and Guidobaldo recognized Galileo as his master.[32] Whether Clavius ever managed to persuade himself of the soundness of Galileo's theorems does not appear from the surviving correspondence. While Italians weighed their judgments, Galileo had the satisfaction of receiving the endorsement of a mathematician from Antwerp, Michel Coignet, who received a copy of the theorems through the mapmaker Ortelius, who had them who knows how. Coignet understood immediately because, he wrote, he had obtained the barycenter of a frustum of a paraboloid in a manner similar to Galileo's treatment of the frustum of a concoid. Galileo's way was more general: "your invention is worthy of acceptance by anyone who cultivates these arts...and we offer you endless thanks for such a favor." In exchange, Coignet propounded a geometrical problem, "which we solve with a little application of the procedures and rules of the great art, or algebra."[33] That would not have recommended it to Galileo.

In order to secure ownership of his unpublished Archimedean exercises, Galileo had them certified by friends of high standing: Bardi, Strozzi, Ricasoli, and Alamanni.[34] The choice confirms that during his gap years Galileo identified more with the interests and members of the Alterati than with mathematical problems and mathematicians. Since the signatories, especially Bardi and Strozzi, had easy access to the grand duke, they may well have signaled Galileo's merits to him and joined with the del Monte brothers in recommending Galileo for a lectureship in Pisa.

2.3 DE MOTU

Borro and Buonamici

When he returned to Pisa as professor in 1589, Galileo became the colleague of his former instructor in physics, Buonamici. Borro too was present, virtually, through his books and polemics, though he had left for a post in Perugia. Neither practiced physics as we do. Insisting on their role as presenters of Aristotle's texts in a form as close to the original as they could get, they did not aspire to develop a new philosophy, or even a new physics.[35] Borro's concept of his job—"effac[ing] myself before those most eminent men in all the liberal arts who speak through me...and from whom I have the science that makes me what I am"—sufficiently indicates the goods he had on offer. "I am not one who thinks himself able to find something new."[36] Like many devoted spokespeople, Borro came to live, breathe, and defend the doctrine that, according to his exegetical principles, he merely transmitted. Michel de Montaigne, who visited Pisa during Galileo's first year there, discovered that although Borro could be good company he was "such an Aristotelian that his...rule of all sound ideas and true knowledge is conformity with the teachings of Aristotle; beyond that, there are only chimeras and nonsense."[37] Borro said as much of himself: "there is no logic above or beyond what Aristotle teaches, nor any other method than what I intend to teach."[38] Careful and responsible exegetes too easily can take as truths of nature what are only accurate readings of texts.

Borro had not made his mind politically correct when closing it. He belonged to the Averroistic line of Aristotelian commentators, who differed over certain points in the capital subject of motion from the Greek line represented by Buonamici. Believing that Averroes, "a philosopher beyond praise," had solved all the little problems that troublemakers had raised against Aristotle's system, Borro saw no reason to read or cite any authority later than the twelfth century.[39] He was prepared to suffer for his faith. One of several dangerous old doctrines he taught held that the rotating sphere of the stars contained the entire universe: nothing lay beyond, no heavenly bodies, no throne of God. The local inquisitor instructed him to introduce a Christian heaven beyond the stars. Borro responded by declaring from his podium, "I have maintained and proved that nothing exists beyond the [stellar] sphere; I've been told to retract; I assure you that if there is anything there, it can only be a dish of noodles for the inquisitor."[40] The Inquisition rewarded Borro's intransigence by maintaining him for a time at its own expense in one of its prisons. Galileo may well have been Borro's student in 1583 when he returned, shaken, from this last of several incarcerations his teaching had earned him.

It is hard to find people nowadays who worry whether "simple light and heavy elements are moved in a straight motion *per se* and by themselves, or *per alia*, by other bodies." For those who do, the Averroistic answer is *per se*, the Greek tradition *per alia*, but in practice both schools took both agencies into account.[41] Fire, the lightest material, proceeds upward, away from the earth's center, and earth downward, by their natures, absolutely; whereas air will fall in fire but rise in water, and water is heavy in air and light in earth. With these obvious things premised, it is easy to get into trouble. Take the question whether air weighs (has a tendency downward) in air. No, because if so the atmosphere would be in perpetual vertical motion; or yes, because an inflated bladder is heavier than an empty one.[42] Borro chose the second alternative and so could solve the puzzle why a big chunk of wood falls faster than a small piece of lead in air but swims in water while the lead sinks. Wood falls faster because the large amount of air it contains weighs something in air; it floats because the same air is light in water; whereas the lead, containing no or little air, is heavy everywhere.[43]

Bodies are moved in natural motion in proportion to their gravity (or levity) whether absolute or relative, and are slowed in proportion to the resistance of the medium through which they move. Hence, as Aristotle taught, all bodies would have the same speed—infinite—in a vacuum. Since that is, or sounds, absurd, void cannot exist in a peripatetic world. Natural motion does not occur at a constant speed, however; heavy objects move faster as they near the earth; light ones as they ascend toward the moon. Why the acceleration? Borro offered several standard explanations and picked "antiperistasis": propulsion by air continually rushing in to fill the space vacated by the moving body in order to prevent the formation of a vacuum.[44] He took up several questions harder even than these, for example, whether another earth would draw ours or vice versa, and how magnetism works. But since, as he says, these are most difficult matters in a subject already sufficiently perplexed, and as Galileo did not follow him there, neither shall we.[45]

During the same year, 1583, when Borro stayed for the last time with the Inquisition, he published in Florence the third edition of his dialogue on the tides, *Del flusso e riflusso del mare*. Galileo knew this book and proposed the same title for the work we know as the *Dialogue on the two chief world systems* (1632). In both dialogues the author cast himself in the lead role and worked his way through an entire cosmological system before arriving at the topic announced in the title—with the difference, to be sure, that Borro wrote to defend, and Galileo to destroy, the traditional cosmology. Borro dedicated his book on the tides to the first wife of Francesco I, Giovanna d'Austria, who had an interest in astronomy. The professor approached the duchess through the Salviatis, a family tied by interest, intrigue, and marriage to the Medici. Borro had worked as secretary for Cardinal Giovanni Salviati and knew the family well; "I was, and will be while I live, and after I'm dead too, if that is possible, [the Salviatis'] most obedient and affectionate servant."[46] Galileo's spokesman in the *Dialogue* on the world systems was another Salviati, Filippo, a shoot of the tree Borro tended.

Borro's dialogue, written in Italian to attract the same sort of audience that Galileo later addressed, takes place in the gardens of the Pitti Palace under the genial direction of the grand duchess. (Galileo placed his *Dialogue* in the palace of a Venetian nobleman who acted as host and master of ceremonies.) The discussion begins with Borro's observation that the grand ducal gardens were much more pleasant in the summer heat than Roman villas with their midges and mosquitoes and brackish water polluted by the lead linings of the aqueducts. The duchess suggests that they find something to talk about that will keep their minds off the temperature. Borro cannot think of anything. Giovanna: nonetheless you are going to talk, in your usual antiquated way. A third interlocutor suggests the refreshing topic of the tides. Giovanna agrees: she has heard much learned talk about them but nothing that made any sense.[47] Borro holds back: he does not know Tuscan perfectly (he came from Arezzo), knows nothing of modern philosophy and only a little of Latin literature, and for relaxation reads only Greek and Arabic. He does not escape. Neither do his auditors.[48] He begins with a brief account of the main features of the Aristotelian world—its three dimensions, its upness and downess, its four elements and effective qualities, and so on—and Aristotle's explanation of them as expressions of perfection and proportion. Galileo was to begin his *Dialogue* in exactly the same place in Aristotle's *oeuvre*, in order to reject explanation by perfection.[49]

Another interlocutor interrupts: what has all this to do with the tides? Borro: "good philosophers always proceed in this way when nothing is known and everything is sought." A lucid account of geocentric cosmology follows, down to the motions of the moon, which brings the discussion to the theory of the tides: they are swellings of the ocean caused by moonlight. The maximum effect occurs around the meridian passage of the moon because then the lunar rays strike more nearly perpendicularly and work most effectively. The company acknowledges that the theory captures the connection between the moon and the tides. But then, they ask, why do tides also occur at new moon and under thick clouds? Anyone interested in Borro's answers will find them toward the end of his treatise.[50]

With these indications that Galileo had learned something from his former teacher, let us return to the perplexing question whether air weighs in air. Borro affirmed that it did. He had thrown a chunk of wood and a piece of iron, which he thought equal in weight but, typically, had not weighed, from an upper window of his house. As often as he did it, he and his students noticed that the wood fell faster than the iron. Buonamici obtained the same result in a similar experiment of greater precision. Pisa must have been a dangerous place when its philosophers were thus philosophizing. Borro explained to survivors that the air in wood counts dynamically but not statically: the wood and the iron could weigh the same on a balance, where the included air behaves as a free sample; but when the wood falls, the conveyed air in it adds its weight to the whole. That a wooden object descended faster than a metal one of equal weight and similar shape was considered a secure phenomenon.[51] Belief in it, as well as mastery of the intricate Averroistic tradition, was one of the many debts Galileo owed to Borro.[52]

Pisa's junior professor of philosophy, Buonamici, was more learned and prolix than his senior, an excellent philologist, and the author of an immense summa, modestly entitled *De motu*, of all ideas ancient and modern about motion. Galileo possessed this compendium and used it as a reference when developing his own quasi-Aristotelian theories. The size, scope, and depth of Buonamici's black hole of 1,000 dense folio pages made an optical demonstration of his precept that the doctrine of motion was the core of philosophy.[53] It also gained him a raise in salary of 15 percent, from 330 to 380 ducats, not a bad return on an academic book.[54] Among the many topics later of intense interest to Galileo treated by Buonamici were the relationship of mathematics to philosophy and the merits of Copernican theory.[55] Buonamici was willing to allow mathematicians a place at the symposium of knowledge. "What is absurd in mathematics [he conceded] is also troubling in philosophy."[56] But only troubling. Take the absurdity of the famous *nova* ("new star") that illumined the heavens and darkened philosophy in 1572. Astronomical observation placed it beyond the moon among the stars and planets. According to peripatetic physics, however, the heavens consist solely of "quintessence," a substance neither light nor heavy, constrained to move in a perfect circle around the center of the universe, and incapable of other alteration. Philosophically speaking, the nova could not be above the moon. Mathematically speaking, it had to be among the stars. To the uncomfortable question, can a mathematical result that "troubles" philosophy kill a physical theory, Buonamici gave an unequivocal "no."[57] Galileo would answer loudly, "yes."

Buonamici dismissed Copernicus' moving earth and central, stationary sun because they violated Aristotle's demand that philosophy be built on confirmed, everyday, commonsense experience. Among the obvious contradictions to experience Buonamici adduced was the thought experiment that Galileo would later claim in favor of Copernicus: dropping a weight from the mast of a moving ship. According to the teacher, the weight would fall to the poop; according to the student, at the foot of the mast.[58] Perhaps neither tried it. Buonamici insisted not only that the earth rest, but also that it be absolutely still, a rhetorical overkill that might have been directed against one of his colleagues, Andrea Cesalpino, a professor of medicine at Pisa. Cesalpino had the pretty taste of decorating the walls of his house with triangles from Plato's creation myth *Timaeus* and balls from the escutcheon of the Medici. Buonamici did not aim his criticism at this curiosity, however, but at

Cesalpino's suggestion that the earth might dance in place to produce some subtle (and spurious) astronomical phenomena deduced by Copernicus and in the process, by shaking the ocean basins, create the tides.[59]

For the rest, Buonamici was as strong a defender of the *libertas philos-ophandi* as Borro and perhaps more anticlerical. He introduced the species "priest" into the Aristotelian classification of sentient life as a link between man and beast. In none of these animals, human or not, did Buonamici allow an immortal soul. A religious friend, a follower of the Thomistic synthesis, asked him if he had read St Thomas. Answer: "I don't read books by priests."[60] Like Borro, Buonamici had many powerful friends in Florence, beginning with Grand Duke Cosimo I. Unlike Borro, he had a strong interest in modern literature. It did not take him far from Aristotle, however, as he won election to the Accademia Fiorentina with a lecture on Aristotle's merits as a literary critic.[61]

Alexander and Dominicus

Galileo's official teaching at Pisa did not go beyond the *Elements* and the *Sphere*, that is, the first five books of Euclid and a commentary on the *Sphaera* of John of Sacrobosco, a short summary of the first principles of astronomy composed 450 years before Galileo began to teach it. He served it up from Clavius' *Commentary* (1581), which, at thirty times the size of the original, was neither brief nor elementary. Of original mathematical work by Galileo during this period there is not a trace. In addition to the *Elements* and the *Sphere*, Galileo taught astrology to medical students who needed it to determine when not to bleed a patient. The important connection between astronomy and medicine is nicely caught by Galileo's request that his father send him his seven-volume set of Galen and his "[armillary?] sphere."[62] The letter containing the request goes on to announce another important connection. "I intend to study with and learn from Sig. Mazzoni, who sends his regards. And having nothing more to say, I close."

Mazzoni, the defender of Dante, inspired the entire university by the breadth of his learning and the size of his salary. For 700 scudi a year he was reworking his earlier contribution to the philosopher's equivalent of the argument over the relative excellence of Ariosto and Tasso: who is to be preferred, Plato or Aristotle? Mazzoni's first answer, published in 1576, was "neither and both"; rightly understood, the Titans of truth had to agree with one another.[63] In his maturity, Mazzoni realized that the Titans conflicted over many matters of mathematics and metaphysics. Was Plato right in affirming that mathematics best described the constitution of matter, or Aristotle in placing quantity among the least important of nature's accidents? Did the philosopher come closer to the truth by exploring the few mathematical ideas he found in his head or the vast range of experience presented by his senses? Mazzoni decided that Aristotle, otherwise reliable always or for the most part, had gone astray by neglecting the quantitative aspect of things.[64]

There exist among Galileo's papers paraphrases of lectures on Aristotle's physics, particularly on questions related to difficulties it raised for Christians, and on Aristotle's logic, particularly syllogistic and mathematical demonstrations, given originally at the Roman College, the central Jesuit university.[65] Neither the date nor the purpose of Galileo's reworking of this material is known, but from the years in which the courses were given and the paper and handwriting of Galileo's paraphrases, very probably they date from the period of his professorship in Pisa. Their purpose? Perhaps Galileo wanted to bone up on Aristotle for discussions with Mazzoni or prepare for teaching philosophy in Mazzoni's concordist manner. That might explain why Galileo turned to Jesuit material, which he might have had through Clavius or the well-read Mazzoni, since, contrary to the teaching of Borro and Buonamici, the material remained within the Thomistic synthesis.[66] Many of the positions represented in Galileo's paraphrases came from St Thomas himself, including the resolution, in God's favor, of the awkward problem of establishing any certain truth about the physical world in the teeth of His omnipotence. For if He can do as He pleases short of contradiction, He might have made the world operate in ways unlike those established by our strongest demonstrations.[67] Galileo would devote much of his time to repudiating this proposition.

In exchange for guidance in philosophy, Galileo could offer Mazzoni help in looking for decisive examples of the errors Aristotle had committed by depreciating mathematics. He did not have far to seek. The tangled problems of motion that had caused Borro and Buonamici to fill the air with missiles offered many attractive targets.

On Motion. A brief visit to Buonamici's black hole will suggest how hopelessly twenty centuries of commentary had perplexed the questions of free fall and projectile motion. Aristotle explained the acceleration of heavy bodies as a natural consequence of their gravity, their tendency to rush toward the center of the universe. His commentators were not satisfied. They proposed that gravity was a consequence of place, or of a pull toward the center, or of a push from the air. Acceleration occurred because gravity continually added new motion to that already present; or because a body, when it began to fall, possessed an unexpressed "impetus" that held it aloft, and gradually died away, like the heat of a pot removed from the fire; or, finally, because of still other principles, the overcoming of obstacles, the nature of the moving body, all of which might accidentally become causes of motion. And free fall was the easy case! Objects thrown, propelled by the wind, extruded like bubbles from water, and so on, did not move freely. In such cases, how does the "violence" of the initial mover survive in the mobile, and how does the original impulse lose its force? Does the air help or hinder the motion? Does the natural motion, free fall and its tendency, act together with the violent motion throughout, or only after the violence has declined to equality with nature, or to zero?[68] It was from these catacombs that Mazzoni hoped to gain release with a little help from mathematics and Galileo.

Something of the content of their conversations can be inferred from Mazzoni's concordist treatise, *In universam Platonis et Aristotelis philosophiam praeludia*, not printed until 1597, Galileo's response to it, and Galileo's unpublished writings "De motu antiquiora." These last consist of scattered notes, a brief dialogue, and two drafts of a treatise, all in Latin. They date from around 1590.[69] Following a hint from Favaro, the dialogue can be read as a record of the lost conversations between Galileo and Mazzoni.[70] One of its interlocutors, Alexander, identifies himself with Galileo by claiming invention of the *bilancetta*. The other, Dominicus, is a composite of the Pisan students and teachers with whom Galileo used to walk and talk, "disputing over many beautiful and pleasant things." We owe this vignette of literal peripateticism to a fellow student, Luca Valerio, the last great representative of the school of Commandino, to whom Galileo later ceded the palm of "the Archimedes of our age" for his prowess in locating centers of gravity.[71] Their center of discussion, however, was Mazzoni.[72]

Galileo's dialogue "De motu" begins as Dominicus literally runs into Alexander. "That trite old adage, 'motion is the cause of heat,' does not hold well for me," he says; his running has not kept him warm. Alexander agreed to test the question further in a brisk walk and talk. *Dom*: "What shall we talk about?" *Al*: "The first thing that comes into either of our heads." *Dom*: "What about what I just mentioned?" *Al*: "And what was that?" *Dom*: "I said, 'that trite old adage…'" A thorough study of heat was too great a subject for a walk, however, as it required frequent reference to Aristotle, whose works on physics Dominicus and Alexander had studied carefully as had Galileo.[73] Even the motion of light and heavy bodies, as treated "most exactly by Girolamo Borro," would have been too much.[74] So Dominicus proposed a few questions raised by his reading. They were:[75]

1. Does a point of rest intervene "at reflection," where a body thrown upwards begins its descent?
2. What reason can you give that a wooden body of equal size with an iron one will fall faster though it is lighter—if you accept that it will?
3. Why is natural motion faster at the end than at the beginning, and violent motion faster at the beginning than in the middle, and faster in the middle than at the end?
4. Why does the same body descend more speedily in air than in water, yet some bodies fall in air but float in water?
5. What reason can you give that a cannon ball carries further in a straight line if fired at an angle than if fired horizontally, although the vertical is more opposed to natural motion?
6. Why do the same guns shoot heavier balls further than light ones, and iron balls further than wooden ones, although the lighter offer less resistance to the "impelling force"?[76]

Dominicus adds the impelling force of flattery to pull a response from Alexander. "You are accustomed to the cleverest, most certain, and subtlest mathematical demonstrations, like those of the divine Ptolemy and the most divine Archimedes."[77]

Subtle Alexander sets forth. "In violent motion, a certain impressed virtue…" "Hold on, hold on," Dominicus objects, "we must go slowly… Aristotle says that this motion is caused by the medium [not by a virtue impressed by the thrower]. Do you think that this opinion of Aristotle is wrong?" What an opening! The sun was now climbing, the weather warming, and the philosophers of motion moved off toward the seashore, for a good lunch of fresh fish, while Alexander multiplied reasons against Aristotle's teaching that a projectile is pushed by air impelled by a virtue impressed on it by the projector. If antiperistasis worked, how could cannon balls carry against the wind? Or a marble sphere spin while the surrounding air remains still?[78]

The discussion took a fresh direction. *Dom*: "I would like to know why nature [places heavy bodies under light ones] and not the reverse." The tentative answer: "Heavier bodies are closer to the center because things are heavier that contain more matter in a smaller space." By "heavier bodies [*graviora*]," Alexander explains, he means denser ones.[79] He rejects altogether the concepts of absolute heaviness and lightness, of *gravitas* and *levitas*; "*gravia* and *levia* can be taken only relatively." Dominicus demands proof. *Al*: We'll be at this forever if I must disprove Aristotle whenever I disagree with him. *Dom*: Our conversation will be brief if you do not. Alexander obliges with a formal definition. "A body is said to be heavier [*gravius*] than another if it weighs more by volume when weighed in the same medium." Dominicus at first does not understand the need to decide relative *gravitas* by weighing equal volumes. Nor does he immediately see the bearing of the medium. To bring him around, Alexander delivers a short lecture on Archimedes' hydrostatics, illustrated, as if by the master himself, by drawings in the sand.[80]

> *Dom*: "These [are] most plain and certain demonstrations. I see that all those philosophers are wrong who say with Aristotle that air is more heavy than light because it helps heavy bodies fall more frequently than it helps light ones rise." *Al*: "Oh ridiculous chimeras!…Immortal gods, how, please, can anybody believe in them since the contrary is obvious to sense?"

Thus prepared to surpass their teachers, our philosophers take up the conundrum that had bothered Borro. Do elements gravitate in their proper place? *Al*: No. Dominicus was not used to such clarity. *Dom*: "Oh what a beautiful answer! How true and beautiful a solution!"

Let us turn to the point of reflection, where a body flung upward begins to descend. There is no rest there, Alexander declares, it is merely the place where the impressed force, the *vis impressa*, of a violent vertical motion equals the gravity of the projectile; and since this *vis* loses its strength in producing its effect, the equality at the top of the path is fleeting. Thereafter the *vis* continues to run out as the body's gravity overcomes it. What appears as acceleration, says the subtle Alexander, is rather progressive loss of the power to rise: if there were a tower high enough that in falling from it the violent *vis* would vanish somewhere, the body would proceed from that point on with constant velocity determined by its *gravitas* and the nature of the medium.[81]

Dom: Very good. Now what about motion in a vacuum? *Al*: No problem, no impediment. A heavy body that moves more quickly in air than in water

would move faster in vacuum than in air. If so, says Dominicus, Aristotle was
wrong again, since he claimed to demonstrate that motion would be instantaneous in a vacuum on the ground that velocity increases as resistance
diminishes. *Al*: Aristotle did not prove the inverse proportion to resistance,
and he would have tried in vain, as it is false. The correct relation is arithmetic: a body's velocity in natural motion is given by the difference between
its gravity and that of the medium. Its speed in vacuum is fast but finite.
Dom: "Oh! Subtle invention, most beautiful thought! Let all philosophers
be silent who think they can philosophize without a knowledge of divine
mathematics!"[82]

Now warmed to his work, Alexander observes that Aristotle's principle
that the velocity of a freely falling body is proportional to its weight conflicts
with his proposition that a body composed of different elements descends
with a velocity slower than the fastest and faster than the slowest component. Take a small lump of lead A and a larger piece B. Joined together A+B
should fall more slowly than B, which would be held back, on Aristotle's
hypothesis, by its companion A. But again, if velocity did depend on weight,
A+B should fall faster than B. If, however, downward tendency depends on
density, the argument fails. In vacuum, where velocity is rigorously proportional to density, "the denser (*graviora*) falling faster than the rarer (*leviora*),"
ten pounds of lead would fall as fast as 100 pounds in vacuum.[83]

A difficulty occurs to Dominicus. Bodies dropped accelerate downward
just as if they had been projected upward and let fall from the point of
reflection. How did they acquire the equivalent of the *vis impressa* without
the preceding violent motion? *Al*: It comes from the shelf or hand that
supported them against their gravity. Thus all freely falling bodies appear
to accelerate initially. However, if you throw an object downward with a
force equal to its *gravitas*, you can kill its residual *vis impressa* and bring it
immediately to its constant terminal velocity. The effect will be hard to
spot, however, because a constant motion can appear to be accelerated. In
Figure 2.8, the body falls in equal times through the equal spaces BC, CD,
DE. Observed from A it seems to accelerate since the eye estimates the
distances covered by the angles subtended, and angles BAC, CAD, and DAE
progressively increase.[84]

On Mathematics. Mazzoni employed the core of Alexander's doctrine—
the arithmetical rule for velocity with the consequence of finite motion
in vacuum, relative weight, and no *gravitas* in place—in pointing out where

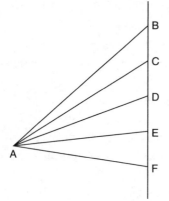

FIG. 2.8 Apparent acceleration of a body moving at constant speed.

Aristotle had propagated opinions "altogether false and absurd...because he did not subject his ideas to the test of mathematics."[85] For example, had Aristotle considered the continuity of motion mathematically, he would have known that there is no rest at reflection. Every astronomer easily could read as much from the usual theory of the planets. In Figure 2.9, P is a planet describing an epicycle around C; the line of sight AP from earth intersects the line connecting the turning points N, U, at T. As P moves from N to U, so does T; and, since the planet's motion is continuous, so must T's be, although Aristotle (and common sense!) might suppose that T rests between changes of direction.[86] Galileo gave a similar argument in his treatise "De motu," with the significant difference that he referred to one of Copernicus' models (without endorsing the Copernican system) and Mazzoni to one by Regiomontanus, the finest Ptolemaic astronomer of fifteenth-century Europe. Further to the attack, Mazzoni criticized Aristotle for teaching that there is no proportion between a curve and a straight line. So did Galileo. And both moderns invoked ancients to make their case: Euclid and Archimedes. The presence in Galileo's "De motu antiquiora" and Mazzoni's *Praeludia* of these two similar, precisely aimed sallies confirms their interdependence.[87]

Fifteen years later Mazzoni sent Galileo a copy of his book containing his exposure of the errors in physics that Aristotle had committed by neglecting mathematics. "I am particularly pleased and gratified [Galileo replied] in seeing that you tend to accept the position I believed to be true and you opposed during the first years of our friendship, when we disputed so merrily. Perhaps you did that [took the part of Dominicus] to display your ready wit, which enables you to sustain false views when it pleases you to do so; or, perhaps, to save in every

FIG. 2.9 Continuity of motion at change of direction.

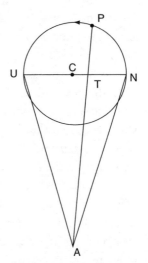

small detail the honor of the teachings of that great master [Aristotle] under whose banner all those do and should march who investigate the truth."[88] Now, the main weapon with which they belabored their master was the arithmetical rule of fall (velocity is proportional to the difference in densities of the body and the medium) that lent itself to physical interpretation via Archimedes' hydrostatics. Very likely Galileo learned the rule from Borro, who knew it from Averroes, whereas Mazzoni, who, unlike Borro, had not stopped his reading in the twelfth century, had it from the court mathematician Benedetti.[89]

The hydrostatic analogy employed by Benedetti, Mazzoni, and Galileo yielded much more than a semi-quantitative rule of fall. It also contained a discovery of utmost importance, the discovery of nothing, that is, nonbeing, emptiness, the Archimedean limit of finer and finer media. Galileo would build his new science of motion on emptiness just as modern cosmology pulls the universe from the "vacuum." The hydrostatic analogy to free fall, however, eventually would have to go. Just as Dante had to part from Virgil, who had piloted him through darkness when the true path was lost, at the top of Purgatory, where he began his climb to the source of all light, so Galileo would have to discard the Archimedean reasoning that took him toward the edge of the Aristotelian cosmos. The extension of hydrostatics, which deals with the equilibrium of bodies at rest, to motion is as difficult to perform as the introduction of a heathen into Heaven.

Besides giving direction to Galileo's thought within the taxingly complex Aristotelian teachings about motion, the discussions with Mazzoni in

Pisa in the early 1590s helped to raise Galileo's sights above the usual business of mathematicians. Mazzoni allowed them not only a say, but even a
veto, in pronouncements about physics made by philosophers. Galileo's
unpublished treatise "De motu antiquiora" responded to this encouragement. It expanded and formalized the ideas developed by Alexander in his
conversation with Dominicus and cleaned up their vocabulary. The treatise recognized that upward motion must always be forced and that the
only natural motion is toward the center of the universe.[90] And the treatise
improved on the dialogue by offering two tests of the theory.

   One concerned a problem that Galileo claimed to be brand new with him:
why heavy bodies moving down planes inclined at various angles to the horizon go more quickly along those most nearly vertical. The answer turns on
an analogy between descent along a plane and the motion of a balance arm.
In Figure 2.10, the arm CD pivoted at A ends in the equal weights c, d. The
weight d tends vertically downward at D along EF with a force measured by,
let us say, the distance AD. Now suppose that d sits at S on the inclined plane
HQ so placed that AS = AD. According to Galileo, d would move along HQ
with the same force that it would exert if suspended from CD at P. Thus the
"force" and so the speed with which d falls along EF is to the force and speed
with which it moves along HQ as AD to AP. But from the similar triangles
APS, HGQ, AS:AP = HQ:QG, or, what is the same, AD:AP = HQ:DF; that is,

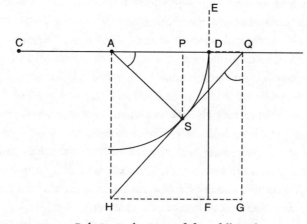

FIG. 2.10 Relative velocities of free fall and motion
along an inclined plane.

the speed of the vertical motion is to that of the oblique motion as the length is to the height of the inclined plane.[91]

Or so it is in the ideal, mathematical case, free from all resistance, either from the plane or from the sliding weight, an unattainable condition, which, as Galileo observed, made experimental confirmation impossible. (Another reason the test would have failed is that Galileo's proposition is false.) In the limiting case of motion along the horizontal, however, the analysis delivered something new and almost true. As the horizontal is nothing but an inclined plane with no inclination, an infinitesimal push should be able to move the weight along it indefinitely. Not quite. Speaking ideally, any displacement along the horizontal implies a rise; but the amount would be very small for any realizable experiment, indeed negligible, just as is the departure from parallel of the directions of fall from the ends of a balance arm. Archimedes had been content to employ such idealizations and abstractions. "He did so perhaps to show that he was so far ahead of others that he could draw true conclusions even from false assumptions... [Let us call it] a case of geometric license."[92] A pity that the geometer cannot correct for the myriad ways in which material bodies escape from his control! That is exactly why philosophers denied mathematics a significant role in physics.

While idealizing and abstracting, Galileo considered the rotation of a sphere whose center coincides with that of the universe. On his principles, the motion was neither violent nor natural, since the sphere's barycenter would not approach or recede from the center to which all bodies tend. Would it spin indefinitely? Galileo raised the question without deciding it. We may infer his likely answer from his teaching that a homogeneous sphere turning around its center of gravity at the earth's surface would rotate forever were there no friction at its bearings or against the air.[93] The natural motion of its falling parts compensates the violent motion of its rising parts. Despite their apparent relevance to the Copernican world picture, these considerations stay within the framework of the Aristotelian system, preserving and exploiting the concepts of natural and violent motion, assuming a central earth, and supposing the heavens to rotate. Galileo's spinning marble ball very probably modeled not world systems but grindstones.[94]

Experiment appeared equally incompetent to test the cornerstone of Galileo's Pisan mechanics, the arithmetic rule for falling bodies. He could find no tower tall enough, even in Pisa, for the experiment. The lightness supposedly acquired by being out of place would not wear away completely during the descent, and the falling body would never reach the constant velocity fixed by the rule. The test was compromised further by the expectation that in free fall wooden balls outpace lead ones at first, a "fact" that Galileo claimed to have tested "often" from a high tower. What could be the reason for so peculiar a phenomenon? Answer: the impressed force is less in wood and at first dies away faster that the corresponding force in a heavier body. "Oh, how readily are true explanations derived from true principles!" But then, why should the greater gravity of lead not kill its impressed impetus as quickly as the lesser gravity of wood does its? "This objection surely has great weight." Did Mazzoni make it? The answer depended on a false dichotomy. The impetus departs, of itself, during the motion, just as hot objects lose their heat when placed in the cold. And as heavy bodies cool more slowly than light ones (as appears from comparing rates at which an oven and a pie removed from it lose heat), so lead retains an impressed motive force more tenaciously than wood.[95] Here the legacy of the Aristotelian doctrine Galileo proposed to better kept him as fully entangled as it did Borro or Buonamici. It would take him twenty years to get free.

2.4 GALILEO AT 25

A wedding in Florence

During May 1589 the Medici threw the grandest party that Florence had ever seen. The occasion was the wedding of Grand Duke Ferdinando I and Christine (Christina) of Lorraine, the granddaughter of Catherine de' Medici, one-time queen of France. The grand duke was then 40 and a duke of only 20 months' standing; previously, for as many years, he had been a cardinal, though without the help of holy orders. His accession was greeted with relief, since his predecessor, his brother Francesco I, had almost all of the vices and none of the virtues of the family. Ferdinando had the opposite mix of virtue and vice and his lavish wedding, in which the entire city participated, cemented his popularity. His bride had the merits in his eyes of advertising Tuscany's liberation from the Holy Roman Empire, to which Francesco had tethered it, and of being uncompromisingly devout (Plate 10). Galileo would take some career steps and missteps based on his calculations of Ferdinando's patronage and Christina's piety.

Galileo's patron Bardi and the architect Buontalenti had overall responsibility for staging the centerpiece of the celebration. This was a seven-hour
theatrical extravaganza: a five-act comedy (the usual form) surrounded by six
intermezzi, each a technological wonder. The poet and architect had worked
closely together before on a similar but lesser performance for the wedding
of a Medici and an Este in 1586.[96] The play for 1589, entitled *La pellegrina*, dealt
with the travels of a pious lady. Bardi supervised the staging, designed the
costumes, and worked out the intellectual program. The symbolism he concocted for the intermezzi floated above the heads of his audience like the
clouds crowded with deities through whom he developed his allegories. Only
an intimate of the Alterati or an unusually informed philologist could understand, though everyone could enjoy, the spectacle. The music too echoed
with a combination of the learned and the emotive, as Bardi fashioned some
of it on the same principles on which that "great genius," his friend Vincenzo
Galilei, had composed the lament of Ugolino.[97]

Buontalenti designed the stage settings, curtains, and backdrops and also
the machinery required to lift mountains and devils through trap doors,
cause the clouds packed with actors to float, and effect instantaneous changes
of light and scene. A hundred men worked the winches, pulleys, and miles
of rope employed during the performance. This, however, represented but a
fraction of Buontalenti's assignments. The theater in the Uffizi had to be refit-
ted, the Palazzo Vecchio expanded and reconfigured, and the gardens of the
Pitti finished and furnished with a temporary water stage for a mock naval
battle.[98] Meanwhile Bardi was setting poems furnished by G.B. Strozzi. These
included several madrigals for the fourth intermezzo, which starred the Prince
of Darkness. It began with the opening of a trap door from which poured
demons wearing silk costumes that looked like snakeskin. Beneath them souls
crackled in flames. Charon the Hellish boatman could be glimpsed as Dante
described him, and Lucifer, standing tall from the waist up above the deep
freeze that contained the rest of him, munched sinners as in Dante's menu.
During the action, the devils on stage sang a dirge, lyrics by Bardi, lamenting
that their business would decline under the gentle reign of Ferdinando and
Christina. When they finished, the whole scene, devil and all, vanished in a
puff of fire and smoke. "The goal of the intermezzi," says Strozzi, "is to stu-
pefy every viewer with their grandezza." An engraving of Buontalenti's stage
set has survived. It was based on Benivieni and Manetti with some advice, we
may presume, from the local expert on infernal cartography.[99]

   Galileo could have participated in the wedding extravaganzas in several other ways as well. Florence had no unemployed artists during preparation for the wedding and no unemployed musicians during it. Ferdinando's staff had to comb the countryside for monks who could sing and gentlemen who could play. Galileo was an expert lutanist. He was also one of those "noble youths" urged to appear as extras, to carry the canopy that shielded Christina and her retinue in the streets, to compete in the joust staged before the church of Santa Croce, or to battle in the naumachia of 18 galleys fought in the artificial basin in the cortile of the Pitti palace. The urging was reinforced by the threat of a fine for absence without a doctor's excuse.[100]

   People from as far as Rome and Venice lined the streets on the morning of Palm Sunday, 30 April 1589, to watch Cristina's progress to the Duomo, which blazed under the light of 38,000 candles. During the ceremony, a cloud filled with angelic musicians descended from the cupola, whose frescoes Cigoli, Zuccari, and others had only recently completed, to rest in front of the astonished bride.[101] All Florence then banqueted, Galileo and his family, parents, brother, and elder sister, as guests of the Ricasolis. We know this particular because during the hearing on Ricasoli's sanity the court asked whether his expenditures in connection with the wedding were excessive.

> *The court*: Is it laudable and proper for a prudent and sound youth, on the occasion of the wedding of his sovereign, as an honor to himself and his family, his city and his prince, to do something out of the ordinary and spend money on literary men and fancy clothes?
> *Galileo, for the literary men*: It is worthy of praise.

The fancy clothes referred to a costume that, if completed, would have cost 300 scudi, or five times the annual salary Galileo would receive at Pisa.[102]

   Dressing up was ubiquitous as well as praiseworthy and an important ingredient of the display of 1589. Chroniclers dwelt on the beauty of the gowns and the women in the audience of *La pellegrina* as Ariosto did on the charms of Alcina and Angelica. All participants in the parades and public events received new livery. The costumes of the gods and heroes who inhabited the clouds were so thickly tricked out in gold and jewels that, according to an eyewitness, "it might well have seemed to everyone that Paradise had opened up, and become the entire stage and setting." The costumes for the intermezzi alone consumed three miles of cloth of all sorts, silk, satin, velvet, wool, which works out to 18 yards each. Many actors and singers would not have been more heavily burdened if they had worn chain mail. The 300 costumes for the comedy and the intermezzi cost 100 scudi on average. Ricasoli was only a little extravagant.[103]

Galileo attended the comedy and intermezzi at the grand opening on 2 May or, perhaps, at subsequent performances (with different comedies) later in the week.[104] Whether employed as an extra or a musician, or seated as a guest of Bardi or Ricasoli, Galileo was present among the Florentine patricians with whom he identified. Two years later, in 1591, a key link to them broke with the death of Vincenzo Galilei, which further compromised Galileo's lifestyle by making him responsible for the welfare of his mother, brother, and sister Livia, and for the balance on Virginia's dowry still owing to her impatient husband, Benedetto Landucci. Galileo's salary of 60 scudi, which, though not high, was well above the lowest in the university, did not suffice to meet his new obligations. Nor had he been able to save anything. During his first year as a Pisan professor he spent so much time in Florence attending to a serious illness of his mother and to the Ricasoli affair that he had to engage a substitute and forfeit pay, which dropped his income from teaching to a little less than the wage of a good tailor.[105] He had made Virginia a substantial gift on her engagement, going, as was his wont, directly to the point: the marriage bed, complete with fringed hangings of silk, purchased and made up in Lucca "for a very good price." Galileo had learned something from his father's brush with trade, although, as a would-be literary man, he denied all knowledge of business.[106]

Tower and toga

We have it on Viviani's authority that Galileo dropped different weights of the same material from Pisa's Leaning Tower to show, "to the dismay of the philosophers," that, contrary to Aristotle, they all fell at the same speed. And he did it not once, or secretly, but "with repeated trials...in the presence of other teachers and philosophers, and the whole assembly of students."[107] Iconoclasts have thrown doubt on this vignette although the tower's tilt made it a perfect platform for the experiment. They objected that no one among the literate throng supposedly present, not even the peripatetic philosophers of motion who went away grinding their teeth, recorded the event. When doubts sprout, the harvest can be bountiful. Iconoclasts have also

savaged the story of Galileo's discovery of the (approximate) isochronism
of the pendulum, which, like the experiment on the tower, exploited a piece
of church furniture. This was the lamp that still sways in the Pisa cathedral.
Its testimony is equivocal, however, because Viviani, from whom the story
comes, placed the critical observation—that the period of the swings is independent of their amplitude—in Galileo's student days, and so before 1585.
The heavy, costly lamp (almost 600 scudi) dates from 1587. Viviani specified
that Galileo determined the isochronism by counting his heartbeats and tap-
ping out a tune, and that he immediately exploited the effect in a pulse meter,
"to the amazement and delight of the doctors of the time."[108] The pulse meter,
or *pusilogium*, a sort of metronome, made its first appearance in print in 1602,
in a book by a Venetian doctor, Santorio Santorio. Previously physicians had
distinguished only 49 different sorts of pulses. With Santorio's compound-
ing of exact pulse rates and imprecise humors, however, they had 80,084,
improving the practice of medicine by 160,000 percent.[109] No more than the
peevish peripatetics, however, did the delighted doctors of Pisa, if there were
any, leave a record of their debt to Galileo.

Owing to the discrepancies in Viviani's account of his master's adventures
with the tower and the lamp, historians take care when he is the only witness. Some have been so bold as to assert that he made things up and that
Galileo experimented only in his head.[110] This was to go too far and, in recent
times, Galileo has become an exemplary, pioneering experimentalist.[111] One
thing for certain can be said about Viviani's stories. Had Galileo stood on
the Leaning Tower throwing down weights and gauntlets before the assembled university, he would have had to wear his toga. The authorities insisted
that professors don their academic gowns when in the town as well as when
discharging their university duties, on pain of the substantial fine of half a
scudo. Galileo so hated this imposition that he wrote a lengthy and irreverent poem against it. The poem tells something about his pursuits and attainments at the age of 25. Many sacred cows came to slaughter by his sharp wit:
university officials, ecclesiastics, academics, philosophers, idiots. And many
youthful preoccupations leave their marks: sex, wine, clothes, money.[112]

Galileo's burlesque follows the style of the critical-satirical digressive
poems of Francesco Berni, who died in Florence long before he could have
regretted seeing his entire *oeuvre* placed on the Index of Prohibited Books.
This did not prevent their publication, however, and Galileo regarded them
highly for their wit and iconoclasm. Although Galileo's "Capitolo contro il

portar la toga" has its zing from his annoyance over the rule of the robe and the stock association of academic dress with pedantry, it was of a piece with his duller sonnets in that it was an exercise to master a literary form. One of Berni's favorite techniques was to treat a common subject in an elevated manner, "in praising [as Galileo characterized the method] the meanest things, urinals, plague, debt, Aristotle, etc."[113] Galileo adapted the technique to his purpose by developing a philosophy of clothes long before Thomas Carlyle's Professor Teufelsdröckh took it up as something new. Also Galileo followed Berni by beginning with the Socratic question, what is the greatest good? Satirists in the Bernesque style enjoyed lampooning the notion of the *summum bonum*, and coupling it with nudity and anticlericalism.[114]

Philosophers have disputed the *summum bonum* for centuries without resolution or imagination, Galileo wrote, because they have not known where or how to look for it.

> The ways of invention are varied, very
> To seize on the good there's but one that has worked
> Look about for an evident contrary.
> That means search out evil, it's easily found
> You've then *Summum bonum*, no trouble at all
> Bad and good are as like as pence in a pound.[115]

If you would know the sinful, look at the good people to whom priests give penance; if you are studying scoundrels and rogues, "it is enough to know priests and friars, who are pure goodness and devotion." In short, to know the best thing in the world, take the easier course and seek the worst. The worst thing in the world is—clothes. Clothes are the source of all deceit. In the good old times, everyone knew what everyone else was good for. A prospective bride could see the equipment of her proposed mate, "See if he is too small, or has French diseases / Thus informed, take or leave him as she pleases." Without clothes, the white, black, and brown robes of the monks would not divide Christians; foremen could not be distinguished from workmen, or patrons from underlings. When people went naked there were no counts, marquises, servants, or paupers. The devil invented clothes as he devised artillery and witches—to the devil with clothes!

If it is God's will, however, Galileo will wear his toga, as if he were a pharisee or a rabbi, "though I am not the least Jewish, even if my name and descent might suggest it." This is a reference to the men of Galilee who stared into Heaven as Christ ascended, not a disclosure that Galileo, like Newton and Aristotle, were Jews. No doubt it is anti-Semitic. Like a rabbi's robe, the academic gown is a cloak for all sorts of skullduggery. Since its wearers feel conspicuous entering a whorehouse, they relieve themselves in other ways more prejudicial to their salvation. To attract less attention and avoid the university's proctor (Cappone Capponi), berobed doctors creep around town on all fours (*carpon carpone*).[116] The toga holds up anyone in a hurry just like (Galileo missed this image) Ruggiero encumbered by his armor.

At work the good doctor is equally ridiculous:

> Why in the world does he not die of shame
> When standing surrounded by eighteen or more
> Bright open-mouthed students awe-struck by his fame
> Looking to all as he parades out before'm
> Like a screech owl among so many robins.

All the screech owls are not equal. Those in velvet gowns are esteemed more highly than those who can afford only worsted. Oh, sighs Galileo, if only men were made like the wine flasks used in taverns, which have so little decoration you can see right through them:

> And yet they are filled with such excellent wine
> It's not surprising if throughout the city
> They say that it's splendid and even divine.
> Bottles, however, straw-covered and pretty
> When opened you'll find to be chock full of sin
> Wine slimy or wat'ry, perfumed or gritty,
> In dec'rative flasks good only to piss in.

Galileo's exercise in the Bernesque manner did not raise his standing at the university. An old professor of medicine, who had joined the faculty at Pisa after many years at Padua, Girolamo Mercuriale, urged him to leave. He should apply for the Paduan chair of mathematics vacant since the death of its admired incumbent Giuseppe Moletti in 1588. Mercuriale had been close to Moletti and to Moletti's friend Giovanni Vincenzo Pinelli, whose library was the cultural heart of Padua. Knowing the territory, Mercuriale also knew, as he wrote Galileo, "the University of Padua is the proper place for your genius." And it paid better. Mercuriale could alert Pinelli and the Venetian overseers of the university to the treasure they might import from Tuscany.[117] The del Monte brothers, always faithful patrons of Galileo, added their recommendations and the hope that Padua would put Galileo in the limelight, "for to say the truth you are known to very few." In September 1592 Galileo paid a successful visit to Venice. The overseers offered him a salary of 180 scudi, three times his Pisan pittance, to replace Moletti. He accepted subject to his obtaining permission from Ferdinando to leave Tuscany. Release was not automatic. Ferdinando's predecessor Francesco had not allowed Galileo's patron Strozzi to leave Florence to serve a Polish prince, presumably because he did not care to lose so valuable a man. Ferdinando had no reason to stand in Galileo's way. There is no record that the university intervened to keep him. A young man with the effrontery to write against his robe of office could be dispensed with.[118] Galileo left Pisa in the fall of 1592 for what he later rated as the best eighteen years of his life.[119]

3

———— ∞ ————

Life in the Serenissima

Galileo spent his first eight years at Padua settling in and setting up, forming important friendships, and living relatively carefree. He kept up his wide interests, learned how to dissimulate, and did nothing significant in science. Around 1600 he burdened himself with debt, began a family, rented a large house, and started a private academy and instrument business. During this second phase, he became subject to a recurrent illness accompanied by a fever that could send him to bed for weeks. These bouts and his many commitments when in good health depleted the time he had to devote to subjects beyond his statutory teaching. Still, with the help of friends, he managed to bring his study of motion to maturity and to develop some ideas about the world system, although nothing important went into print.

3.1 SETTLING IN

A fight for souls

When Galileo gave his inaugural lecture in December 1592, the big man on campus, some fourteen years his senior, was Cesare Cremonini (Plate 7), the junior professor of philosophy. Cremonini had arrived at the university only a year earlier, trailing a reputation as teacher, stylist, writer, and philosopher. He came from the University of Ferrara, where he was an intimate of the duke and a friend of Tasso; he wrote plays and poems, understood mathematics, and could practice medicine.[1] His inaugural lecture at Padua suggests the feelings of a professor newly called from the provinces and the eloquence that quickly made Cremonini a spokesman for the faculty. The theme was hackneyed, the treatment clever.

Each of us is a microcosm of the universe, said the new professor; hence introspection can deliver knowledge of the world as well as of the self; he who knows himself is a natural philosopher! God is a philosopher, the most dedicated of philosophers, as his existence consists entirely in self-contemplation. "You, fortunate youth, have come to the best place to learn to know yourself." The glory that was Athens lives again at the University of Padua through the generosity and perspicacity of the patricians who run and ran the Most Serene Republic of Venice. At great expense and effort they have gathered here the greatest professors from all over Italy, and even from abroad, "to the advantage of all of Europe."[2] Cremonini was one of the great ones, full of his subject, popular with his students, and withering about his colleagues. One of his comedies against a fellow professor hits the level of Galileo's satire on the Pisan toga.[3] Cremonini was the sort of man Galileo delighted to cultivate: thorough scholar, expert logician, ready wit, bon vivant.

The glorious "Bò"—the university's nickname after an inn with the sign of an ox that once occupied its site—was not the only center of learning in Padua. The Jesuit College there had been teaching grammar gratis for over forty years to boys from patrician families when in the 1580s it decided to open its higher courses to the public. Patricians who wanted their heirs taught morals as well as mathematics, and for free, preferred it to the university, where youth risked its soul while learning about it from Cremonini, and at a cost of over 100 scudi a year. In the summer of 1591 some *bovisti* thought to defeat the *gesuiti* by invading a lecture at the Jesuit College and stripping themselves to the naked truth. The Jesuits complained through channels—the Rettori stationed in Padua to oversee conduct at the university and the Riformatori, or general supervisors of education, headquartered in Venice. The strippers received suitable punishment and the Jesuits continued their competition in the false security that the authorities supported them. They reckoned without Cremonini. The arts faculty designated its new philosopher to represent it before the chief executives of the Venetian state, the Doge and the Senate.[4]

The speech Cremonini delivered in December 1591 reverberated throughout Europe. In contrast to his inaugural address, Cremonini now represented the university as declining owing to the operation of a Gresham's law whereby cheap rote learning drives out precious free inquiry. Already Jesuit competition had destroyed the grammar schools of Padua; the same would happen to the university, as it did to the Sapienza at Rome, unless the Senate acted. It is your affair, Cremonini said, switching the burden from school to state: do you want to permit a foreign prince (the pope) to set up an alternative university whose shoddy attractions will draw students from your own?[5] That stung the Senate into action. By the end of the year it had closed the offending courses in the College. The Jesuits in their turn recognized a grave threat to their corporate interests. A long struggle ensued, which lasted until the expulsion of the order from Venetian territory in 1606.[6] The tempest made difficult sailing for Galileo, who wanted to retain good relations with his pro-Jesuit patrons and Jesuit mathematicians and also with his anti-Jesuit patrons and Cremonini's supporters. Navigation required dissimulation. Galileo could not have asked for a better mentor in masquerade than Cremonini. "Think as you please [he liked to say], but behave as expected."[7]

The senior Jesuit in the *contretemps*, and perhaps its instigator, was a man worth knowing, Antonio Possevino, an exemplar of the cultured and obedient, active and passive, generous and ruthless Counter Reformation Jesuit that Galileo could never understand.[8] Possevino had retired to Padua in 1587 after a distinguished career opening and nurturing Jesuit colleges in Northern and Eastern Europe in order to finish an immense *Bibliotheca selecta* of readings suitable for Jesuit savants. In its three editions (1593, 1603, 1607) and many bowdlerizations, the *Bibliotheca selecta* improved on the Index of Prohibited Books by censoring such minutiae as printers' devices and rubricated letters. It did not anticipate the condemnation of Copernicus' treatment of astronomy, however, and recommended it in all three versions. Vincenzo Galilei's books figure in its section on music.[9] Seldom, however, do such free thinkers find a place on Possevino's reading list.[10] To him "liberty" was "license." God's will and order, he shouted, and Mosaic and natural law, condemn freedom of thought as the fountainhead of heresy and atheism.[11] "Oh ignoble and miserable study of philosophy, if by it men are made learned to defend impious ideas and to disdain theology, the mother of the truest doctrines and the true guide to right thinking and living!"[12] A good example of the worst possible sort of philosopher is Averroes, "whom some demented [commentators] make equal to Aristotle and superior to Saint Thomas," though he is absurd, idiotic, and pernicious. Nonetheless, Possevino recommended Averroes' champion Borro, which suggests that, like lesser compilers, he did not look at all the books he cited.[13]

Possevino took fright at the slightest hint of indecency. He recommended that the classical poets be read only in controlled excerpts; fig-leaf ellipses only excite unseemly curiosity. There would have been little left of Ariosto. Nudity drove him crazy, hence, perhaps, the *bovisti* striptease. He held that the reappearance of nudity in art ("the monstrous images of naked women") were the work of the devil in revenge for the evangelization of the New World.[14] "Whoever retains any honesty in his heart hardly dares to look at himself undressed." The man had a problem, exacerbated, perhaps, by seeing a book of erotic and anticlerical poems he had edited for a patron placed on the Index.[15]

Still—we are still with Possevino—just as it was licit for Judith to trick herself out after certain prophylactic measures in order to lure Holofernes to his doom, "so eloquence and the sciences brought by clerics to the rock and citadel of God become shields and weapons to drive away attackers of God's church."[16] Paintings if decorous and historically accurate educate the faithful; architecture builds sound and inspiring churches and colleges; mathematics underpins music and calculations of Easter.[17] In the politics of his order Possevino was liberal enough to get into trouble. He opposed the rigidification that took place under the Jesuit General Acquaviva. He fought, to the injury of his reputation, a new rule forbidding converted Jews from entering the order, and he once acknowledged, again to his detriment, that the Venetians were not entirely in the wrong in the events that led to the expulsion of 1606.[18]

Among Possevino's close friends was Pinelli, the patrician who had urged Galileo's call to Padua and coached him through the hiring process. Since during the first months of his tenure Galileo lived with Pinelli, he had many opportunities to meet his host's conservative friends,[19] for example, Monsignore Paolo Gualdo, perhaps the strongest supporter of the Jesuits in Padua. One of Gualdo's brothers was a Jesuit, the favorite disciple, it is said, of the order's chief theologian Cardinal Robert Bellarmine (Plate 7), and Gualdo almost became one himself. Galileo liked Gualdo's combination of deep religiosity with a taste for banter and a cultivated interest in the arts. Gualdo's partiality to Tasso, whom he knew personally, gave them lots to banter about. Gualdo and Galileo became good friends.[20] Galileo had no desire, and at first felt no need, to ally himself with either the pro- or anti-Jesuit camp. Eventually, however, he would have to choose.

Pinelli had mastered the art of smoothing over differences among savants. His magnificent library, the meeting place of the best-filled and most interesting minds in Padua and Venice, held not only the books recommended by Possevino, but also a quantity of Catholic criticism of the Tridentine spirit, books by heretics, and items prohibited by the Congregation of the Index. Conversation took place in Pinelli's plentiful and peaceful library under the banner of *Patavina libertas*, or local academic freedom. The names of some of the regulars may be gathered from the *Liber amicorum* kept by its sometime librarian, Thomas Segeth.[21] This Book of Friends, compiled around 1599, includes the signatures of Possevino and Galileo and of several other people who will figure later in this story: Jacques Badouère (Giacomo Badovere), a French Protestant student of Galileo's who became a Catholic and possibly a Jesuit; Marino Ghetaldi, a mathematician from Ragusa; Lorenzo Pignoria, Paduan priest and precocious archeologist; Monsignore Antonio Querenghi, Paduan patrician, priest, and patron; and the redoubtable anti-Jesuit polymathic monk Paolo Sarpi.[22] Segeth's bookful of academic friends did not save him from a jail sentence for defaming more solid citizens. Eventually he was expelled from Venetian territory. He made his way to Prague and Kepler, where we will run into him again.[23]

Teaching at the Bò

Pinelli's library included Moletti's books and manuscripts, which Galileo studied while composing his inaugural lecture, delivered to great applause in December 1592.[24] Its text has not survived, but encomiums of it have, from Giacomo Contarini, a Venetian patron of the arts and sciences, then soon to become superintendent of the Venetian Arsenal, and a foreign visitor, a student of Tycho Brahe's.[25] The lecture opened a course on a subject not specified. For the following years, Galileo taught set books in a regular sequence: 1594/5, Euclid and the Sphere, as at Pisa; 1595/6, Ptolemy's *Almagest* (advanced astronomy); 1597/8, Euclid again and the mechanics ascribed to Aristotle; 1599/1600, Euclid and the Sphere. Again as at Pisa, many of Galileo's students came to him to learn the astrology they would need to practice medicine. Galileo also cast horoscopes, some for money, but probably not as cynical calculations.[26]

The policy of teaching set books is a reminder that the purpose of the early modern university was the dissemination, not the creation, of knowledge. The proper approach to any field was to consult the masters. If you want to learn philosophy, Moletti had advised, go to Aristotle; medicine, to Galen; geometry, to Euclid; astronomy, to Ptolemy and his school.

> After you have mastered the way, for a pastime you can see what others have said, if you enjoy leisure or have to dispute or lecture publicly; but not if you teach privately, since then you should just impress on the minds of your students the sound principles of the arts and sciences you teach. Avoid at all costs filling their heads with meaningless questions or diverse opinions...Students [who] have learned nothing systematically dispute...like parrots who speak without knowing what they say.[27]

The statutes of the university for 1607 put the matter more concisely: "All doctors under penalty of losing their lectureships are obliged to and must read and explain clearly and demonstrate the authors they are obliged to teach, *de verbo ad verbum*, word for word."[28]

The system had the advantage that everyone who went through it knew or was supposed to know the same things. And it survives, in the best places, for example, in university courses on Newtonian mechanics and wave optics. The neophyte cannot be exposed to all the arcana at once. The system involves a species of double truth, one for the classroom and one for the laboratory or, in the case of Galileo's colleagues, private conversation or unpublished manuscripts. Double truth can lead to surprising places. By sticking strictly to the regulations, Cremonini gained the reputation of holding the belief, which he taught from Aristotle's *De anima*, that the individual soul dies with the body.[29] Thus his classroom truth. Outside the classroom, however, the truth of the Church held sway and the soul became immortal.

Which truth trumps? Which did Cremonini hold? No one knows. The Jesuits, the Inquisition, and French libertines in search of father figures judged Cremonini a heretic for believing as well as teaching that the soul is mortal, the world eternal, God indifferent, and other Greek aberrations; and they adduced his apocryphal epitaph, *Caesar Cremoninus hic totus jacet*, in confirmation.[30] But these admirers did not have access to his manuscripts, in some of which he strove to prove immortality by the pure light of reason.[31] Prying modern scholars accept his defense against accusations of unbelief (that he delivered only Aristotle's opinions), which, however, he undercut by refusing to rebut his bad pagan texts with good Christian arguments in his books, which were not subject to the statute on teaching. But, did he believe these Greek things? As a good Christian, he said, "it has been and is my intention to…follow the teaching of the doctors and particularly of Saint Thomas."[32] And as a good philosopher? It is hard to unmask Cremonini.[33] What is clear is the counterintuitive proposition that for Cremonini, as for Borro and Buonamici, towing the line was an expression of academic freedom.[34]

Galileo also had a truth for the classroom and another for his manuscripts. The discrepancy did not become public during his tenure at Padua, however, partly because he was not entirely certain of the private truth (the Copernican system) and partly because the relevant public truth (the Ptolemaic system) was generally accepted outside the classroom. An indication of the content of Galileo's public courses, for which neither syllabus nor notes have survived, may be inferred from the texts he prepared for sale to his private students on the same subjects. These concern the sphere and simple machines. The first has the interest that it remains at the level of a thirteenth-century primer, the *Sphaera* of John of Sacrobosco, the second that it builds on sixteenth-century updates of the *Mechanica* then ascribed to Aristotle. Galileo never took ownership of the cosmological "hypotheses" underpinning his *Sphaera* whereas he claimed priority in the formulation of the pseudo-Aristotelian principle on which he developed his account of simple machines.

The world according to Sacrobosco consists of the familiar four-element sublunary kernel surrounded by layered heavens of quintessence bounded by a star-studded spherical shell. There are seven layers counting outward from the stationary central earth, one each for the sun, moon, and planets, arranged in the order of their apparent speeds: Saturn, which moves most slowly against the stars, occupies the seventh sphere; then come Jupiter, Mars, the sun, Venus, Mercury, and, in the first heaven, our moon. Why is the cosmos built like an onion? Interpreting Sacrobosco, Galileo responded as would a modern physicist reduced to fundamentals: symmetry! Why should the universe extend more to the east, or upward or downward, than in any other direction? But we need not speculate. We see the stars go in circles, the earth's circular shadow on the moon, the tops of towers visible from sea before their bottoms, and so on, all Aristotelian arguments for the onion universe; to which Galileo added the characteristic argument, complete with the only figure in his text, that Archimedes required the surface of free-standing water, like our ocean, to lie on a sphere centered on the world's umbilicus.[35]

How do we know that the earth's center coincides with this umbilicus? Among several reasons Galileo offers a "very beautiful observation taken from lunar eclipses." During these extinctions, the earth lies between the sun and the moon. The line joining them can point in any direction, like the diameter of a circle, "wherefore, since...diverse diameters have nothing in common but the center, nor is there another point than the center in common to the diameters, the earth is located at the center."[36] What did Galileo say to the smart student who observed that the same argument applied to a solar eclipse would put the moon at the center? That is not recorded. He did take the trouble to set right those "very great philosophers and mathematicians" like Copernicus who put the earth in motion. No such motion is possible. The earth can not move rectilinearly by nature, since by nature it has to go to the center (a fine *petitio principii*, with which Galileo later taxed peripatetics); nor can it spin in place, for if it did birds would be left behind, buildings would collapse, objects would not fall vertically. Had not Moletti said that the Copernican picture is "a chimera of all chimeras the most chimerical?"[37] To clinch his argument, Galileo observed that a rock let go from the top of a mast of a moving ship hits the deck in the stern.[38] A very pliable demonstration, as we shall see!

After putting the earth to rest Galileo left subjects that belonged to natural philosophy to deliver the useful material he was supposed to teach: definitions of the circles on the sphere (equator, ecliptic, horizon, meridian), coordinates (celestial latitude and longitude), climes (strips between latitudes whose longest days differ by half and hour), causes of eclipses, and, his most advanced topics, the precession of the equinoxes.[39] Though useful as orientation, Galileo's *Sphere* was to state-of-the-art astronomy of his time what physics for poets is to string theory in ours. Accurate predictions of the positions of the planets required mathematical machinery that translated two capital facts unknown in 1600 into geocentric terms: each planet travels in an ellipse containing the sun at a focus; the speed of travel varies so that the line connecting sun and planet sweeps out equal segments of the ellipse in equal times. This clumsy statement, which contains discoveries that Kepler announced in 1609, may be clear from Figure 3.1, where the orbit $\Pi P_1 P_2 \ldots A$ centered at C contains the sun at the focus S (the other focus X is unoccupied). Kepler's "area

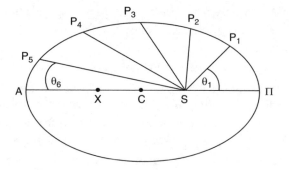

FIG. 3.1 Motion in a Kepler ellipse (S the sun,
P a planet).

law" requires that the planet move on its orbit with varying speed such that, if the areas $\Pi SP_1, P_1SP_2 \ldots P_5SA$ are equal, then the planet takes equal intervals of time to move through the arcs $\Pi P_1, P_1P_2 \ldots P_5A$. Since $S\Pi < SA$, $\theta_1 > \theta_6$ and the planet appears to move more quickly (as seen from the sun) around perihelion (Π) than around aphelion (A).

Ptolemy delivered in his *Almagest* the brilliant simple geocentric approximation to Kepler motion shown in Figure 3.2. We have the same line of apsides $AXCS\Pi$, but the orbit is now a circle and the planet's motion is regulated by the "equant" point X, around which the line PX revolves with constant angular velocity (angle α increases by equal amounts in equal times). Motion on an "eccentric" circle (one not centered on the sun S) regulated by an equant is, for a small "eccentricity" ε ($\varepsilon = CS / C\Pi$), indistinguishable from motion in a Kepler ellipse with the same eccentricity and line of absides. The consequence is that θ behaves in almost the same way in Figures 3.1 and 3.2. Ptolemy incorporated the model of Figure 3.2 into his geocentric system by placing the earth E at S and taking the earth's motion into account by making the planet P rotate around a secondary circle or "epicycle" while the epicycle's center Q rotates around the equant X in the eccentric circle C (Figure 3.3). Galileo had to teach this eccentric stuff in his statutory courses on Ptolemy. He then did his best to forget about it. There are no eccentrics or equants in his later polemics in favor of the sun-centered system. Nor, in his later work, is there an acknowledgment that Kepler motion in a Kepler ellipse is a substantial technical improvement over equant motion in an eccentric circle. It is a great tribute to Galileo's rhetorical skill that he managed to maintain a

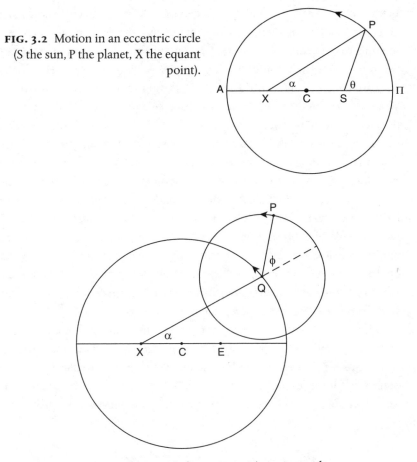

FIG. 3.2 Motion in an eccentric circle (S the sun, P the planet, X the equant point).

FIG. 3.3 Geocentric (*recte* geostatic) motion with epicycle (E the earth, P the planet, X the equant point).

debate on the foundations of astronomy at the undergraduate technical level of Sacrobosco's sphere.

Galileo's treatment of simple machines involved no double truths. He stood behind his version of medieval and ancient mechanics as then recently reconstructed by Benedetti, del Monte, and Moletti. Two of their works, del Monte's *Le mecchaniche* (1577, 1581), and Moletti's incomplete and unpublished *Dialogue on mechanics* (1576), offer instructive comparisons with Galileo's treatment. (A few points in Benedetti's *Speculations* (1585) are also relevant, but as they occur unsystematically, in corrections of Aristotelian assertions, they were not a ready model for Galileo.) Del Monte, Moletti, and Galileo all rely on the pseudo-Aristotelian principle that power to move is proportional to the velocity of the mover. Moletti makes clear the principle, and uses it to derive the law of the balance, by appealing to the wheel-and-axle of Figure 3.4. Since in lifting the weight W through a distance equal to the arc AB the power P must move through a distance equal to the arc CD, velocity of P: velocity of W = arc CD: arc AB = CO: AO. That is just geometry. Now invoke pseudo-Aristotle's fundamental principle, according to which the system is in equilibrium if the velocities are inversely as the weights. Since the velocities also vary directly as the distances, the weights are inversely as the distances: W: P = CO: AO. Behold, Archimedes law of static balance derived from dynamical considerations.[40] As pseudo-Aristotle remarked, the circle is made of contraries (it is both convex and concave, and when spinning moves simultaneously in opposite directions), "[so that] there is nothing strange in the circle being the origin of any and every marvel." There was something bothersome, however, about obtaining a rule for static equilibrium from reasoning about potential displacements.[41]

Moletti claimed jurisdiction over simple machinery, hydraulic engineering, and military technology, and all devices enabling a small force to move a greater. Since every such trick appears somehow to defraud nature, Moletti supposed that the word "machine" derived from "machination." Galileo would be very severe against this way of expressing the productive

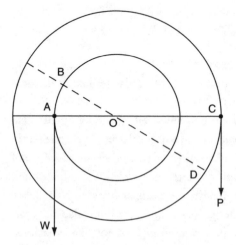

FIG. 3.4 Pseudo-Aristotle's wheel and axle, by which the lesser power P raises the greater weight W.

exploitation of natural powers. Moletti included the theory *de motu* within mechanics and succeeded in persuading himself not only that all objects irrespective of their size and material fall with the same acceleration, but also that Aristotelian physics could not be patched up to account for it. Galileo would take both propositions as fundamental in his revisions of the ideas on motion he had developed in Pisa.[42]

Galileo employed the pseudo-Aristotelian principle of velocities to move in the direction opposite to Moletti, from static equilibrium to dynamical effect. Observing that the slightest additional weight on either side will tip a balance, he proposed to ignore the increment altogether and to set the moving force equal to the equilibrated weight. It followed immediately that the wheel-and-axle in action (without acceleration!) satisfies the relation $OC \cdot P = OA \cdot W$. Now, whereas Moletti emphasized P's greater velocity, Galileo emphasized W's greater sluggishness. In effect, P has to raise the weight W/P P times. Mechanicians had failed to realize (according to Galileo) that their art allows them only to divide up weights virtually and work on them piecemeal. The claim that machines can overcome a large resistance by a small force is not only wrong but fraudulent.[43] A small force moving a larger weight via a machine must either act over a greater distance or over a longer time than a force big enough to move the weight directly. No machine can outwit nature.[44] From this consideration Galileo deduced an explanation for the need for motion in percussion, which he claimed as his own, and the first satisfactory one ever given. The product of the weight of a battering ram and its velocity equals the product of the resistance of the object struck and its velocity. "For things to be otherwise would not only be absurd, but impossible." He would return to this subject in his final contribution to physics.[45]

Like Moletti, del Monte based the theory of simple machines on the principle "learned...from Aristotle, that all mechanical problems and all mechanical theorems are reducible to the [lever and hence the] wheel." In implementing this insight, del Monte followed Archimedes, on whose legendary feats he expatiated, and his own teacher Commandino, "[whose] commentaries smell of Archimedes' own lamp."[46] Thus inspired, del Monte expended fifty pages investigating the consequences of dropping the unphysical restriction, adopted by Archimedes and Galileo, to weightless balance arms.[47] del Monte also made news with his analysis of the pulley and its amplification into the block-and-tackle by showing how to reduce them to a lever. In Figure 3.5,

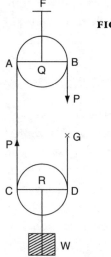

FIG. 3.5 Pulley with a mechanical advantage of 2.

the upper pulley is fixed at F and each of the ropes AC, GD supports half of W if we neglect, as del Monte allowed himself to do, the weights of the ropes and pulleys. He remarked that D may be regarded as fixed (instantaneously) and thus as the fulcrum of the lever CD with force P at C and load W at R. Hence CD·P = DR·W = CD·W/2, or P = W/2: the system allows a force to move twice its effective weight. A third pulley (Figure 3.6) provides a mechanical advantage of 3 or 4 depending on whether it is attached to the top or bottom block. Perhaps of greatest importance for Galileo, del Monte showed that the power must move through a distance $(W/P)s$ if the weight is to be lifted through a height s.[48]

Galileo shortened and simplified del Monte's *tour de force* on pulleys (as did Benedetti) by distinguishing the cases of even and odd numbers of pulleys and generalizing each by appeal to the law of the lever.[49] He followed del Monte closely on the wheel-and-axle, wedge, and screw, showing how all of them too can be reduced to the lever. Galileo took a step beyond his model in a short, clear account of that most "marvelous…miraculous" instrument, the Archimedean screw, a hollow cylinder with blades mounted screw-like on its concave wall (Figure 3.7). The operator dips the screw into a river or well at a convenient inclination, and, by continually turning a crank or gear, lifts water in a continuous flow from the top of the cylinder.[50] In sum, Galileo's *Mechanics* outdid its likely sources

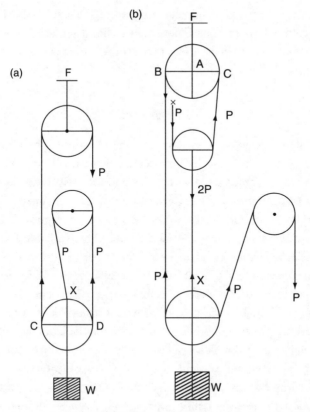

FIG. 3.6a and 3.6b Pulleys with mechanical advantages
of 3 and 4.

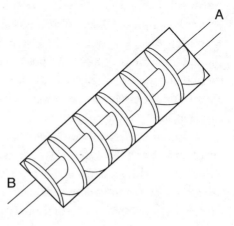

FIG. 3.7 An Archimedean screw
as deployed in Galileo's time.
After Ramelli, *Machines* (1976),
151, 564.

in simplicity, clarity, and coherence.[51] He decided against publishing it for two complementary reasons. There was nothing remarkable enough to make his reputation but enough to make some money when sold in hand-written copies to his students.

Visits to Venice

When Galileo entered its service in 1592, the maritime Republic of Venice was having a hard time on the high seas. Despite the victory of the Holy League, in which Venice acted with Spain and the Holy See, over the Turk-ish fleet at Lepanto in 1571, the Eastern Mediterranean remained a dangerous place. Turkish pirates and their Dutch and English colleagues took increas-ing numbers of Venetian merchantmen. Attempting to mitigate the threat, at least in the Adriatic, the Venetians redesigned their great warships for more speed and maneuverability. Until the battle of Lepanto, Venetian triremes sat three men to a bench, each with his own oar. That suited a regime in which most of the labor was neither convict nor slave. After Lepanto, Ven-ice adopted the practice of other navies, used convict labor, and chained as many as eight men to work a single oar. The system reduced the effects of variations among individuals and suggested the treatment of octets of oars-men as a complex cog in a great machine.[52] Officials at the state arsenal, which made all Venetian warships, wanted to know whether any changes in the length, shape, and support of the oars, and the situation and technique of the rowers, would improve the cog's efficiency.

In 1593, Contarini, newly installed at the Arsenal, asked Galileo how Archimedes would have placed the gangs and their oars.[53] Did it matter where the oarlock was placed? Galileo answered no, since the oar acts like a lever and only the ratio of the distances of the forces from the fulcrum count. Contarini must have felt himself lucky to have such a mathematician to hand, for Galileo, on his own accounting, was perhaps the first person to understand the mechanism of rowing. The standard explanation, given in the Aristotelian *Mechanica*, made the sea the resistance, the rower the force, and the oarlock the fulcrum.[54] In Galileo's version, the ship, acting at the lock, is the resistance and the water, when the oar first attacks it, the fulcrum. Con-tarini replied that there was much more to consider: the height of the rowing bank, the length and counterweight of the oar, the principle that the loom occupied a third of the length of the oar, and the difference in power required at different distances along the loom. Perhaps Galileo tried to take these factors into account. Probably he learned that lengthening galley oars would not work because if made usefully longer they would break unless thickened so much rowers could not move them. The scaling problem again![55] Galileo visited the Arsenal to see the galleys under construction. The operations in the 60-acre plant, which could employ 3,000 men and turn out 50 ships a month, made an enduring impression on him.[56]

While thus mechanically challenged, Galileo applied for a patent on a horse-driven pump to raise a continuous stream of water for irrigation. He obtained a monopoly for 20 years. One such pump was installed in the gardens of the Contarinis. From the same time, 1593/4, date two practical treatises on military architecture, which although derivative in content testify to Galileo's concern to furnish his new patrons with the practical advice they expected from a mathematician worthy of his keep.[57] Galileo continued to have business with the Contarinis through their agent in Padua, who bought wine from Galileo, served as godfather to his daughters, and, perhaps, helped to draw up the marriage contract of Galileo's sister Livia.[58]

Galileo soon found places of resort in Venice more amusing than the Arsenal. One was the palace of the Morosini brothers, Andrea (later an official historian of Venice) and his brother Niccolò, on the Grand Canal near San Luca. The Morosini ran a salon whose roster outdid that of Pinelli's famous library in Padua. Future doges, cultured patricians, university professors, some 20 or 30 strong, came together to talk freely about most subjects.[59] They talked very freely about politics. The Morosini circle included the leaders of the party then largely in control of the state, notably the future doges Leonardo Donà and Niccolò Contarini. These *giovani*, as they called themselves, centered their policies on hostility to Rome, Spain, and the Jesuits, and on promotion of good relations with states beyond the peninsula, particularly France. Their idea of Catholicism harked back to Erasmus, flirted with Protestantism, and opposed the Tridentine spirit.[60] Although the Morosini set were powerful, they did not always prevail when it counted. One of the regulars before Galileo's arrival was the notorious Giordano Bruno, whose dogmatic heresies often built on or incorporated Copernican ideas and other modern notions. Although these notions were not judged heretical in themselves, anyone who urged them could be suspected of guilt by association, and the horror of Bruno's death at the stake in Rome in 1600 recommended caution to innovators. The lesson was the stronger to those

who knew that several members of the influential Morosini circle including
Donà fought Bruno's extradition to Rome but without enough firepower to
defeat the Inquisition.[61]

Another Morosini regular, the Servite Sarpi (Plate 8), would present the
Vatican with a graver threat than Bruno.[62] An anecdote related by Sarpi's
fellow Servite, disciple, and biographer, Fulgenzio Micanzio, who became
a loyal friend to Galileo, suggests Sarpi's redoubtable reputation. One day,
when Pinelli was laid up with the gout, Fra Paolo came to call. In great
pain Pinelli staggered out to meet his guest. His behavior dumfounded
another visitor, the mathematician Marino Ghetaldi. "Why go to all that
trouble for a monk?" Pinelli: "He is the wonder of the age." Ghetaldi asked
in what field Sarpi excelled. "In whatever field you please." Pinelli suggested that Ghetaldi think up a hard problem and spring it on Sarpi at
dinner. Ghetaldi was able and erudite, a former student of Clavius, Coignet, and the algebraist Viète; he had a cornucopia of up-to-date problems
from which to draw a tough nut. Fra Paolo had no trouble cracking it.
Galileo later said of Sarpi that "no one in Europe comes before him in
knowledge of the [mathematical] sciences."[63] He was to serve Galileo as
guidepost and sounding board, and, very probably, as role model. Several ideas akin to Galileo's innovations in physics appear in Sarpi's private
notebooks. Sarpi's cultivation of algebra, however, did not awaken a corresponding interest in Galileo.[64]

Although Galileo did not acknowledge whatever he may have taken from
Sarpi, others were not so reticent. The famous Paduan anatomist Fabricius
of Acquapendente, writing on the accommodation of the eye, acknowledged
that Sarpi, "[a master of] all the mathematical disciplines, especially optics,"
had pointed out the mechanism to him. Fabricius' discovery of the valves of
veins likewise owed something to Sarpi's shrewd observations. Giambattista
della Porta, the Neapolitan playwright and natural magician, considered
Sarpi an authority on magnetism and on almost everything else. "I not only
confess, that I gained something, but I glory in it, because of all the men I
ever saw, I never knew any man more learned, or more ingenious, having
obtained the whole body of knowledge; and is not only the Splendour and
Ornament of Venice or Italy, but of the whole world."[65]

Sarpi was a man of the world as well as its ornament. He had been
to Rome, to represent his order at the Vatican and had come to know at
first hand the most rigid expressions of the Counter Reformation and the

Tridentine spirit, and the most extravagant papal claims to primacy over princes. So extravagant were these claims as put forward under Sixtus V (1585–90), that the pope almost disciplined the Jesuit disciplinarian Bellarmine, a future cardinal inquisitor and now a saint, for the laxness of his views about papal prerogative. In view of Bellarmine's later relations with Sarpi, with whom he was on friendly terms in Rome, and with Galileo, his comment on Sixtus' high-handed policy is high irony: "this is the sort of thing that cost us Germany and one day will cause us to risk losing Italy." Sarpi improved his time around the Vatican by studying Copernicus and della Porta before returning to Venice in 1588 to continue his efforts to reform the Servites. His brethren showed their appreciation by denouncing him several times to the Holy Office for a mixture of silly and serious charges: wearing the wrong cut of clerical clothes, omitting to say a certain prayer, but also frequenting Jews and Frenchmen, and harboring Protestant and even atheistic sympathies.[66]

Around 1600 Sarpi tried to obtain a quiet bishopric as a refuge for study. Pope Clement VIII, whom he had impressed, inclined to grant the request. But Possevino, who was on the spot and supersensitive to the least whiff of Calvinism, convinced Clement to deny it. The advice was good in so far as a post-Trent bishop had more to do than cultivate his mind and disseminate doubt; but it was also bad, as it completed Sarpi's transformation into an enemy of Rome of Lutheran intensity and pertinacity.[67] While he awaited his opportunity, he took daily doses of what he called "moral medicine," or applied hypocrisy. As political science, moral medicine taught that religion is to the body politic what medicine is to the body human: just as a good doctor sometimes deceives his patients to restore health, so the statesman and priest tell tall stories like eternal damnation to secure a peaceful society. As epistemology, moral medicine taught that all knowledge is relative. Prefer the useful or the pleasurable; to every argument there is an objection; respect the foolish views of others for you may come to believe them yourself. As a guide to life, moral medicine recommended disguise. "I have to wear a mask [Sarpi said] because without one no man can live in Italy."[68] Behind his mask, according to his alter ego Micanzio, Sarpi was a composite of the best of Greece: Epicurean, Stoic, Cynic, and Socratic. "The disease of mankind [so fra Paolo liked to say] is the affectation of knowledge."[69]

Galileo's closest Venetian friend, also a member of the Morosini set, was the nobleman immortalized as the independent-minded host in Galileo's most important dialogues. This was Gianfrancesco Sagredo, seven years Galileo's junior, an irreverent, endlessly curious man who studied with Galileo in 1597/8. The two became almost like brothers.[70] A self-portrait, drawn up by Sagredo in 1614, delineates some of the qualities that Galileo admired and perhaps envied in him (Plate 8):

> I am a Venetian gentleman. I never call myself a man of letters. I have an affection for letterati and protect them, but I do not intend to advance my fortunes, or earn praise or reputation from knowing something about philosophy and mathematics. Rather, I rest on the integrity and good administration of my offices in the governance of the Republic, to which I applied myself in my youth following the practice of my elders...My studies turn toward the recognition of those things that as a Christian I owe to God, as a citizen to my country, as a patrician to my house, as a social being to my friends, and, as a gentleman and true philosopher, to myself. I spend my time serving God and my country, and, since I am free from family cares, consume a good part of my life in conversation, service, and satisfaction of friends, and all the rest I devote to my own taste and convenience; and if sometimes I speculate about science, I do not presume to compete with the professors let alone criticize them, but only to refresh my mind by searching freely, without any obligation or attachment, the truth of any proposition that appeals to me.[71]

Among the indulgences that Sagredo permitted himself were Jesuit baiting, art collecting, and womanizing. His best baiting was a lengthy correspondence with a Jesuit in which he pretended to be a widow with theological doubts and an available fortune. His collecting favored "fresh, modern, charming, natural works that deceive the eye, leaving old, smoky, artificial, melancholy and eccentric ones to better minds than mine." That makes a nice transition to his "casino," a private brothel where a Moorish lady chaperoned "another person, very white, eighteen years old."[72] Sagredo liked his wine as well as his women and put his imagination to work on both. For the wine at least he took only the best, and to insure that he quaffed it in optimal condition invented a combination wine glass and thermometer.[73] Most of these tidbits come from letters written after Galileo left Padua. As Sagredo did not start his carousing in middle age, however, we may suppose that he and his brother Galileo sampled the pleasures of Venice together when young men.

Galileo too was a connoisseur of wine and drank more than the usual melancholic scholar of what he called "liquid and light."[74]

Some free thinking

To follow Galileo and Sagredo among the girls of Venice requires a little free thinking. There are some clues. One of the favorite and famous attractions of sixteenth-century Venice was its company of courtesans. Women in the upper levels of the trade, the *cortigiane oneste*, who catered to the patrician class and high-level visitors, possessed qualities of mind as well as body. The better talkers among them charged as much for conversation (so Montaigne learned to his surprise) "as for the entire business."[75] An honest courtesan could sing and play the lute, read and write, and, in some well-studied cases, recite and compose poetry.[76] The best known of them now is Veronica Franco, whose clients included Henri III of France and many lesser dignitaries; whose published poems had some brief acclaim in the society of learned men she frequented; and whose romanticized story is now available to all on DVD.[77] We may reasonably place the likes of Sagredo on the demand side of the business of honest courtesans.

The supply side included girls of good, or, anyway, citizen families like Veronica's, the source of all venetian professionals. Where they obtained their learning in a society in which even noble women were poorly educated unless instructed by a male relative is a mystery. Veronica certainly learned something from her mother, who taught her prostitution, but for most of the rest she probably educated herself.[78] Girls circumstanced like Veronica had few attractive life choices: they could raise a dowry, enter a convent, or go on the streets. The dowry was the key. The larger it was, the higher the prospective bride could aim; without one, she could scarcely aim at all.[79] Galileo's career would be shaped by the inflation in dowries during his young manhood.

Some courtesans saved enough to assemble a dowry and marry themselves off. Perhaps Veronica's mother was such a capitalist. She had enough to provide Veronica with a dowry of 100 scudi in cash and goods amounting to at least that much more with which she could afford to buy a mature doctor for her teenage daughter. (This was about the same amount as Vincenzo Galilei received at the same time to take on his gentle Giulia.) Veronica's marriage was not a success. Re-entering the world dowerless, she quickly earned enough at 2 scudi an engagement to be able to draw up a will (at the age of

18!) making her residual legatee a charity that allotted poor girls dowries of 25 scudi.[80] Galileo later made a similar benefaction and so did Sagredo. A few bequests could not do much, however, against the pressure to sell yourself, and, if you had anything to sell, the desire of frustrated young patricians to buy. The state too had an interest in these transactions since they helped prevent adultery among the upper classes and could be taxed. It was said the tax take on the trade maintained a dozen galleys.[81]

There were around 200 honest courtesans among perhaps 10,000 prostitutes in Venice when Galileo came to town. He was single and likely to remain so; professors, especially in the arts, tended not to marry and, as a citizen of Florence without magnificent prospects, he was not a catch for a prominent Venetian family. Eight years later, he was the father of a baby girl, Virginia, the first of three children by a Venetian woman, Marina Gamba, whom he did not marry. An unfriendly reference to her in 1604 calls her a prostitute; the baptismal certificates of the second child, Livia, and the third, Vincenzo, style her "Maddona," a title sometimes given to, or affected by, honest courtesans.[82]

So much is certain. To fill in the blanks, we may imagine that, to keep Galileo interested for at least seven years (that is, from the conception of Virginia in 1599 to that of Vincenzo in 1506), Marina possessed not only physical beauty but also a talent for music and a taste for poetry. Poetical courtesans tended to favor, and in their writings paraphrase and plagiarize, the *Rime* of Petrarca.[83] Franco submitted her poems in this line to the judgment of the salon kept by her literary advisor, no less a person than the son of a doge, the poet Domenico Venier.[84] Galileo played a similar role (this is an exercise in imagination) toward Marina and her attempts at *rime*. As a conscientious mentor, he entered into the close study of Petrarca evidenced by his surviving annotations, which critics date over a period of time beginning in his early Paduan years.[85] These notes, in contrast to Galileo's comments on Ariosto and Tasso, deal mainly with points of meter, rhyme, and language, or trace standard Petrarchan themes.[86]

Among the themes Galileo marked out, love receives by far the most annotations and exclamations. He flagged every passage mentioning eyes, "eyes that lead the lover by a gentle way to God." He underlined descriptions of feminine attractions, hair in particular, particularly of blondes. He hinted at an internal struggle. "Desire expels love and virtue." "A woman without honor is not a woman."[87] In her last will, drawn up in 1570, honest Veronica made a bequest to a little girl, perhaps an adopted infant, which she left in the care of the "M....i," perhaps the Morosini.[88] As we know, Galileo was on familiar terms with one branch of the Morosini family. He also had ties to the Veniers, two of whom belonged to the Morosini circle, and one of whom, Sebastiano, became a good and useful friend.[89] It is a mere and meaningless coincidence that Veronica's little beneficiary bore the name Marina and that she would have been around 30 in 1600, just the age of Galileo's Marina in that jubilee year.

There exists a long poem attributed to Galileo about a lovers' quarrel for which the swain takes the blame. Critics date it to 1599 or perhaps later. The lovesick poet is not happy with his plight. In one stanza he writes as if he were an astronomer waylaid by physics:

> Heaven raised my face to the stars
> And, along with those eternal beauties,
> Wheeling in the celestial vault,
> Calls me; I do not hear their voices.
> Rather, I gaze upon the earth
> And only attend to human unworthiness.
> A face, a look, a smile,
> I have made my stars and my heaven.
> It seems to me that my soul
> Resides more in certain blonde tresses
> Than in what I can name my own.
> So in my thoughts and not
> In the stars I must wrestle for the truth.

The poor man ends in a standard trope loving his misery:

> I covet my torments
> When I read in my lady's face
> That she desires them.
> No martyrdom but mine will satisfy her.[90]

The undated sonnets mentioned earlier sound the same themes. Here are two of them:

> While displaying in ancient times
> The cruel and impious marks of his madness,
> Amid fires and carnage and brutal massacres,
> The unjust emperor declares:

"My reign heavy with mighty ruins,
With monuments destroyed and temples burned,
Proclaims my greatness in fierce examples
From frozen pole to sunlit beach."
Such another, whose merciless mind is
Defended by impenetrable jasper,
Increases her hardness with my lament.
In a fury, without pity,
Oft she tells me in cruel rough tones,
"In your fire my beauty shines."
. . . .
While love laughs in the quick tremulous
Light of dancing eyes,
A small flame moves in us as from the glow of a dim torch.
But when tears distress love
I feel a true fire burning in my heart.
Oh, the wonder of the strange power
That from tears can draw a voracious flame!
The sun burns as it breaks its potent rays to enter cold pure water
That sparkles between the allurement and the light.
Oh, prime cause of my sweet misery!
To gaze upon those eyes was my destiny.
This is your way of working, and also the sun's.[91]

The optical analogy may limp but the message seems clear. The poor poet armed with natural philosophical images chases a flirtatious blonde beauty with bewitching eyes. Blondes were the preferred type in the *Furioso*, in Galileo's poems, and in paintings of courtesans.[92] Recipes for entering the blonde state were a staple of the natural magician. Della Porta offered several after prescribing a mordant made of honey, the lees of white wine, oil, celandine, madder, cumin seed, saffron, and box shavings. Apply to the hair and let remain for 24 hours. Then rinse with a lye made of cabbage stalks, ashes, and barley straw, "but Rye-Straw is the best: for this, as Women have often proved, will make the Hair a bright yellow." For the dye itself, you will need soap and water, barley straw, fenugreek, quicklime, and tobacco.[93] Good luck.

Galileo wrote about pranks as well as pangs of love during his early Paduan years. He returned to a comedy perhaps begun in Pisa. Its relatively uncomplicated plot stars Orazio, son of the rich merchant Cassandro, and Fiammetta ("little flame"), the daughter of one Frosino, who are in love. Cassandro forbids the match; Orazio, pining, goes away. Three or four years later he returns disguised as a woman, becomes a servant in Frosino's house, "and enjoys the beloved daughter." Frosino conceives a passion for the disguised Orazio; Cassandro, despairing of Orazio's return, decides to marry again; and, of course, he pitches on Fiammetta. She refusing, Cassandro implores her servant, that is, his son Orazio, to plead his case. This produces pleasant complications that can only be imagined since Galileo does not describe them. Finally, Frosino, driven to distraction by the inexplicable refusals of his servant, jumps into bed with her, or rather him. To Frosino's intense disappointment Orazio reveals himself, and marries Fiammetta.[94]

In developing this hackneyed plot during his Padua tenure, Galileo so complicated it that even he could not extricate his hero Ulivetta (as he rechristened the disguised Orazio) from the mesh he wove. At one point Ulivetta had two servant lovers, to both of whom he promised herself, while trying to avoid the attentions of the love-sick Frosino (now Tufano); at the same time he had in hand two commissions to arrange marriages, one of them between his father and his lover Diana formerly Fiammetta. The plot breaks off with over an act to go.[95] One modern critic thinks that if finished it might have made a pleasant fluff, another that it contains nothing in its "cheerful and very licentious pages…for which a comedy comes to be and remains a great work of art."[96]

Galileo may have received the inspiration to revise and expand his Pisan play from della Porta, who was known for his comedies as well as for his cosmetics. They met early in 1593. If that half-mad genius Tommaso Campanella is to be believed, Sarpi and Campanella himself were present too. This formidable quartet had in common a disposition toward free thinking and entanglement in the disciplinary machinery of the Catholic Church. When they met, the Congregation of the Index had recently banned one of della Porta's comedies although the Congregation of the Holy Office had advised him earlier that it preferred his plays to his magic.[97] Campanella had already suffered imprisonment by the Inquisition for having ridiculed a very sound papal injunction excommunicating anyone who removed a book from the convent library without permission. (Some say he was also charged with the lesser offences of reading suspect philosophers and keeping a familiar spirit.) Released, he made his way to Padua, where he was studying at the time the foursome met; he would be in trouble with the Inquisition, and in one or another of its jails, for most of the next 30 years.[98] For the rest, Sarpi would be excommunicated and almost assassinated, and Galileo would earn perpetual detention at the pleasure of the Holy Office. Their luncheon together would have made an Inquisitor's mouth water (Plates 6 and 14).[99]

3.2 STEADY STATE

Family

In 1599 Galileo took a large house with a garden and a vineyard in via dei Vignali (now via Galileo) near the Santo, the great religious complex of San Antonio di Padova.[100] He needed the space to accommodate a coppersmith from the Arsenal, Marcantonio Mazzoleni, whom he engaged as a live-in instrument maker. There were also Mazzoleni's family and a few well-to-do students who came for private courses. As some of them had their own servants and horses, Galileo must have had a full house and fair-size staff when his establishment was in full swing. A letter of 1607 to a prospective student boarder affords a glimpse into his arrangements: the house is neither crowded nor expensive, Galileo wrote, as it then had only one continuing scholar, who had been in residence for four years and would continue another two, the slow scholar's brother, and their page.[101] That was a relative low. For the two years November 1602 to October 1604, Galileo accommodated 16 students with their retinue of 17 servants. Most came for private instruction in the use of the military compass and fortification, and stayed for as long as their intelligence required and their incomes permitted. Perhaps Galileo had as many as ten boarders at a time. In addition to the 16 live-in students of the biennium 1602–04, Galileo had 12 who boarded elsewhere. Most of them bought his compass.[102] Usually the income from room charges, instructional fees, and instrument sales amounted to more than Galileo's salary, which in 1599 was raised from 180 to 320 scudi.[103] The gross for the instrument business in 1599/1600 was 1060 lire or 212 scudi (in Venice five lire made a ducat). Against this he had to take his expenses: 90 scudi for rent, room and board for Mazzoleni and his family, remuneration of servants, heating, materials, and so on.[104] Since as a professor he escaped the tax on wine, he could have lived well enough on his income.[105]

He had other expenses, however. During his first year at Padua Galileo had had to outfit his brother Michelangelo for a job-seeking trip to Poland, provide Livia with a new dress, and meet his brother-in-law's demands for the balance of Virginia's dowry.[106] In 1593 this awkward relative, Benedetto Landucci, threatened to have Galileo jailed if he returned to Florence without the money. Galileo returned and borrowed 200 scudi to discharge the debt.[107] That was small stuff. In 1600 Livia, then living discontentedly in a nunnery, received an offer of marriage. Galileo could do nothing toward a dowry, he told his mother, since once again he had to equip Michelangelo, who had returned to Florence in the hope of an appointment at court, for travel and work. The estimated cost ran to all of Galileo's salary for the year. If Livia did not like life where she was she could change nunneries. Then, when Michelangelo could help with expenses and if she still wished to marry, she could come out and "experience the miseries of this world."[108] Livia soon received another offer and, with Michelangelo's promise to help, Galileo very generously agreed to the extravagant dowry of 1800 scudi. The wedding took place in Padua in January 1601. By November, Livia was pregnant and his new brother-in-law, Taddeo Galletti, exigent. Galileo asked Michelangelo to oblige himself legally to pay his share. For himself, he wrote, he was tightly pinched. Michelangelo could not contribute and Galileo had to borrow another 600 scudi and obtain an advance on his salary to meet his running expenses and the obligations he had assumed to assist his siblings.[109]

Matters scarcely improved even when, in 1606, Galileo's salary increased to 520 scudi. Apparently the Riformatori required some persuasion from Sagredo and Venier, both of whom lent Galileo money and knew his financial position, to grant the raise.[110] Nevertheless, in 1608 he still owed 1400 scudi to his brothers-in-law, one of whom, Galletti, had tried to sue him in Venice, and perhaps as much more to his other creditors.[111] To all of which must be added the cost of Galileo's own family, which, by the end of his Paduan tenure, comprised Marina and their three children, whom Galileo named after his father and sisters: Virginia, born 1600, now famous as "Galileo's daughter"; Livia, born 1601; and Vincenzo, born 1606.

The family did not live together. A professor dwelling in open concubinage would have been anomalous, although during the time that Marina and Galileo were together patricians might live with their partners without suffering social reprobation. The arrangement was not favored by the state, however, as overproductive of bastards, or by the church, as opposed to Tridentine discipline, which prescribed excommunication in persistent cases; and many couples who began irregularly submitted to pressure and married.[112] Marina and Galileo did not take this step, perhaps because of disparity in their social positions. It may be significant that Marina gave birth to Virginia between Galileo's rejection of Livia's first suitor and his acceptance (in January 1601) of Galletti. Galileo's life style might have increased the cost of a respectable husband for Livia. Michelangelo believed that Galileo had sacrificed his chances for a regular family life to establish his sisters. "I know that you would tell me to forget about taking a wife and consider our sisters," he wrote in reply to Galileo's complaint that he had married with 1400 scudi of dowry debt outstanding.[113]

One of Galileo's more lucrative ways of raising money was casting horoscopes, for which he ordinarily charged his students 60 lire (12 scudi). His cashbook for 1603 records five such transactions.[114] His reputation as a practitioner brought requests and no doubt higher fees from cardinals, princes, and patricians, including Sagredo, the Morosini, and someone interested in Sarpi. He corresponded with the grand duke's astrologer, Raffaello Gualterotti, and, in hard cases, with an adept in Verona, a physician named Ottavio Brenzoni.[115] At least once Galileo's astrological predictions reached the accuracy of Copernicus' anticipation of the decline of the Turks and Kepler's forecast of Wallenstein's death. This was the discovery that the planet Venus had a decisive influence on Sagredo's character. Galileo did not need to consult the stars, however, to learn that his friend was "kind, happy, merry, beneficent, pacific, sociable, and pleasure-loving," and a little too devoted to women.[116]

To do astrological computations correctly required determination of the relative strengths of all the planets based on many factors: angular distances from the cardinal points (ascendant, upper and lower meridian, descendant) and from one another; their relations, of which Galileo considered a half dozen types (ruler, sex, sect, face...), to the zodiacal signs; their motions, whether swift or slow, ascendant or descendant, direct or retrograde; and their placement within the "houses," sections of the zodiac cut off by the horizon, meridian, and certain auxiliary circles, and associated with various aspects of life.[117] It took time, labor, and knowledge to cast and interpret a geniture. That Galileo took this trouble even when no one paid him for it suggests that he ascribed some value to the art. Three of these gratuitous calculations concerned his daughters and himself.

Galileo calculated his birth chart for the afternoon of 16 February 1564 at least twice, and once for the 15th, if the chart he designated "Georg: Giocomi[o]" ("George Myass") should also be his. All three charts agree in what matters: Galileo's rising sign is Leo, his sun is in Pisces (in one case, Aquarius) in the eighth house (death) and Saturn and Jupiter are in conjunction and retrograde in Cancer in the eleventh house (animosity). Even a novice astrologer can see that there are some bad influences at work here. No wonder Galileo was melancholic. We do not have his elucidation of the significance of the conjoined retrograde planets unfavorably placed in the Crab, of the combative Mars glowering at the horoscope from the midheaven, or of the sun and the inferior planets drowning among the descending Fish. However, an ingenious recent astrologer has interpreted one of Galileo's sonnets as a poetical résumé of the chart's meaning.[118] Here it is:

> Darting forth flames, the celestial lamp
> Burning in the lion ascends into the sky.
> Every zephyr ceases, and the world kindles
> Under a blaze from the ardent south.
> The earth below and the heavens above grow hot
> Striking with a double heat to inflame the air.
> And the fish hardly can escape the
> Heat that reaches the sea by diving to the depths.
> The birds, the beasts, and the tired vanquished flock
> Seek grottoes, caves, dark valleys
> That protect from the withering rays.
> But you, poor heart, in your burning breast enflamed,
> What protection can you find?
> I believe it can be nothing but cold death.[119]

On this reading, what appears to be a love lament in Petrarchan mode is a commentary on the aspects and prospects of a native born on a very hot day in July. Despite that happy hit of the uncomfortable fish, the poem seems unlikely to represent red-hot influences on the infant Galileo, who was born in February.

We know exactly how Galileo interpreted charts with the sun rising in Leo, as both of his daughters were born in August. Here is what the stars said about Virginia:[120]

Virginia's temperament:

P[rim]o, although Mercury and the Moon are separated and have no aspect to one another, nonetheless they indicate a certain discord between the rational and the emotional powers. Because Mercury is very strong and in a ruling sign [Virgo] and the Moon is weak and in a subject sign, reason will dominate desire.

Saturn the signifier of character is very strong, which tends to bestow a correct and severe temperament although mixed with some nastiness. This, however, is moderated and tempered by beneficent Jupiter in a robust sextile aspect. This makes her patient in work and trouble, solitary, taciturn, frugal, concerned with her own comforts, jealous, and not always truthful in promises.

The Sun also well placed [in its ruling sign Leo] bestows a certain pride of reason and character.

Spica in the ascendent adds charm and religion. The rising human sign Libra lends humanity and gentleness.

Her aptitude:

As to her natural genius, Mercury, so well placed on several counts, promises intelligence and the association of Jupiter increases wisdom, prudence, and humanity.

Saturn, also fruitful and powerful, will strongly assist memory.

Libra, rising with several planets, favors her talents, and...[the MS breaks off].

A year later Galileo had to repeat the operation:

Livia's temperament:

Mercury and the Moon in separated signs indicate a certain discord between the rational power of the soul and the disposition of the senses: however, the very well placed Mercury [in Virgo, conjoined with Jupiter and in sextile to Venus] overcomes the weak Moon so that the emotional side is completely subjected to the rational. Mercury here is the significator of character, conjoined with Jupiter and both in benign sextile aspect to Venus, gives a disposition with very nice and praiseworthy traits.

Also Spica, preceding Mercury, combines charm with grace and religion: thus she will be very keen in mind, gentle, cautious, talented at everything, poet, mathematician, learning on her own, a good companion, accommodating to every occasion and person.

Her aptitude:

Mercury being very strong in the ascendant shows a natural genius adjusted to all things; and by its association with Jupiter adds knowledge, probity, simplicity, erudition, prudence, humanity. Venus being in sextile wondrously increases quickness and grace of speech and character.

There is this caveat, however, that because of the unfortunate position of the Moon, she might understand fully but resolve badly, and deliberate well for others but poorly for herself.

Although Galileo worked out the significant aspects of both charts in detail, his interpretations turned almost entirely on the positions of the planets relative to the horizon and one another, and on the standard association of Mercury with reason and the Moon with emotion. In both nativities Mercury, being in the ascendant in an equable sign (Libra, house of the autumnal equinox) and accompanied by other planets, is well placed to imprint his rationalizing spirit, more strongly and pleasantly in Livia's case, where his companions are beneficent (the sun, Jupiter) than in Virginia's, where they are trouble makers (Mars, Saturn). In both cases, the separation between the rational signifier Mercury and the emotional Moon is around 90 degrees, which confers some opposition and suggests a little disharmony. But Livia did not have much to worry about because her bad Moon, being under the earth and in opposition to its house, is very weak. Variations in the Moon's place and in Mercury's partners account for the differences in the girls' characters.

This analysis comes straight from the most authoritative of astrological guides, Ptolemy's *Tetrabiblos*. "Of the qualities of the soul, those which concern the reason and the mind are apprehended by means of the condition of Mercury...and the qualities of the sensory and irrational part are discerned from the...Moon." The stronger Mercury is, the keener the intellect; the more prominent the Moon, the more the feminine traits of character express themselves. The effect of good company appears from Ptolemy's description of a Mercury allied with Jupiter: natives result who are "learned, fond of discussion, geometricians, mathematicians, poets, orators...good

natured, generous...philosophical, dignified." In good aspect with Venus,
Mercury produces "artistic, philosophical [people], gifted with understanding, talented, poetic, lovers of the muses...sagacious, resourceful, intellectual, intelligent, successful, quick to learn, self-taught..." On the other hand,
Saturn and Mercury at best make "meddlers...fond of the art of medicine,
mystics...shrewd, bitter, accurate, sober, friendly, fond of practical affairs,
capable of gaining their ends"; and, at worst, "frivolous talkers, malignant,
with no pity in their souls, given to toil, hating their own kin, fond of torment, gloomy..."[121]

Galileo's interpretation omitted many of the questions customarily asked
of genitures, like length of life, fortune, and illnesses. It was one thing to
suppose that planetary influences could imprint the fresh soul of the newly
born, quite another to derive from the heavens foreknowledge of events.
Astrology could be helpful without making predictions. Does your geniture
indicate sensitivity to the sun? Then retire to a cool place when the sun is in
Cancer. "And in this way [thus Galileo's predecessor Moletti] judgments of
astrologers should be understood, and not in any other form."[122] The degree
of Galileo's personal commitment to astrology is as hard to divine as his
politics. That he subscribed to it in the form in which he deployed it in his
daughters' nativities seems likely. That he sold or offered astrological advice
for another 20 years or more is incontestable. The accuracy of his predictions, even when given "almost as a joke," could astonish their recipients.[123]
Sed contra, Galileo's role model Sarpi, having studied astrology carefully and
sympathetically, had declared it rubbish. "I hold to very few things about
which I am not prepared to change my mind; but if I believe anything for
sure it is this, that judicial astrology is pure nonsense."[124]

Academy

Pinelli's death in August 1601 deprived Padua of its literary center. He left his
collections to a Neopolitan relative, Duke Cosimo d'Acerenza, who also suffered from bibliophilia. The duke engaged three ships to transport his inheritance to Naples. Pirates captured one and being indifferent readers vented their
frustration by throwing a third of the books into the sea. Fisherman netted
some rarities and sent them on to Naples. Meanwhile the duke had died. His
widow sold the lot at auction in which the Society of Jesus and the Cardinal
Archbishop of Milan, Federico Borromeo, fought it out. Borromeo outbid

the Jesuits and brought his purchase safely to Milan as the nucleus of a library named for the city's one-time bishop Saint Ambrose. This became the famous Ambrosiana, which still houses the part of Pinelli's library that survived the bombs of the Second World War.[125]

A new Pinelli was already on hand when the old one died. This was Antonio Querenghi, who had been at the university with Sarpi and returned to his native Padua in 1597 as a cathedral canon. While on the road he had made a friend of the future Clement VIII and other influential churchmen and had begun to assemble a library. He was a classical scholar, poet, theologian, and rhapsodic mathematician, and, like his good friend Paolo Gualdo, fond of jokes and Jesuits. "Monsignore is a man who loves peace and quiet, prizes his bed as if it were the fifth element, and dislikes peripatetics; he thinks of nothing else than how to recover Archimedes' trick of pulling a boat along the ground so that he can go in a gondola overland whenever he pleases..."[126] Pinelli's circle became Querenghi's; Gualdo, Sarpi, Galileo, and Pignoria were members.[127] Querenghi's library was not quite half of Pinelli's (3,000 against 6,500 volumes) but larger in doubtful works: foreign imprints, magic, and astrology, with emphasis on the works, published and manuscript, of the consummate astrologer Girolamo Cardano. No doubt it also contained a copy of Sixtus V's bull of 1586 "against judicial astrology and every other sort of divination." Galileo would have found in Querenghi's library everything he needed to cast the birth charts of Virginia and Livia including Cardano's commentaries on Ptolemy's *Tetrabiblos*.[128]

Querenghi and his friends also met at the Accademia dei Ricovrati, a learned society set up in 1599. Its 26 founding members listed themselves in order of merit as determined by its founder, the young abbot (and monsignore) Federico Cornaro. His qualification? He had been educated at Padua in "all the sciences most fitting to a true gentleman." Cornaro put Galileo in 15th place, eight below Cremonini.[129] The group spent its first two years making itself interesting and conspicuous. A solemn public inauguration attended by all city officials, professors in the university, and, exceptionally, their wives, followed a high mass at the Santo. There were few speakers and much music. Then the academicians got down to the serious business that would occupy them for months: the design of *imprese*, the emblems and mottoes peculiar to each member and the corporate logo peculiar to them all. At a very private session on 23 May 1600, Galileo received the sensitive assignment of reviewing, revising, and, when necessary, rejecting his colleagues' *imprese*. Since the shield and slogan were to demonstrate the learning, cleverness, and artistry of its owner, Cornaro evidently valued Galileo's scholarship as well as his *buon gusto* and, what may be unexpected, his diplomacy.[130]

Galileo also helped with the corporate logo. What would be appropriate for a group calling themselves *ricovrati*, that is, inmates of a (mental) hospital? Taking the association seriously, Galileo chose as his academic name *Abba-tutto*, "depressed," perhaps with his financial and familial circumstances in mind. As the academy's *impresa* he favored a two-mouthed cave bearing the legend *bipatens animis asylum*, "an asylum for souls with two entries."[131] Clever people might have recognized the words as a paraphrase of a line from Boethius's *Consolation of philosophy* and the cave as an illustration of a scene in Homer's *Odyssey*. Boethius had written, *Hoc patens unum miseris asylum*: "Here may a wretch have refuge from his pains." The wretchedness in question is that of everyone ensnared by the senses, in all senses: addiction to material things, trust in sense impressions, ignorance of the good and the true.[132]

The airy cave belonged to the naiads, a species of water nymph. It had an involved symbolism. One entrance, to the North, received humans; the other, to the South, gods only. The cave contained a loom of stone and a hive of bees.[133] It appears that the Accademia dei Ricovrati was a naiads' cave in which Galileo and his fellow nymphs could be cured of their illness by hard work and divine enlightenment. Querenghi specified the nature of the hard work needed. It begins in a more famous cave, that of Plato's *Republic*, where people imprisoned by their senses perceive reality only dimly, via shadows on the cave walls. The road out begins with arithmetic. "Let no one ignorant of mathematics enter here," read the doorposts of Plato's Academy; "let no one ignorant of geometry leave here," read the exit sign from Homer's cave.

Galileo remained active in the Academy of Inmates for a few years, rising in 1602 to the office of censor of books submitted for its imprimatur.[134] But he, Cremonini, and other serious savants drifted away as their colleagues took up the sort of literary questions that Galileo had enjoyed during his gap years: What is the best way to woo? Can you love more than one woman at a time? Does Tasso's Rinaldo act more from love or honor?[135] Galileo had more serious claims on his spare time: private teaching, instrument business, spasmodic research. One of the spasms concerned magnetism. Sarpi seems to have started it. As an expert on magnetism and a friend of heretics, he was perhaps the first Italian reader of the inspirational *De magnete* of Queen

Elizabeth I's physician William Gilbert. This book, published in London in 1600, reports the first sustained, systematic experimental inquiry into any branch of natural philosophy. Many of Gilbert's experiments used a "terrella," a simulacrum of the earth made from a spherical lodestone, with which he demonstrated among many other things that a magnetic needle able to rotate around a horizontal axis dipped beneath the horizon when placed in the magnetic meridian. To show the phenomenon, he moved a "versorium," a short compass needle appropriately pivoted, around his terella. Measurement of the angle of dip required in addition that the lodestone be perfectly spherical, the poles well defined, the needle very short and perfectly magnetized, and the versorium placed exactly in the magnetic meridian. Then the great question was to find a rule that related dip angle to latitude.

The angular displacement μ of the versorium at a latitude ϕ with respect to its orientation at the equator equals ϕ plus δ, the measured dip (Figure 3.8). At the poles, $\mu = 180°$ because each of its constituents is $90°$. Gilbert ascribed the phenomenon to a new physical agency, a special magnetic virtue, a "disposing and rotating influence."[136] Sarpi liked the mystery of magnetism, which he often used to illustrate the epistemological difficulty in reasoning about physical entities we cannot detect immediately by sense. (We can see a lodestone draw iron filings and feel one dropped on the toe but we cannot detect magnetism directly.) To natural philosophers who held that magnetism was an irreducible occult quality (which meant that it owed its action to a hidden cause), Sarpi replied that it is occult only in the sense that a blind

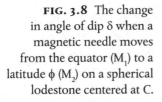

FIG. 3.8 The change in angle of dip δ when a magnetic needle moves from the equator (M_1) to a latitude ϕ (M_2) on a spherical lodestone centered at C.

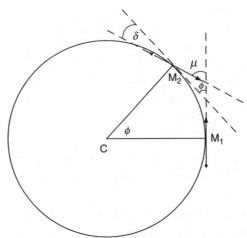

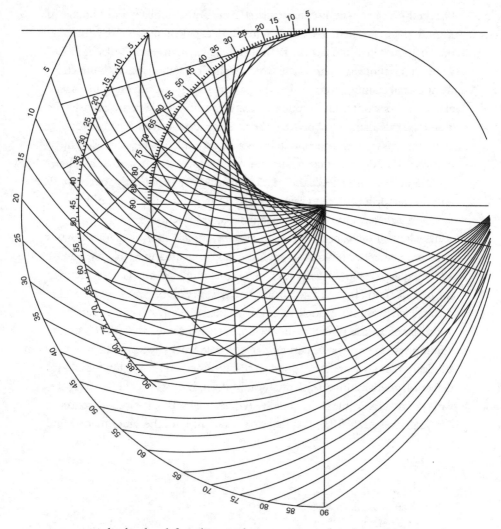

FIG. 3.9 A display that defeated Sarpi. The innermost circle is the earth; the straight lines (apart from the vertical) dip, and the figures latitude. The construction divides μ into ϕ and δ, or would if μ were known. From Gilbert, *De magnete* (1600), 201.

man cannot see.[137] With enough information, the mind's eye might perceive its deeper nature. Gilbert's dip pointed to deep things; Fra Paolo turned his powerful mind upon it.

At the equator and the poles, everything seemed clear. And in between? Sarpi raised the question with Galileo in conversation and then by letter.

Gilbert had given a complicated geometrical rule for obtaining μ and thence δ (Figure 3.9). "How did he find his method, by experiment or reason?" Sarpi answered his own question: not by experiment, since that would have required either traveling all over the world or having a versorium far more sensitive than any procurable; nor by reason, since Gilbert could not explain the principles underlying his exotic diagram. Why did he draw those circles and divide them as he did? "Please think about my difficulties and supply what is wanting in our author, who has withheld the causes of the most obscure things." Galileo made versoria for experiments and no doubt also for sale. "Please tell me how to make them," Sarpi asked, "how to fix the pivot...everything.[138]

Sagredo joined in the inquiry and Galileo sent him some magnets.[139] One of them, "a real Rodomonte," had great potential. To realize it, to fit an armature, Sagredo needed Galileo's help.[140] Soon he had a double Rodomonte, wherewith he experimented diligently with no known advantage to science. He gave it to Galileo for sale to Grand Duke Ferdinando, who wanted one that could pull a sword from a strong man, or, at a minimum, support its own weight of iron. During the dickering over price, Galileo busied himself enhancing the magnet's attractiveness. When he received it, the double Rodomonte could hold five pounds, a little more than its weight.[141] By resourceful application of Gilbert's hints about improving performance and his own stubborn ingenuity, Galileo had managed almost to double the weight the magnet could lift when, in the spring of 1608, he dispatched it to Florence.[142] Galileo included in his shipment a few instruments and magnetized bits of iron to demonstrate the magnet's strength "and some other stupendous discoveries I have made with it." The stone deserved a place in the grand duke's cabinet. Galileo suggested that it bear the emblem, *vim facit amor*, "love makes power," to indicate the jurisdiction given by God to just and enlightened rulers. This was a typical progression for Galileo: from episodic engagement in a problem of physics stimulated by questions from friends or students through the development of instruments for business, presents, and research, to a few respectable results, "miraculous, marvelous," bestowed where they might do a poor professor the most good.[143]

The students in Galileo's private academy included many foreigners, several from the land of Copernicus, and some hangers on, rather like postgraduates, who had reasons to remain in Padua.[144] Among the latter, two monks from the Benedictine monastery of Santa Giustina became disciples.

Dom Girolamo Spinelli would collaborate with Galileo and Querenghi in 1605 in a pseudonymous reply to a philosopher who had had the effrontery to impugn mathematicians. Dom Benedetto Castelli (Plate 9) became almost a son to Galileo, a trusted adviser, personal agent, and, in his turn, a pseudonymous collaborator. Austere but not humorless, pious and fiercely loyal, he once proposed the monkish remedy of burning the books of Galileo's enemies.[145] First among these malefactors was another hanger-on, Baldassar Capra, whose sin was to publicize the most lucrative of Galileo's instruments, his geometrical and military compass, without troubling to disclose that he, Capra, had not invented it. The instrument enabled adepts to calculate many useful things quickly. To protect his monopoly, Galileo had instructions for its use copied out by hand for each purchaser. Because the instruction booklet contained no illustrations, it made little sense without the instrument. When equipped with both, "anyone of average intelligence" could master it.[146] Let us see.

Galileo's compass consists of two flat arms striped with radial scales and hinged together at one end (Figure 3.10). It performs calculations by proportion according to the angle of opening. The prototypical case requires the solution of $x:a = b:c$. The operator picks out the distances b and c "lengthwise" along the two symmetric scales, or "arithmetical lines," marked on the compass arms and opens the compass until the distance a just fits "crosswise"

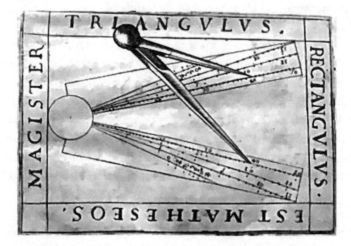

FIG. 3.10 Galileo's proportional compass showing crosswise fitting. From Capra, *Usus* (1607). Courtesy of the Museum of the History of Science, Oxford.

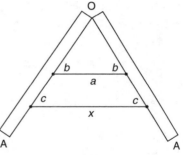

FIG. 3.11 Galileo's compass opened to find $x = ab/c$ using the "arithmetic" (linear) scales.

between the marks for b. Then the distance crosswise between the marks for c is x (Figure 3.11). Galileo explained how to apply the technique to map scaling and currency conversion (Venetian ducats to Florentine lire in his predictable example), and compound interest (by repeated application of the process).[147]

The concept of a pocket calculator in the form of a compass was not new with Galileo. Among his predecessors, both authors and transmitters, were Pinelli, Moletti, Giacomo Contarini, Coignet, and, above all, Guidobaldo del Monte. They shared a widespread interest in measuring and calculating instruments for surveyors, builders, and military men; an interest that, according to a bibliography published by Galileo's former student Levinus Hulsius in 1604, had produced seventy pertinent books during the previous half century.[148] The particular merits of Galileo's version lay in the nature and arrangement of its scales, which underwent several modifications between its earliest form described to Pinelli in 1597 and the definitive one of 1604. The arithmetical lines were the last added, apparently because problems involving areas (surveying) and weights and volumes (gunnery) were more pressing and difficult. The arithmetical lines transformed a specialty instrument into a general calculator.[149]

To handle areas, Galileo inscribed "geometrical lines" scaled to the square roots of integers from 1 to 50. With them finding the mean proportion between a and b, that is, the solution of $x^2 = ab$, could not be easier: make marks at $\sqrt{b}$ and $\sqrt{a}$; open the compass so that a fits crosswise between the marks $\sqrt{a}$; x is the line that runs between the marks at $\sqrt{b}$ (Figure 3.12). Say you have Q soldiers to arrange in a rectangle whose front is to its flank as 5:3. How many soldiers wide and deep should you make the army? Galileo reduced the problem to finding the value of x for which $15x^2 = Q$. In his example, Q = 4335, which he rounded down to 4300 and wrote as 43·100.

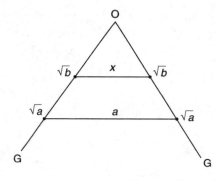

FIG. 3.12 The compass opened to find $x = \sqrt{ab}$ using the "geometrical" (square-root) scales.

Then $5x = 50\sqrt{(Q/100)} \cdot \sqrt{15} = 50\sqrt{(43/15)}$. The front $5x$ therefore is the crosswise distance between the marks for $\sqrt{43}$ when the compass is open so that 50 fits between the marks for $\sqrt{15}$. The front works out to 85; with 30 in place of 50 crosswise, the flank is 51 and the army 4335.[150]

We rise to "stereometric lines," scaled to the cube roots of the integers up to 148. The operations of enlarging or reducing a solid, of taking a cube root, and of finding mean proportionals brings nothing new.[151] Next to the stereometric lines Galileo's compass carried his most important help to the artist of war: "metallic lines," scaled as the inverse cube roots of the densities of the materials. Do you wish to know how much a silver statue will weigh equal in size to a marble one weighing Q pounds? Then you must find x such that $x{:}Q = \delta_{Ag}{:}\delta_{Ma}$. Do you want to know the weight of a cannon ball of any material and any size from knowledge of the size and weight of a standard cannon ball, and hence the relative charges needed to fire them?[152] The calculation will not be pursued here lest it fall into the wrong hands.

Galileo's compass had a reverse side, with lines for geometrical work, such as finding the side of a regular polygon for inscription in a given circle, squaring a circle or rectangular figure, and, what was new, "additional lines" for finding areas of segments of circles. Having mastered all this, the student was only half way through his course. He could acquire an additional device from Galileo to turn the compass into a surveying instrument good for taking distances to enemy locations and fixing the elevations of muzzles. This device was a quadrant similar to ones long in use but adapted to fit onto the compass with its arms open to 90°. Galileo's instructions included directions for finding heights and distances; here he provided diagrams, presumably because he had nothing to add to the usual techniques and repertoire.[153]

Galileo did not date his instructions nor did Mazzoleni sign or date the compasses. This oversight, combined with the fact that Galileo did not invent the instrument from whole cloth, made an irresistible opening to academic buccaneers. In 1603 he learned that a Dutchman, Jan Eutel Zieckmesser, was showing a compass similar to his. That was bad for business. Galileo challenged Zieckmesser to a duel with compasses drawn at the palace of the Cornaros. The witnesses present agreed that in most respects the weapons had been fashioned independently, and declared a draw.[154] Then came Capra. This scapegrace had received lessons in the use of the compass from Galileo, who thought he had the makings of a good astrologer, and copies of the booklet and the instrument from Giacomo Alvise Cornaro, at whose house the lessons had taken place. Soon Capra also had an authoritative printed copy of the instructions. Galileo had run the risk of publishing them to secure the patronage of the Medici crown prince, Cosimo, whom he had tutored in the use of the compass during the summer of 1605, and to whom he dedicated the instruction booklet and the instrument. He minimized the risk of unauthorized access by printing only 60 copies, without illustrations, on a press in his own house. These precautions did not defeat Capra. He reissued the booklet in a Latin paraphrase with a very nice drawing showing crosswise fitting (Figure 3.10).[155]

With equal effrontery, Capra's quack physician father gave a copy of the plagiarism to Cornaro. Alerted by him, Galileo collected testimonials from Sarpi, Querenghi, and others to whom he had shown versions of the compass over the preceding decade, and arranged to examine Capra on his knowledge of compass principles before university authorities. Poor Baldassar! He failed the test. The authorities confiscated his book and threw him out of the university.[156] That did not satisfy Galileo, who published a brief, brutal rebuttal of the goat (la capra) who had got his goat.[157]

The ferocity of Galileo's counterattack betrays his need for the money the compass earned him. By 1610 he had sold some 300 instruments. At 150 lire each, they brought him 450 lire a year for over a decade. If expenses for materials and Mazzoleni amounted to half of that, he cleared say 2,000 lire or 400 ducats a year, about half his academic salary over the decade. And he anticipated that his compass income would rise. Military men as well as poets sought his advice about their art. From Florence came a request from a general just returned from a victorious siege for help with artillery. From Venice came a request from another soldier, Pietro Duodo, for a mathematical curriculum for a new Venetian military school called the Accademia Delia. Instruction would include the use of the military compass. Curiously, the first Delian lecturer in mathematics, Ingolfo de' Conti, had not been able to afford to take private lectures from Galileo.[158] But, although the stakes were high, pecuniary interest alone seems insufficient to explain the nastiness of the counter to Capra that Galileo himself later admitted was excessive. As will appear, the missing motive must be sought in the stars, and an affair of honor.

Church

There was another traitor in the Scuola Galileiana. On 21 April 1604, Silvestro Pagnoni, who had worked for Galileo for a year and a half as a copyist, presented himself "spontaneously," which often meant, as in this case, on the urging of his confessor, before the nearest inquisitor. His conscience forced him to disclose that his former employer cast horoscopes. Pagnoni volunteered further that during his 18 months in Galileo's house he saw his master go to mass only once, and then to look up his ecclesiastical friend Querenghi. "I never knew him to go to confession." Question: "With whom does Galileo associate?" Answer: "With Cremonini, almost every day, and [what was not compromising] with the illustrious Giacomo Alvise [Cornaro]." The rascal then allowed that Galileo had never mistreated him and that, in his opinion, Galileo, though a poor Christian, believed in "matters of faith." Inquisitor: "You have said that Galileo gives certain judgment about the nativities he makes: this is a heresy; how could you then say that he believes in matters of faith?" Pagnoni took the hint and hit again. "I know that I said this and that he does give certain judgment about the nativities, but I did not know that this had been declared heretical."[159]

Pagnoni then enriched his testimony with some precious details about Galileo's dealings with his mistress and his mother. During a visit to Galileo in 1603/4, Giulia had enlisted Pagnoni to spy on her son. He duly relayed her confidences to the inquisitor:

> I've understood from his mother that he never makes confession or takes communion, and she asked me to find out whether he went to mass on feast days. I observed that instead of going to mass he went to the house of his Venetian prostitute Marina....I believe that his mother went to the Holy Office in Florence against him and [or because?] he

called her terrible names, *puttana, gabrina*. And she said that the Florentine Holy Office issued him an admonition to him.

The first bit is no doubt true. Probably the Florentine incident deserves credit
too.[160]

The Venetian authorities had no trouble seeing that Pagnoni acted through
spite and that his charges were "very frivolous and of no importance." Galileo had to teach astrology; there was no reason that he should not practice
it also; and Pagnoni certainly had not shown that Galileo interpreted the
charts fatalistically.[161] Galileo's private life was no business of the Inquisition.
The weakness of Pagnoni's charges is perhaps their most interesting feature.
Despite Galileo's compromising friendships with Cremonini and Sarpi, his
earlier exposure to Averroistic interpretations by Borro and Buonamici, his
school of foreign heretics, and his dissolute life style, Pagnoni could not find
anything derogatory to say about Galileo's religious beliefs. Galileo had many
friends and supporters among the clergy and among laymen close to the
Jesuits. He detested true libertines like Cremonini's student Antonio Rocco,
whose licentious defence of homosexuality, *L'Alcibiade fanciullo a scuola*, published in 1650 but no doubt long meditated, teaches that Moses invented the
story of Sodom to stop men from preferring boys to women grown ugly
after 40 years of wandering in the desert.[162] Galileo liked ribaldry but not
blasphemy.

The inquisitors did not need Pagnoni's hint to interrogate Cremonini, who
had already come to their attention as a corrupter of youth and promoter of
disbelief. The view was widely held even outside the thought police. Sagredo's father thought that Cremonini spread atheism and alumni like Rocco
gave color to the charge. Former students popped up now and then to accuse
Cremonini of leading them astray.[163] In 1604 one of his colleagues, Camillo
Belloni, professor of moral philosophy at the university and a fellow Ricovrato, took up the cause. Belloni exploited the occasion of a Lenten sermon,
in which a Jesuit preached against discussing the doctrine of the mortality
of the soul in courses on Aristotle, to denounce Cremonini and relieve his
own conscience.

The Riformatori knew perfectly well that Belloni and Cremonini had
carried academic dispute to the point of fisticuffs. They ordered the Rettori
to squelch the processes against the greatest philosopher and the primary
mathematician of their university, "it being altogether improper that people

should make use of the services and respectable office of the Inquisition to settle personal quarrels."[164] The inquisitor agreed, no doubt encouraged by the knowledge that Cremonini took the sacraments at the recommended intervals, taught religiously at Santa Giustina, and regarded astrology as nonsense.[165] Galileo's file was archived so imaginatively that it took almost 400 years to resurface. Cremonini's had been sent to Rome, where it joined a growing mass of denunciations.[166] Contact with the Inquisition was an ever-present hazard for scholars in Counter-Reformation Italy. Often it resulted in a warning, a reprimand, or a dismissal, and sometimes in a brief stay in jail. Many charges had little to do with faith or morals, but indulged jealousies or settled scores. As usual in Italy, outcomes depended on family connections (inquisitors had relatives) and influential contacts (inquisitors listened to princes) as well as on the acumen and judgment of the bureaucrats of correct thinking. The Inquisition was a fact of life, of many peoples' lives, a sort of low-level background terrorism, and they learned to live with it according to their circumstances.

The terror increased for everyone living in Venetian territory in April 1606 when Paul V excommunicated the most serene Doge and Senate and put the entire population of Venice under interdict. Paul expected that his order, which prohibited clergy in Venice from performing mass or serving sacraments, would force a hasty surrender from a government threatened with the unrest of Christian subjects deprived of the solace of marriage and burial. The immediate cause of Paul's wrath was the Republic's claim of jurisdiction over two priests accused of serious crimes and its clampdown on giving or willing landed property to the Church. Venetian efforts at negotiation, in which Galileo's military friend Pietro Duodo played a conspicuous part, failed to move the pope. Paul had not taken the measure of the men he dealt with. After recalling Duodo, the excommunicated officials ordered the secular clergy in Venetian territory to ignore the interdict; they ejected religious orders identified with the Counter Reformation, first the Jesuits, then the Capuchins and Theatines; and they took every opportunity to display respect for the Catholic religion. Attendance at mass shot up when the worshipper could cleanse his soul by spiting the pope.[167]

Galileo watched the evicted Jesuits, each carrying a candle and a crucifix, leave Venice, "to the pain and sorrow of many women devoted to them."[168] It was not a matter for mockery. The interdict was to last a year and the expulsion almost 50. Although neither the Serene Republic nor the

Apostolic See was a major military power, their conflict had the potential of escalating into a war on Italian soil involving France and Spain and of driving Venice into the Protestant camp. In the event, hostilities took place mainly on paper. The greatest champions were men of central importance in Galileo's story: for the Venetians, Fra Paolo, newly appointed to canonist and theologian of the Republic; for the Romans, Cardinal Bellarmine, admirer and advisor of Paul V, and longtime professor of controversial theology at the Jesuit College in Rome. They set out clearly the opposition between a world order based on an outmoded conception of Christendom and a world disorder based on the Machiavellian concept of *raison d'état*. "It was the function of the interdict, assisted by the growing inclinations of Europeans in the period to deal in rigid absolutes, to classify, systematize, and at last bring fully to the surface the antithesis between the political and cultural achievements of the Italian Renaissance and the ideals of medieval Catholicism, now invigorated by the Counter Reformation."[169] Galileo would expose a similar antithesis by insisting on the independent authority of his physico-mathematics.

Among the eloquent supporters of the state was Cremonini, who, in speaking once again for the faculty, praised Doge Donà for defending true religion from those who had "dared to poison the faith at its very source."[170] Cremonini's reference to pollution of the pure water echoed a leitmotiv in Sarpi's arguments. The Pope must be opposed in the interest of religion. Princes consecrated by God had the right and duty to intervene to save the church from error and presumption. It is a heresy, Fra Paolo advised, to declare as necessary to the faith a belief that is not so and then to force others to accept it. The fewer the dogmas and the fewer the Jesuits, the better the system. Sarpi appealed to history to make the case that the Roman church had not been faithful to its founders.[171] Bellarmine replied by pointing to the lengthy official Catholic history created by his close friend Cardinal Cesare Baronio, which argued the improbable thesis that the church had not changed in any significant way since its earliest days. To deny or challenge this history, wrote Bellarmine, "can be called heresy in history and temerity in theology, because it is repugnant to all the histories and sacred canons." As Sarpi said, Counter-Reformation Rome hinted at new heresies all too easily. He proposed as a partial defense introducing the critical study of history into the university, and showed the way himself, by writing an unflattering account of the Council of Trent.[172]

Paul withdrew the interdict in April 1607. The Venetians surrendered little in return apart from the two criminal clerics whose case had prompted the affair. They went to the King of France to dispose of as he wished. The laws against alienating property remained on the books. Soon, however, dissensions suppressed by the need for concerted action against Rome stirred. Lucrative papal appointments beguiled prominent members of the Donà-Sarpi camp. Others came to see that Venice's continuing rapid loss of commerce reduced its independence of action and made cooperation with Rome mandatory.[173] As the resistance to Rome softened, the Inquisition and censorship returned to normal. In 1609 Sarpi complained that the Venetian printing presses put out nothing but books of devotion and that papal agents opened his correspondence outside Venice.[174]

On 5 October 1607, five assassins tried to dispatch Sarpi. A crowd forced the bunglers to take refuge in the house of the papal nuncio. They left one of their daggers buried in Fra Paolo's face. "I recognize the *stilo romano*," he is said to have quipped after its removal. He received the solicitous attention of Donà and other magnates, the good wishes of the people, and the care of the great Acquapendente. He recovered. Other attempts on his life were made, but in the end the Holy Office had to be content with burning his books rather than his person.[175] By 1613 Sarpi's communication channels with the wider world had silted up. Although he continued to hold his appointment as state theologian, he devoted his time to writing history and to thinking through a world order determined by the struggle between "a few lonely champions of virtue and piety and a handful of wicked men at the Curia."[176] That was the way that Galileo came to see himself in relation to ecclesiastical authority.

Close observers of Sarpi might have concluded that an accomplished savant strongly supported by a vigorous and virtuous prince could be an influential public servant and an effective reformer of clerical abuses. They would also have recognized that in a busy republic, where many voices must be heard and the span of attention is short, "everyday new things appear and take away the importance of the old." Even a strong prince like Donà could not do what he pleased against the legitimate interests of his fellow patricians. And, of course, observers of Sarpi would find confirmation, if any were needed, that Counter Reformation Rome was quick with the charge of heresy even where matters of faith did not enter, and did not scruple to destroy people it considered threats to its interests. Sarpi-watcher Galileo could see himself as the accomplished savant advising the virtuous prince, and perceived that a single-minded tyrant might be easier to serve than a multiheaded republic. He seems to have missed the possible application to himself of the Roman hierarchy's inclination to hint at heresy to silence people it found threatening. A good test for a free thinker thinking to relocate from Padua to Florence in the early seventeenth century was the differential reception in Italy of a book on the divine right of kings by no less a prince than James I of England. Venice read it, Rome prohibited it, and in Florence Grand Duke Ferdinand had his confessor burn it.[177]

4

———— ⚬⚬⚬ ————

Galilean Science

4.1 RELUCTANT ASTRONOMER

Although his job description did not oblige him to create new knowledge, Galileo used his last years at Padua—the years of relatively stable life with Marina—to arrive at a new, fundamentally important treatment of motion. He also made some progress toward a new cosmology. He worked harder and more seriously than he had at Pisa. He could have said with Shakespeare's Lucentio, "I have Pisa left / And to Padua come, as he that leaves / a shallow plash, to plunge him in the deep." An indication of what he found in the depths follows. It may not be for all tastes. As Lucentio's servant reminded his master, poetry, rhetoric, and music were amusing pursuits, but not so Aristotle. For the rest, "The mathematics and the metaphysics / Fall to them as your stomach serves you / No profit grows where no pleasure ta'en."[1]

The Copernican confession

It was fortunate for Professor Galilei that astronomy could not get at the truth. He could teach Ptolemy and prefer Copernicus with a clear conscience. He had formed his preference while teaching at Pisa, or perhaps earlier, as appears from the long letter he wrote to Mazzoni in 1597:

> Turning…to the agreement of your opinions with those I regard as true even if they differ from ordinary ideas, I confess that I hold more strongly to them than I did at first, when I did not think that I had such strong support. To say the truth, however, as much as I was certain also about the other conclusions, so much was I confused and timid at

the first attack, seeing you so resolute and openly in opposition to the opinions of Pythagoras and Copernicus on the place and movement of the earth, which I held to be more probable than the opinion of Aristotle and Ptolemy.[2]

Now bolder and firmer, Galileo told his former mentor that an argument Mazzoni had devised against Copernicus had no force.

Among Aristotle's sins against mathematics that Mazzoni picked on was a report that the sun shines on Mount Caucasus for a third of the night. Let AI (Figure 4.1) be the mountain, DH the horizon at I, B the sun's position at last visibility, $\alpha = 60°$. Since the earth's radius r is negligible in comparison with the solar distance BC, angle BCJ is almost a right angle and angle ACJ = α = 60°. From $\cos 60° = \cos\alpha = r/(r + h)$, h must equal r, that is, 3,579 miles. Quite a mountain! On the advice of the mathematician Scipione Chiaramonti, who would become an antagonist and target of Galileo, Mazzoni allowed that Aristotle might have meant that a third of the mountain glowed when the lower two-thirds had gone dark. Then, taking $\alpha = 6°$ as the extent of twilight, $2h/3 = 20$ miles and $h = 30$ miles, "which does not seem impossible."[3]

From this considerable height an observer would see a conical horizon 192° in extent. It occurred to Mazzoni that a Copernican observer would have an even wider panorama, since his platform would be much farther from the center of his world than a Ptolemaic observer on the top of Mount Caucasus would be from the center of the earth. But our horizon does

FIG. 4.1 Last light on Mount Caucasus (AI) from the set sun at B; α = ∠GCB = 60° = ∠GCK/3 or one-third of the sun's nocturnal path.

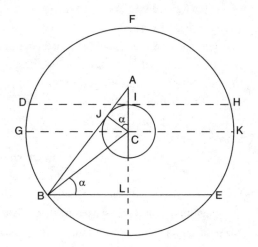

not usually exceed 180°. "Therefore Copernicus' placement of the earth is false and impossible."[4] Galileo answered that the analogy did not hold: the operative parameter in the first case is the distance above the earth, in the second, distance from the center. A prettily packaged piece of geometry made clear the distinction. In the same figure, let L be the place of the earth on the Copernican hypothesis and the arc BFE the extent of the heavens visible from it at noon; in this case the sun is at C. A Ptolemaic observer at A who saw the same arc BFE above him would stand on a mountain $r(\sec\alpha - 1)$ miles high. Using standard Ptolemaic values $s = 1216r$, $R = 45,225r$, and $r = 3035$ miles, where s and R are the radii of the sun's orbit CL and of the stellar sphere CB, respectively, Galileo found $h = 1.10$ miles, which, to be sure, would give a detectable effect, since $\alpha = 1°30'$. But for Copernicans, Galileo added, R is very much larger than $45,000r$ and, consequently, the line BLE in practice coincides with DIH, $h = 0$, and the horizon in both the Ptolemaic and the Copernican system split the heavenly sphere into apparently equal halves. "Please tell me if you think Copernicus can be saved in this way."[5]

A few months after this private assistance to heliocentrism, Galileo received a challenge to declare himself publicly. Johannes Kepler, then teaching mathematics precariously at a Lutheran secondary school in Catholic Austria, sent two copies of his new and mysterious *Mysterium cosmographicum* to Italy via the music teacher in his school, Paul Homberg. Since Kepler had not heard of "Galileus Galileus," whose redundant name later amused him, Homberg must have picked Galileo as a worthy recipient of the Mystery; or, better, Venice put Homberg in Galileo's path by its enlightened policy of encouraging foreigners and even heretics like Homberg to study in Padua.[6] The episode was important because Galileo, who seldom read contemporary works in mathematics or philosophy unless given them or attacked in them, probably would not have stumbled on Kepler's book on his own. He was to take something from it that decisively influenced his thought.

Like many people who receive unexpected books, Galileo thanked the author immediately so as not to have to comment in detail. He had had time only to read the preface, he said, from which he gathered that congratulations were in order, not to the writer, but to the reader, for "having acquired such a lover of truth as an ally in the search for truth." Kepler had found some choice things, which Galileo promised to study, "and that the more willingly since I adopted Copernicus' opinion many [five or six?] years ago, and deduced from it the causes of many natural effects doubtless inexplicable on the ordinary hypothesis. I've written out many reasons for it and many responses to reasons against it, which I have not dared to publish as I've been deterred by the fate of our master Copernicus. For although he has gained immortal fame among a few, he has been ridiculed and derided by countless others (for such is the number of fools). I would venture to disclose my thoughts if there were more like you; but as there are not, I will forbear."[7]

Kepler tried to stiffen the backbone of his shy ally. "I was very pleased to receive yours of 4 August, firstly because of friendship begun with an Italian and secondly because of our agreement about Copernican cosmology." Mathematicians everywhere (Kepler continued) side with Copernicus and calculate according to his principles. If we all speak out together, people ignorant of mathematics will have to take our word for it. "If I'm right, not many good mathematicians in Europe will wish to differ from us; *tanta vis est veritas*, such is the power of truth. If Italy is not a suitable place for publication, and if you encounter other difficulties, perhaps Germany will grant us this freedom…Have faith, Galileo, and go forth."[8] To this pep talk, and an appended request to make a certain astronomical observation in the common cause, Galileo did not respond at all. There were good reasons for his silence. To the practical Catholic Galileo, the rhapsodic Lutheran Kepler was no closer to the truth in astronomy than he was in religion.

Kepler could be faulted not only for pretending to rethink the thoughts of God, but also for having been a student of the detestable Michael Mästlin, professor in the Lutheran stronghold of Tübingen. The Index had condemned all of Mästlin's books, which he richly deserved for his vicious pamphlet war against good Father Clavius over the Gregorian reform of the calendar. Now Mästlin's protégé, the author of the *Mysterium cosmographicum*, had set up as a sort of prophet, indeed, as an answer to the Lord's prayer. "God Himself has waited for 6,000 years for someone to study Him [properly]."[9] (Galileo used much the same formulation when he set up as a prophet 15 years later.) Kepler had demonstrated a profound harmony, and perhaps more, between the sun, the fixed stars, and the intermediate space where the planets play, on the one hand, and God the Father, God the Son, and God the Holy Spirit, on the other. This vision, though pale in comparison with Bruno's kaleidoscopic cosmology, nevertheless coincided with it in key points that interested the Inquisition: the Copernican system, which, if taken literally, conflicted with Scripture, and the identification of the Trinity with the

Cosmos, which, if pursued logically, could end in pantheism.[10] When Galileo came out of the closet, some sharp-eyed commentators, including Kepler, would perceive similarities between his worldview and Bruno's. It was not a desirable connection.

The cosmic mystery Kepler claimed to have cracked was the puzzle why God created only six planets and gave them the speeds and distances they display in a Copernican universe. The clue to the right answer came to Kepler during a lecture on an astrological theme and led to the exotic geometry of the Platonic solids. Of these rare objects—each of which consists of the same regular polygons meeting at the same solid angles—there can exist only five even in the heaven of geometry. In an extraordinary feat of imagination, Kepler saw these five solids nestled inside one another so that a spherical shell inscribed in one would circumscribe another. The five solids therefore defined six spherical shells, to each of which Kepler assigned a planet; Saturn circulated in the shell circumscribing the cube, Jupiter in the shell inscribed in the cube, and so on.

Kepler chose the order of the solids to get the best fit. Once established, the sequence fixed the relative values of the mean distances of the planets from the sun without wriggle room for the mathematician. To determine the thickness of each shell, which represented the difference between the maximum and minimum distances of the planet from the sun, Kepler resorted to observation and left their derivation from fundamental principles to another occasion. Still, he had seen to the heart of the cosmic mystery. "Never could I express in words the joy that this discovery gave me. I no longer regretted lost time. I no longer felt disgust at the toil, I no longer shirked the most laborious calculation, and I spent days and nights computing until I could see if my opinion...agreed with the orbits of Copernicus or if my joy would dissipate in the winds."[11]

Kepler must have been disappointed that Galileo did not mention the great discovery of the Platonic spaces when indicating his reasons for preferring the Copernican theory. "From it I have found the causes of many natural effects that doubtless cannot be explained on the ordinary hypothesis." Kepler supposed that Galileo had the tides in mind. This idea was not hard to find. One of Kepler's correspondents had suggested winds, tides, and ocean currents as phenomena favoring a moving earth and, as we know, Galileo's colleague at Pisa, Andrea Cesalpino, had proposed earth shakes as the cause of tides.[12] Sarpi's notebook for 1591 has a similar but subtler tidal generator

in the small excursions of the earth's center of gravity from the center of the
universe occasioned by evaporation and precipitation, the descent of rivers
to the sea, rock slides, and so on. Four years later he noted that tides might
be caused by the combination of the diurnal and annual motions supposed
by Copernicus.[13]

The theory Sarpi then sketched invokes the same principles that Galileo
later made the foundation of his proof of the Copernican system. Sarpi
observed that the two ends of a long seabed running east–west (think of
the Mediterranean) would have different velocities, as indicated in Figure 4.2.
The plane of the paper is that of the celestial equator; A and B are points in
space with which the ends P, Q of the semicircular canal PQ coincide when
we begin to watch; CD is the direction of the earth's revolution around the
sun projected onto the equator; and the arrow at C is the sense of the earth's
spin. The velocities of spin and revolution reinforce at A and oppose at B.
The canal around P must therefore decelerate, and that around Q accelerate,
during the twelve hours in which the spin moves the canal to the opposite
side of the earth and the annual revolution brings the earth's center from C
to D. This alternate acceleration and retardation creates tides because the
water, unable to keep pace with the acceleration of its bed, heaps up where
the canal slows down and falls away where the canal speeds up. Since, however, in real life the sloshing depends on the size, depth, and location of the
seas, the theory cannot predict the timing of the tides observed. The question whether this fateful and fallacious theory was Sarpi's or Galileo's has
been answered variously.[14] Whoever deserves the credit, the theory contains

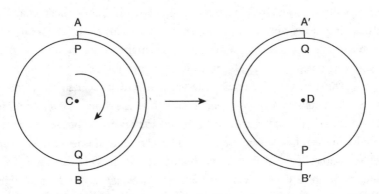

FIG. 4.2 Sarpi's tidal theory (1595), which moves the waters by a combination
of the earth's diurnal and annual motions.

an idea basic to Galileo's later physics: the concept of the composition of velocities, the ability of the same body to enjoy two different natural motions at the same time.

The insight that the tides have nothing to do with the moon may have sealed Galileo's commitment to the Copernican theory. His idea of "neutral motion" in "De motu antiquiora" and his deduction that a marble sphere centered on the center of the world could spin forever "naturally" had made the *diurnal* motion physically possible; the derivation of the tides reported by Sarpi now seemed to make the *annual* motion necessary. Hence by the time he wrote Kepler in 1597, Galileo was a Copernican of at least two years' standing.[15] Kepler would have preferred an ally who relied on solid metaphysical rather than dodgy physical arguments; as he rightly observed, no theory of the tides that omitted the moon could correspond to the nature of things.[16] But the theory reported by Sarpi corresponded perfectly to the nature of Galileo. It was witty and economical, requiring only motions presupposed for other reasons, and purely kinematical, eschewing ad-hoc, unintelligible, spooky influences of the moon upon the waters. And, also characteristic of many of Galileo's simplifications, it was wrong.

The single subject that took Galileo's fancy as he glanced through Kepler's *Mystery* came at its end, where Kepler made a preliminary guess at God's scheme for relating the periods τ of the planets to their distances a from the sun as determined by the Platonic solids. The guess failed quantitatively and also, in Galileo's view, qualitatively. Kepler related the periods to the presence and activity of the sun.[17] That was to introduce too much physics, or the wrong physics, for Galileo. His first try at relating τ and a invoked what we may call the Pisan Drop in memory of Borro and Buonamici's theorizing by throwing heavy objects from upper-storey windows. The Pisan Drop has God release the planets from a point in the firmament whose whereabouts a good mathematician could discover on the assumption that the linear velocity of a planet in its orbit equals the vertical velocity v acquired in its fall. Galileo's calculation has very considerable interest if only for showing that astronomical problems intruded into, and may perhaps occasionally have guided, important developments in his ideas about motion. We saw another likely example in the spinning marble sphere at the earth's center, which could represent a Copernican problem or a misplaced grindstone.[18] Here, however, the connection is clear. Galileo knew that v is twice the average velocity acquired during the fall, from which he gathered that if the time of fall equaled the period τ, the orbit's circumference would be twice the distance of fall.[19]

Let Saturn be the reference orbit and b_s its height of fall. Can Jupiter's observed period τ_J be obtained by a drop from the same creation point, that is, through a distance $b_J = b_s + (a_s - a_J)$? Galileo arranged the calculation to minimize reliance on the relatively unknown radii a. Since $v_J\tau_J$: $v_s\tau_s = a_J$:a_s, $(b_J$:$b_s)(\tau_J$:$\tau_s) = a_J$: a_s provided that, as Galileo then believed, velocity increases with distance fallen in natural descent. And since $b_s = \pi a_s$, τ_J:$\tau_s = \pi a_J$:b_J. If Galileo had solved this equation for a_J taking a_s as known, he would have learned that it did not come close to the Copernican value. In the calculation as preserved, he proceeded by successive approximations, altering a_J and b_J so as to come as near as he thought necessary to the observed value of τ_J. That gave him some confidence in the Pisan Drop, to which he would return again and again, as an indication that the motions of rocks and planets follow the same rules.[20]

Galileo's first pursuit of a relation between τ and a probably dates from 1602/3, if not earlier. The dating rests on letters from Edmund Bruce, like Segeth a Scot in Pinelli's circle, to Kepler, whose cosmology Bruce much admired. Bruce squealed that Galileo taught secrets from the *Mystery* as if they were his own. It is charitable to suppose that Bruce had in mind Galileo's attempt to implement the Pisan Drop, inspired by Kepler but not stolen from him. Bruce's charge did not bother Kepler.[21]

Infrequent communion

Kepler sent a copy of his *Mystery* to Tycho Brahe, then the prince of European astronomers, at his island research headquarters off the coast of Jutland. Brahe was not there to receive it. A new Danish king had decided that an astronomical price was too much to pay for an astronomer and Tycho received the unknown Kepler's little book while on the road looking for a wealthier despot. Tycho replied that the trick with the solids was impressive and the fit with Copernican data better than might be expected; and also that Kepler's enterprise was fatally flawed by assuming a sun-centered universe and relying on Copernicus' data. Tycho had a cure for both evils. In the true world system, as Tycho conceived it, the earth remains at rest in the center in accordance with good philosophy and common sense, the moon and sun and stars revolve around the earth, and all the planets circle the sun. As for data, Tycho had a castleful, collected by himself and his assistants or serfs in the first continuous systematic planetary observations ever undertaken.[22] Hoping to get his hands on data that could confirm God's deployment of the five solids and fix the widths of the spherical shells, and finding Graz increasingly inhospitable to Protestants, Kepler negotiated for a job with Tycho, then newly re-employed as mathematician to the eccentric Holy Roman Emperor, Rudolf II. Tycho foresaw that the brilliant intense mathematician who could think like God would be able to perfect the Tychonic system. The union was not made in heaven, however, since Kepler had his own agenda and Tycho would not allow him free access to the precious data. The mathematician's star proved the stronger. In 1601, after two years of uneasy collaboration, Tycho died. Kepler eventually got the data and also Tycho's post as Imperial Mathematician.[23]

Tycho had a mixed reputation in Padua around 1600. Pinelli had corresponded with him, owned his books, and understood his claims to glory; but as Tycho became more interested in promoting himself than his science, Pinelli increasingly doubted his reliability. As part of the promotion Tycho sent his student and son-in-law Franz Tengnagel to Italy to try to win the financial support of the Venetian Senate and the Grand Duke of Tuscany for an observatory in Egypt to extend Tycho's catalogue of stars southward. The cause did not carry. Nor did the Tychonic system win many friends in Italy since the Jesuits, particularly Clavius, opposed it.[24] Against this background, Tycho tried to get in touch with Galileo. Pinelli urged Galileo to reply; Galileo preferred silence. In 1600 Tycho tried again through Pinelli. "Please give my greetings to your most excellent professor Galileo Galilei on the grounds of our common subject of study, and tell him that if he writes me…he will not find me slow to respond."[25] Silence again.

Tycho stooped to writing directly. The occasion, he said, was a conversation in which the ambassador of the Grand Duchy of Tuscany to the Holy Roman Empire sang Galileo's praises. "I could not help but write you to build a basis for friendship and further correspondence." Then came a more plausible reason:

> I learned from my assistant Franz Tengnagel on his return from Padua that you have examined the first volume of my *Epistolae astronomicae* [1596] and found some things in it about which you wished to consult me…[I]f there is something that you want to discuss, I welcome it, and

you will find me a most faithful correspondent. Perhaps it concerns my hypothesis of the celestial revolutions...; there are some details that neither the Ptolemaic nor the Copernican system can handle so well. Perhaps you want to discuss fixed stars, or comets..., or something else mentioned in my book. Do it frankly, as you judge best. I in turn will tell you my opinion and not neglect to confer with you about astronomical matters.[26]

Certainly a civil letter.

Why Tycho wanted Galileo's correspondence is not known for certain. At the time he sought a biographer in Italy and at all times he strove to control the field by collecting astronomical observations. Possibly he was misinformed by Kepler, who inferred from Galileo's endorsement of Copernicus and his station as a mathematician that he actively pursued astronomical research. Whatever Tycho had in mind, it failed. Galileo could not have liked what he read in Tycho's *Epistolae astronomicae* against Copernicus and had no interest in setting the heretic right. He did not answer Tycho's civil letter.[27]

Nevertheless, he carried on a private debate with Tycho that sometimes spilled over into caustic references in freewheeling discussions with his Paduan friends. Perhaps he laughed at Tycho's compulsion to collect data: "it is no demerit for a mathematician not to have seen the star at its first appearance, as if he were obliged to stay up all night for his entire life staring at the heavens to see if he can spot a new star."[28] Or, he might have depreciated Tycho's planetary system for its bad taste, its awkward compromise of Ptolemy and Copernicus. These or similar belittlements came to the ears of Tengnagel through the generosity of Galileo's rival Magini. "I'll not dignify these emulators and calumniators of Tycho," Tengnagel replied, "these dwarfs who rant in private and from the lectern in Padua against whomever they please," by mentioning them in connection with Tycho's glorious name. "The truth cannot be expressed or repressed by owls hidden in the shadows (I mean those proto-mathematicians, that celebrated, if it please God, professor of mathematics and his Venetian brother in ignorance [Sarpi? Sagredo?]). Being incapable of publishing anything on their own, they envy the immortal, super-Herculean labors of others."[29] Galileo had not published anything because he had nothing to say that could withstand his own criticism.

It took nature itself to force Galileo into broadcasting, if not printing, some of his unripe cosmological opinions. The cause was public interest in the nova of 1604, first noticed in Padua, on 10th October, by Capra and two associates. Capra told his patron Cornaro, who told his protégé Galileo, who caught up with the spectacle on the 15th. The university turned to its mathematician for an explanation. The three lectures Galileo devoted to the nova in November and/or December drew large audiences eager to know its importance for philosophy and astrology. "Grave questions, these," said the professor, "and worthy of your minds. If only I could give answers that corresponded to the importance of the matter and your expectations...! I will take up only a single point, which pertains to my particular competence, the evidence for its motion and its position."[30]

Galileo explained the meaning and relevance of parallax, reported that the nova displayed none, and concluded, as a certainty, that it lay beyond the moon. Here he might have stopped, having dispatched his single arrow. Instead he sketched a theory that ruined the Aristotelian cosmos: the nova very probably consisted of a large quantity of airy material that issued from the earth and shone by reflected sunlight, like Aristotelian comets. Unlike them, however, it could rise beyond the moon. It not only brought change to the heavens, but did so provocatively by importing corruptible earthy elements into the pure quintessence. That raised heaven-shattering possibilities. The interstellar space might be filled with something similar to our atmosphere, as in the physics of the Stoics, to which Tycho had referred in his lengthy account of the nova of 1572. And if the material of the firmament resembled that of bodies here below, a theory of motion built on experience with objects within our reach might apply also to the celestial regions. "But I am not so bold as to think that things cannot take place differently from the way I have specified."[31]

The nova took 18 months to fade away. Galileo observed its protracted death in the hope of detecting a constantly diminishing parallax. If the nova rose from the earth into regions beyond the moon, perseverance in its original direction might plausibly be the cause of its dwindling intensity. Suppose, then, that the nova began to fade when not far past Saturn and that it moved steadily on along a line that passed through the earth's center. Its changing parallax might be discoverable if the universe were Copernican. The maximum distance measurable by sightings from two stations depends on their separation. That is the principle of binocular vision. For Ptolemaic observers, the maximum possible separation of the stations is the earth's diameter; for Copernican observers, it is the diameter of the earth's orbit (Figure 4.3).

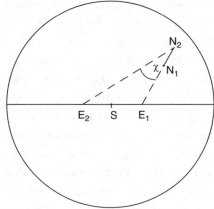

FIG. 4.3 The semi-annual parallax
χ of the comet supposed to move
from N_1 to N_2 as the earth moves
from E_1 to E_2.

Nevertheless, the nova refused to show parallax. Perhaps failure to detect the parallax undercut Galileo's confidence in the Copernican theory.[32]

The menacing question whether a mere measurement could kill established physics again featured high on the philosophical agenda. The several solutions suggested by friends and opponents of Galileo will indicate the ingenuity, alarm, and wishful thinking that the recent nova inspired among the committed learned. Clavius belonged to the wishful. He accepted that parallax measurements placed the recent nova beyond the moon, just as Tycho's numbers did the nova of 1572; but after watching the sky he buried his head in the sand, and declined to be drawn on the physical consequences of the placement.[33] Guidobaldo del Monte doubted the observations that established the null parallax and made some himself, with little luck; however, he would not have his brother mathematicians pollute the immaculate heavens and walk away from their dirty work. "The mathematicians will soon agree among themselves that it is a star. But they will not know how to reply to the philosophers' demonstrations that the heavens do not corrupt and cannot suffer this novelty…It is against all philosophy…I cannot understand why knowledgeable people want to make the heavens corruptible in order to be able to pronounce the nova a star."[34] Del Monte too thought wishfully. "[The nova] does not bother me for I believe it is a comet," and hence, for an Aristotelian, quarantined beneath the moon. He recalled that Mazzoni knew so much about comets that he had managed to persuade the Grand Duchess Christina that she had nothing to fear from them. "Would that Mazzoni were still alive!"[35]

A nest of mathematical philosophers in Verona accepted that the new star lived beyond the moon and offered various solutions to the implied problem in physics. Ilario Altobello, who was a Minorite monk and a minor poet as well as a mathematician, wanted to cause the maximum damage. Noting that the nova first appeared near and during a conjunction of Saturn and Jupiter, he supposed it an offspring of their union, "a most marvelous marvel of the heavens, given as the last light in this last of the penultimate age of the world, to make us understand the contrivances and truths of celestial nature." Owing to its placement relative to Jupiter and Mercury, the star would expose the truth, which is that the world is Tychonic.[36] Other correspondents from Verona ascribed the nova to sunlight reflected from relatively dense quintessence created either by chance overlap of denser parts or by compression of rarer parts by Jupiter and Saturn.[37] The authoritative Kepler denied that the nova owed anything to the grand conjunction, or Europe's problems anything to its fading, but to say the truth there was no reason to believe him.[38]

Among the purer philosophers, two have a claim on our attention as Galileo replied to one, Antonio Lorenzini, and had a running battle with the other, Ludovico delle Colombe. Lorenzini was a freelance, the author of a philosophy of laughter chuckled over as far away as Germany, and a friend of Cremonini, who probably contributed the more telling passages in Lorenzini's assessment of the nature of the nova. Padua's primary philosopher had taken pains to observe the "exhibition in the sky made to drive peripatetics crazy," and rebuked through Lorenzini mathematical cobblers who did not stick to their last. Galileo's well-attended performances had trespassed beyond the disciplinary boundaries regulating public lectures and, with the authority of the podium, had planted doubt about the general reliability of the received world picture. And that on the basis of a number or two obtained from instruments operated at or beyond their limits of reliability! Philosophers drew their conclusions from the "principles of natural things known through the senses, observed at an appropriate distance, and confirmed by induction," whereas mathematicians were literally out of their senses as the things perceived lay very far away.[39]

The argument has its merits. To defeat it Newton would posit a special "rule for philosophizing," which requires that qualities we find in all the bodies we can experiment with, like extension and impenetrability, should be taken as qualities of all bodies universally.[40] Cremonini would have added, provided that the extension does not upset the physics we have developed from experience with objects we can handle. From this point of view, abandoning Aristotle, not cleaving to him, would have been irrational.[41] Lorenzini did not pursue the argument at this level, however, but tried to discredit parallax measurements on technical grounds that he did not understand. That opened him to easy dismissal by astronomers. Capra replied immediately, with an annoying boast that he had seen the nova five days before Galileo. The gadfly thus stung the professor on his most sensitive spot, his private and professional honor. The consequent grudge, amplified by the plagiarism of the compass, precipitated Galileo's intemperate attack on Capra mentioned earlier.

A Latin edition of Lorenzini's *Discorso* fell into Kepler's hands. He read it as an affront to astronomy. "What do you say to [this fellow], Italian mathematicians, you Clavius, Guidobaldo, Magini, Galileo...? Why do you shut your eyes and show such patience before something so dishonorable?"[42] When he received this call to arms, printed in Kepler's account of the nova, Galileo had already taken steps to level Lorenzini in a way, however, that Kepler would not have understood. Galileo hid behind others, a feint foreign to Kepler, who was candid to a fault. And instead of writing a serious reply in Latin or Italian, which might have served as a purgative to scholars who had swallowed Lorenzini, Galileo and his collaborators wrote a lampoon in the unintelligible dialect of the Paduan countryside.

Galileo and his friends did not take Cremonini–Lorenzini as seriously as Kepler did. They laughed at the philosopher of laughter.[43] The mockery did not bother Cremonini, who had learned not to get upset when opposed. "Disagreement about learned matters does not make enemies of honorable men, rather, it is a sign of great ignorance to dislike people who do not accept your opinion."[44] Historians sometimes miss this point. Like his diatribe against wearing academic gowns, Galileo's put-down of Cremonini–Lorenzini was a farcical exercise in a popular comic form, in this case the rustic dialogue of the sixteenth-century writer known as Ruzzante. Galileo guffawed at Ruzzante's earthy humor and the rough expressive dialect of his characters, and later kept his Florentine friends rollicking with his readings in Padovan.[45]

The collaborators took the corporative name Cecco di Ronchitti, who claimed to be from a village in the domain of Querenghi, and to have learned about the nova from his master. In addition to Querenghi, Castelli's Benedictine brother Spinelli and probably Castelli himself, who was a great fan of

Ruzzante, had their hands in the composition.[46] Galileo apparently devised the story line and Spinelli wrote it up. It stars one Natale (Nale), who knew Lorenzini's opinions, and Matteo, who knew Galileo's. After Cecco insinuates in Bernesque fashion that a gown makes a doctor, his bumpkins are discovered under a walnut tree pondering the cause of the drought that afflicts them. Nale confides that a doctor in Padua traced it to the new star.

> *Mat:* "A pox on those great turds at Padua." Anyway, it's so far away they cannot know where it is.
>
> *Nale:* A philosopher there says it's much closer.
>
> *Mat:* "A philosopher is he? What has philosophy to do with measuring...? It's the mathematician you've got to believe. They are surveyors of empty air, just like I survey fields."
>
> *Nale:* The philosopher says that mathematicians do not understand the problem, that they think the sky can be created and destroyed.
>
> *Mat:* "Now, where do mathematicians talk that kind of reasons? If they just stick to measuring, what do they care whether or not something can be created? If it was made of polenta, couldn't they still see it?

Nonetheless, just like Galileo, Matteo speculates about the star's nature and settles on Galileo's theory as the best available. Having turned philosopher, he has a go at Aristotle:

> *Mat:* "I think that in the sky there are hot and cold, wet and dry, just like down here. Why? Thick and thin are seen there, and so are light and dark. Those are contrary things. What more do you want? That star could have been there but it wasn't, and now all of a sudden it is. Isn't that a switch?"
>
> *Nale:* "But he goes on to say that if earth and air and water and fire were up in the sky, we could not see it the way we do, because it would be thick and dark...Also he says that [if made of the four elements] the sky couldn't go around because the elements go up or down, but not around."
>
> *Mat:* "He leaves out writers who say that the earth goes around like a mill...Plague take me if that fellow, Doctor though he is, wouldn't look the same as anyone else with his clothes off."

> *Nat:* "Llisten to this one. He says that the mathematicians have some good instruments and solid arguments, but just don't know how to use them."

Our rustics have now come to the point. Matteo the surveyor explains how to measure parallax and that, despite the cavils of the doctors, astronomers could find it accurately enough to know whether an object is under the moon or beyond it. Matteo then runs Nale around the countryside sighting trees from various uncomfortable positions to teach him and any "learned dwarfs" who could read Padovan the practical meaning of parallax and its determination.

> *Nale:* "I've got it, and it is as plain as a cowshed." Now here's another thing—the author of this precious book says that unless it is at the zenith, the moon cannot cover all of the sun.
>
> *Mat:* "The Hell you say! The idiot must think that the moon is like an omelette. What an ass!...Is there more?"
>
> *Nat:* "Sure. What's the meaning of 'gralaxy'?"[47]

Cecco's farce had a second printing later in 1605 significantly different from the first in the two places in which Matteo alludes to Copernicus. The allusions in the earlier edition, printed in Padua, are friendly: "there's a lot of people, and good ones too, that believe the sky doesn't move"; "he [Lorenzini] leaves out writers who say the earth goes around as a mill." The later edition, printed in Verona, refers to "the lie that the sky didn't move" and the "false witness" of the Pythagoreans and Copernicans, and dismisses the writers who suppose the earth to spin as "wild-eyed." Although Galileo must have been disappointed that after losing half its brilliance, and when seen from opposite ends of a diameter of the earth's orbit, the nova showed no change in parallax, it is unlikely that he authorized the changes.[48]

Early in 1606 a more formidable opponent than Lorenzini entered the lists. It was Ludovico delle Colombe, a Florentine of about Galileo's age, who, having accepted the mathematicians' location of the nova, undertook to place it in the heaven of the philosophers. On principle it could not be newly created, but only newly visible. Assume then that it is at so great a distance as to be invisible without the sort of lens that extends the vision of myopic people. We need only to find a lens that we can all use at once. Delle Colombe spied it in the transparent heaven whose slow turning causes the precession of the equinoxes. A thickening in this heaven coming between the viewer and the otherwise invisible star might reveal it, without doing any violence to philosophy, theology, or astronomy.[49] One Alimberto Mauri, of whom nothing is known, replied to delle Colombe point by point. Perhaps Mauri was partly or wholly Galileo. In support of the identification, Mauri often throws numbers at delle Colombe (distances and sizes of stars, rates of precession and other processes), puts the terrestrial elements in the sky, and refers occasionally to Cecco.[50] Against the identification are Mauri's dullness and, what would have been foolhardy on Galileo's part during the great Venetian interdict, the dedication of the book to Paul V's treasurer, Luigi Capponi.[51] Delle Colombe, who wanted to know his antagonist, at first identified Galileo as both Cecco and Mauri, but soon accepted that he was only Cecco. On this understanding he wrote most civilly to Galileo, whom he held in honor and affection, he said, and was most eager to serve.[52] But not for long. Ludovico and his brother Raffaelo, a Dominican preacher, would make serious trouble for Galileo in Florence even before he returned home in 1610.

4.2 MOVER AND SHAKER

An uncertain trajectory

Galileo did not date his manuscripts or, usually, indicate what problems they concerned, and if he published the results recorded in them, he did so decades after obtaining them. These sloppy habits vex and inspire the serious student of his mathematics of motion. Since the 1970s, when Stillman Drake proposed an ordering of the manuscripts based on watermarks, inks, and orthography, the problem of what Galileo knew and when he knew it has become deeper and darker. No single ordering uniquely makes logical and chronological sense.[53] What has been gained is an appreciation of Galileo's skill as an experimenter. The manuscripts contain many numbers and some diagrams that suggest the experimental arrangements that produced them: pendulums, inclined planes, water clocks, free drops. Several modern Galileians have repeated these experiments and, after much cut and try, have reproduced the numbers.[54] Their success has only enlarged the domain of mystery.

Did experiment drive theory or theory experiment? Sometimes the one and sometimes the other, but also, and not rarely, neither. Galileo could stick to an attractive theory in the face of overwhelming experimental refutation. During his period of greatest creativity in the science of motion, from 1602 to 1609, he probably jumped from theory to experiment and from one idea to another, circled back and forth, inventing the form of a descriptive mathematical physics, guided often enough by little more than his *buon gusto*. The principal outcome was a somber, limited, exact science, and a few striking results, advertised as more exact than they were, to serve as a replacement for one chapter of the vast, colorful, diffuse library of Aristotelian philosophy.

Galileo's exegetes have thought this dating game worth the candle for several reasons. Drake thought it important to show that hard-headed experiment and calculation preceded the theories on which modern physics stands. In his reading, the practical man, not the philosopher, was the proto-scientist.[55] For others, knowledge of the pathway to discovery, with its jumps forward, backward, and sideways, is of surpassing interest, both in its details and in its indications of the modes of human creativity.[56] Still others want to know how Galileo's ideas related to Aristotelian commentary, medieval geometry, and practical mechanics.[57] Here the vast potential field is narrowed appreciably by the content of Galileo's "De motu antiquiora," which framed his work in Padua.

While at Padua, Galileo had several interlocutors who played the roles that Mazzoni had played at Pisa: sounding board, source of ideas, and, perhaps most important, propounder of problems. One of the interlocutors, the late Moletti, communicated through his manuscripts, which Galileo consulted in Pinelli's library; in them Galileo would have found or confirmed such essential ideas of his mechanics as the composition of velocities and the retardation of descent along inclined planes.[58] Guidobaldo del Monte's notebooks also contain considerations very close to ones Galileo later published as his own. Some historians read these considerations as records of conversations with Galileo.[59] It is safer to attribute them to "Galileo's group," a set of friends with whom he discussed all sorts of things freely and openly, and to regard them, as the group seems to have done, as common property. Sarpi did not name Galileo in his notebooks and Galileo did not credit Sarpi with any of the Galileian ideas found in them.[60] Galileo thought that Mazzoni had derived much of his treatment of Aristotle's failings from their exchanges during walks around Pisa, but Mazzoni did not think an acknowledgement necessary. Del Monte did not mention Galileo when describing ideas very similar to ones Galileo later used as if his own. Nor, to be fussy, did Galileo claim to be the author of the main new ideas about motion presented in his published work. He remains fastidiously off stage while his spokesman describes inventions revealed in conversation with him. "Our academician told me that he has found out…" Even in this form, the claim is about knowledge, not about priority.

By 1610, Galileo possessed virtually everything he later made public in the *Two new sciences* (1638). He, Sarpi, and at least one or two others then knew that in free fall bodies increase their distance from their starting point in proportion to the square of the time elapsed, and their velocity as the time linearly; that the trajectory of a body thrown or shot obliquely is a parabola; that violent and natural motion can coexist in the same body from the onset of the violence; and that an impressed velocity does not waste away of itself. What seems essential to know is not who made what discoveries when, but why Galileo came to see them as fundamentally important, as the basis of a new science of motion. Why did he reject the questions about motion as well as the answers given to them by Aristotle and his commentators, among whom we should number Galileo's former self, the author of "De motu antiquiora"?

To obtain an answer in this muddle of uncertain dates, perplexed attributions, and reciprocal influences without claiming greater certainty than our information allows, a Galileian dialogue, which permits digressions and approximations, might suit.[61] Its participants are Alexander, Galileo's alter ego from Pisa, who since parting from Galileo in 1592 has cultivated algebra, and Galileo himself. The conversation takes place during the summer of 1609, when Galileo, according to his custom, was spending part of his vacation in Tuscany.

An imaginary reconstruction

Al. What have you done with the subjects we talked about at such length in the old days? You almost finished a formal discourse and drew it up in Latin before you left, but you have not published a word.

Gal. The truth is, I've been working hard to make ends meet and attend to three children and several clever friends. Actually, that's not the reason. The truth is that I found serious difficulties in our dear old "De motu" and I have not been able to solve them within the old framework.

Al. If you can't solve them, no one can.

Gal. You recall that our Pisan mechanics turned on an extension of Archimedes' treatment of floating bodies to falling ones. That led us to discard levity, except when laughing at Aristotle, as unsuitable to science, and let relative gravity, the tendency of all bodies to go to the center of the universe, do all the work. Of course, as Aristotle dimly perceived, heavier bodies possess this tendency more powerfully than light ones: but he erred in measuring this tendency by gross weight rather than by weight per unit volume, or density. With this important clarification, we agreed that, *grosso modo*, the world is constructed with the heaviest, that is, the densest, matter closest to the center, and the lightest, that is, the rarest, at the top, under the sphere of the moon.[62]

Al. Yes, we thought that this was a secure move away, but not far away, from Aristotle, as the theory allowed bodies to come to their proper places without suffering the impossibilities of motion that Aristotle put on them.

Gal. You mean his crotchet that in moving freely or naturally a body assumes a velocity determined by its weight and the resistance of the medium through which it falls. We had a good laugh at that and at his deduction from it that vacuum cannot exist. Since then I've come to see that his desire to write the medium into his theory responded to something profound. It is after all the separation of a body from its natural place that somehow provides the cause and occasion for it to move. Separation implies the existence of an intermediate entity or ground, since otherwise the current and future, or actual and potential, positions of the body would coincide, which is absurd. Aristotle went wrong in estimating the velocity of return to natural place by the ratio of the weight of the mobile to the resistance of the medium.

Al. We showed that this relationship fails before experiment as well as logic, and replaced it with the relation we derived from Archimedes: velocity of return is proportional to the difference in densities of the mobile and the medium.

Gal. From that we easily showed that in a vacuum a body would move with a finite speed proportional to its density. That annihilated Aristotle's argument against the void. However, his insight that something must provide the separation between present and future positions of a moving body remains: our "vacuum" is not nothingness, but a ground of unresisted motion.

Al. You are becoming very generous in your maturity. I sense that acquaintance with your clever friends in the Veneto has blunted the know-it-all attitude that endeared you to the faculty at Pisa.

Gal. Moderation and common sense lie deep in my character. Anyway, I came to see that we did exactly what Aristotle did. We began from sound principles but generalized prematurely. The reliance on Archimedes was a mistake. His theory relates to static equilibrium; its generalization to bodies moving continuously in natural motion is not obvious and may not be possible.[63]

Al. We generalized by adding the old concept of a self-consuming or wasting impetus to explain why a body thrown vertically upwards gradually slows, ceases to rise, and falls back faster and faster as the wasting impulse dies away. If the body can fall far enough that the impetus wears off completely, it will proceed with a characteristic velocity proportional to the difference in densities of the body and the medium. Until that point, it appears to accelerate but in fact is just using up the impressed impetus. As Aristotle perceived (to talk in your nice new way), the final or true velocity in natural motion is a constant. The common view that natural fall is accelerated is false.[64] The hand or shelf holding an object before it drops supplies the same amount of wasting impulse it would have retained if thrown there from the surface or maybe the center of the earth. So in this case too, acceleration is only the apparent result of the decay of the imposed impulse. And to finish the main traits of the theory, in projected or violent motion at an angle α to the horizon the onset of the downward motion owing to gravity occurs variously depending on the value of α. At one extreme, vertical projection ($\alpha = 90°$), gravity does not kick in until the wasting impulse has vanished; in horizontal projection ($\alpha = 0°$) it starts immediately; in between, it happens at times and places closer to the starting point the smaller the value of α.

Gal. Your summary is admirable. Unfortunately, the theory is completely wrong. We might have been able to guess that, since our results never agreed with the few experiments we may have made. We had to invoke accidental factors like friction to explain the disagreement.

Al. Yes, certainly, that's the way Aristotelian theory works. Measurements may confirm it but not kill it.

Gal. You do not have to make mathematics a murderer to see that the concept of a wasting impulse dependent on the nature, position, and direction of projection of a moving object is logically inconsistent. For we say, on the one hand, that a body released from rest has a residual impulse that is used up in the fall; and, on the other hand, that a body projected upward begins to descend where and when its wasting impulse expires. In the first case gravity begins to act immediately and acceleration occurs through the wasting of the impetus; in the second case, gravity kicks in only at the top of the rise and acceleration occurs through the action of gravity. I began to see that wasting impulse was vapid and that natural motion truly is accelerated at the same time that I recognized that, at least for projection away from the vertical, natural motion downward begins as soon as the object—rock, cannon ball, arrow—is released from its mover.

Al. When did you tumble to this?

Gal. On my first journey from Pisa to Padua I spent a few days with that wonderfully wise and helpful man, Guidobaldo del Monte. Besides being a good mathematician, he was the inspector of fortifications for Tuscany and delighted in the flights of cannon balls. For him the question when natural motion begins to act in shots inclined at various angles had practical importance. We happened to be discussing it while drinking wine in his garden near a shed with a sloped roof. I may have mentioned the trick of slowing natural motion with the help of an inclined plane. Perhaps it came to both of us at the same time that a ball rolled upward along the sloped roof at an angle to the horizontal would simulate a cannon shot in slow motion. We did the experiment using an inked ball on a plane more convenient than the shed roof. Let me draw it for you at a representative angle α (Figure 4.4). As you see, the ball traced out a line resembling a parabola, hyperbola, or, as we

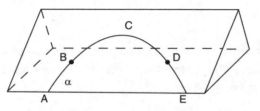

FIG. 4.4 The inked-ball trajectory as it may have been observed by Galileo and del Monte.

fancied, an upside-down version of the curve formed by a chain hanging from two widely separated supports.

Al. How extraordinary! But why refer to the chain? Does that not reintroduce the old confusion between the static and the dynamic?

Gal. Yes, but the analogy between the simultaneous actions of the stretch and weight of the chain, on the one hand, and of violent and natural motions, on the other, may give an opening to a solution to this important and problematic connection. The essential point is the symmetry of the path. As Guidobaldo wrote in his account of the ball's trajectory, "it rises as it descends, which is reasonable because the violence it acquired going up causes it to behave similarly coming down, where the natural motion then dominates. The violence that directed the motion from B to C is conserved, making CD equal to BC, and, as it gradually lessens during the descent, makes the path from D to E equal to BA. There is nothing to show that from C toward E the violence vanished completely [*a fatto*]. For although it diminishes continually toward E, yet enough remains that the ball never travels in a straight line there."[65]

Al. I see that you or both of you were still uncertain whether the violence continues unabated or dies away in time.

Gal. The answer of course lies in the symmetry of the path. But it would be after some years and extensive conversations with Sarpi that I realized that the violent velocity must be conserved throughout, apart from the inevitable loss in overcoming the resistance of the air. You see, what misled us was that on the way up the slope of the curve initially follows the direction of projection, more or less, over AB, whereas over DE the slope is reversed, as if the impetus were running out and could no longer call the tune. Also, to be frank, I've rather exaggerated the symmetry of the curves traced by the inked ball.[66]

Al. Very often pictures are misleading. I am too much of a gentleman to point out that misleading ideas from geometry can be rendered harmless in the course of algebraic abstraction.

Gal. There is something in that. I got into a dreadful muddle through what appeared to be a very apposite drawing. It depicted the fall from A to B

(Figure 4.5) and the muddle arose from supposing that in the acceleration velocity increases as the distance fallen. Or, in geometric terms, v_C: v_B = CG:BJ, where v_C, v_B are the degrees of velocity acquired in passing over AC, AB. And I should add in partial exculpation that I also had vaguely in mind that the distance a hammer can drive a nail at a single blow depends on the hammer's velocity.

Al. I see Aristotle behind this formulation. He always took velocity as his main measure of motion, in preference to, say, the time required to cover a particular distance. Greek mathematicians had trouble with the concept of velocity because, as a mixed ratio of distance and time, it did not have the dignity of a Euclidean quantity. In my opinion, you will never create a satisfactory dynamics of motion if you do not relax about these Greek demonstrations. To pick one example: your concept of indifferent motion derived from the relationship v_h: v_k = k:h, where h and k are the height and length of the same inclined plane; and from this you concluded that on a plane of zero slope a body would be indifferent to motion. But v_k in this case is not indifferent but indefinite because on a plane of zero slope k:h = 0:0.[67] What do you say?

Gal. I'll have to think it over. Meanwhile I'll continue the confession of my embarrassment at how long this erroneous association of velocity and distance dominated my thinking. I've only freed myself from it during the last year or two. Liberation came with my discovery that the path of a cannon ball is in fact a parabola. In looking back I can see that I arrived at this

FIG. 4.5 A useful but misleading diagram of velocity against distance in free fall.

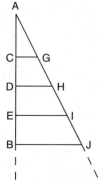

enlightenment by swinging back and forth between good and bad ideas and better and better experiments.

Al. You always liked pendulums.[68] But I must say that the use that doctors have made of your pulse meter would not convince anyone that numbers are relevant to reality.

Gal. During my last illness Santorio declared that my pulse rate made me a hypochondriac of the 97th class. That contented him but not me. Doctors should know and treat the causes, not the numbers, of diseases.

Ant. Perhaps we should keep that in mind. Have you had any success in proving your claim that a pendulum performs its oscillations in the same time regardless of the amplitude of its swings?

Gal. No, but I've made a lot of progress in trying. Early on I tumbled to a relevant and very beautiful proposition deducible from our old Pisan mechanics: the time a ball takes to roll down a chord to the bottom of a vertical circle is the same for every chord including the vertical diameter. It follows from the rule that the velocities of descent along inclined planes of the same height are inversely as their lengths. Let us draw (Figure 4.6). ADB is the vertical circle centered at C. The very beautiful proposition ("VBP" for short) is that descent from D to A takes as long as descent from B to A no matter where D lies on the circle. To see this, extend CB and AD to their intersection at F. From the rule from "De motu antiquiora," we know that $v_{BA}: v_{FA}$ = FA:BA. We want to show that $t_{BA}: t_{DA}$ = 1. Well, $t_{BA}: t_{DA}$ = (BA:AD)($v_{BA}: v_{FA}$) = (BA:AD)(FA:BA); and,

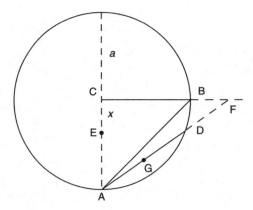

FIG. 4.6 Galileo's very beautiful proposition (VBP).

if you remember your Euclid, you know that AD·AF = 2AC² and, therefore, that $t_{BA} = t_{DA}$.[69]

Al. Very nice, very nice indeed. And I see how you might hope to get from here to the isochronism of the pendulum by filling the arc with little chords and trying to get the total time by Archimedean exhaustion.

Gal. I've managed only for two chords but the result is encouraging: the time for the broken journey SI, IA (Figure 4.7) is always less than the direct one SA, from which I infer that descent along the arc SA takes less time than any other route from S to A and, further, that this minimum is the same for all arcs.[70]

Al. That is quite an extrapolation. Even if it is true in mathematics, I doubt that you would ever be able to show it for real pendulums.

Gal. You must have been talking with Guidobaldo. I sent the chord theorem (the VBP) to him seven years ago. He replied that he did not believe it or the consequences I wanted to draw from it in favor of the isochronism of pendulums. I told him that I had demonstrated the isochronism with two pendulums of equal length swinging in widely dissimilar arcs, which stayed synchronized through hundreds of oscillations. Perhaps I exaggerated. He could not duplicate the feat.[71] Nor did he expect to be able to since he thought as we did once that experiment could not confirm theoretical results precisely because of inevitable and uncontrollable disturbances.[72] I am almost willing to return to this position myself. I have not been able to

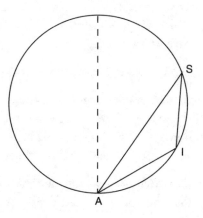

FIG. 4.7 First step in an Archimedean exhaustion of motion along an arc of a vertical circle.

confirm by experiment my elegant theorem of the isochronism of descent along chords. It works well enough for real chords, that is planes, with slight inclines but the time over these is about 15 percent more than the time along planes close to the vertical.[73]

Al. Maybe the VBP is wrong. In any case, your demonstration of it doesn't hold up. You took the velocity of descent as constant, as we did in the old motion theory. It is in fact accelerated, as you since conceded.

Gal. I can fix that up. The average velocity of descent along inclined planes of the same height is half their final velocity, since, as I postulate, velocity increases uniformly in natural motion; so everything goes through as before with half velocities in place of whole ones. Consideration of average velocities also yields the nice useful result that the times of descent along inclined planes of the same height are as the lengths of the planes. Let's give it a name, say the average time theorem (ATT).

Al. Nonsense, the average velocity over DA cannot be equal to the average velocity over FA.

Gal. Before you go too far, let me tell you a second beautiful result that followed from the VBP and the ATT. The derivation is very easy. The ATT gives $t_{CA}: t_{BA} = CA:BA$ and $t_{EA}: t_{DA} = EA:DA$. The relationship I'm after, the ratio of the time of fall from C to that from E, is then $t_{CA}: t_{EA} = (CA:BA)(DA:EA)(t_{BA}: t_{DA}) = (CA:BA)(DA:EA)$. A touch of the VBP removes the ratio of the times.

Al. I see that the next step, to express AD:BA in terms of CA and BA, is a piece of cake. By Pythagoras, $AB^2 = 2a^2$ and $AD^2 = ED^2 + AE^2 = (a^2 - x^2) + (a - x)^2 = 2a(a - x) = 2aEA$. So $DA:BA = (EA/CA)^{1/2}$, and $t_{CA}: t_{EA} = (CA:EA)^{1/2}$. Hence, if you imagine the diagram reversed and the fall to begin at A, you have $AE:AC = (t_{AE}: t_{AC})^2$.[74]

Gal. You did that more elegantly than I did.

Al. The result—that the distances fallen go as the squares of the times elapsed—is refreshing. However, it appears to rest on your dodgy derivation of the VBP. You must do better if you are to persuade the Sarpis and del Montes, let alone the philosophers.

Gal. I told Sarpi about the times-square rule and the pretty consequence that the spaces passed over in equal times in free fall are as the odd numbers 1,3,5 ... He had already absorbed ideas similar to those in "De motu antiquiora."[75] Like you, he wanted a better principle than the VBP and the ATT, which he said were just disconnected bits, like my odd-number rule. I rose to the challenge—and fell on my face.

Al. Ah! You went back to the velocity–distance concept. I don't blame you. It is so natural an assumption. As we observed earlier, the point of natural motion is to overcome distance or, better, separation, and velocity measures the effectiveness with which the process occurs.

Gal. This is what I wrote Sarpi:[76] "I suppose (and perhaps I'll be able to demonstrate it) that the naturally falling body constantly increases its velocity according as the distance increases from the point of departure ... that the degree of velocity ... at E is to the degree of velocity at D as EA is to DA ..." (Figure 4.5). Then I show that the sum of all the velocities with which the body passes through AD is to the sum with which it passes AE as ΔADH is to ΔAEI, that is, as $AD^2:AE^2$, which, of course, is the same as $DH^2:EI^2$.

Al: I do not see what significance to assign to the concept of "sum of all the velocities."

Gal. The sum of the degrees or moments of velocity, that is, of all the lines that can be drawn parallel to DH and EI. I blush to say that I took the concept and the analysis from a medieval technique called the "latitude of forms." That was not very clever of me since the scholastic philosophers were interested in making a picture, not in calculating the accidents, of motion. But I had known the technique for a long time and seized upon it to answer Sarpi's demand for a principle.[77] The appearance of the square of the distance, or, better, the square of the velocity in the formulation seemed promising; for if you take "sum of velocities" proportional to the square of the "latitudes" CG, DH ..., and also proportional to the distance fallen AC, AD ..., you have, with s as distance and v as the velocity achieved over the distance, $s \propto v^2$. Now, as I continued in my letter to Sarpi, velocities go inversely as times, so $v \propto s{:}t$, and $s \propto t^2$, as I claimed.

Al. This is worse than your error about average velocities in deriving the VBP. You have made s proportional both to v and v^2. Even Aristotle would not have made such a blunder.

Gal. That's what Sarpi said too. Nonetheless, I confirmed the t^2-rule by experiment. I've rolled balls down inclined planes of small slopes and timed their passage over measured distances with a water clock and by beating time. I never would have found the rule by experiment, but the measurements come close enough to the theory, and could be made closer if anyone thought it worth the trouble.[78]

Al. That would be a waste of time.

Gal. As for the principle Sarpi demanded, I found it at last, after twenty years' study of motion. It is that in natural motion, not distance or velocity, but time is the ordering principle. The axiom that I should have offered to Sarpi is that in natural motion "velocity is proportional to the square root of the distance fallen." This picture of velocity against distance is a parabola, not a triangle.[79] In fact, it was while playing with parabolas that I came to see the truth.

Al. Before going there, I cannot help pointing out that the way you arrived at your t^2-rule should comfort philosophers who think that mathematics only confuses physics. Just look at the range of conflicting assumptions you have used to get the same result: $v \propto s$, $v \propto \sqrt{s}$, $v \propto t$, "total velocity" proportional to s^2, natural motion both accelerated and unaccelerated, and so on. There may be a thousand theoretical assumptions from which the t^2-rule follows, and each of these assumptions may imply a different causal mechanism and so a different physics.

Gal. So what? Let us first obtain exact descriptions and then worry about qualitative explanations if we still want them. In the process, we can eliminate many possibilities, just as measurement of parallax has negated much of Aristotle's physics.

Al. Although in that case measurement is used qualitatively, to show whether an object is closer or further than the moon. The exact numbers do not matter.

Gal. I'll tell you what I told Fra Paolo when he raised similar points about experiments and causes. He objected to my conclusion from the symmetric curve that Guidobaldo and I had studied that an object shot upward would return to the point of projection with its original velocity reversed in direction. He remarked that an arrow that could pierce a board when shot at it directly at short range scarcely dents it when descending from the height to which the bow propels it vertically. I replied that we cannot remove the resistance of the air, which slows the arrow in both directions to an extent we cannot calculate. I made the same observation to Guidobaldo when explaining to him that, notwithstanding his troublesome experiments, my chord theorem is too beautiful not to be true.[80]

Al. So would you say that the jury is out on the value of experiment? Or is the correct position that it is conclusive when it confirms a quantitative rule, but only an inconvenience when it does not?

Gal. Perhaps something like that. Let's go on to causes. The inexhaustible Fra Paolo asked me why objects of different material, like gold and silver (they are the first things a Venetian thinks of), receive different amounts of impetus from the same mover.[81] The more I considered this question, the more I convinced myself that it is not possible to say anything exact about it. From then on, except for discussions with Sarpi, who likes hopeless causes, I have not worried about such questions.

Al. This is to leave the field to the philosophers.

Gal. Judge after I've told you how I came to know that $v \propto \sqrt{s}$ and therefore that $v \propto t$ and the tremendous discovery, indeed, my best discovery in mechanics, that I made along the way.[82] You know that I've had financial trouble largely because of the incessant demands made on me by my—or should I say our—siblings.

Al. And partly because of your self-indulgence.

Gal. I've supported these little indulgences and the family in large part by teaching military engineering. Gunnery is an essential part of it. My business and theoretical interests thus came together in the weighty matter of the trajectory of cannon balls. Gunners don't care whether or not a useful mathematics of trajectories rests on a solid philosophical foundation. That made things simple. I could ignore the annoying relations among experiment, mathematical theory, and explanatory causes that worry philosophers.

Al. So you went directly to your workshop. How did you model gunshots at various elevations?

Gal. I turned the problem upside down. I rolled balls down the plane GA (whose slope I exaggerate in the drawing (Figure 4.8)) and turned their velocities so that they leapt into space at A. I have an adjustable board that can be placed at BE, CF…I observed the places B, C…where the ball struck the board. The distance AE, AF…always came out as the squares of the distances BE, CF…, not like the thousand synchronic swings of the pendulums through different arcs, which, to be sure, was rhetoric, but truly and neatly.[83] Now $BE^2 \propto AE$ is just the symptom of a parabola.

Al. That should interest bombardiers. They might see the connection with your earlier formulations more easily if you were to use the double-distance theorem that follows from the proposition that the average velocity of descent is half the final.[84] If then you plot points horizontally at distances $2k$, $4k$, $6k$…and vertically below them points at pk^2, $4pk^2$, $9pk^2$…, p a constant, and join the points you will have a parabola (Figure 4.9).

Gal. To get the parabola, I had to assume that the horizontal motion, the motion of projection, and the motion of fall occur simultaneously and independently.[85] Here again Archimedes showed the way, with his method of generating a spiral as the locus of a point that moves radially and tangentially at the same time. I'm often surprised by the way that art anticipates nature. Come to think of it, old Cesalpino used to point out that rain can be driven violently by the wind as it continues to fall naturally.[86] And my predecessor Moletti had worked it out, as I've seen in his manuscripts. The concept of

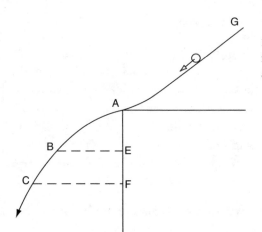

FIG. 4.8 A roll down an inclined plane GA turned into a reversed gun shot ABC.

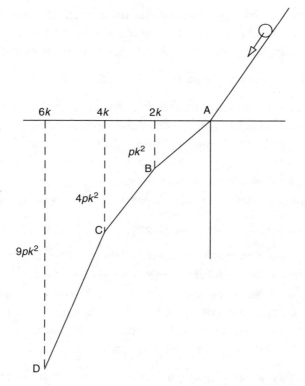

FIG. 4.9 Construction of the parabolic path of a projectile launched horizontally at A.

mutual disturbance or struggle belongs on the same scrap heap as positive levity and wasting impulse.

Al. I see how to get the t^2-rule and its principle, $v \propto t$. The horizontal spaces $2k$, $4k$..., are proportional to t, the vertical spaces pk^2, $4pk^2$..., therefore, to t^2. And from $v \propto t$, achieved at last with the help of heavy artillery, $s \propto t^2$ implies $v \propto \sqrt{s}$. That the velocity of fall should be proportional to the square root of the distance fallen does not appeal to most people's intuition.

Gal. That is the chief reason for mathematizing physics—to correct the misleading intuitions arising from everyday experience.

Al. Where do you go from here?

Gal. Remember that we agreed that Copernicus' geometry of the solar system is more plausible than its rivals? Since then I have become convinced that he is almost certainly right and that, in consequence, no theory of motion based on gravity toward the center of the universe can be right—with the exception of an old crotchet of mine about the creation of the universe.

Al. That seems out of character. What is it?

Gal. An idea that came to me while leafing through Kepler's *Cosmic Mystery*. It is that all the planets fell from one and the same creation point as far as they needed to acquire the velocities we observe them to have. I tried out the idea using $v \propto s$, which, of course, did not get me far. With $v \propto \sqrt{s}$, I can work out a common drop point for Saturn and Jupiter, or for any other pair of planets, and preliminary calculations suggest that one point may do for all.[87]

Al. What else do you have for philosophers?

Gal. The composition of velocities and the idea of indifferent motion, which may permit a spinning sphere to spin forever, equip anyone to defeat the usual physical arguments against Copernicus. With the composition of velocities, the persistence of indifferent motion, and the t^2-rule, I can derive the trajectories of cannon balls and contrive a Copernican theory of tides. Naturally, that does not prove Copernicus right.

Al. Philosophers abhor a vacuum more than nature does. How can you hope to defeat a physical theory or rather an entire physics without one to put in its place?

Gal. As our friend Mazzoni used to preach, philosophers should test their theories against quantitative arguments. I admit that this use of mathematics as a sieve for removing crude ideas does not meet my ambitions. Mathematicians should not merely enforce right thinking among philosophers, they should initiate and guide it.

Al. Have you the firepower to begin such a program?

Gal. Not yet. I only have some charming rules about motion on inclined planes and a theory of projectile paths that would never serve to aim a gun. And my refutations of physical arguments against the Copernican system are not compelling to those who do not incline toward or accept it already.

Al. What are you waiting for?

Gal. A gift from God.

5

⌥

Calculated Risks

5.1 STARRY MESSAGE

Florentine foreplay

On 12 May 1590, an hour after nightfall, Christina, Grand Duchess of Tuscany, fulfilled the promise of her extravagant wedding and produced a son. Jupiter, blazing in mid-heaven, "looked down from that sublime throne...and poured forth all the splendor of his profuse magnificence into the purest air." The recipient of this wholesome radiation was Cosimo II, who, twenty years later, having risen through Jovian influences and the death of his father to the rank of grand duke, would add Galileo, the author of the preceding astrological effusion, to his court.[1] When the prince approached the age of 12, Galileo's reliable advisor at Pisa, Dr. Mercuriale, suggested to Christina and Ferdinando that Cosimo should learn mathematics. Simultaneously the matchmaker advised Galileo to perfect his military and geometrical compass.[2] The compass showed the way. Several Florentine gentlemen, who had acquired one by gift or purchase, mentioned to Christina Galileo's desire to present a suitable example to Ferdinando during a visit in the summer of 1605. Let him send it, Cristina replied, and when he comes to Florence "he will be seen in accordance with his merits."[3]

And so he was. Christina invited Galileo to the Medici villa in the cool hill town of Pratolino, "with a good room, modest table, good bed, and warm welcome."[4] Like most Jovians, Cosimo did not excel at mathematics and Galileo would have to return for several more summer tutorials. During and between these visits, the Medici and the mathematician bonded. Ferdinando helped to resolve some of Galileo's legal-financial problems and provided enough black satin to make a respectable gown for the hefty lampooner of the Pisan toga.[5] Galileo commissioned two compasses in silver to point the royal road to computation and adopted the style, or mask, of an angling courtier.[6] "[I am] one of those most faithful and devoted servants, who think it the grandest grace and glory to be born subjects." Nothing would please this subject more than to spend his life teaching Cosimo's mathematics. "Unfortunately, great intrigues and enterprises are not for me." To which the savvy princeling replied, "either you dissimulated about knowing your own worth or think that I do not recognize it."[7]

Galileo used the implied intimacy to recommend his Paduan doctor, Acquapendente, as successor to Mercuriale, who died late in 1606. The recommendation incorporated an argument Galileo was to recycle on his own behalf three years later. "Having acquired everything he can hope for here in reputation and authority, and having reached an age unsuited to the daily toil of caring for many friends and patrons, [Dr. Acquapendente] wants a little quiet to conduct his life and finish his work. He needs nothing to fulfill his ambition but the titles and degrees that others of his profession have obtained but which he cannot have unless they are given by a great absolute prince."[8]

Ferdinando had need of a doctor. As he sank into sickness, Christina took charge of whatever she could, including Galileo. In the spring of 1608, in a fit of fever and melancholy, he gave her an opportunity to measure his ambition and vanity. He had intended to spend that summer in Pratolino by invitation from their highnesses. He therefore rejected an invitation to reside instead in Florence. Feeling snubbed, he declined to go at all. "I would serve my prince even at the cost of my life," Galileo wrote the Tuscan state secretary Belisario Vinta, but otherwise, as he had no business in Florence, he would not come. Christina understood: "Tell Galileo that since he is the first and most prized mathematician in all Christendom the Grand Duke and I want him to come this summer even if it is inconvenient in order to instruct our son the prince in mathematics..." Galileo set out for Tuscany as soon as his doctor and his health permitted travel.[9]

The summer school had a newcomer that year, the prince's clever close friend Giovanni Ciampoli. Like Galileo, Ciampoli came from an old unwealthy Florentine family and gained entrée to higher society through his confident, aggressive ability with words. He versified spontaneously aided by a memory able to retain a sermon at a hearing. His important patrons included Strozzi, who took him into his household and gave him an allowance, and the grand duke, who employed him even as a teenager to versify for Medici events. Persuaded by Cosimo's coach to study Euclid, Ciampoli quickly progressed to astronomy; when Galileo returned to Florence in 1610, Ciampoli assisted him in the observations of Jupiter's moons.[10] He knew how to please.

Ciampoli's friend Cosimo had more to worry about in the summer of 1608 than mathematics. The ailing Ferdinando decided to procure a wife for him. He pitched on a 21-year old archduchess, Maria Maddalena of Austria, a pious young lady given to hunting. Galileo did not neglect the opportunity provided by his early knowledge of the betrothal. A great expert in *imprese*, he suggested a theme for the party the Medici would throw to solemnize and energize the marriage. The theme should be a universal one indicative of the groom's "celestial piety." Galileo proposed a magnet (Cosimo) supporting iron fragments (his subjects) with the motto he had suggested for Ferdinando's lodestone, *Vim facit amor*. There was even more to the metaphor: since our globe is a giant magnet and Cosmo means world, "the very noble metaphor of the spherical magnet could be understood as our great Cosimo." A coin with this *impresa* on one side and Cosimo's effigy on the other with the tag *Magnus Magnes Cosmos* would make a fine souvenir. The magnet's mother did not like the idea, however, and Galileo put the connection Cosmos-Cosimo in the cupboard for another occasion.[11]

The wedding of Maria and Cosimo rivaled the spectaculars staged for Christina and Ferdinando 20 years before. Bardi, Strozzi, and, especially, Cigoli, again played prominent parts and another of Galileo's friends, Michelangelo Buonarroti, the grandnephew of the artist, wrote the comedy. Christina also participated. With a deadly mixture of bossiness and incompetence, she vetoed parts of the intermezzi that Bardi and others thought integral to their artistic integrity.[12] The damage would have been difficult to discern, however, after the week of intense partying that began with the bride's arrival in Florence on 18 October 1608 proceeded by five companies of cavalry and four of musketeers. On entering the city through an elaborate triumphal arch designed by Cigoli, Maria Maddalena exchanged her army for an escort of 52 noble youths carrying a baldachino under which she proceeded to the cathedral. There Cosimo beheld her for the first time and happily found her less plain than reported; nonetheless, her squat figure and square face, inherited from the Habsburgs and the positions of Venus and Libra in her geniture, would become more pronounced with time and child bearing. But she was radiant as she entered the cathedral filled with light and flowers, silk hangings, and little angels; and no doubt her satisfaction was as great as her surprise when a cloud carrying a chorus of singing saints descended over her as she knelt to pray.

The following day, a Sunday, brought a ball and a banquet, for which artists created in sugar forty famous Florentine sculptures. The festivities ended at seven in the morning. The following week witnessed, among much else, Buonarroti's play and the intermezzi; a ballet executed by horses; and a representation, on the Arno, of Jason's theft of the Golden Fleece. This adventure featured Cosimo as Jason commanding an *Argo* filled with soldiers and musicians, a true float. He wore gilded arms, a magnificently plumed helmet, and a mantle of gold reaching to the ground. Amid fireworks and dragons Cosimo, scarcely mobile under his weight of gold, carried off the fleece and presented it to his bride. More usefully, he gave dowry money to 200 poor girls.[13]

The party was too much for Ferdinando. He took to his bed. Would he recover? A reliable answer involved inspection of his nativity. Christina asked Galileo to work out which of two dates assigned to Ferdinand's birth was correct. He obliged. "It seems to me [he wrote the anxious Duchess] much more in conformity with the rules to believe that His Highness was born on 30 July 1549 than on 19 July 1548." Ferdinando would not face his "climacteric year" for some time. Therefore his current illness would not prove fatal. Nonetheless, Ferdinando died.[14] The erring astrologer lost no time in offering condolence and congratulations, and in subscribing himself his former pupil's "most humble servant and vassal."

The vassal was not so humble as to hide what he wanted from his new lord. "I've worked for twenty years, and those the best of my life, dribbling out at everyone's request that little talent that God and my own efforts have given me in my profession." Galileo needed leisure and quiet to finish three great works he had in hand. He would not willingly leave Padua. "Still, the freedom [there] is not enough, since I must spend hours every day, and often the best, responding to the demands of one person or another. It is not customary to enjoy a stipend without public service in a republic, however splendid and generous, because to get anything useful from the public you must satisfy it, and not just one individual; and while I am capable of teaching and serving, no one in the republic can exempt me from them and leave me my salary; and in sum, I cannot hope for such an opportunity from anybody but from an absolute prince." Galileo understood that the prince might want something in exchange. He would give it willingly, he said, provided he did not have to suffer republican servitude to all comers. "I would not abhor service to a prince or grand seigneur...but rather desire and covet it."[15]

Although several of Galileo's Florentine patrons worked on this project during the spring of 1609, Cosimo was not yet willing to burden himself with his old tutor.[16] He already had enough authority figures to please and battle with: his mother, his late father's ministers, and his wife. Nor, perhaps, now that he had to pay the bills for his extravagant wedding, did he want to increase the expenses of the court beyond the necessary new staff to see to Maria, who required, besides the usual attendants of a princess, her own cook, a Jesuit confessor, and a handful of huntsmen to see to her horses and hounds. While awaiting more definite news, the hopeful vassal continued to teach, run his boarding school, peddle his instruments, and carry on an expanding correspondence that dealt with poetry as well as mathematics. From Rome Margherita Sarrocchi, known for her learning and her beauty, "and for a none too excessive puritanism," sought his advice about her heroic poem *Scanderbeide*. He sent her some critical advice and also copies of his old theorems on barycenters for her partner, his old friend from Pisa, Luca Valerio. He revealed to one of the Medicis that projectiles fired from the same place to the same height have the same time of flight over level ground irrespective of their range.[17] He had other things in stock too, many marvelous things, "perhaps the greatest curiosities that so far have been sought out by men," but he did not have time to tell anyone what they were.[18] Nor did he have time to go to Florence in the summer of 1609. That was lucky for him.

The Dutch gadget

"In Italy there is nothing new except the arrival of that spyglass [*occhiale*] that shows distant things, which I much admire for beauty of invention and level of art; but as for use in war on land or sea, I think there is nothing in it." This was Sarpi's considered opinion in the early summer of 1609 when he had known for over six months of the claim of Dutch spectacle makers to a gadget that made distant objects appear near.[19] He had not believed it, partly because he had thought of something similar in theory when young, but had not followed it up "owing to the recalcitrance of matter," and partly because he had deemed the report exaggerated, "as rumors usually become when traveling."[20] Galileo reacted similarly. He did not see the possible importance of the gadget for his flirtations with Florence or for his instrument business. He made no recorded effort to reproduce the Dutch performance although he dealt in spectacle lenses and kept a selection of them in his shop.[21] Perhaps Galileo and Fra Paolo discounted the rumor as just another in a long series of stories about magnifying mirrors or lens-mirror combinations. Natural magicians like della Porta taught that a perfect artisan could accomplish wonders with optical setups like those installed at the ancient Lighthouse of Alexandria, by which the curious could see from one end of the Mediterranean to the other.[22] But neither he nor any other modern magician had explained how to do it.

Those who know how to read Scripture can find in it references to everything. Father Benito Arias Montana, who attended to the publication of the polyglot bible sponsored by Phillip II of Spain, was perforce an expert reader. In Luke 4:5, which reports that "the Devil, taking [Jesus] up into a high mountain, showed unto him all the kingdoms of the world in a moment of time," he perceived a reference to "the perspective or optical art, which the devil knows; as by the same art we make *inspicilla*, which bring very distant objects very exactly before the eyes."[23] The text dates from 1575. No Christian dared to develop the Devil's device for another three and thirty years. Then several bold artificers in the low countries produced versions that magnified by a factor of two or three. One came into Sarpi's hands in July 1609. Having examined it, he could advise the Senate not to buy it from a traveling salesman who had offered it, together with its "secret," for 1,000 scudi. By then, August 1609, the secret was out.[24]

Sarpi's knowledge of optics gave him confidence that the gadget could easily be bettered, and his knowledge of men assured him that Galileo was the one for the job. Not that Galileo had any particular expertise in optical theory beyond perspective; neither Sarpi nor Kepler thought that he understood the principles of the instrument and Sagredo, who applied to him several times for explanations, judged none of them satisfactory.[25] Sarpi would have expected Galileo to proceed as he had in improving Sagredo's magnet, by clever convergent variations of known methods of cut-and-try; and, once he had found the method, to work doggedly until he could produce a saleable instrument for his inventory of military accessories.[26] To make things easier, Sarpi informed Galileo that the gadget consisted of two lenses, one on either end of a tube four feet long. Perhaps Sarpi helped in initial explorations as he certainly did in subsequent improvements. The Dutch gadget became the Italian telescope through the efforts of "the mathematician [Galileo] and others here [in Venice] not ignorant of these arts."[27]

By the end of August 1609, Galileo had a 9x (9-power) instrument, magnifying three times as much as the original gadget. Following his bent, Galileo would have kept his secret and sold his version to a market already developed by news of the Dutch invention. With his head start and compulsive industry, he could have expected to stay ahead of his competitors for some time. He chose another way, almost certainly counseled and directed by Sarpi. On 24 August Galileo wrote to Doge Donà to offer his invention gratis to the state:

> Galileo Galilei, most humble servant of Your Serenity, constantly and vigilantly alert not only to fulfill his duty as lecturer in mathematics at the University of Padua, but also by some useful and noteworthy discovery to provide an extraordinary benefit to Your Serenity, now comes forth with a new artifice of a spyglass [*occhiale*] derived from the most recondite speculations of perspective, which brings visible objects so close to the eye, and presents them so large and distinct that, for example, what is distant nine miles appears as if it were only a mile away: a thing of inestimable value in all business and every undertaking at sea or on land...And therefore, thinking it being worthy of being received, and judged as most useful, by you, he has decided to present it to you and to leave it to you to determine whether and how this invention should be manufactured. And this the said Galilei presents with every affection for Your Lordship as one of the fruits of the science that he has professed at the university for 17 years, with the hope of being able to offer you better ones in the future, if it pleases God and Your Lordship that, in accordance with his desire, he spends the rest of his life in Your Lordship's service.[28]

The phrase about the theory of perspective betrays a concern to reassure Venetian admirals and generals by rooting the novel contrivance in the soil of science.[29] The Republic accepted this precocious product of academic science for military purposes and rewarded Galileo, perhaps by prearrangement, by increasing his salary to 1000 scudi and granting him tenure for life. That seems a reasonable recompense. If Galileo had marketed his spyglass as effectively as he did his compass, it would have bought him 400 or 500 scudi

a year for as long as he controlled the market. The increase in salary and life
tenure provided the same sum forever with no expenditure of effort. Galileo
informed his brother-in-law that he had shelved his plan for repatriation and
would continue in Padua.[30]

The Florentine agent in the Serenissima, Giovanni Bartoli, kept the Tuscan
court informed about the sensation caused by Cosimo's vassal in Venice.
With Sarpi's help, according to the gossip, Galileo had analyzed the gadget,
found the "secret," and offered his improvement to the state; in recompense,
the Senate had raised his salary to 1000 scudi, "with the obligation of remaining in his lectureship permanently."[31] The day this report went out, Cosimo
asked Galileo for a duplicate of the instrument he had just given to the Venetian state.[32] The vassal could not oblige. He could neither sell the *occhiale*
nor tell how to make it, and had to construct a dozen for the Venetians; his
gift had made him a servant. Secretary of State Vinta asked Bartoli to seek
something that might satisfy Cosimo on the open market. Bartoli located a
Frenchman who claimed to sell instruments as good as Galileo's; but neither
through it nor any other gadget he examined could Bartoli keep a distant
object in view. "To my eye they do not show so many marvels."[33] He would
have said the same about Galileo's version, which had an even smaller field of
view than less powerful gadgets. In instruments of Galileo's design, the field
diminishes as the magnification increases; at 10x, about 20′, two-thirds the
size of the full moon; at 20x, no more than 14′; at 30x, under 10′.[34]

Around 1 December 1609, Galileo raised his best telescope, then of 20x,
to view the most conspicuous object in the night sky. The moon was four or
five days old when its pockmarked face came fuzzily into view: like lesser
beings, Cynthia, the lunar goddess, lost her smooth complexion on a closer
view.[35] As the moon grew, Galileo could see less and less of it at a single
peek, since his field of view encompassed less than a quarter of a full moon.
From his partial views of the lunar surface, he made composite portraits of
Cynthia in several phases—portraits, impressions, not exact mappings. The
six that survive in wash drawings are spectacular works of creative imagination (Plate 5).[36] They called on Galileo's knowledge of perspective, the joint
property of mathematicians and artists; on his own exercises in drawing and
water-coloring; on his keen eyesight and trained eye; and, perhaps, on his
addiction to Ariosto, who had described just such a moon as Galileo drew
when relating Astolfo's search for the wits of the half-cracked champions of
Charlemagne.[37]

Galileo's special preparation for lunar viewing may be suggested by the moon map drawn by the English mathematician Thomas Harriot shortly before Galileo turned his telescope on what he interpreted as its mountains. As a mathematician both pure and applied Harriot was Galileo's superior. If he had published his discoveries and theories instead of leaving them scattered through 10,000 pages of manuscript, he would have driven Galileo crazy, for he anticipated much of Galileo's mechanics, including parabolic trajectories, and explored the heavens almost as fruitfully as Galileo did. With instruments of 6x and perhaps as much as 10x, Harriot saw enough of the moon to map it in the style of a surveyor, a meager thing without a hint of landscape (Figure 5.1a). One of his collaborators, Sir William Lower, viewing independently, remarked that the terminator looked like a coastline and the full moon like a meat pie; neither of them recognized the smudges as shadows cast by craters.[38] After reading Galileo's description of the moon, Harriot looked again and drew a big crater along the terminator (Figure 5.1b). Probably the instruments through which he first sighted the moon revealed more evocative detail than he had recorded.[39] In his later lunar explorations he used a telescope of 30x made by himself and his friends. The transformation of the Dutch gadget into an instrument powerful enough to detect novelties in the heavens did not require a Galileo. His unique strength lay in interpreting what he saw.

In their printed form, Galileo's renditions move from portraiture almost to caricature.[40] The gigantic crater cut by the terminator at quarter moon

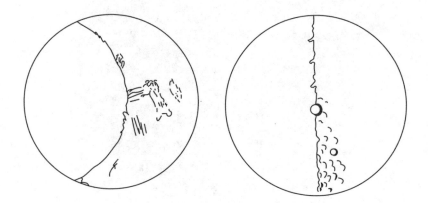

FIG. 5.1 A surveyor's moon. Harriot's depictions before (5.1a) and after (5.1b) seeing Galileo's *Sidereus nuncius*. After Bredekamp, *Galileo* (2007), Abb. 214, 215.

in Figure 5.2 would be large enough, if it existed, to be seen by the naked eye.[41] Galileo compared it in size to Bohemia, whose rendering, on Ortelius' widely used *Atlas*, may have guided the engraving of the great crater.[42] The flecks of light on the moon's night side and spots of light on the day side also indicated structures of vast scale, for, as Galileo explained, they recorded the illumination of mountain tops and the shadows cast by crater walls: and from these appearances he inferred the colossal scale of lunar features. In the remarkably favorable circumstance depicted in Figure 5.3, the top of a big mountain AD on the moon's limb is just touched by the sunray GCD; otherwise night prevails throughout the hemisphere CAF. Galileo estimated CA or, what is close to it, the line CD, at one-twentieth of the moon's diameter and CF at around 2,000 miles; he then worked the Pythagorean theorem and found that the mountain was almost five miles tall. "It is therefore manifest that lunar peaks are loftier than terrestrial ones."[43]

While following the moon, Galileo noticed a host of new stars previously invisible or blurred together. The Milky Way was not a complex terrestrial exhalation, as Aristotle would have it, but a congeries of stars.[44] The excitement of discovery kept Galileo observing well into the night, against his

FIG. 5.2 The half-moon featuring an exaggerated crater along the terminator. Galileo, *Sidereus nuncius* (1610). After *SN*, 45.

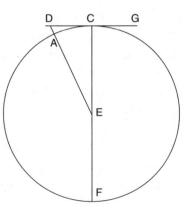

FIG. 5.3 Galileo's method for determining the height AD of a moon mountain. Galileo, *Sidereus nuncius* (1610). After *SN*, 52.

custom, in the chill December air, squinting through his imperfect glass at patches of the night sky almost unresolvable by the unaided eye. With one of his 30x instruments, he and his group—Sarpi, Micanzio, and others—could "see only a hundredth of the moon at a time, but of such a size as the whole moon appeared through the first telescope, the cavities are conspicuous and seen so exactly, that it is truly astonishing."[45] Galileo understood the significance of what he had seen, and was grateful. "I thank God from the bottom of my heart that he has pleased to make me the sole [!] initial observer of so many astounding things, concealed for all these ages."[46]

Galileo's move from 10x espionage to 20x moon studies occurred just after his lunatic mother had concluded a visit to Padua. Her departure may have released his creative energies as it certainly revived his spirits. "Don't forget to write me [Giula wrote] and to fill a page with the contentment and delight caused by my leaving." The addressee of this directive was Galileo's servant Alessandro Piersanti, whom Giulia tried to engage as a confidential agent. She had good reason to want to know what plans her son had for her granddaughter Virginia, whom she took back with her to Florence.[47] But she also wanted Piersanti to spy and steal. She provided a sure route for correspondence and sealed the pact by signing herself with the fearsome words, "affectionately, as if your mother." Piersanti was to report the goings-on in Galileo's household, insure that some linen Giulia had bought did not fall into the hands of "the lady you work for," and purloin three or four spectacle lenses from Galileo's store. Giulia knew exactly what she wanted, "not concave lenses for short vision…but plain ones that go under the tube, that is, those at the bottom through which you can see to great distances." Apparently she had followed the fortunes of the telescope without understanding that the secret lay in the lens combination. "Put [the lenses] in the bottom of a little box and fill it up with those pills of Acquapendente that I [like]." Giulia destined the stolen lenses for Landucci. "I ask you to do this because Galileo is so ungrateful to this man who has always cherished him"—when not suing him.[48] Giulia's spy ring had not informed her that Galileo had asked Christina for a modest post for Landucci and Vinta for help in recovering a loan made by Piersanti, who was then seriously ill, unable to work, and dependent on Galileo's charity. Naturally Piersanti turned over Giulia's letters to his employer.[49]

Had Galileo drawn up a balance of blessings during the Christmas season of 1609 he would have had to set discontent and uncertainty in his personal life against satisfaction over confirming Ariosto's lunar landscape and detecting myriads of unsuspected stars. His manipulative mother and aging mistress were at stiletto points; his preteen daughters needed his attention; Sarpi's intervention had locked him into a lifetime job in Venice while he angled for a place in Florence; his brother Sagredo, still abroad as the Republic's consul in Aleppo, could give neither comfort nor advice. Although Galileo's discoveries were notable, they did not break the bounds of earlier speculation. The existence of stars dimmer than the dimmest we can see, and the interpretation of the mottling of the moon as earth-like features, had been canvassed by the ancients. Galileo did not rush to publish his portraits of the moon or the news about the stars. He hoped for something never conceived before.

Medici stars

Far above these earthly cares, in the crisp evenings of that Christmas-tide, an ascendant Jupiter poured his rays toward Padua. Fate had placed the planet there, alluringly bright and close, a perfect and convenient object for telescopic examination. It presaged a brilliant future for astronomy, although, being near its closest approach to earth it was retrograde, a hint that the future might not be cloudless for the first astronomer to detect its secrets.[50]

Sometime before 7 January 1610, when Galileo described his lunar discoveries to Antonio de' Medici, he noticed through his 20x telescope that Jupiter had lined up along the ecliptic with three little stars. The first sighting of this striking alignment, perhaps incorporating only two starlets, took place it seems at Sarpi's monastery in Venice.[51] Agostino da Mula, a Venetian patrician, an old friend of Galileo's, and an enthusiastic student of light and vision, claimed the honor. Or so we read in a report by another member of Sarpi's group, Giovanni Camillo Glorioso (Plate 6), who would succeed Galileo at Padua against the algebraist preferred by Sarpi. Glorioso suspected that Galileo had blocked his earlier opportunities and, indulging his resentment and his jealousy, claimed that Galileo had taken the starlets from Mula as he had the telescope from the Dutch. If Mula did prompt Galileo's work during January of 1610, his intervention only makes Galileo's follow-up the more extraordinary and admirable. Galileo alone would recognize Mula's starlets as elements of a miniature solar system. That took immense skill and application; or, as Glorioso explained it, "the carefulness and industry of the Florentines."[52]

We can follow this care and industry day by day in Galileo's drawings of the changing configurations of Jupiter and the starlets. The triplet viewed on 7 January divided two to the east and one to the west of the planet (Figure 5.4a). On 8 January, the planet appeared to the east of all three starlets (Figure 5.4b). Galileo inferred that the standard tables erred in listing Jupiter as still in retrograde since it evidently had moved east with respect to the starlets, which he naturally supposed fixed, like all other stars. He expected to find Jupiter still further east of them on the 9th, but clouds veiled the scene; on the 10th, he was startled to find the planet west of the configuration and only two starlets in evidence (Figure 5.4c). On the 12th there were again three, one now smaller than the others, but with Jupiter among them, as in Figure 5.4a. So rapid an alteration of retrograde and direct motion could

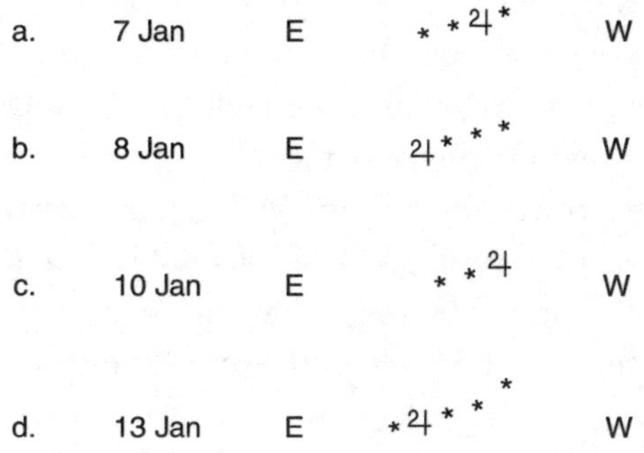

FIG. 5.4 Jupiter's starlets as they appeared to Galileo early in January 1610.

not occur on either the Ptolemaic or the Copernican theory. Galileo inferred that the starlets accompanied Jupiter in its rounds and performed some complicated minuet in their travels.[53]

On 13 January he had another stupefying surprise. A fourth starlet had appeared, not quite in line with the others and further than any of them from the planet (Figure 5.4d). On the 15th, Jupiter was again east of them all, on the 16th of three only, one having disappeared; on the 19th there were only two; and so on. What made the flighty starlets stick to Jupiter? Galileo gave an answer for which there was no precedent. The starlets were not stars at all, but moons. The supposition that they circled Jupiter gave a qualitative explanation of their appearances and disappearances, and of their changing distances from the planet. Galileo devised a second argument to enroll the starlets among planet-like objects that would recur in his polemics over telescope viewing: all stars twinkle; the starlets do not twinkle; the starlets are not stars. This syllogism contributed to the solution of the agitated question whether stars and planets shine by their own or by borrowed light. Kepler's opinions may suggest the range of respectable options in 1610: stars twinkle by reflection of sunlight; Venus, because she showed no phases, must have some light of her own.[54]

Our moon made a special case because of the dim glow of its dark regions prominent around new moon. Astronomers disputed over whether the glow arose from sunlight seeping through the moon's body, from native light, or from sunlight reflected from the earth's surface. Galileo took this last alternative as his own, and wrote as if he had invented it.[55] Several astronomers had already traced moonglow to earthshine, notably Kepler and his teacher Mästlin.[56] And Sarpi had puzzled it out before either of them.[57] The theory made the moon earth-like by depriving it of indigenous light and the earth moon-like by making it a reflector. Galileo also denied to planets any light of their own. That suggested the fallacious syllogism, planets do not twinkle, the starlets do not twinkle, therefore the starlets are planets.[58]

On 30 January 1610 Galileo informed Vinta that he was in Venice printing his account of the marvels God had vouchsafed unto him, of which the greatest were the "four new planets...that move around a very large star, like Venus and Mercury, and perhaps the other known planets, do around the sun." He felt free enough from his state service to promise to send a good telescope with which their highnesses could see the marvels he had found.[59] Their highnesses expressed their enthusiasm. Galileo wrote Vinta again, under strictest secrecy, that he would seize the opportunity—the first since Augustus named a star, or, as it turned out, a comet, after Caesar—to promote his patrons to the heavens.[60] Would Vinta please advise whether *pianeti cosmici* or *medicea sydera* would suit the situation better? Vinta replied that "cosmic" did not refer unambiguously to the modest Medici and that the attribution of one starlet to Cosimo and one to each of his three brothers would answer perfectly.[61] Thus Galileo's account of his discoveries, rushed into print early in March 1610 under the title *Sidereus nuncius*, included the fanciful, fateful designation of Jupiter's moons as Medici stars. "The Maker of the stars himself," Galileo wrote in his dedication to Cosimo, "admonished me to call these new planets by the illustrious name of Your Highness." Then, in the fawning style of the times, Galileo recalled that he had long bathed in "the rays of Your [Highness'] incredible clemency and kindness…Night and day [I reflected] on almost nothing else than how I, most desirous of your glory (since I am not only by desire but also by origin and nature under your dominion), might show how very grateful I am toward you. And hence, since under your auspices, Most Serene Cosimo, I discovered these stars unknown to all previous astronomers, I decided by the highest right to adorn them with the very august name of your family."[62]

This nonsense, however much enjoyed in Florence, did not go well in Venice. Galileo had obtained a nice promotion by assigning to the Republic the invention he had made under its auspices as professor in Padua; and now he was claiming that the astronomical discoveries he had made with the same instrument, also in Padua, had occurred under the auspices of the Grand Duke of Tuscany. Furthermore, he presented his novelties, from moonglow to Jupiter's moons, as exclusively his own observations, interpretations, and discoveries. In announcing the new heaven, Galileo lost sight of the help he had received on earth from his fellow observers Sarpi, Micanzio, da Mula, and also Pignoria, who participated gamely though worried, rightly, that "burrowing into the secrets of heaven might be reckless."[63] And Galileo neglected to mention the capital importance of the testimony of trustworthy Venetians able to certify that the discoveries announced in *Sidereus nuncius* were not optical illusions. Acknowledging his debts to his Venetian collaborators would not help him secure the position he designed for himself at Cosimo's court. His chief concern was not the feelings of his friends but whether their Tuscan highnesses would manage to see their stars through a telescope operated by unpractised hands. Should Galileo run down for Holy Week to insure successful viewing? He was scheduled to return to Florence for the summer; but that would be too late, for by then Jupiter and his brood would have disappeared behind the sun. But then the trip was long and tiring and he would need a litter from Bologna. But again, the business was of the utmost importance; another such occasion to show his devotion might not recur.[64] He was becoming hysterical.

Cosimo resolved his vassal's indecision by sending a horse-drawn litter to fetch him to Florence during Holy Week. Galileo's command perform- ance would dispel all doubts about the marvels. "As God has privileged you with this most singular discovery, He will also provide you with ingenious and judicious and fluent eloquence and expression needed to present [the subject] persuasively to everyone."[65] The stakes were high. Cosimo hoped to see a new miracle, to the confusion of the doubters, and to insure that his subjects who had any knowledge in the matter would confirm by their own experience the existence of the Medici stars.[66] Galileo's stars did not desert him. He was eloquent and persuasive and left Florence assured of the good will of their highnesses and the experts he had convinced.

Galileo now faced the problem of persuading people outside the circles of his Venetian colleagues and Florentine patrons that his discoveries had the status appropriate to a premise for a *demonstratio potissima* in physics. The program faced the significant impediment that far too few competent observers had access to the means of certification to permit an appeal to the repeated sense experience of rational animals. And even with this confirma- tion there might remain uncertainty over whether things seen at a great dis- tance through a glass darkly could have the same epistemological standing as things seen clearly, naturally, and close at hand. The obvious strategy was to distribute telescopes, directions for their use, and copies of *Siderius Nuncius* as guides to viewing. Galileo pushed Cosimo to organize the distribution without, however, having anything to distribute. In the spring of 1610 he had lenses enough for only ten instruments and they were the acceptable residue of 100 tries.[67] There was also the problem whether he had the right to choose their recipients.

The right to distribute would cease to be a problem if Galileo entered Tuscan service. Consequently he told Vinta that he would not give out instru- ments or disclose their "secret" without the grand duke's authorization,

and pressed hard for a final decision about his position.[68] Galileo reminded
Vinta that in Padua he had a lifelong salary of 1,000 scudi a year, which he
could double by private teaching and taking boarders. His official teaching
was not burdensome and otherwise "I am entirely my own master," apart
from the private students and the boarders. "If I am to return to my native
land, I desire that the primary intention of His Highness shall be to give me
leave and leisure to draw my works to a conclusion without teaching." He
would replace the foregone income by writing books "dedicated always to
my lord," and by perfecting inventions, "for of these I have a great many...
[on which] His Highness may rest assured that he will not be wasting his
money." Galileo went on like a peddler or a della Porta: "Particular secrets,
as useful as they are curious and admirable, I have in great plenty...Great
and remarkable things are mine..." On his writing table were two books on
the universe ("an immense conception full of philosophy, astronomy, and
geometry"); three on motion ("an entirely new science...discovered by me
from its very foundations"); three on mechanics ("what has been done is not
one-quarter of what I write, in quantity or otherwise"); and books on sound
and the voice, vision and colors, ocean tides, continuous quantities, military
matters..."I must get rid of distracting thoughts that retard my studies." As
for his title, an important consideration at court, Galileo asked that it include
"philosopher" as well as "mathematician"; "for I may claim to have studied
more years in philosophy than weeks in mathematics." The letter ends with
the truest statement in it: "to me [this] is the most serious matter that exists,
concerning as it does the continuation or the complete alteration of my
present way of life."[69]

Cosimo quickly agreed to everything and sent 200 scudi to help with
the expenses of sending out telescopes. Galileo would have a salary of 1000
scudi for life, paid out of the funds of the University of Pisa, of which he
would be the primary mathematician without obligation to teach; he would
enjoy every aid to his work, access to their highnesses, and the title of Mathematician and Philosopher to the Grand Duke of Tuscany.[70] The duke's new
star accepted the generous terms and added two requests. One was for a
two-year anticipation of his salary, to be repaid over four years, to relieve
himself of debt. The other was for a further public testimonial. Cosimo had
had a gold medal engraved with the Medici stars. Galileo asked that it be
displayed with an attribution of their discovery to him. The House of Medici
had been very fortunate. "No other heroic deed approaches the [discovery

of Jupiter's moons] in nobility or praiseworthiness." Let Cosimo show his gratitude for his quartet of stars. "Fortune reserved them for him alone and took them from everyone else: and I have now begun to suspect that there are no other planets."[71]

Toward the Demonstratio potissima

To whom should the first instruments go? Mathematicians? Philosophers? They would merely quarrel and raise objections until, when they accepted the inevitable, they would claim priority in its discovery. It would be better to give them to princes, who could instruct their mathematicians to take care how they dealt with a prominent courtier of a fellow prince, and who would want to possess the instrument that showed (or made!) so many marvels. Convince the princes and the rabble will follow. *Cuius regio, eius religio.*[72] Galileo began with princes close to home, notably cardinals—his reliable patron Francesco Maria del Monte, and cardinals Scipione Borghese and Odoardo Farnese. Borghese preceded the Holy Roman Emperor in Galileo's parochial politics because he was the pope's nephew, and Farnese, though only the son of a duke, came before the Queen of France even though she was a Medici.[73] When the emperor discovered that a mere cardinal had bumped him, he observed that Galileo should value his work more highly. Marie de' Medici appealed on the ground that her late husband, Henri IV, assassinated on 10 May 1610, had wished to commission Galileo to find him a new star or two. Still she had to wait. No doubt the emperor and the queen would have shown their gratitude most royally, outdoing Borghese, who gave Galileo a gold chain, and del Monte, who offered a quantity of indulgences blessed by the pope.[74] The Elector of Bavaria, Maximilian I, having received a copy of *Sidereus nuncius* through his lutanist, the author's brother Michelangelo, expressed his enthusiasm volubly, "which is not a little…as he is a man of few words," and, of course, wanted a telescope. Although he was prepared to reward Galileo handsomely for one, he dwelt too far from Florence to have priority. Michelangelo saw only the money motive. "Even if I am not a prince able to remunerate you, I am at least your brother, and so it seems strange that you do not want to satisfy my desire [for a telescope]."[75]

Perhaps the best indication of the burden of proof Galileo had to lift in the spring of 1610 by the adroit bestowal of telescopes is Bartoli's report of the view from Venice:

Everyone is reading and thinking about Galileo's book in which he makes a show of having found four new planets with his *occhiale*, and seen another world in the moon, and similar things, which open a delightful pastime to professors of these sciences, especially because of the title of *Siderea medicea*. I can not refrain from saying that many of these gentlemen now think that he has made fun of them when he gave out as a secret the common spyglass [*cannone*] on sale in the street for four or five lire, of the same quality, it is said, as his. They laugh at [these discoveries], and call them rash, while he tried to make them a great feat, and has done so, and gained an increase in salary of 500 fiorentini. I understand that he really is a most worthy man and a friend of F. Paolo...[76]

Certification stumbled not only because of a dearth of good telescopes, but also from an abundance of bad ones and, whether good or bad, from inept operators. Recipients of Galileo's handiwork did not always know how to correct for its adjustment to his peculiar sight, keen and clear on the right, weak and fuzzy on the left.[77] And every user of a high-powered telescope of Galilean design faced the problem of mounting it so that it would be both maneuverable and fixable.

Galileo unintentionally demonstrated the problems of his equipment himself during a stop off in Bologna when returning to Padua after his triumph at Cosimo's court. Neither Magini nor any of the other twenty learned men present saw Jupiter's moons. One observer, Magini's myopic assistant Martin Horky, advised his patron Kepler and other German mathematicians that the Medici stars did not exist, and went after the sideral messenger and his message in small book published in June 1610.[78] Denounced in turn by Magini, Horky fled to Milan and his good friend Capra, whom he trusted to distribute his book where it would do the most harm, and then on to Prague, where Kepler told him that he had made an ass of himself.[79]

Horky may have been a better observer of Galileo than of the stars. Here is his portrait of our hero after the debacle in Bologna:

His hair hung down; his skin, in its tiniest folds, is covered with marks of the *mal français*; his skull is affected, delirium fills his mind; his optic nerves are destroyed because he has scrutinized minutes and seconds around Jupiter with too much curiosity and presumption; his vision, hearing, taste and touch are shot; his hands have the nodules

of the gout because he has stolen physical and mathematical treasure;
his heart palpitates because he has sold everyone a celestial fable...
Fortunate and thrice happy the physician who returns the sick messenger to health.[80]

Some of this agrees with Galileo's own descriptions of his physical condition.
He began to suffer severe rheumatic attacks at the age of forty, in 1604/5. He
shivered in the summer of 1608 with a persistent fever. He was bed-ridden
during most of the winter of 1610/11 with severe miscellaneous pains, sleeplessness, discharges of blood, and depression (melancholy).[81] Horky noticed
delirium, arising from the French disease. That was probably a good hit; the
Galileo–Sagredo life style almost guaranteed a dose of syphilis, whose symptoms can mimic those of other ailments. No doubt Galileo had overstrained
his eyes portraying the moon and chasing the Medici stars. The gout goes
well with Galileo's indulgence in wine and with his tendency to rheumatic
ailments. As for palpitations, Galileo may already have been suffering from
the irregular heartbeat of which he later complained. Although we can safely
reject Horky's etiology of moral faults, his description may show us something of the true Galileo striving desperately, when ill and overworked, to
persuade influential mathematicians and philosophers of his trustworthiness. Heroes must be tried.

The first important astronomer to endorse Galileo's findings unequivo-
cally was Kepler. He did so after reading a gift copy of *Sidereus nuncius* forwarded by the Florentine ambassador to Prague, Giuliano de' Medici, with
the request that Kepler write a commentary on it. Kepler was elated by the
book, though not for a reason easily guessed. He had been distressed by
rumors about Galileo's discoveries because there was no place in his universe for even one more planet. So he had conjectured that Galileo had seen
moons, one around Venus, Mars, Jupiter, and Saturn, and once again was
right for a wrong reason.[82] It followed for Kepler that if the earth had a single
moon and Jupiter four, Mars should have two and Saturn eight. He urged
Galileo to look for them. If found, they might explain why the dimensions of
the planetary system as worked out by Platonic solids did not agree perfectly
with the observations of Tycho Brahe. There had to be space to accommodate the moons![83] Galileo looked and was pleased not to see another moon.
The Medici stars remained unique in the planetary system—provided that
the earth was not a planet.

Kepler did not risk much in accepting the lunar landscape or the multitude of moons.[84] He accepted the main obstacle to belief, the Medici moons, for a hard-headed romantic reason. As he wrote Magini, Copernicans had no reason to mislead one another deliberately.[85] Galileo would not risk his good reputation (and his prince's!) on claims that anyone with a proper telescope could prove false.[86] In September 1610, having put his hands on the telescope that Galileo had presented to the Prince-Elector-Archbishop of Cologne, Ernst of Bavaria, Kepler confirmed by telescope what he had endorsed in Copernican fellowship. Both he and Thomas Segeth, who served as a literal eyewitness, could see everything advertised in *Sidereus nuncius* except one of the four Medici moons. They deemed that sufficient to corroborate all of Galileo's claims.[87] During the same month Galileo took up his position as the Grand Duke's Mathematician and Philosopher.

Galileo's desertion of the Venetian Republic aroused perplexity as well as anger.[88] Why was he so eager to place himself with a mediocre prince and a priest-ridden court? The reasons he gave Vinta—the desire to free himself from teaching and from the demands of his various patrons—perhaps were the weightiest. He saw other advantages as well. He knew that to conquer Italy he would have to convince the Jesuits to abandon Aristotle's cosmology. This monumental task could not be shouldered in Venice, where citizens still were prohibited formally from corresponding with the Society of Jesus and where Galileo had its *bêtes noires* Sarpi and Cremonini for friends. In Florence, interaction with Jesuits was only too easy. Counterintuitive as later events may make it appear, Galileo probably regarded easier access to Jesuit mathematicians and philosophers as an attraction in 1610. One of the first letters he wrote after moving to Florence reopened his correspondence with Clavius, interrupted, as Galileo wrote, for "reasons I need not detail to your perceptiveness."[89]

Then there was Galileo's quixotic desire, compounded of nostalgia, ambition, and wishful thinking, to serve Cosimo. He had been at the fringes of the court since boyhood through his father's musician friends, the literary cardsharps around Ricasoli, the court mathematician Ricci, and, latterly, his summer tutee, the heir to the grand duchy. Galileo had been impressed by the magnificence of the Medici in action, their ability to mobilize their wealth for weddings if not for warfare, and their undemocratic favoritism of those they deemed worthy of it. A correspondent put his finger on it. "I knew that your devotion to your prince was enough to make you leave something better [to serve him]."[90] Although Cosimo was no Charlemagne, Galileo would be his Orlando: and just as a single great champion of yester-year could turn a battle against the odds, so Galileo, secure in the support of his prince, would rout the armies that opposed him. Or, to change the metaphor, the Florentine court would become the Archimedean place, and the telescope the lever, from and with which Galileo would move the Aristo-telian system to the garbage heap of history.

On a lower level of novelty though perhaps not of priority, moving to Florence implied parting from Marina. There was no room for her, spatially or emotionally, at the Tuscan court, and scarcely more for their children.[91] Vincenzo, then not yet five, would remain in Padua with Marina; Livia would come with Galileo to Florence to join Virginia. One of Galileo's first cares was to find a nice nunnery where the girls could be taught whatever they needed to know. Already he had made arrangements to place Virginia in a convent at a cost of 42 scudi a year plus expenses for a bed and other necessi-ties.[92] And he would pay for Vincenzo's upkeep and perhaps also something toward Marina's. Galileo did not shirk the financial, only the emotional, responsibility of maintaining his children.

And so, having completed "the best eighteen years of my life," Galileo took up the career of a courtier. From a lowly professor he had risen to a high-class jester, expected to help relieve the dull and punctilious court rou-tine by producing an occasional wonder.[93] He arrived in September 1610, after a brief stay with Magini. He had to come by litter as he was too weak to ride a horse.[94]

5.2 CELESTIAL MESSENGER

A new Columbus

Galileo's startling revelations upset those who pretended to know the cos-mos and delighted many who were content to know only the world. Tricks they might be, or perhaps prodigies, "forse prestigi / son questi o di natura alti prodigi."[95] Philosophers in the Aristotelian tradition had either to reject Galileo's findings or patch their philosophies to accommodate an earth-like moon and a secondary center (Jupiter) for lunar motions. The curious edu-cated received the celestial message as an announcement of a new America and, dependent on temperament, rejoiced in or despaired of an age so full of novelties. Here are some strains from the joyful. "You were born to honor our country."[96] "Your Lordship! In you God not only united all the gifts that formerly He had spread among other men, but also vested...the revelation of new heavens, of which he made you the new Atlas! He has led you along paths never trod before by the human mind, like a new Columbus!"[97]

Comparison of Galileo with Columbus became a refrain. Italy can now boast "the discoverer of a new part of the world as well as the discoverer of a new world of stars." Only Italy could produce a Columbus of the cosmos, Dutch barbarians could never have done it. Thomas Segeth: "Columbus gave us lands with much bloodshed / Galileo stars with no injury. Which is greater?"[98] Could there be a doubt? Galileo's is the greater achievement "in the same proportion as the heavens are more noble than the earth." Greater than Columbus, greater than Tiphys (the pilot chosen by Athena to steer Jason to the Golden Fleece) is the celestial navigator: "You, greater than the one and the other / Looked into the world of the stars / In inaccessible regions far away / And interned within these recesses unknown / In their profundity you knew how to find / New orbs new light new movement."[99]

This is dull stuff. Let us allow ourselves some enthusiasm. Ilario Altobello: Galileo has eclipsed the glory of the astronomers of old, of Hipparchus, Ptolemy, Copernicus, the Egyptians, the Chaldeans. Kepler: Galileo's very name implies divine things, prophetic things. Was it not asked of Christ's apostles, "Ye men of Galilee, why stand ye gazing into heaven?" They had stopped to watch their lord ascend before they turned to spreading His gospel; and soon after they set to work, they had turned everyone who spoke about higher things into a Galilean. Dom Castelli, the modern-day man of Galilee's chief apostle, read the short new gospel, *Sidereus nuncius*, ten times, soaking himself in the "deep learning, high thought, learned speculations, and...the marvelous harmony and unity of it all." To which the prior of Castelli's abbey in Brescia added, "How happy is our age in which Sig. Galilei has made such stupendous discoveries."[100]

Galileo and probably also the Medici solicited poetical accolades for the Italian edition of *Sidereus nuncius* that never appeared. A folder of these profusions survives. It contains 40 hexameters, 10 Sapphic odes, two epigrams, and four distiches, all in Latin, all by Jesuits. Perhaps the earliest and best of them, by the Neapolitan Costanzo Pulcarelli, installs Galileo as a new Atlas, whose brilliance frightened the heavens into switching on new stars to compete with him.[101] The poets were to serve Galileo as newspapers do modern scientists, advertising and celebrating discoveries before the experts had their say.[102] Starring in heroic poetry can go to a man's head. Galileo began to identify himself with his message. *Nuncius* can mean "message," as Galileo first intended, or "messenger." Kepler and, following him, most modern translators, take *nuncius* in this second sense. Although wrong as to Galileo's original intention, they catch his subsequent conviction that he was an agent of the stars, and even of God their creator.[103]

After his quick survey disclosed no moons around the other planets, Galileo set himself to determining the periods of the revolutions of Jupiter's satellites. The problem, "truly Atlantic," that is, worthy of an Atlas, took him a year and more to solve. It required distinguishing the moons and inferring their orbits in space from their changing angular distances from Jupiter and the intervals between their occultations. Everyone knowledgeable hailed it as a feat: and indeed, the accuracy of his determinations, to within five minutes of time, amazes modern connoisseurs as it did Galileo's contemporaries. In his combination of stubbornness and dexterity, intuition and geometry, acuity of mind and sight, he surpassed all the observers of his age.[104] What made him persevere in a task that Kepler had declared virtually insoluble? "I rely on the Dear Lord, who, having had the grace to make me the only one to detect so many and new wonders from his Hand, will permit me to find the absolute order of their revolutions."[105] Perhaps the worry that the Devil might help Magini, who opposed Copernicus and loved to calculate, strengthened his resolve.[106]

When the moons became unobservable in the summer of 1610, Galileo looked carefully for other secrets that God might have reserved for him. He turned his glass on his fellow melancholic, Saturn. Another marvel! The old boy was not his former circular self, but three-bodied, "accompanied by two attendants who never leave his side." The news was too hot to conceal and too useful to reveal. Galileo announced it as if he were a prophet speaking in tongues: "smaismrmilmepoetale," quoth he, "umibunenugttauriras." He dispatched this news to Prague with the hint that it concealed a celestial message. It drove Kepler to distraction. Even he could not extract "altissimum planetam tergeminum observavi" from the scramble.[107] No more could Harriot, who made ten or maybe fifty tries at it. Galileo decoded it around the end of December.[108]

Saturn's deformity worked wonders for astronomers. Frightened by it, Emperor Rudolf commanded his mathematician Kepler to check it out,

CALCULATED RISKS   167

gave him 200 scudi for his trouble, and promised to pay the arrears on his stipend, which by then amounted to many thousands. Rudolf also promised to square accounts with Magini, whom he owed for a burning mirror. (Both Kepler and Magini had thought to improve their financial position by trying for Galileo's chair in Padua.) The three-bodied Saturn settled a third account of greater interest to Galileo than the income of his colleagues. Horky threw in the towel. He would give two quarts of blood, he said, without specifying whose, not to have written against a man who could pull such wonders out of the sky.[109]

Galileo did not keep his eyes on Saturn for long. Venus, leaving her bath of solar rays, emerged into the evening sky in June 1610. She had a great secret to reveal. Though on both the Ptolemaic and the Copernican systems Venus should show phases, she nonetheless appears with unchanging brightness to the unaided eye. Copernicus had signaled the importance of the missing phases of Venus and, as we know, Kepler had allowed Venus some light of her own to account for her constant intensity.[110] Much depended on a better answer. If the telescope could untangle the counterbalancing effects of distance and phase, which counterfeit constant intensity, it could indicate whether Venus revolved around the sun or, as in the Ptolemaic system, she spent all her time either above or below it. In 1610, the transition to half-planet (phase 5 in Figure 5.5b), which is not a possible Ptolemaic form, occurred in December.[111]

Despite the known importance of the question, there is no word in Galileo's correspondence or entry in his notebooks about the phases of Venus before 11 December 1610. On that day he wrote to Prague, briefly, inquiring what Kepler thought of the triune Saturn and transmitting another torment for the poor fellow, *Haec immatura a me iam frustra leguntur o.y.* The plain text, apart from its hint at Jewish ancestry, means "I am now bringing these unripe things together in vain, Oy!" It concealed the discovery of the long-sought Venusian phases. *Cynthiae figuras aemulatur mater amorum*, the decrypted text reads, "The mother of love [Venus] copies the forms of Cynthia [the moon]." On 30 December Galileo gave two different accounts of the observations underlying this important finding. One, addressed to Clavius, placed their start when Venus was first visible after nightfall. Then at its greatest distance from the earth and near the sun, it appeared very round and small (Figure 5.5b, phase 4). The other letter, addressed to Castelli, who had reminded Galileo some three weeks earlier of the question of phases, stated that the

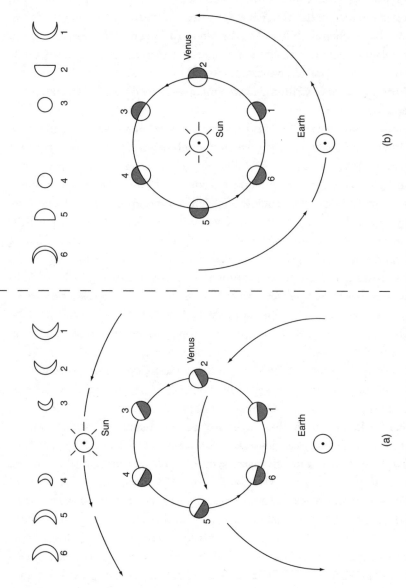

FIG. 5.5 The phases of Venus on the Ptolemaic (a) and Copernican (b) systems. After Van Helden, in SN, 108.

observations began three months before the time of writing. Both accounts then describe the change from the small round phase to a gibbous one, increasing in size as the planet drew closer to the earth until, at maximum elongation, only half was visible (phase 5). Continuing toward lower conjunction, its magnitude growing as its lit portion diminished to a crescent, it would, as Galileo guessed, become bigger, thinner, and more horned (phase 6) until it again disappeared into the sun.[112]

The coincidence in timing between Castelli's letter of 5 December, pointing out that detection of the right phases should "convince any mind obstinate against Copernicus," and Galileo's letter to Prague of 11 December containing the encrypted discovery, has aroused suspicion among exact chronologists.[113] The interest in the suspicion does not lie in the charge that Galileo portrayed Castelli's suggestion as his own. We know that often he came to think of ideas he had developed and enriched as originating with him, as exemplified by the compass, telescope, and pusilogium, and the appropriations from del Monte, Sarpi, and the Venetian telescope group.[114] The interest in the suspicion lies in the question why, if he did not undertake serious study of Venus until prompted by Castelli, or, as seems more likely, until October 1610 as indicated in his reply to Castelli, he had delayed so long in pursuing an inquiry so interesting to Copernicans. One plausible explanation is that he was too deeply engaged in patron politics and Jovian studies to undertake anything less promising.[115] Another is that he inclined toward Kepler's view that Venus has an intrinsic brightness that swamped the phases.[116]

A third possibility is that he did not regard detection of the phases as determinative. The limited elongation of Venus (and Mercury) showed clearly enough that they circled the sun, as several ancient astronomers had taught, and Galileo had taken for granted in *Sidereus nuncius*. When decoding his Cynthia riddle, he said that the phases showed beyond doubt that Venus shines by reflected light and also, of course, circles the sun, "as do Mercury and all the other planets." Pythagoras had believed as much, and so had Copernicus, but only Galileo had managed to "prove by the senses" that the superior planets behave like the inferior ones.[117] Galileo might have added Tycho to the list of believers, because as he well knew and one of his supporters reminded him, putting the planets around the sun did not entail putting the earth in motion.[118] Until Galileo added, "and the earth is a planet," to the proposition that all planets revolve around the sun, or endorsed the full

Copernican theory, his formulation of the conclusion forced by the sequence
of Venus' phases agreed perfectly with the Tychonic system.[119]

The shapes of Venus astonished astronomers as much as the mountains
on the moon. She seemed so bright and round! None of Kepler's eight tries
at deciphering the anagram ending in "oy" ("you see what misery I suffer
from your reticence") concerned Venus.[120] In Venice and Padua, where the
Galilean faithful had cudgeled their brains over the riddle, there was satisfaction as well as amazement that their master had done it again.[121] Magini
sent congratulations from the Bologna group, now convinced of the inadvis-
ability of doubting Galileo's claims. And an old Italian hand, Mark Welser,
a prominent banker in Augsburg close to the Jesuits who was to play an
important part in circulating Galileo's ideas, applauded the detection of the
phases but failed to see why they implied that Venus goes around the sun.[122]
To make clear how the phases of his Venus differed from those expected on
the Ptolemaic system, Galileo would have had to adopt openly either Tycho's
or Copernicus' model. And that he was not prepared to do before the Jesuits
were on side.

Roman holiday

In his revived correspondence with Clavius, Galileo expressed an intention
of visiting Rome to show the Medici stars to the astronomers of the Roman
College. Cigoli urged that he come soon, as the Jesuits could not find their way
around heaven without help and Clavius himself thought the *sidera medicea* a
hoax.[123] Christoph Grienberger, soon to be Clavius' successor, rated Jupiter's
companions and Venus' phases as optical illusions, and lunar mountains as
entirely imaginary. Then one of their colleagues, Giovanni Paolo Lembo,
made a telescope that resolved the stars in Orion and the Pleiades, and eventually Clavius himself procured a Galilean telescope from a cardinal to test
the rest of *Sidereus nuncius*. By then, the fall of 1610, the world had freed itself
from Galileo's optical monopoly. Antonio Santini, a merchant of Venice,
succeeded in September in seeing the Medici stars through a telescope he
had made; not finding the making difficult, he deduced that Galileo's opponents were either incompetent or obstinate. In October he delivered this
insight, and soon also lenses, to Clavius; and on 4 November he could report
to Galileo that the *clavisti* had seen the Medici stars.[124] Meanwhile Galileo
had worked on the Jesuits within his reach. All those stationed in Florence

and many who passed through had the opportunity of seeing the disputed starlets under the direction of their discoverer.[125]

Clavius' group, which included Grienberger and Odo van Maelcote, had not found the observations easy even with a good telescope. "No doubt this instrument would be of inestimable value, if it were not so difficult to use."[126] Galileo reassured Clavius that all his group needed was practice and crowed to others that, at last, the Jesuits had recognized the truth and "confessed it."[127] Delighted at the news, Cosimo enthusiastically supported Galileo's proposed trip to Rome. It would be done in style. The grand duke supplied a litter, paid the expenses of Galileo and a servant, and procured rooms for them at the main Medici villa in Rome. Illness and bad weather prevented departure until spring. Galileo arrived in Rome on the Tuesday of Holy Week, 29 Mar 1611, carrying a letter from Cosimo asking Cardinal del Monte to favor Galileo's enterprise, "for his merit (a born Florentine), for the public interest, and for the glory of our age."[128]

The day after his arrival Galileo visited the Roman College. He found the fathers chuckling over an attack on him made around Christmas 1610 by a reckless young man named Francesco Sizzi, whom Kepler placed on the same mental level as Horky. The title of Sizzi's Christmas present to the learned world, *Dianoia astronomica, optica et physica*, intimated that it would defeat an opinion with facts. The opinion was Galileo's assumption that he could see accurately to Jupiter. Against it Sizzi marshalled the "facts" that refraction distorts the more the further away the object viewed; that the telescope is unreliable over a distance greater than a sixth of the earth's radius; and that Jupiter is distant 17,615 terrestrial radii. Galileo's assertion implied a reliable scale-up of the same order of magnitude and incredibility as his conclusion to the stability of the Inferno from a model 200,000 times smaller. What effrontery! "And who are you, who presumes to descry / Such distant things a thousand miles off / Glancing from your bench with a feeble eye?"[129] Galileo did not join wholeheartedly in his hosts' merriment over Sizzi's pseudo-quantitative rhetoric and his objections based on scripture, which also made them smile; for Sizzi was a Florentine, who deserved consideration from an official representative of the grand duke. Moreover, he had the protection of Don Giovanni de' Medici, to whom Sizzi had dedicated his dianoia.[130]

The astronomers of the Roman College could laugh at Sizzi because they had managed to see everything that Galileo had reported from the skies. With their confirmation, the rejection of Jovian planetoids, Venusian phases, and so on as artifacts of the telescope lost all plausibility. Even in Belgium, so wrote a former student of Galileo's stationed there, the truth would out. Welser in Augsburg accepted the judgment of the Roman College and read it as a God-sent lesson in humility, "that we may recognize how little we know in comparison to what we do not know."[131] Even the Inquisition accepted the substance if not the spirit of the lesson. Its vigilant theologian, Bellarmine, who was quite prepared to discover things in the heavens unsuspected by Aristotle, asked Clavius' group for an official evaluation of the telescope and the observations Galileo claimed to make with it. On all points the *clavisti*— Clavius himself, Grienberger, Maelcote, and Lembo—approved the findings, with the notable exception that Clavius, for whom things were moving too quickly, thought that the irregularities seen on the moon might arise from differences in density, rather than of height, of moon stuff.[132] On hearing this Cigoli scoffed that the dean of Jesuit mathematicians did not understand perspective. Still, the approval of the Roman College was a great victory, personal as well as scientific. Piero Dini, a Florentine man of letters who had moved to Rome to serve his cardinal uncle, and who would become a confidential agent for Galileo, saw the mathematicians' reply to Bellarmine. The bottom line, in Dini's judgment: Galileo and the Jesuits had become "great friends."[133]

There remained the omen of Clavius' reluctance to allow mountains on the moon. He owed this resistance to his conscience and to his colleagues of the Roman College, who had followed his lead over the forty years since he had first published his view that Copernican theory opposed sound philosophy, standard astronomy, and Holy Writ. The inference from an earth-like lunar landscape to earth-like moon stuff, and thence to the destruction of the capital distinction on which the Aristotelian cosmos turned, was only too obvious. The cool cheese of the ancients had become a hot potato for the moderns; or, to improve the metaphor, Galileo's mountains were philosophical as well as physical monsters. To Galileo's growing impatience, the Jesuit mathematicians on whom he had counted declined to admit his or embrace another interpretation of the lunar surface.[134] The philosophers of the Roman College did not shrink from the task, however. In 1612 the professor of physics, Girolamo Piccolomini, whose cosmological views differed little from his predecessors' a generation back, told his students that the lunar features probably arose from shadows cast by rarer parts near brightly illuminated denser parts, "as we see in little spheres made of amber or crystal that appear spotted because they are not uniformly dense and lucid."[135] The idea that the blotches Galileo saw lay within a perfectly spherical, opalescent or transparent moon was not hard to find. It occurred to Ludovico delle Colombe.

Galileo made good sport of the argument in a lengthy letter, doubtless intended for circulation, to the factotum of Cardinal Joyeuse, Gallanzone Gallanzoni. What is so splendid about sphericity? A man would not be more perfect for being spherical, nor the earth either, since God had had the opportunity to make it so, round and even, at the Creation and again at the Deluge, but had decided that it would better suit the non-spherical life forms here if rough and jagged.[136] You could take the earth together with its atmosphere to be a perfect sphere or believe in invisible moon mountains thousands of miles high. But then, Galileo continued, you would not be doing astronomy, which is not a game of the imagination but a science based on "sensory experience and necessary demonstration." This phrase, which became a battle cry, related to the *demonstratio potissima*.[137] Despite their reservations about the mountains, Galileo interpreted the Jesuits' endorsement of his findings as a major step to the completion of the consensus required to establish a premise in physics.

On 18 May Galileo returned to the Roman College to be celebrated before all its students and professors, and also princes, prelates, and cardinals. Maelcote gave an oration on the work of "the nuncius," the sidereal messenger there present, including his three-bodied Saturn and many-splendored Venus.[138] Naturally Galileo also attended literary assemblies. At one of them, held in the palace of Cardinal Giovanni Battista Deti, he heard his old friend Strozzi give a talk on the timely topic of pride; and a very good talk, Galileo reported to Vinta, which made him proud to be a Florentine. Everywhere Galileo was the lion of the season. He added several purple supporters—Cardinals Ottavio Bandini (Dini's uncle), Tiberio Muti, François de Joyeuse, and, with the help of a letter from Buonarroti, who had become a proselytizer for Galileo's celestial discoveries, Maffeo Barberini.[139] Thus began a friendship that would end in an enmity of world-historical importance.

Barberini came from a noble commercial Florentine family that marked its upgrading in the 16th century by replacing the three silver wasps or (according to detractors) horse flies on its escutcheon with as many golden bees. Maffeo, who was to plaster the bees all over Rome, studied law at Pisa after the usual education under the Jesuits in Florence and Rome. At

Pisa he lived in the same house as Buonarroti, whom he had known since childhood. They slept on broken mattresses in the same room. "If I had only known / that some day Maffeo would be the Pope / My bed I would have laid upon a stone / or on a rough tile way down the roof's slope." Thus Buonarroti, a poetical prodigy, who had been elected to the Accademia fiorentina at the callow age of seventeen.[140] Barberini too was a poet. Most of what he wrote around his time at Pisa, which coincided with Galileo's teaching stint there, was in the mildly erotic Petrarchan mode. A specimen reads:

> Now your lips of coral bring back to me
> Cheeks suffused with carmine, a milk white breast
> Now you show your lovely eyes like twin stars
> Your blonde tresses and your sparkling white hands
> A mind all yours has long roamed within me
> Why then, Portia, have you so many snares?[141]

The Portia picked out for Maffeo did not like the match. Rebuffed, he entered the Curia through the good offices of a rich uncle able to buy him a good place in the bureaucracy. He rose quickly, to special envoy, and then nuncio, to France. His four years in and around the Court of Henry IV inspired a Francophilia that would have significant consequences for foreign policy during his papacy. Paul V made him a cardinal at the time of the great Venetian interdict, and then, in the year he met Galileo, legate to Bologna. In the interim Barberini had set up a magnificent establishment in Rome with a legacy from his uncle and began to practice the nepotism for which he later became a watchword. Maffeo Barberini was shrewd, handsome, quick, attentive to business and the arts, but cautious, distrustful, stubborn, easy to anger, and indifferent to the opinions of others (Plate 10). In short, a Florentine like Galileo.[142]

Galileo made other important contacts during his Roman holiday, including Paul V, who received him with the honor due a savant with the wisdom to leave Venice for Florence.[143] At the salon of Margherita Sarrocchi, he met many other well-wishers.[144] Most important for Galileo's future, indeed for Italian cultural history and the history of science, was his meeting with Federico Cesi (Plate 11), a prince in waiting who enjoyed the favor of the pope. When only 18, Cesi had had the extraordinary idea of forming a learned society, which, according to its statutes of 1605, would study natural science and mathematics, pursue new knowledge and publish discoveries. Since he had no

scientific standing himself and knew no one who did, Cesi's first lynxes, as he
called his academicians, were odd animals. Besides himself, the initial group
consisted of two other young noblemen and a Dutch doctor whom Cesi
extracted from the jail in which he languished for murdering an apothecary.
The ceremony in which Cesi initiated the first three members of his broth-
erhood captures the sense of Christian knighthood with which he infused
his Order. On Christmas day 1603, dressed in a long robe, he called each
brother in turn, read him the rules, swore him to obey them, and gave him
a gold chain bearing the image of a lynx. The murderous medic Johannes
van Heeck (Eck) understood the relationship of the brothers to the founder
as one of liegeman to lord: "we are only brothers, but you, Oh Lord, are our
prince." "Love [of you] belongs to us," declared Francesco Stellutti, the most
faithful and competent of the original group, "to you, the empire to which
you were born, and to which you were destined from heaven."[145]

Cesi's father Federico, Marchese di Monticello, Duke of Acquasparta, took
alarm at his son's involvement in a homoerotic band inclined toward mysticism and melodrama, organized like a religious order, and perilously close to
heresy.[146] He quickly put an end to it. Eck drifted to Prague. Cesi hastened to
Naples, to master mysteries under della Porta. They got on well as each had
a use for the other—Cesi needed the cachet that association with so great a
luminary would bring, and della Porta saw promise in patronizing the heir
to a great title and fortune. Cesi learned many "secrets" to convey to his little
band and della Porta established himself as its primary advisor. In 1610 Cesi
was able to revive his almost extinct lynxes. They then numbered three of
the original four: Cesi himself, Eck now back in Rome, and Stelluti. In July
Cesi made della Porta a lynx and eventually his vice president and head of a
satellite den in Naples. Stelluti became general business manager in Rome.
Cesi too was on the rise. In 1610 he was but a marquess; in 1613, by special
nomination of Paul V, he became Prince of St Angelo and St Polo, so titled
after two of his village holdings.[147]

Cesi thus was prepared to bag the lion of the season when he met Galileo
in Rome in the spring of 1611. Their collaboration began with a dinner party
Cesi gave for a dozen other guests at a villa on the top of the Gianicolo from
which many prominent buildings in the city could be seen. The party began
before sundown to allow viewers to read the letters on the distant façade of
San Giovanni in Laterano through Galileo's telescope. After dinner everyone tried to see the moons of Jupiter, with mixed success. The participants

all agreed, however, that the suggestion of the mathematician to the Duke of Gonzaga, John Demisiani, to call Galileo's instrument "telescope," was a happy one.[148] Eleven days after the party, on 25 April 1611, Cesi made a lynx of the lion. That gave him two big beasts, Galileo and della Porta, both of whom claimed the invention of the telescope. Since Cesi's academy very seldom met, the two could co-exist in it without deciding which if either of them was the inventor.[149]

Although Cesi supported a range of interests as befitted an academic impresario, he recognized the special importance of astronomy. He followed and admired the work of Tycho and Kepler. It was he who pointed out to Galileo the importance of Kepler's *Dioptrics* (1612), which gave the world the theory of the telescope and Kepler as much pleasure (Kepler supposed) as the discoveries announced in *Sidereus nuncius* gave Galileo.[150] The first publication by a lynx concerned the nova of 1604. It was the work of Eck, then in Prague, suffering from the intellectual companionship of Kepler's entourage. He agreed with them that the nova resided above the moon, but not that the heavens could change unless by special act of God; and he filled his little essay with nasty remarks about the heretical antiperipatetics among whom he had the misfortune to find himself. Since Cesi wanted to forge ties with Kepler and believed that the heavens do change, he edited out Eck's vitriol and changed his text to favor a more progressive astronomy.[151] Cesi later wrote Galileo that he had shared the ideas of Tycho and Kepler—the fluidity and alterability of the heavens—attacked by Eck and that since then he had come to accept the elliptical orbits Kepler had introduced in 1609.[152] In this he was more advanced than his new Florentine academician.

Galileo placed great store in his membership in Cesi's academy. He identified himself as a lynx on the title pages of his books and referred to himself in his dialogues as "the academician." Why? No doubt he thought that the young, energetic aristocrat with excellent ecclesiastical connections would help him maintain the interest and patronage he had won in Rome. In addition, Cesi's concept of a band of scholars pursuing free inquiry under the beneficent eye of a prince appealed to Galileo's romantic as well as to his practical bent. Although Cesi had outgrown some of his adolescent romanticism by 1611, he retained the concept of a band of chosen servitors. All lynxes, he wrote Galileo, had the true nobility of independent minds. They owed allegiance to their prince, and so could not enter a religious order; otherwise they were free, or, rather, obligated, to think as they pleased.[153] Galileo liked the idea of being the Orlando or Ruggiero of such a group, which soon added Valerio and Welser to its roll, and the novelty of having the intellectual support of a prince. With this encouragement he could risk accepting Kepler's old challenge to come out strongly for Copernicus.[154]

The Roman holiday was a great triumph. Had it been ancient Rome, so Cardinal del Monte wrote Grand Duke Cosimo, the city fathers would have erected a statue in Galileo's honor in the capitol. Something potentially more useful was created at the Roman College. Shortly after his return to Florence Galileo received a very friendly letter from Grienberger conveying to him in Florence the warm greetings of all the Jesuit mathematicians in Rome.[155]

5.3 MORE RABBITS FROM THE HAT

Sinking bodies

Filippo Salviati was the sort of man Galileo delighted to know. He was as noble a Florentine as could be found, related to all the families that counted, generous, open-minded, clever, and rich, *richissimo*. He had been raised for war and courts, enjoyed the similar pursuits of jousting and dancing, and, like Galileo, had a compulsion to serve his prince. The knightly virtues of Ariosto's paladins would have had a late flowering in Salviati if ill health had not militated against a military career. In 1606, at the age of 24, he decided to retool himself for a quieter life. He learned Latin and Greek, advanced to Aristotle and the philosophers, and had reached mathematics when Galileo returned to Florence. They became teacher and student, client and patron, friends. Between 1611 and 1613, Galileo spent many months at Salviati's villa Le Selve outside Florence recovering from his various ailments, writing, and disputing with the learned guests that Salviati patronized.[156] The delights of Le Selve included many varieties of creature comfort. Salviati employed two-dozen servants and four soldiers for a domestic establishment that consisted of himself and his wife. In wealth, connections, and, above all, intellectual compatibility, Salviati was another Cesi.[157] Galileo had him enrolled among the lynxes.

On a summer's day in 1611, when Galileo was at Le Selve enjoying the fresh air and an argument with two Pisan philosophers, the talk turned to the nature of condensation and rarefaction. One of the philosophers adduced ice as an example of a body heavier than water. The lynx pounced. Nonsense, he said, ice floats on water. On the contrary, replied the philosophers, ice, being a solid, is evidently denser; a chunk of ice floats because its broad flat surface prevents it from cleaving the water. Nonsense again, said the lynx: water has no resistance to the separation of its parts, as is plain from the facts that a submerged piece of ice rises when released and the smallest particles of mud settle, albeit slowly, to the bottom of a well. Water resists motion but not parting. Whether a body sinks or swims depends solely on its specific gravity.

Some days later Galileo learned that his tireless critic Ludovico delle Colombe claimed to be able to show that a body that sinks when spheroidal floats when formed as a lamina. He offered ebony as an example. Galileo declared the demonstration invalid. He had had in mind bodies wetted by water; a piece of ebony whatever its shape if once submerged stays submerged. His adversaries refused the condition; they had exhibited a clear case where shape determined buoyancy. The parties fought over the admissibility of delle Colombe's evidence with the tenacity of lawyers in their own cause. A showdown scheduled at Salviati's house fell through when Cosimo informed Galileo that, as a court ornament, he was not free to sparkle in public arguments with noisy professors. It would be better, the grand duke informed his former tutor, to write out your arguments and dispute on paper.[158] That enabled Galileo to turn a likely loss into a qualified win.

A contest by pen, Galileo wrote Cosimo, would give him a great advantage. The enemy would have to write too, and so either abandon or expose their customary cavils, chimeras, distinctions, sophisms, and contradictions. They had already abused the spoken word, he claimed, shouting from piazzas, chiesas, gondolas, that he could know nothing of physics because he was a geometer. But they did not know their man. They could no more "choke me off and exterminate me from profaning [Aristotle's] sacred laws" than Ariosto's savages could discipline Orlando for freeing the succulent maiden they had reserved for their orc. And just as Orlando despised them as a bear does yapping dogs, and killed thirty of them with ten blows of his telescope—or rather sword—so Galileo, with no armor but the shield of veracity and the protection of his prince, would dispatch "every madman who irrationally assaults the truth."[159]

Curious to know what had ignited his mathematician, Cosimo caused the then new professor of philosophy at Pisa, Flaminio Papazzoni, to represent the peripatetic position against Galileo in a debate before the grand-ducal family. Such performances often occurred after meals as a substitute for television and Galileo was obliged by contract to participate in them when requested. He then played his part of high-powered jester, answering questions put to him and defending himself as wittily as he could. As recreation, postprandial disputes did not require a declaration of victory or an admission of defeat. In the case before us, Galileo easily won—not because Pappazoni was incompetent but because, as he owed his professorship to Galileo's recommendation, he had no interest in exerting himself.[160] Cosimo so enjoyed this version of Androcles and the lion that he restaged it in the presence of two visiting cardinals, Barberini and Gonzaga, who joined in the fun by supporting, respectively, Galileo and Papazzoni. By then—October 1611—Galileo had developed enough material to make a little book and a theory general enough to attack the peripatetic position on several fronts. Stricken again with illness or hypochondria, he retired to Le Selve to write *On floating bodies* (1612). It is an exemplary Galilean text. It secures a new neat Archimedean result by applying pseudo-Aristotelian mechanics to hydrostatics; clutters the application with repetitious geometry; introduces a sharp observation; bases upon it a physical principle that literally does not hold water; throws out hints at a non-Aristotelian physics; and spices the whole with references to the idiocy of his opponents.[161]

The crisp result concerns a prism of wood floating in a rectangular vessel. Let the original level of the water in the vessel be I and the level after immersion of the prism be II (Figure 5.6). Galileo's Aristotelian mechanical principle is that the moment (weight x velocity) of the wood must equal that of the water in any small vertical displacement from equilibrium. Press the

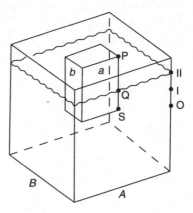

FIG. 5.6 A prism in free float in a basin with rectangular cross-section *AB*.

prism down a distance x; the water level simultaneously rises by y where $x : y = (A - a)(B - b) : ab$. If the motion takes place in the small time t, then, according to the principle,

$$\text{(weight of prism)}(x/t) = \text{(weight of water)}(y/t).$$

Let the specific gravity of the wood be δ relative to that of water taken as unity. Then the preceding equation can be written

$$\delta(abc)x = (A-a)(B-b)(c-z)y,$$

where $z = PQ$ is the height of the prism above the final water line II. After substituting for $x : y$, we arrive at $z = c(1 - \delta)$. Since in the case of wood, $\delta < 1$, z is positive and the prism floats. If $\delta > 1$, the body sinks. Period. Except that, as delle Colombe showed, a flat ship of ebony will float although its specific gravity is greater than 1.

The concise conclusion $z = c(1 - \delta)$ has several implications besides sinking or swimming. They may be seen more clearly in the unrealistic picture Galileo drew of the situation (Figure 5.7), from which the inspiration for the mechanical analogy he used leaps to the eye. Galileo portrayed the prism as lying against three walls of the basin so that $b = B$ and the problem became two-dimensional. He conceived the water prism QO and the wood prism PR as two weights in equilibrium; pressing down on PR raises QO, etc., and of course leads to the relation $PQ = c(1 - \delta)$ or $c\delta = c - PQ$ or, what we want, $PS \cdot \delta = QS \cdot 1$. The only geometrical quantities in this equation are the heights of the wood and the water: the log will float irrespective of its size and the amount of water in the vessel provided there is enough to wet the wood and to fill a prism of height QS and base OS that can be almost as thin as you please. This is the celebrated hydrostatic paradox, which was (and is) useful

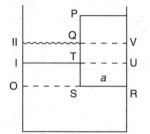

FIG. 5.7 A prism constrained to float against the side of the basin of Fig. 5.6.

for lifting weights and astonishing neophytes. Galileo celebrated the paradox in his usual subdued way as revelatory of "the causes of some admirable and almost incredible events, as that a very small quantity of water may raise up and sustain with its small weight a solid body that is a hundred or a thousand times heavier."[162]

The celebrant called attention to another marvel implied by his analysis: Archimedes made a mistake or, at best, claimed as general a principle of limited applicability. Only when the size of the floating object is negligibly small compared with the width of its basin is it true that the volume of the body immersed equals that of the water displaced. For in Figure 5.6, the displaced volume IIQTI, which must equal the volume TURS of the immersed prism below the original water line I, is evidently less than the total immersed volume QVRS. But as the width of the vessel increases, as OS $\to \infty$, QT $\to$ 0, I $\to$ II, and the displaced volume equals the volume immersed. At all times, however, the weight of water that could be accommodated in the submerged part of the solid equals the weight of the entire solid: QS = δPS, aQS = $a\delta$PS, Q.E.D.[163] It appears that Galileo persisted in the same error as Archimedes until he began his calculation and that discovery of the fault prompted him to invent the analogy to the virtual motion of the lever, with which he always felt comfortable.[164]

After these hors d'oeuvres, Galileo faced up to the question why ebony chips can float. First answer, a quibble: such chips are not in water but on it; if thoroughly wetted, they go to the bottom and stay there. Second, a paralogism: a cone of wood or wax submerges partly whether placed base or apex down on the water; hence water offers no resistance whatsoever to being parted by any body of any shape. Third, a contradiction: water acts as if it had a tough skin not easily penetrable by flat bodies of densities not much greater than its own. This skin explains the behavior of the ebony chips. Regarded closely, they lie in a little cradle, surrounded by ridges or banks of water that rise above them to the average water level (Figure 5.8). What contains and restrains the fluid in the ridges against its gravity? Galileo offers no explanation: it is a fact. Likening the cradle and the chip to a ship and its cargo, Galileo argued that the relevant specific gravity was that of the combination chip plus air. Thus we have flotation, in agreement with Archimedes' principle.[165]

To these ingenious speculations, Galileo added exercises in geometry to bamboozle peripatetics who had skipped mathematics. Close observation

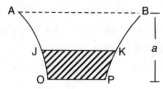

FIG. 5.8 An ebony chip JKOP lying in a water cradle of height a; air occupies the space ABKJ.

suggested that no ridge could be deeper than the depth a. What thickness x of a piece of ebony of base A could such a cradle support? Or, as Galileo put it, what is the ratio of empty to filled space in the cradle? Answer, $(\delta - 1):\delta$. He went on to investigate how the space and equilibrium changed with the shape of the floated objects, notably cones, cones on cones, and so on, to the admiration of the men of Galilee and the confusion of the pharisees.[166] Cardinals del Monte and Capponi were enthusiastic; Cardinals Bellarmine, Deti, and Gonzaga indifferent; the Paduan and Pisan professors foxed; Welser and other loyalists pleased and perplexed until, after sufficient rereadings, they had convinced themselves that Galileo had resolved all the paradoxes he had started.[167]

In fact, Galileo had resolved little but unleashed a lot. Under his floating bodies he put a medium that extruded objects of lesser density and parted, without resistance, for objects of greater density. This was Galileo's first published denial of levity and his hint in print at his new theory of motion. He added a hint at an anti-Aristotelian concept of matter. Water is a pile of atoms without mutual coherence or attachment, whence its ease of parting. But just as detached persons in a crowd slow the progress of individuals pushing through it, so the atoms resist the passage of bodies through water. None of this accounted for floating ebony chips. To the end of his life Galileo puzzled over the "obscure behavior" of fluids. Yet in the heat of the debates of 1611 he affected to understand them well enough to subject them to geometry. His opponents might not have understood the mathematics but they got the message, and replied. Galileo was a category error, a mathematician who had forgotten his place, a philosopher who did not play by the rules. "Mathematical proofs and propositions cannot grasp the true cause of natural phenomena." "Those who want to demonstrate natural accidents through mathematical methods are delirious." Not delirious mathematicians, delle Colombe added, but sober philosophers, had the job of judging how to use mathematics in physics.[168]

Galileo did not condescend to reply to his opponents under his own name. Instead he provided the ammunition for a counter attack by Castelli published in 1615. By then the battlefield had many fortified places. Castelli required 18 salvos to answer 25 objections to Galileo's hints on atoms fired off by a single opponent, Giorgio Coresio.[169] Although insignificant in their detail, these maneuvers taken together had the important consequence of increasing and hardening the opposition to Galileo in Florence and Pisa already flourishing under the gentle heat of jealousy. His highly paid sinecure, charged on funds reserved to the University of Pisa from tax collected from ecclesiastical holdings, excited envy and resentment among courtiers, clerics, and professors.[170] And his supercilious refusal to name his opponents in his replies (to do them a favor, he said), his sending his squire to rebut them, and his cocksure insistence that physics must bow down to mathematics irritated even people inclined to support him. Campanella may speak for all of them. Writing from his dungeon in Naples, he upbraided Galileo for freighting floating bodies with propositions neither known to be true nor easily defended, "so that you give your enemies an opening for denying all the celestial things you pointed out...O Dio what a pity it was to humble the immense pride with which you could have gone forth so happily revealing to mortals so many great things!"[171]

Spots on the sun

As a facilitator in the Republic of Letters, Welser liked to put his friends in contact with great men. In 1611 he sent Galileo an attempt by Georg Brengger to lower the lunar mountains. It met a polite but firm refutation.[172] In January 1612 Welser tried again with a more powerful challenger. He called himself "Apelles" after the Greek artist who hid behind his paintings to hear what viewers said about them. To answer this cautious contender, Galileo embarked on an extensive new line of investigation that would change his public persona from Assassin of Aristotle to Champion of Copernicus.

An adept mechanician, Apelles made himself a telescope of almost 30x. In March 1611 he looked through it and heavy mist to measure the diameter of the sun. He noticed what seemed to be spots on its surface but, not wishing to injure his eyes, he put off further examination until he had leisure, stained glass, and a colleague, Johann Baptist Cysat, fit for the purpose. They resumed the observations in October 1611 to discover that the spots moved slowly across the solar disk, changing shape as they went. When plotted at daily intervals on a series of circles, they appeared to come into existence (or into view), mutate as they moved almost in parallel lines, and die (or disappear behind the sun) within two weeks or so. After satisfying himself by astute tests that the spots were not artifacts, Apelles proposed to explain them in a way that did least damage to the traditional cosmology.[173] For Apelles was a Jesuit, Christoph Scheiner by name, professor of Hebrew and mathematics at the Jesuit college in Ingolstadt, a clever man of 35 already launched on a distinguished career (Plate 14). He would become the world's expert on sunspots, an observer noted for his improvements in instrumentation, a useful theorist on light and vision, and an important Jesuit, advisor to his general and confessor to a prince. To make a small world of it, Scheiner's confessant was Archduke Karl Josef of Austria, a brother of Cosimo's archduchess Maria Maddalena and of Galileo's admirer Archduke Leopold.[174]

Scheiner's first contributions took the form of three letters written in November and December 1611. Welser published them in January 1612 with a nice plate roughly depicting the spots on smallish circles representing the sun. The size of these representations (almost all 2.5 cm in diameter), which had little space to indicate detail, supported Scheiner's suggestion, in keeping with his obligation to stay close to Aristotle, that the spots were bunches of tiny stars circulating around the sun. That would explain their coming to be and passing away, their motion in roughly parallel lines, and, because of the supposed opaqueness of the little stars, their appearances, which Scheiner perceived as uniform darkness against white spaces. He thus assimilated the spots to miniature Venuses, on the theory that Venus, if she revolved around the sun like the spot starlets, would reveal the fact by showing herself near inferior conjunction as a moving mark on the solar disk. In keeping with this model, he expected that the sequence of spots would recur as the orbiting starlets returned to the configuration in which he first sighted them. In an inspired extension of the model, he guessed that the coming and going of the Medici stars and the bumps on Saturn might also be owing to the antics of starlets.[175] These conjectures did not please Galileo.

He took almost four months to reply to Welser as a client and to Scheiner as a patron, that is, patronizingly. "It seems to me that Apelles, being of a free and not servile mind and quite capable of sound knowledge…cannot yet totally free himself from those fancies previously impressed on him," that is, Aristotelian physics.[176] Scheiner deserved a better reception from the sidereal messenger, as he had in effect founded sunspot studies; but Galileo could not concede even that little. He believed that he had priority. He had noticed them first, he told Welser, around November 1610. Perhaps he saw them even earlier, for Micanzio recalled many years later that Galileo had shown them to Sarpi before he left Padua; which, however, would not have given him priority, since Gualterotti had described clearly, in print, a spot he had seen without a telescope in 1604; Kepler did the same, mistaking a spot he had observed·in 1607 via a *camera obscura* for Mercury; and Harriot had detected and followed many solar blemishes with a telescope in 1610, as usual, however, without publishing them.[177] The first published telescopic observations of the sun, the work of a German student using a Dutch gadget, appeared in time for the Frankfurt book fair of 1611.[178]

The earliest secure date for Galileo's observation of sunspots is the spring of 1611, when he showed them to people in Rome around the time that Scheiner noticed their existence.[179] According to Galileo's game rules, it did not matter when Scheiner first saw the spots or that he established some of their properties before Galileo did. That gave him no priority. Why not? Because he misinterpreted what he saw.[180] The lion's share of the discovery of the spots belonged to their first correct interpreter and, as we know from Aesop, lions do not share.

The lion began his response without a roar, circumspectly, he wrote Welser, because the notoriety of his previous discoveries had recruited a band of implacable opponents. "For the enemies of novelty, who are infinite in number, would attribute every error, even if venial, as a capital crime to me, now that it has become customary to prefer to err with the entire world than to be the only one to argue correctly." Still Galileo would hazard some conjectures. Everything that Apelles wrote about the spots apart from their existence was wrong. They are not dark, only less bright than the sun. There is nothing star-like about them. And they might well reside on the solar surface despite Apelles' argument that they cannot be carried around by a hypothetical rotation of the sun. This last point was capital. Apelles had observed that a spot that crossed the sun's disk without vanishing took around 14 days to do so; if the sun did rotate, a hypothesis that Scheiner rejected, and if the spin caused the motion of the spots, then the same ones should return to the same positions once a month. They do not. Therefore… Galileo replied that the argument would be compelling if we knew that the spots were permanent features. But we know they are not. Therefore…[181]

Poor Scheiner failed even in endorsing the great truth that Venus and Mercury circle the sun. For he supposed that their transits might be visible as big spots moving across the solar disk at speeds calculable from the planetary models. In fact, Galileo wrote, they would be far too small to distinguish themselves and—here is the point—no further proof of Venus' place was needed beyond her display of phases as revealed by the sidereal messenger. Venus travels around the sun as do all the other planets according to Pythagoreans and Copernicans. If the spots are not star bunches, then what? If there is anything they resemble within our experience, replied Galileo, that thing is a cloud, which comes to be, passes away, expands and contracts, and moves almost contiguous to the earth's surface. If we could look at the earth from the sun, "and if the earth rotated on its axis," we would observe in the clouds the phenomenon known as sunspots.[182]

Galileo concluded this first of his three portentous letters on sunspots by giving another of his hostages to fortune. Apelles had had the effrontery to suggest that the Medici stars resembled sunspots in their sudden appearances and departures, and might therefore be only flights of starlets. It followed that sometimes Jupiter might well show more than four (or perchance no) satellites—Scheiner thought that he had found a fifth, which he named after Welser—and that the other planets probably had companions now and again too. Galileo countered that he had shown that each Medici star had its own orbit and period. Next, Apelles tried to enroll Saturn among the sites undergoing only apparent alteration. Saturn did not look three-bodied to him, but like a soup tureen with lid and handles. Galileo assured Welser that either Scheiner's eyes or his telescope had failed him. Galileo's own occasional inspections showed absolutely no alteration in the appearance of Saturn, "and reason itself, based on the experience we have of all the other motions of the stars, can render us certain that likewise none will take place." Galileo based this extrapolation on the supposition that any relative motion among the three bodies should have been detected over the period he had watched them.[183] It is a good measure of his self-image. He had come to think that the revelations that God had reserved to him for ages could be established definitively in a year and a half.

It was just this rapid pace that prevented him from writing the big book on cosmology that everyone expected of him. Or so Galileo said in the introduction to *On floating bodies*, in which, to show his ongoing participation in astronomical discovery (and to claim his priority), he incongruously presented his first values for the periods of the Medici stars. When he sent *On floating bodies* to press in March 1612, he had not pronounced in favor of the cloud formations on the solar surface, or, probably, decided its capital relevance to Copernican theory.[184] Galileo continued to update his astronomy via *On floating bodies* by inserting in its second edition, published in Autumn 1612, the bulletin that the spots adhered closely to the sun's body and were "carried around by rotation of the sun itself, which completes its period in about a lunar month—a great event, and even greater for its consequences."[185] The announcement of an unprecedented gyration of the sun in a book about experiments with bits of ebony suggests and symbolizes the fruitful symbiosis between Galileo's ideas about terrestrial and celestial motion, and his conviction that, in a few cases at least, he had discovered the physical truth about astronomical bodies.

Support for the conclusion that the sun rotates came from observations by Galileo in Florence, Cigoli in Rome, and astronomes as far away as Sicily organized by Castelli. The group began its work in mid-February and kept at it into August 1612. Their drawings, informed by knowledge of perspective and practice with washes, rendered the ambiguous shapes of the spots more like clouds than like the hard-edged opaque forms drawn by Apelles. Ideas of the physical character of the spots developed with the methods of depiction. Since the location of the spots was a matter of stellar importance, "[artistic] style was a carrier of knowledge of truly cosmic proportions." In early May Galileo began to use Castelli's method of throwing the sun's telescopic image onto a sheet of paper fixed to a board to make large scale maps of the spots rendered in ink and wash. In contrast to Scheiner's 25 mm sun, Galileo and Cigoli entered their observations on circles 125 mm in diameter, making manifest the foreshortening at the limbs and allowing for fine shading elsewhere. Gradually Galileo's and Cigoli's drawings for the same days became very similar although they did not, as Galileo boasted in his second letter to Welser, "fit exactly with mine."[186]

In this second letter, Galileo argued from the thinning of the spots and the shortening of the distances between them as they moved toward the sun's limb that they must be on or near its surface. Taking, as he did, the sun's axis as perpendicular to the ecliptic, a spot seen from the earth's center would appear to move along a straight line AB parallel to the ecliptic and perpendicular to the line of sight (Figure 5.9). By hypothesis, the spot moves in fact along the semicircular arc ACB, so that when at X (or Y) it

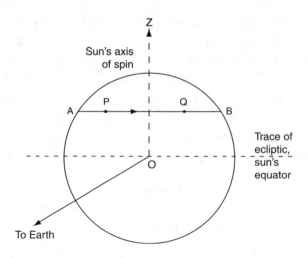

FIG. 5.9 Apparent path (AB) of the sun spots P,Q projected onto the solar disk by a distant observer in the plane of the ecliptic. The plane of the drawing is perpendicular to that of the ecliptic.

appears at P (or Q), as in Figure 5.10. Now let the spot describe a circle of radius b greater than the sun's radius a. Then by hypothesis it will be at X', Y' when seen at P, Q. As b increases, X'Y' tends toward equality with PQ. But XY > PQ. At any radius $b > a$, X'Y' < XY. The test is made most conveniently when the line of sight bisects X'Y' and XY, since the expected value of XY can be calculated from the earlier measurements of P and Q (on the assumption that the spots lie on the sun's surface) and compared with the measured value of X'Y'. The parallel or orthographic projection that Galileo employed in this analysis was a staple in the theory of proportion.[187] The result of the calculation and comparison, Galileo claimed, ruled out spot orbits with radii much different from a; but as the sample he gave contains a small miscalculation, it might not have persuaded anyone who bothered to work it out.[188] Those who accepted the conclusion, however, and the underlying premise that the sun turns on an axis through its center, gained something. They had attained a new perspective on the universe.

The Copernican theory does not merely exchange the positions of the sun and the earth. It also endows the earth with the role previously assigned to the sphere of the stars. A spinning earth? What could keep so sluggish and heavy a body in motion? And the clouds, would they not be left behind, not

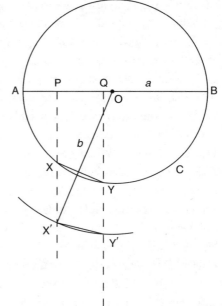

FIG. 5.10 True positions of the spots (X,Y or X',Y') projected by the observer of Fig. 5.9 onto the sun's disk at P,Q. The drawing is in the plane through AB parallel to the ecliptic.

to mention birds and anything else not bolted down? To these objections Galileo had no good replies before his study of sunspots. But now he had at last an object that behaved exactly like Copernicus' earth, spinning on its axis while describing its yearly Ptolemaic circuit, and leaving not a cloud behind. To be sure, it was no proof, but it showed that, just as Jupiter's traveling with the Medici stars strengthened the possibility that the earth could move without losing the moon, nature could accomplish things undreamed of in Aristotle's philosophy.[189] Galileo gave the sun a little physics to help it move. Just as a weight can rest or stray on a horizontal plane (that is, a spherical shell concentric with the earth), having neither desire nor repugnance to a motion that takes it neither toward nor away from the center, so the sun, a spherical body, has no reason to prefer resting to spinning.[190] The spinning marble sphere that Galileo had considered in his Paduan mechanics had become not the earth but the sun.[191]

Galileo did not specify the origin of the impulse that set the sun spinning. What counted was that the medium in which it revolved be sufficiently fluid that the sun might continue to spin until God dissolved the world. Galileo had just issued proof that water, and a fortiori the air and the interplanetary medium, offered no resistance to parting. He had only to endow the heavens with this yielding fluidity to persuade anyone who accepted the paradoxes of *On floating bodies* that the sun can spin without meeting resistance.[192]

Galileo did not claim to be the first to melt down the solid spheres that faithful Aristotelians like Cremonini invoked to portage the planets. Tycho, whom as usual Galileo did not mention, had made the heavens fluid to accommodate the comets he placed beyond the moon and the intersecting orbits of the sun and Mars implied by his planetary system. Tycho was right. Peripatetic philosophy was no more solid than its celestial spheres. Or so Galileo wrote triumphantly to Cesi on 12 May, shortly after beginning his large-scale solar mappings, "[sunspots] should bring about the funeral or rather the extreme and last judgment of pseudo-philosophy." To Welser he confided that the frequent creation and destruction discovered on the sun signaled still another marvel to add to those he had found in the sky: the obliteration of Aristotle's heavenly city.[193]

Meanwhile Apelles had corrected and extended his first hasty letters with three more directed to Welser, who published them in September 1612. In this *More accurate inquiry into solar spots and Jupiter's stars*, Scheiner again argued that sunspots are shadows cast by dark matter, to which he now ascribed a slight translucence to account for his more careful observations of the change in shapes and play of shades as the spots crossed the sun's disk. Again he supposed that the same dark matter came together and parted to create the illusion of moons around Jupiter and lumps around Saturn; and he hinted that moonglow, which he attributed to sunlight seeping through the lunar body, indicated that our moon might also be made of it. By 25 July, when he dated his sixth letter to Welser, he had read Galileo's first one. His balanced response expressed pleasure that his observations agreed with those of a "witness greater than all," the starry messenger. It also contained a lengthy account of sources of observational error, admitted the earth among reflecting bodies, and called for a new cosmology incorporating the dark matter he supposed to fly around the planetary system.[194]

Galileo had Scheiner's *More accurate inquiry* before him when, on 1 December 1612, he sent Welser the last of his sunspot letters. His friends Cesi and Cigoli had advised him to go easy, which he did, to spare Welser.[195] Still, he allowed himself a jab at Apelles and at philosophers in general for preferring the authority of thousands to the "spark of reason in a single individual" who saw that the spinning sun confirmed Copernicus.[196] That went too far for Welser, who shrank from publishing Galileo's letters to him ostensibly because of the expense of reproducing the large drawings in their original size. Cesi was happy to take on the task, however, and to edit as well as to publish Galileo's contributions to solar physics. Editing involved negotiating with the Inquisition over certain passages and in negotiating with Galileo to tone down his rhetoric. Cesi took particular care with the engravings, about which he consulted Cigoli; and Galileo's handsome book, *Istoria e dimostrazioni intorno alle macchie solari* (*Observations and demonstrations concerning sunspots*) came from the press in 1,400 copies, half of which also contained Scheiner's letters to Welser.[197]

The title and language are significant. Unlike *Sidereus nuncius*, *Observations on Sunspots* was not merely an account of the phenomena (*istoria*) but also an indication of the true character of the universe (*dimostrazioni*).[198] And unlike the earlier work, the *Letters* appeared in the vernacular, in answer to Cigoli's plea for works accessible to people unpracticed in academic Latin and to the great inconvenience of their most interested reader, Scheiner, who did not know Italian. The choice of language shifted the grand drama of the solar spots from the European stage to Galileo's theater of operations. Making philosophy accessible to young Italians who think that "those miserable [Latin] pamphlets that contain great new things...remain way over their heads" was but one of Galileo's objectives in writing in the vernacular.[199]

It was as well that Scheiner did not understand the preface that Cesi saw fit to add to Galileo's *Istoria*. It breathed the fire of the small devoted band of impressionable liege men (Cesi, Stelluti, and Angelo de Filiis, the lynxes' librarian and author of the preface) eager to promote their champion Galileo over all others. The horned Venus, lobed Saturn, spotted sun, the instruments for discovering them and the ability to interpret them, "all this was reserved for Signor Galilei alone." He had been as one against the multitude. At first he met denial and rejection, then envy and jealousy; he and only he, guided by God and Nature, had the "firmness of judgment, perspicacity of discourse, integrity of mind, [and] nobility of spirit" to have brought the novelties to light, all of them, on his own.[200] And this formulation of the case was a shortened and softened version of de Filiis' original draft, which even Galileo could see would give unnecessary offence.[201]

Galileo chose Welser, newly lynxed in 1612, as the recipient of the announcement of a double marvel: a new production in the heavens and a mistake by Galileo, "an unexpected wonder…which has recently come to disturb me."[202] Despite Galileo's assurances, Saturn's fortunes had changed drastically: he was now again solitary. Should the enemies of reason take heart? Galileo predicted that the strange and unexpected event would be undone after the winter solstice of 1614 (as it was) and that, no less than Venus' phases, it "agrees in a wondrous manner with the harmony of the great Copernican system, to whose universal revelation we see such favorable breezes and bright escorts directing us, that we have little to fear from darkness or crosswinds." Apparently Galileo assimilated Saturn's blobs to Jupiter's moons and estimated their period from the interval between his first view of them and their disappearance. Calculation of the date of their reappearance involved the relative motions of the earth and Saturn as well as the supposed period of the blobs and so related to the Copernican system.[203]

Just before this declaration of faith, as he was objecting to Scheiner's application of the word "stars" to spots, Galileo recalled the lunar explorer Astolfo. The connection he made was a far-fetched allusion to the stars, in this case to the eyes of the evil sorceress Alcina, who had the bad habit of turning disused lovers into myrtle trees. That was Astolfo's plight when Ruggiero, dismounting from his flying horse on Alcina's island, stopped to chat with him. The siren came to Ruggiero that evening, wrapped in nothing much and smiling through "joyful twinkling stars" that made him feel as if hot sulfur coursed through his veins. Fortunately he had an antidote to her meretricious paralogisms. He had Angelica's ring, the Ring of Reason, with which he disengaged himself after a most enjoyable evening. Escaping from Alcina and her peripatetic followers, Galileo—that is Ruggiero—ran to the far side of the island, where her sister, a rival sorceress, cultivated sound philosophy.[204]

5.4 FAMILY AFFAIRS AND FORMER FRIENDS

Family affairs

Knights-errant travel light. Galileo divested himself of his daughters as soon as they became nubile. In 1613 he placed them in a nunnery of the Poor Clares in Arcetri just outside Florence, where they would spend the rest of their lives. It had not been easy—not because the separation pained him particularly, but because the girls did not meet two principal criteria for perpetual incarceration. For one, they were under age; for another, they were sisters. He could solve the problem of age by placing the girls first as boarders. The prohibition against establishing natural siblings as religious sisters in the same monastery could only be set aside by dispensation from Rome. Through his contacts with cardinals including the reliable del Monte, and, it is said, the decisive intervention of Castelli, Galileo obtained his dispensation. After three years of monastic life, at the age of 16, Virginia took her vows and the name Maria Celeste (Plate 9); Livia did the same, a year later, to become Suor Arcangela.[205]

Galileo did nothing unusual in disposing of his children. The cost of marrying off two illegitimate daughters to families good enough for a grand ducal philosopher would have been prohibitive. Galileo may well be criticized, however, for not allowing them to investigate and choose their convent, as his new friend Barberini recommended to a niece contemplating the contemplative life. Galileo did not have Barberini's means, but he had a good income and was almost or entirely out of debt when he had to place the girls. He had shown himself generous, overly generous, in providing for his sisters. Very likely the Poor Clares recommended themselves because its abbess, the sister of Galileo's friend Belisario Vinta, pressed Galileo to have his daughters take the veil during her term of office. Life among the Poor Clares was hard. Virginia, who had a calling for it, passed to a better one at the age of 33. Livia, who chafed under monastic discipline, had to bear it longer. She outlived her father, who left her a very small legacy to buy some comfort in her last years.[206]

Galileo was no misogynist. He thought that women could be as intelligent as other people. "Women have excelled in every art / in which they've carefully taken part." Thus Ariosto, praising women's ability at arms as well as with the pen, preparatory to describing Astolfo's adventures with some Amazons. Galileo did not try to develop his daughters in either geometry (which, since it defeats many, can stand for war) or poetry, although while he was considering a convent he had the example of Margherita Sarrocchi, who exceled at both, before him.[207] That was a pity since at least one of the girls, Virginia, had a talent as a writer now obvious to everyone.[208] The monastery was the easier option. Galileo made it available not only to his daughters but also to his great niece, Virginia Landucci, who thanked him most fulsomely for it. But then she did not choose the Poor Clares.[209]

The story of this niece well illustrates the mix of generosity and hard-ness in Galileo's character. He had maintained a poor girl, Anna di Cosimo Diociaiuti, in a convent. When his nephew Vincenzo Landucci proposed to marry Anna, Galileo agreed to continue her subvention of 6 scudi per month to help support the couple. He did not want to continue it after Anna died in 1633; but Maria Celeste, who supervised the payments, allowed them to go on as charity to the Landucci family. When she died in 1634, Galileo stopped the money; whereupon Vincenzo, as litigious as his father, sued and won on the ground that Galileo had encouraged the marriage and had a responsibil-ity for the consequences. A few years later, Galileo converted the debt by paying Landucci 50 scudi down and defraying Virginia Landucci's convent costs. That was an entirely satisfactory solution to him. As we know, the maintenance of poor girls in nunneries or through dowries was an estab-lished form of charity.[210]

Marina Gamba did not live to see her daughters settled. She died in 1612. Galileo then prepared confusion for historians by placing his son Vincenzo with a married woman named Marina. Lorenzo Pignoria made the arrange-ments and acted as a conduit for Florentine scudi for the boy's care and Vene-tian pills for Galileo's pains. Vincenzo was to do well in life. Legitimized by Cosimo, educated in medicine at the University of Pisa (thus fulfilling his grandfather's plan for his father), Vincenzo became a responsible Florentine citizen, and, less successfully and noisily than his father, a man of science and letters. He would be a comfort to Galileo in his old age.[211] In that he differed from his uncle Michelangelo, who continued to drain Galileo's resources. It will cost you nothing to help out your brother, Michelangelo wrote in 1611, recalling that he had promoted Galileo's glory by advertising the sidereal dis-coveries around Munich.[212] He did not appeal in vain.

Among the inconveniences Galileo could not leave behind was his poor health. He was ill for most of the winter 1610/11 and again in the summer after returning to Florence from Rome.[213] What was his problem? His friends advised him to cut down on wine (none except with meals said Cigoli, only with moderation added Sagredo and Castelli), and no doubt overindulgence in food and drink did its damage.[214] Whatever the cause of his recurrent fevers, depressions, pains, and palpitations, they altered his physical appear-ance. The Venetian ambassador to Rome, Simone Contarini, who had known Galileo during Paduan times, was struck by the deterioration in his appearance when they met in 1615. Contarini linked the evident decline in health with an increased apparent religiosity. "He frequently takes the sacraments and is very much changed from what he was."[215] The probable etiology of this increased observance will soon be clear.

The only member of his extended Venetian family who relocated near Galileo was his quasi-son and disciple Benedetto Castelli, who had left Padua in 1607 for a teaching post from which he developed the first Galilean apostolate. "My students venerate Your Lordship's rare virtues, which are unique in our time." We know Castelli's enthusiasm over *Sidereus nuncius* and his suggestion about the phases of Venus. His place, as he saw it, was with the celestial messenger. He requested a transfer to the Benedictine abbey in Florence.[216] He arrived there in April 1611 and immediately made himself useful in compiling data about the Medici stars. Soon he was helping to meet objections to *On floating bodies* and to draw the trajectories of sunspots. In 1613, he became professor of mathematics at the University of Pisa, where, in time, he had Vincenzo Galilei and two of Galileo's nephews as students, and trained two excellent mathematicians in the Galilean cause, Bonaventura Cavalieri and Evangelista Toricelli.[217] The post gave him access independently of Galileo to the Tuscan court. In 1626 Castelli entered the service of the Barberini as advisor on hydraulics to the pope and tutor to one of the papal nephews. Despite his many successes as administrator and expert, Castelli tended to sell himself short where Galileo was concerned. "I know very well that everything I am I am through you and the reputation I have of being your disciple."[218]

Old friends

Cremonini and Galileo at first stayed in touch through Paolo Gualdo, whose reports may reflect his taste for practical jokes. In May 1611 Gualdo wrote that Cremonini spoke most affectionately about Galileo though preparing to publish a treatise *De caelo* ("On the heavens") against him. Galileo replied expressing feigned trepidation, to which Cremonini answered that he had no need for concern since the new *De caelo* would not mention his discoveries. According to Gualdo, Cremonini refused to look through a telescope. This behavior, made famous in Galileo's later lampoons, has become the trademark of the purblind philosopher. But Cremonini's reason for not trusting the telescope was that he had looked through one. The image confused and dizzied him. He inferred that only people with quirky eyesight and unrestrained imagination could see what Galileo had claimed to see.[219]

In any case, for Cremonini as for many others, Galileo's observations could not make an effective challenge to Aristotle's physics. Cremonini reduced the epistemology at stake to a single question. If the moon is a big rock, what keeps it from falling down? Aristotle's system might be wrong, and Galileo's observations indicated problems in it; but unless Galileo could provide a physics adequate to account for standard phenomena, the inconvenient evidence from the telescope would remain just an inconvenience. No astronomical system, including Ptolemy's, satisfied Cremonini's strict demand for a science of first principles.[220] Galileo appreciated the difficulty raised by Cremonini, which amounted to finding the physical principles on which to anchor a *demonstratio potissima* of the Copernican system. He could not do it. And without a replacement for Aristotelian physics he had no chance of bringing philosophers with him. "Yea verily [wrote Campanella], you cannot philosophize without a true and confirmed system of the constitution of the universe. We [it is Campanella in papal mode] expect Your Lordship to provide one."[221] Philosophers will not leave one bone until tossed another.

Cremonini's solution to the problem of a realistic astronomy, which was to reassert Aristotle, brought him once again to the active attention of the Inquisition. Revisions were required to his *De caelo*; the pope, still Paul V, demanded to see the revisions before publication and ordered Bellarmine to dictate to the Venetian inquisitor the sorts of corrections required. Cremonini procrastinated. A new pope, Gregory XV, a friend of the Jesuits, was less patient than Paul. He ordered that all Cremonini's writings be placed on the Index, but his decisiveness did not bother or deter the errant philosopher.[222] While scrutinizing Cremonini, the Holy Office did not overlook his friend Galileo. During his visit to Rome in 1611, the Inquisition, alerted perhaps by Bellarmine to errors in faith that might arise from the new discoveries, decided by formal decree to see whether "Galileo Professor of Philosophy and mathematics...is named in the case against Dr. Cesare Cremonini."[223] "Oh, how much better Galileo would have done [Cremonini sighed to Gualdo] if he had not taken up these caprices and left the freedom of Padua."[224]

Where a new idea appears in Galileo's writings it is wise to look for it in Sarpi's notebooks. Fra Paolo had considered the problem of solar fuel before Galileo took it up in connection with sunspots. Following a Stoic idea, he had likened the sun's physical economy to the burning of oil by a lamp.[225] Galileo thought that the spots might be the food, if not the excrement, of solar consumption, and spied the food's provenance in the planets. Sunspot activity appeared confined to a narrow band around the ecliptic, which suggested an exchange of solar light for planetary exhalations. This ingenious theory, proposed in the third letter to Welser, did not make it into the printed edition of *Observations on sunspots*, which had problems enough without it in passing the censorship.[226] Galileo did not drop the idea, however. He coupled the notion of a material bond between sun and planets with the discovery of the solar spin to obtain the motive power to sweep the planets around. Apparently he had in mind something like Kepler's anima motrix, by which the turning sun drives the planets as the beams from a lighthouse would ships—if the beams were made of steel.[227]

Galileo informed Sarpi early in 1611 of his latest findings: the phases of Venus and their implication, the discovery of a way to obtain the periods of the Medici stars, the unchanging appearance of Saturn and his companions. The signs in the Roman sky, as he read them, were favorable. The Jesuit mathematicians, "forced by the truth...have confessed and admitted everything." Among the holdouts, the most obstinate were the philosophers of Padua, that is, their mutual friend Cremonini and his followers. Continuing in this paranoid or melancholic vein, Galileo feared that the philosophers would try to "exterminate mathematics" at Padua, that they would choose a cipher as his successor, "so that if ever something beautiful and true were discovered, it would be suppressed by their tyranny."[228] Sarpi answered through Micanzio: what you have done is very fine, but enough; stop fretting and return to your important work, which is perfecting your (our?) science of motion.[229] A satisfactory theory of motion would challenge received physics at its core. There was where the principles of a new philosophy lay. To Sarpi, who continued to ask after him, and to Gualdo, Sagredo, and other Venetian friends, Galileo's Copernican campaign was a quixotic sideshow.[230]

From the Venetian point of view, the ongoing celestial discoveries weakened the case for them all. Was there not something almost magical and even dishonest about this incessant revelation of distant things?[231] Certainly, they enriched the range of hypotheses for academic discussions; but they did not establish heliocentrism as a demonstrated realistic theory to be foisted on philosophers and theologians. In its state and the world's state, it was not worth a major fight with the duplicitous occupants of the Vatican. Sarpi acted on this conviction in 1616. At the time, Venice was at war with Austria and Sarpi had several works in hand demonstrating the machinations of the papacy, including his devastating *History of the Council of Trent*, published anonymously in England in 1619, and an exposé of plots against Venice by Rome.[232] So in 1616, when the Congregation of the Index condemned certain Copernican works, Sarpi advised that, stupid as it was, the Serenissima should accept it. It did not threaten the Venetian printing industry or anything else with a significant constituency. "Since so few people apply themselves to astronomy, there is no reason to fear a scandal."[233] Galileo and Sarpi were no longer in touch.

Sagredo could not understand his friend's fancy for Florence any more than he could his Copernican compulsion. "Where will you find freedom and self determination as in Venice...? Serve your prince [if you will]; but here you had command over those who order and govern others, and you did not have to serve anyone but yourself, as if you were monarch of the universe." At court jealousy is always ready to tear down virtue. "Who knows what the unfortunate and incomprehensible accidents of this world can effect? Especially when exploited by the deceits of evil and envious men who sow and cultivate false suspicions in the mind of the prince."[234] Though hurt by Galileo's desertion, Sagredo maintained his end of their weekly correspondence. The letters are jovial, licentious, scientific, gossipy, and opinionated. The friends exchanged gifts, medicine, plants, *objets d'art* from Sagredo, dogs, wine, and sausages from Galileo. They discussed heat, thermometry, lenses, telescopes, magnetism, comets, and the calculation of Easter.[235] And they assassinated characters. Two of Sagredo's judgments are arresting. Kepler was not as good a mathematician as Galileo, indeed, not good enough for a call to Padua; and "Della Porta has the same place [among the learned] as church bells among musical instruments."[236]

Sagredo's death in 1620 cut Galileo's last material tie to Venice. It was in any case as fragile as glass. Galileo had dropped his mechanic Mazzoleni as brusquely as his mistress Marina, and had to depend on others to procure Murano glass for his telescopes. He does not seem to have installed in his dwellings in and around Florence any of the grinding machines he had in his house in Padua but depended largely on Sagredo for finished lenses.[237] Most of the artisans Sagredo commissioned were mirror makers or polishers of *pietra dura* by trade and had neither the skill nor the interest to develop reliable techniques for freeing optical glass from twists and blemishes. Yields were low. In a representative try, a friend of Sagredo's cast 300 lenses, of which he deemed 22 suitable; of these Sagredo took only three and none of them was perfect. He then commandeered a glassworks and tried to forge lenses of rock crystal. In optical properties they outdid glass but the yield of good ones was if anything poorer. Often the best lenses came from broken pieces of mirrors. From these Sagredo fashioned opera glasses, which, he said, made women short distances away look more beautiful. Fuzziness has its uses. Galileo tried in vain to attract specialists from Murano to Florence. Cosimo did better. He obtained a few to run the glassworks he set up in 1618, which also kept Galileo supplied with lenses. It appears that he had a business in telescopes for twenty years or more after leaving Padua.[238]

The petty court jealousies of which Sagredo had warned Galileo may be illustrated by a spat between Salviati and a low-level Medici over whose carriage had precedence. The Medici won, the insult stung, and, despite attempts at mediation by Christina and cardinals Bandini and Barberini, the quarrel smoldered for over three years until in January 1613, in the presence of the grand duke, the teapot tempest petered out. Toward the end of the year, having patched up his relations with the court and nursed Galileo through the writing and publication of the *Observations on sunspots*, Salviati set off to see the world. He went first to Venice, where he may have met Sagredo, and to Padua, where he talked at length with Cremonini, who again expressed his affection for Galileo, "except in theory." He then took ship to Spain. The country did not agree with him. He died suddenly in Barcelona in 1614, not yet 32.[239]

6

⬯⬯⬯

Miscalculated Risks

Galileo's published attacks on traditional physics worked most powerfully and transparently where they destroyed distinctions between celestial and terrestrial affairs. *Sidereus nuncius* and the *Letters on sunspots* integrated the earth into the heavens and earthed the heavenly bodies: the moon's surface copied the earth's with rocky irregularity, the sun's atmosphere ours with intermittent clouds; Venus showed herself as opaque as the moon, and Jupiter more earthy than earth in the number of his satellites. The distinction between a heaven made of incorruptible perfectly spherical bodies imbedded in frozen quintessence, already weakened by earlier detection of novas and comets without parallax, now could survive in its original purity only in minds not authorized to think for themselves. *On floating bodies*, which did away with levity and reduced the list of factors relevant to buoyancy to a single entry, specific gravity, was to terrestrial physics what *Sidereus nuncius* was to cosmology.

Although obliteration of the distinction between celestial mechanics and terrestrial physics destroyed the wall on which the traditional world picture hung, it did threaten the structure of received religion in principle. It did so in practice, however, owing to the interdisciplinary fusion of Catholic doctrine with Aristotelian philosophy sanctioned by the Council of Trent. The upper hierarchy of the church had been trained in this fusion, and many of the men still in charge in 1615, notably Bellarmine and Paul V, were rigid defenders of Tridentine dicta and papal prerogatives. Still, squabbles among the professors that did not touch on matters of faith generally did not excite the censors; and just as Platonic and other deviant philosophies escaped without official censure, so, no doubt, the great questions of the fluidity of the heavens and the rockiness of the moon would not, in themselves, have commanded the urgent attention of the Holy Office.

What aroused its vigilance was the effort to make Copernicanism the framework of an alternative world system. The challenge of the sun-centered universe to the traditional cosmos was not the heart of the matter, although no doubt it meant much to people who had professed the old way for decades. Nor was the Copernican geometry the main point: only mathematicians had cared about the ways in which astronomers arranged orbits, spheres, epicycles, equants, and eccentrics in their calculations. What concerned the Holy Office was the formal opposition to a stationary sun extracted from such passages in scripture as Joshua 10:12. Why should a man informed by God command the sun to stand still if it had not been moving? But even this consideration might not have forced the issue had not some volunteers, including Galileo, undertaken to twist the plain words of scripture to allow, and even to favor, the sun-centered universe. The volunteers ignored a primary decree of the Council of Trent. "In matters of faith and morals pertaining to the edification of Christian Doctrine, no one, relying on his own judgment and distorting the Sacred Scriptures according to his own conceptions, shall dare to interpret them contrary to the sense which Holy Mother Church, to whom it belongs to judge their true sense and meaning, has held and does hold, or even contrary to the unanimous agreement of the fathers."[1]

6.1 FREELANCE EXEGETES

Galileo was alerted to opposition to his as yet unpublished Copernicanism during his wooing of the Jesuits in 1611. The future lynx Welser, peering from Augsburg, knew that Galileo had been talking up Copernicus in Rome. He so informed Gualdo, who in turn warned Galileo: "Your actions are observed minutely, and are being published around the world."[2] Galileo understood that his enemies might push theological objections to heliocentrism and wrote Cardinal Carlo Conti, whom he had met in Rome, for an assessment of the significance of the scriptural argument. The cardinal answered that a heaven subject to destruction fit the bible better than eternal hard quintessence; a spinning earth presented few problems, a revolving one and a stationary sun many more. Conti advised Galileo not to force the issue.[3] There were other opinions. "If I were God I would not suffer such a race of human ignoramuses [as your enemies] to live," Galileo's former student Daniele Antonini, now well launched on the military career that would cost him his life, wrote from his barracks in Brussels. "I guess that the dear Lord permits them to act as buffoons for mother nature."[4] Galileo settled on putting his *Letters on sunspots* through the press shielded by Conti's powerful scriptural argument for the corruptibility of the heavens.

Despite Cesi's standing in Rome and his impeccable religiosity, he could not obtain permission to publish the *Letters* as Galileo wished. Galileo wanted to begin with Jesus' rebuke to those who have ears but have not heard, "the Kingdom of heaven suffereth violence, and the violent take it by force" (Matthew 11:12). The application to world systems was too transparent. Galileo responded to the censors by removing the biblical context and tried, with success, the paraphrase, "Already the minds of men assail the heavens, and the more valorous conquer them."[5] Galileo wanted to claim "divine goodness" as the ground of his success; he had to settle for "favorable winds." That was the easy part. Galileo desired also to argue that the bible agreed better with Copernicus than with Aristotle or Ptolemy. On these points the censors refused to negotiate. Although Cesi had strengthened the argument that the bible favors a corruptible heaven with passages from scripture and the fathers, he had to concede. Scripture could not be mentioned, nor Conti's argument.[6] The *Letters on sunspots* consequently came out without scriptural confirmation of the Copernican confession with which Galileo ended them. The censors did not object to the confession. They were not concerned with astronomical systems but with biblical interpretation.

Hermeneutics in Florence

The matter of the compatibility of Copernicus and the bible returned, in a manner Galileo believed he could not ignore, late in 1613. Just after taking up his professorship at Pisa, Castelli received a friendly admonition from the archbishop–overseer of the university not to "enter into opinions about the motion of the earth, etc." To which Castelli replied that he planned to follow the example of his master Galileo, who had not mentioned it during 24 years of teaching in Pisa and Padua. The reassured archbishop allowed that if Castelli wished he could touch on similar questions as "probable."[7] A straw in the November wind. A month later the new professor lunched with the grand ducal family and, among others, a Pisan philosophy professor, Cosimo Boscaglia. The conversation turned to Jupiter's moons. Christina muttered that they had better be real. To be sure, she had seen them in the heavens and also on earth, in the form of four Florentine gentlemen who played the Medici stars in an extravaganza held in 1612 in honor of the Duke of Urbino.[8] Nevertheless, she entertained doubts because Galileo linked his discoveries to Copernicus' possible heresy. How could anything be true that implied false doctrine? Christina asked Boscaglia for his opinion. The philosopher replied, to the whole table, that the new phenomena were genuine, and to Christina, privately, that the Copernican implications were worrisome. After the meal Christina summoned Castelli to her rooms. Cosimo, Maria Madd-alena, Boscaglia, and Galileo's supporters Antonio de' Medici (a natural son of Francesco I and his second wife Bianca Capello) and Paolo Giordano II Orsini (a future patron of Scheiner) also attended. Christina put the case that scripture rendered the earth's motion impossible. Castelli in reply "play[ed] the theologian with…finesse and authority….Only Her Ladyship contra-dicted me, but in such a way that I thought she was doing it in order to hear me [out]." Boscaglia remained silent. Antonio de' Medici assured Castelli of his good will toward Galileo. The tale appeared to have a happy ending.[9]

Thinking that his disciple might again be called upon to defend the faith, Galileo sketched out a general method of meeting biblicist objections to the conclusions of "sensory experience and necessary demonstrations." The method would become standard Catholic hermeneutics long after he had suffered for proposing it.[10] Many scriptural passages are silly, Galileo wrote Castelli, even blasphemous if taken literally, like those assigning God arms and legs, and all-too-human emotions. In such cases another interpretation should be sought. The same consideration applied to the bible's occasional observations touching natural phenomena; and in these cases, where "sen-sory experience or necessary demonstrations" can settle the facts, the testi-mony of scripture "should be reserved for the last place." We must keep in mind that the Holy Spirit accommodated its message in the Old Testament to the limited capacities of an ignorant people. Today, however, a few people are clever enough to use God's gifts to see things as they are. "I do not think it necessary to believe that the same God who has furnished us with senses, language, and intellect would want to bypass their use and give us by other means the information we can obtain with them." To place scripture first in matters not pertaining to faith is foolish as well as "disorderly." It incurs the risk of binding faith to assertions about the physical world that may turn out to be false.[11]

If God made the world heliocentrically, Catholic exegetes better informed about the world system than Hebrew savages should be able to interpret biblical references to astronomy more plausibly on Copernicus' model than on Ptolemy's. "It being obvious that two truths can never contradict each other," the Joshua story correctly understood convicts Aristotle's cosmology of falsity and absurdity. Really? Well, according to the Greeks, the sun's diurnal motion is caused by the rotation of the sphere of the stars. Arresting the sun could only stop its annual motion; and that, being from west to east, could only shorten the day created by the sphere rotating from east to west. And if God had stopped the sun while allowing the stars to continue to move, He would have altered the relationship between sun and stars forever. In Galileo's Copernican world, however, God could shut down the system and prolong the day by stopping the earth's spin without introducing any confusion when normal operation resumed.[12]

This accomplished hermeneutics revealed that Joshua's words could have been spoken by a Copernican biblical literalist and that theologians would not require supernatural insight to reinterpret standard texts. Galileo accepted that a satisfactory interpretation was both possible and necessary, for he believed as surely as Bellarmine and the majority of Catholic exegetes of their time that every statement in scripture is in some sense true. Although the Holy Spirit had to talk down to reach Moses' uneducated masses, unlike many popularizers He never descended to patent untruths. As humankind progresses, the correct meaning of His utterances gradually comes to light. "The Lord is subtle but not malicious."[13] Owing to God's decision not to confuse the bible's moral message with an exposition of the world system, the efficient philosopher investigates nature without reference to scripture. After efficient philosophers and mathematicians have arrived by sensory experience and necessary demonstrations at true propositions, an authorized exegete can reinterpret the bible and hail its unveiled teaching as confirmation of the true propositions. Thus, having demonstrated that Joshua 10:12 indicated a Copernican universe, Galileo appealed to Joshua 10:13 ("the sun stood still in the midst of heaven") to confirm the central position of the Copernican sun. To the believer, this circle was not vicious. It brought together two truths that had to be one.[14]

Like many men of science, Galileo did not bother with administrative matters. Who was to work out the correct exegesis when philosophers and mathematicians had determined that the literal word of God could not be defended? Certainly not the man of science eager to assert his independence from scripture! The crux, as Borro and Cremonini had maintained, was to separate philosophy from theology and keep them apart. That was Galileo's goal too. But as he thought he knew some truths about nature, he could not forebear to show how the bible should be read when it conflicted with them. His concordist effort in effect redid the Thomistic synthesis with different ingredients and, in so far as he practiced it, returned Galileo to the prison from which the argument for independence might have freed him.[15]

Galileo's *Letter to Castelli* came into the possession of a group Galileo derided as "the pigeon league" because Ludovico delle Colombe ("of the doves") helped energize it. Among its members were several Dominicans including Ludovico's brother Raffaelo, the leading preacher in town. The delle Colombe brothers, who together represented the interdisciplinary Tridentine ideal, had been antagonists of Galileo since Ludovico had opposed him over the nova of 1604. In 1610/11 they struck both philosophically and theologically. Ludovico proved the immobility of the earth on physical principles and Raffaelo declared, on mere principle, opposition to novelties of all kinds and the inquisitiveness that created them. "They are the most dangerous expressions of human pride." Among the instances he mentioned were Galileo's old lectures on Dante, then still occasionally discussed at the Accademia Fiorentina. Not even Hell was safe from the unwholesome curiosity of mathematicians.[16]

Truly unwholesome. The speculations of mathematicians are a sort of drunkenness. "[They] cannot grasp even close-up and very simple things...What is clearer than *Deus firmavit orbem terrae qui non commovebitur*, and still the Copernicans say that the earth moves...Therefore if he who drinks the wine of the science of the world does not temper it with what is written, *Aqua sapientiae salutaris potabit illum*, he will fall into delirium and madness."[17] Did anyone in Florence know a drunken, delirious, crazy astronomer? Raffaelo gave the answer himself, from the pulpit in the Florentine Duomo. The madman in question claimed to see spots in the sun. "Did not the ancients say of a man who seeks defects where non exist, *Querit maculam in sole*? The sun has no blemish, nor does the mother of the sun, the Virgin Mary."[18]

The informal pigeon league had the important support of the Archbishop of Florence, Alessandro Marzi Medici, of Don Giovanni de' Medici, and, probably, of the Grand Duchess Christina. The archbishop and the grand duchess also conspired in the wider project of strengthening clerical influence in the grand duchy by eroding the power of patrician officials like Vinta.[19] Among the clerics they favored the Dominicans Niccolò Lorini, a familiar figure around the court, and Tommaso Caccini, an ambitious preacher, who, though only 35, had been a hound of God for almost 20 years. (Plate 12). He took on Galileo in 1611, perhaps at the suggestion of his patron the Archbishop. Tommaso's yapping bothered his brother Matteo, the head of the Florentine House of Caccini, more than it did Galileo. With the help of the Salviatis and Maffeo Barberini, Matteo managed to get his noisy brother out of town for a while. Egged on, probably, by Raffaelo delle Colombe, and in any case unrestrainable, Tommaso returned in 1614 to deliver the Advent sermons in Santa Maria Novella.[20]

Having decided on his own that the propositions of a moving earth and standing sun were heretical, Caccini gathered what evidence he could, including the *Letter to Castelli*, that Galileo and his disciples believed them and, indeed, many things worse. His research methodology featured eavesdropping. One day he overheard a private discussion between a student of Galileo's, a young Florentine noble named Giannozzo Attavanti, and a Dominican master, Ferdinando Ximenes, who instructed Attavanti in casuistry. They were discussing in the hypothetical mode whether God is substance or accident, and whether He has human senses and emotions. Listening at his spy hole, Caccini heard enough to infer that Attavanti held the opinion he defended hypothetically (an accidental, sensual, emotional God) and had learned it from Galileo. On a previous occasion, Caccini had heard Attavanti and Ximenes discussing Copernican theory. He had then rushed into Ximenes' cell and declared, "it was a heretical proposition to assert that the sun stands still and its center does not move." He further declared that he wanted to preach about it.[21] He took the book of Joshua as the text for his Advent sermons. When he arrived at Joshua 10:12, he barked at Copernicans for promoting a theory in flagrant contradiction to Holy Scripture.[22]

Caccini's virulent sermon made a scandal. The Dominican General wrote Galileo to apologize; so did Lorini; and Matteo blasted Tommaso in terms that can only be used between brothers. You fake religion and zeal to cover your animosity, Tommaso. Or was the attack your idea? "How thoughtless you have been to let yourself be taken in by those pigeons, or idiots, or doves!" You have ruined your reputation. "A grave mistake! A gross stupidity!"[23] Matteo redoubled his efforts to remove Tommaso from Florence. He need not have bothered. Exactly a month after his nasty sermon against mathematicians and Galileists, Tommaso Caccini showed up in Rome to testify before his fellow Dominican, Michelangelo Seghizzi, the Commissary General of the Holy Roman and Universal Inquisition. He deposed as follows. "With the modesty which befits the office I hold," he had preached to the multitude that the Copernican view was "discordant with Catholic faith," indeed, heretical, as it ran counter to the opinion of all scholastic theologians and all the Holy Fathers. "After this discussion I counseled them that no one was allowed to interpret divine scripture in a way contrary to the sense on which all the Holy Fathers agree."[24] Asked by Seghizzi about Galileo's reputation regarding matters of faith, Caccini replied, "[by some] he is regarded with suspicion in matters of faith because they say he is very close to Fra Paolo, of the Servite order, so famous in Venice for his impieties; and they say that letters are exchanged between them even now." Galileo also corresponded with people in Germany, as did other members of an academy he belonged to, a further indication of his unreliability.[25]

Caccini knew more, and worse. Galileo's letter to Castelli contained "questionable doctrines in the domain of theology." Some Galileists believed that God was an Accident and that the saints had not worked miracles. Pressed by Seghizzi for his sources, Caccini replied that Galileo's Copernicanism was notorious in Florence and could be found in his book on sunspots, and that Attavanti (Caccini could name no other) was the misled Galileist. As for the Castelli letter, Caccini volunteered that he had seen a copy in the possession of Lorini, from whom he had learned about Galileo's ongoing correspondence with Sarpi.[26] Lorini had not neglected to inform the Roman authorities about the Castelli letter. Early in February he sent a copy of it to the Dominican prefect of the Congregation of the Index, Cardinal Paolo Camillo Sfrondati, for delivery to the Inquisition. That was only to do his duty, he said, to act as a hound of the Holy Office. All of his brothers at S.M. Novella who had read the letter agreed that it contained many rash or suspect propositions, to wit: Holy Writ sometimes speaks inappropriately; scripture holds last place in disputes over natural effects; biblical exegetes often err; scripture pertains to faith and nothing more; and "in questions about material phenomena, philosophical or natural argument has more force than the sacred

and the divine one." Unpardonably, like so many Protestants, the Galileists
want to interpret scripture in their own way. "They speak disrespectfully of
the ancient Holy Fathers and St Thomas." "They trample underfoot all of
Aristotle's philosophy, which is so useful to scholastic theology." All true. So
evident were Galileist errors and pretensions that Lorini felt no need to add
more. "As St Bernard said, [in Rome] the holy faith has lynx eyes."[27]

Knight-errantry in Rome

The Holy Office sent Lorini's version of the *Letter to Castelli* to a consultant,
who saw nothing that "diverge[d] from the pathways of Catholic expression"
except three short phrases in the Castelli letter, two of which must have been
written by a negligent or malevolent copyist, as they are much stronger than
Galileo's wording, and the third of which offers to "assume and concede"
Joshua 10:12.[28] The inquisitors perceived that Lorini might not have sent a
faithful copy of the letter and wrote to Florence for one. Since the original
was less damaging than the copy, it seemed that the storm might subside if
the parties stayed calm.[29] Galileo opted for a pre-emptive strike. He asked
Piero Dini, then serving in the Vatican bureaucracy and living with his cardinal uncle, Ottavio Bandini, to forward correct copies of the *Letter to Castelli*.
"The most immediate remedy would be to approach the Jesuit Fathers, as
those whose knowledge is much above the common education of friars."
After all, Galileo wrote, he strove to save uninformed decision-makers
in Rome from falling for the frauds of his detractors; his motive was not
personal but patriotic. Surely the Jesuits would help him to overcome the
"wickedness and ignorance of my opponents." Galileo always had trouble
distinguishing the ignorance of his opponents from their wickedness. To set
them right, he was putting the finishing touches on a formal statement of his
hermeneutical methods and conclusions, which would circulate as a letter to
the Grand Duchess Christina.[30]

   The Jesuits declined to come to the rescue. Grienberger saw something
sophistical in Galileo's reasoning and Bellarmine doubted that humankind
could reach truth in astronomy. He did not foresee a ban against the Copernican theory, he told Dini, but, perhaps, a directive to regard it as a hypothesis, that is, a useful fiction. Early in his career, Bellarmine had concocted a
cosmology from the bible and the fathers, the only way, according to him,
to reach truth in the matter. A stationary sun seemed irreconcilable with the

bible. The most difficult text for Copernicans, he confided in Dini, was the beautiful psalm that begins, "The Heavens declare the glory of God." The difficulty follows a few lines later: "In them hath he set a tabernacle for the sun which is as a bridegroom coming out of his chamber, and rejoiceth as a strong man to run a race." Dini replied that the Galileists could interpret it favorably. Bellarmine urged restraint. No one should rush to conclusions, he advised, and offered to read whatever Galileo had to say. "As you can see, things are rather unclear because everyone is on guard about a business of such importance." Dini returned to giving copies of the Castelli letter to friendly cardinals including Barberini and del Monte.[31]

Bellarmine's apparent flexibility increased Galileo's missionary zeal. He replied to Dini that Copernicus would not have labored so valiantly to demonstrate a mere hypothesis. He thought his system true and we, who have even better reasons to believe it than he did, should either accept it as true or probable, or dismiss it altogether. Furthermore, it is nonsense to say, as Bellarmine did, that because astronomers employ concepts like eccentrics and epicycles all their constructions are fictions. For epicycles, interpreted as orbits around a body other than the earth, do truly exist, as we can deduce from the moons of Jupiter and the phases of Venus. As for eccentrics, Mars' path around the earth most certainly is not geocentric for the planet is twice as distant at conjunction with the sun as at opposition.[32] This clever and misleading response, which implicitly takes force-free epicyclic rotation as the principle of planetary motion, helps explain why Galileo could not accept Kepler's ellipses.

There remained Bellarmine's prize text. To neutralize it, Galileo developed the celestial physics he had begun with his paradoxical identification of sunspots as excrement and continued in his explanation of Joshua's miracle. We know from Genesis that God made light before he made the sun to be its vessel. From the sun and the universal background radiation left over from the initial *fiat lux*, all bodies in the universe obtain their fertilization and growth, and the planets have their principle of motion. The bridegroom running his course is the radiating sun, which, like a bridegroom, is eager to fertilize, rejoicing in his power of penetration...To reduce the offense of this exegetical caricature, Galileo followed the advice of Roman friends to qualify it, and his essays in theology generally, as conjectures submitted to higher authority with "the humility and reverence due to the Holy Church and all its very learned fathers, whom I respect and honor and to whose judgment

I submit myself and every one of my thoughts." In practice, however, most of the higher-ups were ignoramuses. Hence God had sent Galileo for their instruction. "Divine love may sometimes deign to inspire humble minds with a ray of His infinite wisdom, especially when they are full of sincere and holy zeal."[33]

Prophets do not deal in hypotheses. Not for them the magic spell of the four-letter talisman, the tetragrammaton HYPO, which converts a true, manly thesis into a tranquilizing hypothesis. Only the uninformed or the malevolent could think that Copernicus' theory was a hypothesis. Moreover, taking heliocentrism to be hypothetical demeaned the human mind as perfected in mathematician-philosophers. When by sensory experience and necessary demonstration they reach certainty, they cannot amend their convictions; "it is not within the power of the practitioners of demonstrative sciences to change opinion at will." Exegetes must give way. They can change their interpretation; the philosopher cannot change nature.[34] If they ducked these plain facts, Bellarmine, Grienberger, and the rest would desert the truth as well as Galileo.

On 7 March 1615, the very day on which Dini reported that everyone, including Bellarmine, was stepping gingerly around the Copernican question, Cesi sent Galileo a pamphlet that prompted its premature resolution. Its author, Paolo Antonio Foscarini of Venice, was Principal of Calabria for the Carmelite order and a professor of theology then lecturing in Rome. He was also the author of an encyclopedia. Putting together its section on cosmology, he had recognized that Copernicus might have been closer to the true system than Ptolemy. Further thought, closer acquaintance with Galileo's work, and the mistaken idea that all lynxes were Copernicans, convinced him that heliocentrism was probable and that the church might err grievously by condemning it. Like Galileo, he wrote to protect the church from itself. Unlike Galileo, he believed that a close reading of scripture could reveal the truth about the natural world. He was a true concordist.[35] Had he only known how to derive the Copernican theory from the bible! He contented himself with the lesser task of showing that every known biblical passage apparently opposed to heliocentrism could be construed either neutrally or in its favor.

Like Galileo, Foscarini exploited the principles of accommodation and of priority (demonstrable science has precedence in matters not touching faith and morals) and, as Galileo would do, ran into trouble specifying usable tests

for propositions amendable to demonstration. Cesi rated Foscarini's *Letter*—
the pamphlet had the form of a letter to the General of the Carmelites—a
timely godsend, "unless," he added, "it be harmful to increase the anger of our
adversaries, which I do not believe." He decided to bring out a Latin edition
with more passages and construals that would convince everyone. "[Coperni-
cus'] opinion will be permitted and approved so fully that anyone who wants
to hold it can do so freely, as in matters that are only physical or mathematical." Other lynxes urged caution.[36] Although Foscarini presented his views
with due humility and in the spirit of charity, the Tridentine decrees applied
more directly to his published *Letter* than to Galileo's semi-private musings.
Moreover, Foscarini was a monk subject to discipline. His letter required an
answer.

Galileo had Foscarini's pamphlet with its rich citations to Augustine and
other saintly exegetes before him when revising his wordy, worthy letter to
Christina.[37] Very likely he was familiar with the arguments, since the theology of Sarpi's circle inclined toward the Protestant elements in Augustine's
thought. So prepared, and appealing to the military spirit of the devout duchess, Galileo presented the struggle over Copernicus as a battle between good
and evil. The pigeon league and birds of similar feather, ignorant, incompetent, embittered, obstinate, self-interested, vindictive, malicious men, unable
to wield the instruments of science, that is, disciplined observation and necessary demonstrations, have drawn "an irresistible and fearful weapon," the
sword of scripture.[38] Against them stand diligent philosophers, acute mathematicians, famous exegetes, and the good, holy, fair-minded fathers in Rome
who, with the help of God and Galileo, will guard the church from reckless
condemnation of the truth.[39] Citing authorities from Augustine to Trent, and
emphasizing that Copernicus had had the support of popes and bishops,
Galileo tried to convince Christina that the church would expose itself to
mockery and its sacred documents to scorn if it opposed truths obtained by
properly ascertained natural knowledge.[40]

In astronomy especially, scripture cannot stand up to "sensory experience and natural demonstration" (SEND). The writers of the Old Testament,
which contains all the passages to which cowardly philosophasters and the-
ologasters appealed, did not see fit even to mention any planet but Venus.
The sluggish sun of Joshua and the bounding sun of David were not introduced to instruct moderns in cosmological theory. Similarly the unanimous
agreement of the fathers, who never considered carefully the constitution of

the world system, weighs nothing against SEND. To declare the Copernican version heretical on the basis of their accumulated *obiter dicta* would be a usurpation of the prerogative of popes and councils to declare heresies as well as a gross and ridiculous error.[41] All of which Galileo submitted to those to whose superior judgment on matters so far from his profession he was always ready, he said, to defer.[42]

In order to gain a fair hearing for his application of his exegetical principles, Galileo made the damaging concession to Christina that when SEND did not hold, when the reasons given for a particular physical proposition are only probable, then the literal reading of scripture should prevail.[43] On this test, Joshua should trump Copernicus; for, although SEND established essential and growing support for Copernican theory, it gave no unimpeachable proof. Nonetheless, in Galileo's reckoning, Copernicus trumped Joshua. The resolution of the contradiction lies in an equivocal distinction between propositions likely forever to elude SEND (whether stars are animate) and those likely at some time to submit to it (whether the earth moves). Galileo gave no criteria for deciding to which category a given case belonged. The distinction was less a guide to epistemologists than a sop to theologians; and, unintentionally, a demonstration that the SEND test set a standard higher than Galileo's cosmology could reach. However, he could reinterpret biblical passages to agree perfectly with Copernican theory, indulging in the sort of paradoxes he loved, and he ended his letter to Christina with his compromising explanation that Joshua lengthened his busy afternoon by stopping the spinning of the earth.[44]

Fatal interdiscipline

The unidentified censor to whom Foscarini's letter went for review found that it "excessively favors the rash opinion" of Copernicus, allows it probability though it contradicts the obvious meaning of scripture, twists the sacred texts, "and explains them contrary to the common explication of the Holy Fathers, which agrees with the more common, indeed the most common, and most true opinion of almost all astronomers."[45] The charge of rashness greatly offended Foscarini, who dug three senses of "rash" from standard theology and showed that none of them applied to him. He reemphasized that the bible spoke to common people and that the Fathers knew nothing about astronomy. He protested, exactly on target, against hitching scripture to secular ideas that might prove false. "[I]n matters pertaining to the sciences acquired by human effort, no one ought to be so addicted to a philosophical sect, or to defend some philosophical opinion with such tenacity, that he thinks that the whole of sacred scripture should henceforth be understood accordingly…Therefore we should not be so tenaciously committed to the philosophy of Aristotle or to Ptolemy's world system…Nor should the passages of scripture be interpreted according to the meaning of these philosophies only."[46]

Foscarini sent a copy of this sensible defense and of the original letter to Bellarmine, who replied in his kindly, tough-as-nails way, and signed himself "as a brother." He congratulated Foscarini and Galileo on their prudence in speaking about Copernican theory only as a hypothesis, as he thought that Copernicus had done. (We should regard this praise as Bellarmine's advice rather than his belief.) To interpret it as true would be very dangerous, "not only because it irritates all the philosophers and scholastic theologians, but also because it is damaging to the Holy Faith by making the Holy Scriptures false." The septuagenarian Bellarmine followed this opening shot with a fusillade that showed, if nothing else, that the Curia should have a mandatory retirement age. He argued as if his opponents were the Protestants against whom he had done battle in the previous century. The Council of Trent had forbidden interpretations contrary to the consensus of the Holy Fathers. "Ask yourself then how could the Church, in its prudence, suggest an interpretation of scripture which is contrary to all the Holy Fathers and to all the Greek and Latin commentators." It will not do to say that their agreement about matters irrelevant to faith and morals does not count. There are no such matters in scripture. Every word is an article of faith, if not in substance then by source. "Thus anyone who would say that Abraham did not have two sons and Jacob twelve would be just as much of a heretic as someone who would say that Christ was not born of a virgin."[47]

From this frozen fundamentalism Bellarmine thawed to admit that should SEND of the earth's motion and sun's rest ever be found, the church would have to proceed cautiously in interpreting the apparently opposed scriptural texts. However, no such demonstration existed. "And in case of doubt one should not abandon the sacred scriptures so interpreted by the Holy Fathers." Did not Solomon himself write, "The sun rises and sets, and returns to its place," Solomon, "inspired by God," and also wiser than anyone else in the human sciences? "All this wisdom he had from God."[48] Foscarini planned a reply but did not live to deliver it. Death took him unaware, in 1616 or 1617, only two-sevenths of the way through the encyclopedia that had stimulated his contribution to the Copernican cause.

Although the Holy Office conducted its business in secret, Galileo did not have to guess that it took an interest in him. On its orders, the Archbishop of Pisa asked Castelli for a correct copy of Galileo's letter and admonished him, for his own good, to give up belief in the earth's motion. "Overcome by such benevolence [Castelli wrote Galileo], I could do no other than reply that I was eager to comply with his suggestions, and that it only remained for me to accommodate my mind to the reasons that I might hope [to receive] from his profound wisdom and sound learning. He took for me but a single reason from his stock...: since all created things are made for the service of man, it clearly follows as a necessary consequence that the earth cannot move like the stars...He went on to say...that it was soon to be made known to you and His Serene Highness and to everyone that these ideas are all silly and deserve condemnation."[49]

This bulletin roused the bull. Galileo's friends directed him to keep his mouth shut. Cesi: do not try to vindicate yourself against Caccini, hold your peace about Copernicus; Bellarmine thinks that the theory is heretical since the earth's motion opposes scripture. Barberini: be more cautious, do not stray beyond physics and mathematics, allow theologians their territory, give no opening to the enemy. Anything you say will be distorted, the cardinal said; you put mountains on the moon and soon people say you put humans there too, and ask how they could have arrived on the ark.[50] Grienberger: prove the earth's motion before trying to square it with scripture.[51] Ciampoli: relax, your enemies though noisy are few. Galileo could not relax. He longed to rout the obstinate, instruct the ignorant, and slaughter the slanderous. "[B]ut my mouth is shut and I am ordered not to go into the scriptures." It was unreasonable, intolerable! "This amounts to saying that Copernicus' book, accepted by the church, contains heresies and may be preached against by anyone who pleases."[52] Galileo suffered this frustration until November 1615, when, perhaps having wind of the Florentine inquisitors' interviews of Ximenes and Attavanti, he gained Cosimo's permission to go to Rome to "defend himself against the accusations of his rivals." Cosimo behaved throughout the affair more as an indulgent and supportive friend than as a prince concerned about his standing. He directed his ambassador in Rome to accommodate Galileo, a secretary, a valet, and a small mule. His secretary, Curzio Picchena, wrote the superintendent of the Medici Villa, Annibale Primi, that Galileo was to have everything he needed for his mission and wanted for his wellbeing.[53]

The lesser and the greater mule arrived in Rome on 10 December 1615 to the annoyance of Ambassador Guicciardini. He knew that Galileo meant trouble. Galileo had stayed with him for a few days in 1611 while celebrating with the Jesuits and (Guicciardini recalled) alienating consultors and cardinals of the Holy Office. Bellarmine had told him at the time that if Galileo stayed much longer the Inquisition would have had to take a position about his opinions. To Guicciardini's good advice that he lower his profile and expectations, Galileo had paid no heed then. Now the stakes were higher. "I do not know whether he has changed his mind or his attitude, but I do know that some Dominicans prominent in the Holy Office and others think ill of him; and this is not a place to come to dispute about the moon or, at this time, to uphold or bring new ideas."[54] Galileo had not changed. At 52 he was more obdurate than ever, impatient, and dismissive of advice; "he is fired up over his opinions, and has a passion about them that he lacks the strength and prudence to curb." Rome would not be taken by so mad an Orlando. "The Prince here [still Paul V] abhors literature and these clever ideas, and cannot listen to novelties or subtleties." Almost everyone, the sympathetic and informed as well as the ruthless and ignorant, followed the pope's lead.[55]

Galileo had two goals in mind. One was to exculpate himself from charges of impiety and recklessness; the other, to demonstrate that his opinions were not erroneous. Towards exculpation, he brought with him the potent stain remover of Cosimo's testimonial to his character. "I know him well," Cosimo wrote the cardinal nephew Scipione Borghese, "he is a good man and very observant and zealous in religion." No one could doubt the sincerity of a courtier so close to a pious prince. Cosimo appealed to Cardinal del Monte to introduce Galileo to "intelligent and discreet people" to whom he could demonstrate his "correct and pious intentions."[56] That required many visits to cardinals and other important individuals, and the increased expression of religious sentiment noticed by the Venetian ambassador. Galileo bore the fatigue with the help of the "loving care" of Superintendent Primi and the prospect of "securing and increasing my reputation."[57] Early in the new year Galileo could report that he had penetrated the snares set for him by his persecutors. A month later he could announce that he had finished his first task:

the people who counted now recognized his "frankness and integrity" and the "diabolical malignity and iniquitous intention of my persecutors."[58]

There remained the bigger task, shouldered "as a zealous Christian and Catholic," the task of saving the church from the serious error into which prelates deceived by Galileo's enemies might lead it. He promised himself a threefold increase in reputation when he succeeded.[59] His method was the Ancient Mariner's. He talked and kept talking, brilliantly; he took on all comers, like a chess master playing a dozen opponents simultaneously, or like a cat among pigeons. He played with and humiliated his victims by sharpening their arguments apparently to invincibility before annihilating them. God Himself assisted in these parlor tricks. "I am consoled in seeing how much the blessed Lord enjoys the integrity and purity of my mind." "I trust in God that, as he gave me the grace to uncover the frauds of [my enemies], he will give me the means to foil them, and prevent any decision that would bring scandal to Holy Church."[60] An eyewitness to these performances, Galileo's old friend Querenghi, probably spoke for many in admiring Galileo's manner while rejecting his message. "Outside the world of these mental fireworks, we can stand firmly in our place without flying away with the earth like ants on a rising balloon."[61]

Pressed by both friends and adversaries for a demonstration that would make the Copernican world plausible to them, Galileo wrote out early in January 1616 the kinematical theory of the tides adumbrated in Sarpi's notebooks for 1595. When in his extremity Galileo had to deliver a proof of the Copernican theory, he could proffer one that had passed the scrutiny of Fra Paolo.[62] Galileo's "Discourse on the tides" has the form of a letter to Alessandro Orsini, who then was about as old as the theory and a cardinal of exactly 18 days standing. Apparently Galileo could not enlist any of the senior cardinals he cultivated to receive and disseminate this *demonstratio potissima*. Untested but enthusiastic, Orsini undertook to explain to the pope that the tides are to the seas what sloshing is to water in a barge, if the earth-barge sails on a Copernican itinerary.[63]

If the inquisitors saw Galileo's sloshbucket tidal theory, they were not taken in by it. It agreed neither with received philosophy nor with the physics Galileo had invented to treat sunspots (clouds rotating with the sun) and earth spin (the atmosphere turning with the earth). The theory could not account for observations of tides in particular places because of the complexity of the secondary causes. In fact, it misfired altogether, as it seemed to work as well without as with the annual motion. On Galileo's reckoning, the spin alone introduced a disparity of motion between any pair of diametrically opposite points on the earth's equator. The annual revolution merely picked out one particular diameter—that along the line from the sun to the earth—against which to reckon the "acceleration." It was, moreover, a very peculiar acceleration as it required a sea bottom to suffer the "curious effect of having its parts moving at different speeds at different hours of the day."[64] But in his emergency, Galileo clung to his and/or Sarpi's theory and, as was his wont when challenged, never gave it up.

Galileo ended his discourse on the tides with a pungent and even bitter reference to its purpose. What would happen if Copernicus' "hypothesis, previously corroborated only by philosophical and astronomical reasons and observations, were declared fallacious and erroneous by virtue of more eminent knowledge"? Obviously, the discourse on tides would be declared null and void. There would then be three paths to follow. The eminent authorities could demonstrate where the discourse failed philosophically and astronomically; or they could say that the matters treated were among those God wanted to keep secret; "or, finally and more advisedly, remove [me] from these and other fruitless inquiries."[65] The stakes were high. Young Orsini went on his mission. Paul sent for Bellarmine. They decided that Copernicanism was erroneous and perhaps heretical.[66]

The eleven theologians empanelled by the Holy Office to evaluate Copernican theory returned their unanimous verdict on 24 February 1616 after five days of deliberation. They judged the assertion that "the sun is the center of the world and completely devoid of local motion" to be "formally heretical since it explicitly contradicts in many places the sense of Holy Scripture, according to the literal meaning of the words and according to the common interpretation and understanding of the Holy Fathers and the doctors of theology." The proposition of the earth's motion qualified for the lesser censure of "at least erroneous in faith." Moreover, the propositions failed, this time equally, to conform to the world system to which the theologians had yoked their doctrine. "[They are] false and absurd in philosophy."[67] There was no room in this formidable interdiscipline for a salutary opposition between science and religion. The consultors had no reason to ponder the accommodationist argument: in their view, the findings of philosophy concurred perfectly with the results of traditional exegesis.[68]

   The cardinals of the Holy Office accepted the advice of their consultors. The pope thereupon ordered Bellarmine to warn Galileo to abandon his opinions; if he should not accede to this friendly warning, Bellarmine was to issue a formal "precept" or injunction against him "to abstain completely from teaching or defending this doctrine and opinion or from discussing it." If he did not acquiesce to the injunction, he would go to jail. The following day, 26 February 1616, Galileo appeared before Bellarmine and Seghizzi. To the confusion of subsequent history, the unsigned minute describing the interview does not agree with the papal order. Bellarmine duly warned Galileo that the "abovementioned opinion" conflicted with scripture and advised him to abandon it. Then, before Galileo could express his voluntary acquiescence, Seghizzi proceeded, *succesive et incontinenti*, to the second step and, before Bellarmine and other witnesses, "ordered and enjoined the said Galileo…to abandon completely…the opinion that the sun stands still at the center of the world and the earth moves, and henceforth not to hold, teach, or defend it in any way whatever, either orally or in writing." Galileo accepted the injunction and agreed to obey it.[69] He had no attractive alternative.

   It was easier to condemn Copernican work than to acquaint the faithful with the news. The Index of Prohibited Books came out irregularly, at long intervals; the latest edition available for consultation in 1616 was Clement VIII's of 1596. The right to publish interim condemnations belonged to the Master of the Sacred Palace (as chief censor in Rome) and to the Secretary of the Congregation of the Index. The first to act in 1616 was the master, whose order, printed but never published, concerns only three items: Foscarini's *Letter*, banished outright for its attempt to show that Copernican ideas "agree with the truth and are not contrary to scripture"; Copernicus' *De revolutionibus* and a commentary on Job that interpreted the verse, "[H]e shaketh the earth out of her place, and the pillars thereof tremble," in a Copernican sense, suspended "until corrected."[70] The order mentions neither heresy nor Copernican works in general. The Congregation of the Index persuaded the pope to suppress the master's version in favor of a sterner prohibition. Its decree disposes of the three items on the master's list as he did; bans all books teaching Copernican theory; and refers to the offending doctrine as "the false Pythagorean doctrine, altogether contrary to the Holy Scripture."[71] The Congregation of the Index announced this ruling—the only official document concerning the decision of 1615/16 published at the time—on 5 March 1616.[72]

Cardinal Bonifazio Caetani's assistant Francesco Ingoli, who had disputed publicly with Galileo in 1615/16 and earned appointment as consultor to the Holy Office for his performance, received the assignment to purge Copernicus. He recommended leaving the core of the work intact despite its reliance on the earth's motion because of its utility for calendrics, and cleansing it for Christians by crossing out passages where Copernicus spoke of his hypothesis as a "theory" or otherwise asserted the truth of his system. The Congregation of the Index received Ingoli's recommendation in 1618 and, after the mathematicians of the Roman College had signified their approval, published them in 1620. They did not become generally available before the publication of the second edition of Clement's *Index* in 1624.[73] It appears that only Italian owners, and not all of them, marked their copies as ordered. The entire business shows an enviable faith in the effectiveness of the verbal distinctions beloved in the schools and in confessional disputes. Anyone could read the corrected condemned doctrine and anyone but Galileo could discuss it, in principle, provided only that he called it a hypothesis.[74]

Copernicus' *De revolutionibus* may have been the first book devoted almost 100 percent to what we would call science to be condemned, even temporarily, by the Congregation of the Index. Earlier, natural science had been hit sometimes as parts of books banned as magic, divination (necromancy, geomancy, fatalistic astrology), or heresy (mortality of the soul), or in books judged obscene (gynecology), or in writings by proscribed authors (Protestants), but only as collateral damage.[75] Since the condemnation of Copernicus would not only break new ground but also ban a book permitted for 70 years, Galileo thought he had reason to believe that his reason would prevail. He misjudged the fleeting conjunction of personality and doctrine in the Rome of Paul V. He also neglected the lesson of his own experience. Of the twelve Catholic authors of books touching natural knowledge who were questioned or detained by the Inquisition before 1616, Galileo knew or just missed knowing at least half: Borro, Bruno, Campanella, Cremonini, della Porta, Giulio Libri.[76]

Ambassador Guicciardini informed Cosimo about the failure of Galileo's mission. "Everyone feared that his coming here would be prejudicial and dangerous."[77] And everyone perceived the failure after the publication of the decree of the Index—everyone except Galileo, who represented it as a justification of his efforts to serve the church. Caccini and his followers had labored to have Copernican theory declared heretical; Holy Church decided on the lesser fault that it disagreed with Holy Writ; Galileo had won. Only Foscarini was banned. Copernicus' book would lose a few phrases where it claimed compatibility with Scripture. "I am not mentioned, nor would I have gotten involved in [the business] if…my enemies had not dragged me into it." Galileo's behavior throughout had been exemplary; "a saint would not have handled [the affair] with greater reverence or with greater zeal toward Holy Church." Saints spend time wrestling with the devil. On this score Galileo too could claim sainthood. "My enemies…have not refrained from any machination, calumny, and devilish suggestion."[78]

Five days later Galileo wrote home again. He had had a long audience with the pope. "Timid with equals, ungrateful to benefactors, supercilious with inferiors, and passionately fond of money," Paul V had orchestrated the inquiry into Galileo's beliefs, ordered Galileo alone among mathematicians and philosophers to receive instruction directly from Bellarmine, and persecuted Galileo's friends Cremonini and Sarpi.[79] Galileo applied to his persecutor for protection. Paul replied that neither he nor the Holy Office would listen to slanders and that, as long as he lived, Galileo could feel safe.[80] The unheeded slanders continued and rumors that Galileo had been disciplined or made to recant started up. Castelli and Sagredo wrote in alarm. In response, Galileo did something prudent. He applied to Bellarmine for a certificate of good standing. Bellarmine generously wrote that Galileo had not abjured or received penance, but was informed that the Copernican theory "is contrary to Holy Scripture and therefore cannot be defended or held." Perhaps this certificate, which has become famous, was as much a favor to Cosimo (Bellarmine was a Tuscan) as to Galileo.[81]

The decisions of 1616, though not enlightened, were not immoderate. The misfortune for the church was that the pope, Bellarmine, and doubtless others at the Holy Office believed that they had to act. From their Tridentine perspective, Galileo's attempt to set up an independent school of cosmology and biblical criticism looked like the budding of a new head of the Protestant hydra; and (to stay in the animal kingdom), once the pigeon league had pecked out the danger, the thought police had to look into it. A coincidence decided the timing of the showdown: the simultaneous activity in Rome of Caccini, unburdening himself to the Inquisition, and Foscarini, trying to prevent it from heeding him.[82] The consultants to whom the Holy Office submitted the question of the tenability of Copernican theory had little leeway in their considerations. We can appreciate as inevitable their qualification of a stationary sun and moving earth as absurd in philosophy. That was precisely what Galileo thought. But whereas the consultants foisted the absurdity on Copernican theory, Galileo placed it on Aristotelian philosophy. Had the expert consultants and their inquisitorial clients only understood the direction of the argument, they could have achieved the uncoupling of theology from philosophy desired as heartily by the Borros and Cremoninis as by the Galileists.

The promiscuous use of the charge of heresy by the Roman hierarchy, on which Sarpi liked to dwell, confused the status of Copernicanism as decided in 1616. The consultors to the Holy Office went too far in qualifying as "formally heretical" the proposition that the sun does not move unless they meant no more by it than "contrary to the literal meaning of scripture." They had no power to declare any belief heretical. Neither did the cardinals of the Holy Office, who ought to have struck out the reference to heresy in their consultors' report before adopting it. But they were no more indulgent than they were alert. Three of them—Bellarmine, Sfrondati, and Ferdinando Taverna—had participated in the proceedings that had condemned Bruno. Their intention or oversight was compounded by the presence of the pope at their decisive meetings. Since popes might have the power to declare heresies, Paul's involvement and approval have supported the opinion that in 1616 the Roman Catholic Church declared Copernican cosmology a heresy.[83]

This does not appear to be the way in which either the Master of the Sacred Palace or the Congregation of the Index interpreted the situation. Neither referred to heresy in describing the offense of Copernican writings. One member of the Congregation, Maffeo Barberini, may have opposed imposing any sanctions at all. "That was never our intention," he said later in his papal we, "and if it had been up to us, that decree [of the Index] would not have been made." Barberini looked at the Copernican system from a point of view that made it ineligible to be a heresy.[84] He owed this insight or its confirmation to "his faithful Bellarmine," his theological advisor Agostino Oreggi. Like Bellarmine, Oreggi believed that when the bible spoke about the natural world its word was law. Where it did not, as in the disposition of the light that shone before the creation of the sun, we are free to speculate, but no more. God did not consult man about creation. "Where wast thou [Job] when I laid the foundation of the earth? declare if thou hast understanding, who hath laid the measures thereof, if thou knowest. Whereupon are the foundations thereof fastened?"[85] Oreggi–Barberini understood that we cannot decipher God's plan any more than Job could. In His omnipotence, He could make and remake the world in a thousand ways we cannot even imagine. We can never know for certain how he did or does it.

Barberini, who was a lawyer but no theologian, adopted this excessive voluntarism (so called for stressing God's will, *voluntas*) as a sort of legal restraint on philosophizing. The doctrine was not fresh, as it goes back at least to the time of Adam, who had it from the Angel Raphael. Shut up the book of nature, Raphael had advised Adam, after you have reckoned the months and years and seasons, for God will not reward further researches.

> From man or angel the great Architect
> Did wisely to conceal, and not divulge
> His secrets to be scanned by those who ought
> Rather admire: or if they list to try
> Conjecture, he his fabric of the heav'ns
> Hath left to their disputes, perhaps to move
> His laughter at their quaint opinions wide.[86]

Conjectures that provoke divine chortles do not make heresies. Barberini's voluntarism did not allow Copernicanism to be true or false. Both Galileo and the consultors to the Holy Office erred in taking it to be more than a mere hypothesis.

Barberini knew that Galileo held dangerous views on this subject. In *On Floating Bodies*, he would have found that Galileo sought the "true, intrinsic, and entire cause" of natural phenomena.[87] If he had also read his friend's *Letters on sunspots*, he would have learned that not even Omnipotence could have made the spots appear as they do by a mechanism different from Galileo's, all others being "manifest impossibilities and contradictions."[88] In Christian charity, Barberini undertook to inform Galileo about the hypothetical character of all claims to natural knowledge. Their conversation, dating, probably, from 1615 or 1616, and reported over a decade later by Oreggi, went something like this:

> *Bar.* Even if the Copernican theory fit all relevant phenomena perfectly, would you deny that God might have constructed the universe on principles entirely different from those Copernicus adopted? I do not mean whether another universe might

be possible, but whether, all appearances remaining the same, ours might be produced by many different means. Rephrased in Aristotelian terms, can the "accidents" of bodies—their colors, shapes, motions—remain the same if their essences change?[89]

Gal.   It is not my business to inquire into all the ways that God might have done something, but the way that, in His wisdom, he chose to do it.

Barb. How do you know that in His wisdom he does not now and again transubstantiate earth from platform to planet, and the world from Tychonic to Copernican, or vice versa, without our knowing it; or annihilate the whole works while continuing their appearances; or expand or contract the universe and everything in it by the same proportion, without changing anything as perceived by our senses?[90]

"Having heard all this [so Oreggi concluded his anecdote] the very learned mathematician fell silent. For which he earned praise for [his] intelligence and his behavior."[91] Silence is not acquiescence, however, but "the best composition and temperature [so advises Francis Bacon]...if there be no remedy."[92] Galileo could not concede that the grail he sought lay beyond the reach of humankind.

From later events it appears that Barberini did not know of the special warning given Galileo by Bellarmine. The probable reason for the unusual admonition, which would not have been necessary if everyone understood that belief in a stationary sun was a heresy, was that Galileo had tried to set up a critical philosophico-theology in place of the uncritical interdisciplinary Thomism he thought he had destroyed. If that is a correct reading, Galileo did well to escape relatively unscathed from the predicament in which he had put himself. As he said, he was not mentioned by name. Had he published his *Letter to Christina*, it would have been condemned along with Foscarini's Copernican apology. Galileo's devoted enemies in Florence, especially Archbishop Marzi Medici, thought that he had got by too lightly, and commissioned new fire-breathers to preach against him.[93] To be sure, the admonition against teaching Copernicus, if taken literally, would have been a bitter pill. Galileo sought ways to avoid swallowing it while Bellarmine and Paul prepared for a better life. They both entered it in 1621.

6.2 POETICAL INTERLUDE

Guicciardini was delighted to see his troublesome guest leave for Florence. Galileo would not listen to him or to Cardinal del Monte or other friends, but remained belligerent. "He [was] in a mood to castrate the friars." Guicciardini had to live another two months with his firebrand before he could persuade Cosimo to order Galileo home. No good could come of his belligerence, Guicciardini predicted, correctly; "he can only lose if he fights the *fratri*."[94] There was a second reason that the ambassador wanted to rid himself of the father of modern science. Galileo and Superintendent Primi sponsored "strange and scandalous doings" when they got together, "a riotous life" for which the ambassador had to pay.[95] No doubt the bibulous evenings included the literary types among whom Galileo found many of his patrons and supporters.

Literary lynxes

In 1620 the literary world had the thrill of reading the *Poemata* of Maffeo Barberini, published in Paris by the French polymath litterateur Nicolas Claude Fabri de Peiresc. The poems Peiresc printed included verses on Mary Magdalene, Saint Louis IX, and Galileo. The lines regarding the last of this odd trio appear as illustrations in a poem on the variety of human life and its deceptions.[96] People go off in many different directions, the cardinal sang, looking here and there,

> Some at the Scorpion's heart, or into the Dog's face,
> Others stare rather at Jupiter's Clan,
> And his father Saturn's, discovered,
> Learned Galileo, with your glass.

Yet, things are seldom what they seem:

> Not always does a thing that blazes
> Externally also shine within: in the sun we see
> Black spots (who would believe it?) uncovered
> By your art, O Galileo.[97]

Placing the spots within the sun was to go a step beyond Galileo.

Barberini sent Galileo a copy of his *Poemata* with a warm inscription warmly reciprocated. The two Florentines could acknowledge and appreciate one another's merits as their unfettered ambitions had not yet brought them into competition. Like Galileo's literary criticism, Barberini's poems were products of leisure hours.[98] And like Galileo's poems, Barberini's do not please modern taste. "Barberini's *Poemata*..., [are] desolately empty of any poetic inspiration. They are a string of commonplaces of insupportable banality."[99] Good or bad—and many discriminating contemporaries discovered nuggets among them—they were praised mercilessly after Barberini became pope. As obscure early works of winners of the Nobel Prize for Literature shoot up in value retrospectively, so Barberini's poems rose into reprints immediately upon his elevation and saw a dozen or more revised and enlarged editions during his lifetime, some set to music. The resultant accolades—"a new David...a new Apollo"—helped puff up a vanity already greater than Galileo's.[100]

Among Barberini's discriminating contemporary admirers were several lynxes. Although not of Cesi's band, the future pope was close enough to them to be regarded, with a little poetic license, as a patron and fellow traveler. Around 1618 he developed an important connection with two young literary lynxes, Strozzi's protégé Ciampoli and Cesi's cousin Virginio Cesarini, both of them then on the threshold of spectacular careers at the Vatican and devoted service to Galileo (Plates 7 and 11).[101] After a year spent in Padua at Galileo's suggestion, Ciampoli had freed himself from the *libertas patavina* to spend time with Barberini, then, in 1611, legate in Bologna, and with the bibliophile Cardinal Federico Borromeo, still Archbishop of Milan. They represented different but complementary approaches to Catholic renewal, Borromeo the old-testamentary and disciplinary, and Barberini the uplifting and poetical.[102] This, anyway, was the conceit invented by Ciampoli, who presaged a glorious future for the church under Moses' staff wielded by Great Federico and David's lyre strummed by Great Maffeo. This future Ciampoli opened to Borromeo (whose chances of becoming pope he evidently estimated as higher than Barberini's) in a *Poemetto sacro* dispatched in 1616. "To live a king, 'tis but one way / For to this truth the soul doth swear / 'A kingdom's thrall, beauty decays / Virtue alone dominates fate'."[103]

While Ciampoli sought to follow both Moses of Milan and David of Bologna, Galileo recommended him to the Lynx of Rome. Ciampoli quickly

jumped into an orbit around Cesi, in whose cousin Cesarini he found someone to whom he could devote himself entirely. Cesarini, five years Ciampoli's junior, had everything this world has to offer but good health. The
son of a duke, closely related to the Orsini and the Farnese, rich, brilliant,
sensitive, he had improved his natural gifts through study with the Jesuits.
At the Roman College he impressed not only Barberini but also Bellarmine
as a prodigy in erudition. The great controversialist saw in the young man
"another Pico"—at the age of 23 the first Pico (Pico della Mirandola) had proposed to defend 900 theses on all subjects against all comers—and tried to
enlist him in the pious work of destroying the teachings of that pest from
Padua Cremonini. When he met Ciampoli, Cesarini was deeply committed
to Aristotelian philosophy as mediated by the Jesuits and to a moralizing
poetics in the tone of Ciampoli's *Poemetto*.[104]

Although Cesarini had been intrigued by the *Starry message* when his
teachers at the Jesuit college in Parma had enthused over it, he did not take an
interest in astronomy until Galileo came to Rome. The brilliant disputations
for Copernicus and against all opponents that Galileo conducted there had
the same effect on Cesarini as the mathematical lessons Galileo gave Cosimo
had had on Ciampoli, and with the same result. The learned youth underwent a sort of baptism (washing out his mind, he said, as mineral waters
purged his body) and opened the pages of Euclid. Soon, despite a lingering
adherence to Jesuit philosophy, he became lynx material, or so the Roman
lincei, including Galileo, decided at a meeting in the spring of 1616. In 1618
Cesarini and Ciampoli, who shared the homoerotic orientation of the founding *lincei*, were admitted together. Cesarini quickly reached the academic van
as vicar of the Roman *lincei* during Cesi's protracted residence at his family
seat in Acquasparta between 1618 and 1621.[105]

Ciampoli and Cesarini enrolled in Galileo's campaign against received
philosophy at first in so far as they saw themselves engaged in a similar
fight against authority in poetry. Although Galileo liked their fighting spirit,
he could not have approved their targets, which included the classic Italian poetry he loved. They wished to write in Pindaric odes, at once pious
and lyrical, moralizing and inventive. In this they claimed Barberini as their
cicerone, "the great Maffeo, who joined together the strings of David with
the Argive harp."[106] In March 1619 Cesarini sent Barberini a recipe for an
extract of cedar, which, when mixed with a little limoncello, was very useful
to poets. "It is a most potent diuretic." Perhaps it was responsible for the

effusion with which Cesarini acknowledged one of Barberini's inspiring productions, worthy, he wrote, of the happy times before Cesar. "Although I'm sick in bed, I do not want or dream of any other brook to slake my thirst than the inexhaustible fountain of your genius."[107] Ciampoli and Cesarini's heroic-moralistic poetry in turn inspired Barberini.[108] Their output has won only a modest place in literature. A standard authority dismisses Ciampoli's work as conventional blather about sacred, moral, and funereal subjects, or about the vacuity of the human condition, "not deserving the name of poetry." Cesarini's moralistic productions receive higher marks owing to their occasional expressions of honest melancholy inspired by his debilitating illness. The moral verses of their appointed leader Barberini rank as more skillful than heartfelt, "dignified but not lyrical."[109]

Though writing uplifting verses in a Grecian style for his friends and patrons remained Ciampoli's ambition, he needed greater security than poetry and Strozzi's continuing subvention could bring. His active networking brought him into the entourage of Alessandro Ludovisi, who succeeded Paul V in 1621. The new Pope, Gregory XV, was the first incumbent of St. Peter's chair trained by the Jesuits. Among his accomplishments during his short time in office were the canonization of the Jesuit saints Ignatius Loyola and Francis Xavier, the establishment of the Congregation for the Propagation of the Faith, and the appointment of Giovanni Ciampoli of the fluent pen as his Secretary for Latin correspondence. Ciampoli introduced Cesarini into the pope's household as a gentleman in waiting.[110]

Meanwhile the Roman lynxes had commissioned a piece of literature that made Cesarini famous. Together with Cesi, they pushed Galileo to answer a counterattack on him by a Jesuit he had provoked under circumstances soon to be related. The three lynxes decided that the answer, entitled the *Assayer* (*Saggiatore*), would appear in the form of a letter to Cesarini; and Galileo wrote him into the work as if he were collaborating in judging the Jesuit, who, to add to the fun, was a distinguished member of the college that had praised Cesarini as a new Pico. Without the editorial help and frequent goading of the Roman lynxes, Galileo might not have written the *Saggiatore*, in which he embedded, among much unfair criticism, some beautiful expressions of his method of philosophizing.[111]

The production of the *Assayer* and subsequent interactions among Ciampoli, Barberini, and Galileo, illustrate with particular clarity the importance, indeed, the necessity for innovative natural science to link itself to literature in Italy during the early modern period. As shown by Ciampoli's career, ability to write poetry gave entrée to academies and powerful patrons, and might end with a position of influence and a command of patronage within and outside the omnipresent church. As we know, Galileo first made contact with patrons like Strozzi through Florentine literary academies.[112] And, as we also know, among the first to extol his discoveries were poets. This mode of promotion did not cease with the decree of 1616. A poet could sing things a philosopher could not say. Two examples must suffice.

The first comes from the *Adone* of the bold and verbose Neapolitan Giambattista Marino, at home equally in courts and prisons, who stuffed his poems with sensual images, clattering metaphors, and the stories, mythologies, tropes, clichés, and paradoxes of all the humanist poets from Petrarch to Tasso. The sensuality and subject matter, which made Marino anathema to Barberini's circle, might have recommended him to Galileo except that Marino liked the sort of word play (32 consecutive oxymorons in one case), foreplay (Venus' endless chase of Adonis), and neologisms that Galileo had lambasted in his comments on Tasso.[113]

Marino's Adonis takes the opportunity to examine the dark regions of the moon while rising in Dante's manner toward the stars. He considers several explanations of the mottling, including delle Colombe's polished crystal with mountain's inside, before his better informed traveling companion tells him that if his sight were stronger he would see seas and rivers, cities, hills, and plains. That will come to pass. "You'll be able to shorten the longest spaces / with a little tube and two crystalline lenses."[114] The engineer of this miracle is a new and better Columbus:

> Exposing the bosom of the deep sea
> Not without violence and anger
> The Ligurian Argonaut discovered in this lower world
> A new clime and a new land.
> You [Galileo], second Typhus not of sea but of heaven,
> Spying out what turns and what holds
> Without any risk, for all peoples everywhere,
> Will discover new truths and new things.[115]

The risk-free environment for new truths may exist only in the poetical imagination.

The future tense ("Galileo will discover") belongs to a gimmick, often employed by Ariosto and Tasso, whereby a past seer predicts the completed future. It allowed the Venetian poet Giulio Strozzi, who had seen Galileo in action in Rome in 1611, to move from discovery of the Medici stars to revelation of the Copernican system. Strozzi's seer is the wizard Merlin, who invented the telescope and discovered Jupiter's moons, Apollo's spots, and Saturn's companions. With these clues, Merlin worked out that the planets revolve around the sun. Being wise as well as clever, he kept his discoveries from people too ignorant to appreciate them. That was everyone. Eventually, he prophesied, the time would be ripe for rediscovery and revelation. "The time will come...when the finest mind of Tuscany will renew my famous glasses."[116] The arts and sciences would then flourish with a race of Merlins inspired by Galileo.

Literary lion

Poetry is a medicine on its own as well as a sugar coat for unpleasant truths. "Therefore [Sagredo instructed Galileo] continue to read Berni and Ruzzante, and set aside Aristotle and Archimedes for a while." Galileo never lacked poems to read. Poets good and bad submitted their work for his judgment. Not infrequently he himself was their subject, "Galileo, who first opened the closed gate beyond the earth."[117] Buonarroti and several promising younger poets met in an informal poetical seminar in Galileo's house in Florence.[118] Occasionally he attended meetings of the Accademia Fiorentina, of which he was elected consul in January of 1620.[119] Rhetoric rather than geometry would be the weapon of his missionary struggle.

Before 1616, the books that Galileo published under his own name were reports of new phenomena (*Sidereus nuncius*, *Letter on sunspots*) or developments of old theory (*On floating bodies*). Although the Italian works sometimes stooped to sarcasm and insult, their purpose was not to offend but to present discoveries and persuade readers of their reliability and of Galileo's right to claim them. The three books Galileo wrote between 1616 and 1632 deliver little new in science, but many innovations in the art of persuasion. The first of these books, *Discorso delle comete* (*Discourse on comets*), was published in 1619. Tellingly, a literary man, Mario Guiducci, unveiled it, in a lecture to the Accademia Fiorentina. It has two off-stage participants, Galileo and Orazio Grassi, the Jesuit mathematician under attack. The second, the

Assayer (1623), again has two off-stage personae, "Lothario Sarsi," the pseudonym under which Grassi replied to "Guiducci," and Cesarini, to whom
Galileo turns frequently for a laugh at "Sarsi." The third book, Galileo's rhetorical masterpiece or, to use Campanella's words, "philosophical comedy,"
the *Dialogue on the two chief world systems* (1632), brings the three interlocutors
of the *Assayer* on stage, with "Salviati" as Guiducci, "Sagredo" as Cesarini, and
"Simplicio" as Sarsi.[120] Galileo remains in the wings, to manipulate the actors
in their oppositions and digressions (comedies within a comedy, as Salviati
put it) and to receive their praises.

We have a commedia dell'arte, a series of duels among masked men. Sarsi
(Grassi) attacks Guiducci (Galileo); Galileo as assayer defends Guiducci, that
is, himself, against Sarsi; Salviati sometimes wears the mask of Copernicus
(he says so himself) while also masquerading as Galileo; Sagredo, while
pretending to be a neutral commentator, is another avatar of Galileo. Only
Simplicio is what he appears to be, a gentle idiot, stuffed with the learning
of the schools, unable to think for himself, as immovable as the Aristotelian
earth, in short, a comic caricature of all the philosophers who had had the
effrontery to dismiss the revelations of Galileo. Participants in a masquerade
can speak and act more extravagantly than they do in ordinary life. Galileo
took full advantage of the license, asserting, via Guiducci or Salviati, many
doubtful and outlandish things, and not a few white lies. His many exaggerations of the instances and accuracy of his measurements fit well here. In
his philosophical comedy, bodies of different sorts all fell at the same speed,
"within a hair's breadth," in hundreds and even thousands of trials; pendulums of different amplitudes swung in unison, "within a heart's beat," after
hundreds or even thousands of oscillations; an inked ball described a perfect
parabola when rolled on a sloped plane; and so on.[121]

And what shall we make of the extravagance of the Pisan Drop, the creation point of the planets, said to answer perfectly to calculations? Or of the
bizzarria (oddity), to which we will return, that a freely falling body does not
accelerate but only changes its direction of motion? Salviati takes the trouble to demonstrate this *bizzarria* and Sagredo accepts it as a marvel. Modern
readers also wonder at it. Was Galileo, the master of experiment, the facile
geometer, the slayer of Aristotle, dishonest, as Arthur Koestler would have
him, or just a charlatan, as Paul Feyerabend preferred? Neither. Galileo as
stage manager is the creator of ingenious fancies, mathematical caprices, an
epic poem, a set of stories. Sagredo asks Salviati to describe the curve of a

freely falling body in space; the masked Galileo replies with the clever non-sensical *bizzarria*; nature too is part of the masquerade.[122] Galileo's comedic talent reached full strength in the barbs, jokes, word plays, paradoxes, irony, satire, and gross caricature of the *Assayer*.[123] As if to signal its epistemologi-cal level, Galileo inserted into it more quotations from Ariosto than he did in any other of his books.[124]

The poetical lynxes grew more important for Galileo as some of the more scientific ones distanced themselves from him after 1616. Johann Schreck, astronomer and botanist, to Johann Faber, head of lincean botany: "I'm amazed that Galileo is pushing the motion of the earth so hard, as if it were not enough to say that it is an hypothesis useful in astronomy, whatever its truth. To my great inconvenience the edict will prevent me from using it to calculate eclipses for the Chinese."[125] A one-time lynx, Schreck had to resign his membership when he joined the Jesuits and their mission to China. Luca Valerio also resigned. As an employee of the Vatican compromised by his intercession with his cardinal patron on Galileo's behalf and behest, Valerio thought it best to cancel his connection with a group identified with his old friend. Like Kepler, Valerio blamed Galileo for bringing on the condemna-tion of 1616.[126]

That left more room for poets.

6.3 ILL OMENS

The edict of 1616 hit the Jesuits as well as the Galileists. The enthusiasm of their mathematicians for telescopic astronomy ran against the tendency of their general, Claudio Acquaviva, to see the fight over novelties in the heav-ens as a second front in the struggle against heretics. It is truly written that generals always fight the last war. In 1611 Acquaviva had demanded that his troops, particularly those in the combat zone of the classroom, stick to Aris-totelian philosophy as corrected by St Thomas. Disappointed over the lax observation of this command by the rapidly increasing and diversifying body of Jesuits, which grew from 5,000 to 13,000 during his reign, he renewed it in December 1613 in even stronger terms: "whoever teaches views contrary to St Thomas or who introduces new things into philosophy on his own ini-tiative or from obscure authors is ordered to retract them immediately." As appears from the case of Grienberger, this order made it very difficult for the

Jesuits of the Roman College to support Galileo's views on floating bodies, let alone his Copernican religion.[127] The decree of the Congregation of the Index further restricted the adventures of Jesuit astronomers by discouraging positive allusions to the Copernican theory or its defenders. Informed professors who felt a responsibility to their subject found themselves in a painful conflict. Between 1616 and 1620, when they issued their first *Sphaera* after Clavius' last, the Society's mathematicians sought ways to remain true both to their general and their pedagogy.[128]

Giuseppe Biancani, the author of the new *Sphaera mundi* (1620), tied his book to Clavius' farewell injunction: "in view of what Galileo diligently and accurately set forth in the *Nuncius sidereus*, astronomers should see how the celestial orbs are to be constituted so that these phenomena can be accounted for." Biancani's solution, worked out partly in correspondence with Grienberger, was to adopt Tycho's system and justify it and other statements about the constitution of the world by appeal to "the best astronomers" or "the common opinion of astronomers." The closer to physics, the more diffident the statements. Are the stars carried by a rigid firmament, like rivets in steel? Probably. Do the planets move through the heavens like fish in the sea or birds in the air? "*Incompertum mihi est*, I do not know."[129] Under this cover, Biancani delivered a very good textbook, filled with information indifferent to the choice of world systems (calendars, eclipses) and descriptions of the new phenomena—the mountainous moon, spotted sun, horned Venus, companions of Jupiter, bumps of Saturn. The sun sits in the middle of the other planets, except for the moon, as if their Lord, "according to the common opinion of astronomers." Copernicus, Tycho, and Kepler all teach that the planets circle the sun as the moon does the earth.[130] The statement was strictly true and patently false, a perfect, even a Jesuitical equivocation, since the earth is a planet to Copernicus and a unique something else to Tycho.

Biancani ended his celestial survey with comets. Again he follows Tycho in placing them in solar orbit above the moon. The arrangement agreed with observations by Jesuit astronomers in many parts of Germany and Italy of the brilliant comet first seen in late November 1618. Analysis carried out at the Roman College showed that it sailed along a great circle in the sky, around though not centered on the sun, at a distance to be determined. There was this difficulty, however, that if circumsolar in the manner depicted by Tycho (Figure 6.1), comets should show phases like Venus. Why do they not? Furthermore, the traditional cosmology had no place for them, their

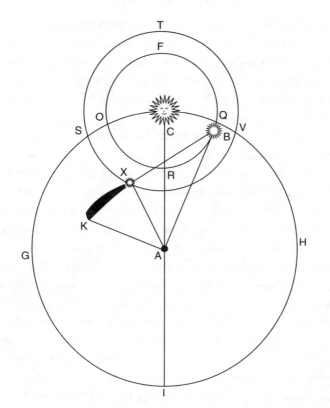

FIG. 6.1 A typical cometary orbit according to Tycho Brahe. After BH, *Disc.*, 438.

sudden appearances, slow declines, bizarre shapes, and free motion beyond the moon. Could cosmology be patched to suit? "You see, dear reader, there are many important questions to answer."[131]

Baleful comet

The great comet of 1618 had given Grienberger and his colleagues the opportunity to engage the entire Roman College in the problems presented by the new astronomy. During the Christmas vacation of 1618/19, they arranged for four lectures on the comet, one each by a theologian, a mathematician, a philosopher, and a rhetorician. The philosopher tried to relate the phenomena to Aristotelian principles, but showed some flexibility toward them, which, to be fair, Aristotle did as well, since the best that he could say about his account of comets was that it was not impossible. The theologian also showed openness toward Galileo and his ideas.[132] The mathematician, Orazio Grassi, deduced the whereabouts of the comet from parallax measurements and did not scruple to place it beyond the moon. A student of Maelcote and Grienberger, Grassi was far different in real life from the fat pig, rash idiot, and utter fool Galileo made him out to be in the *Assayer*.[133]

By 1618, when he was 35, the utter fool had succeeded Maelcote and received a charge from Acquaviva to set up a school of architecture. Grassi's fencing with Galileo took place in odd corners of a religious life devoted to turning out mathematicians and churches for the glory of God. This is not idle rhetoric. We have from Grassi an extended metaphor linking the Virgin Mary and the Stella Maris, on the one hand, to salvation and mathematicians (as navigators) on the other. Galileo's mathematics reached only to the heavens, Grassi's to the Heaven that counts. In 1622, as Galileo put the finishing touches on the *Assayer*, Grassi staged an operatic ascent to paradise.[134] The occasion was the canonization of Saints Loyola and Xavier. In Grassi's libretto, carloads of actors representing the various Jesuit provinces from Italy to China brought gifts to the celebrants while a chorus of angels sang and danced in clouds. The spectacle, set in ancient Rome, demanded the building and burning on stage of a large pyre in accordance with the rite that made gods of dead emperors. Grassi designed all the machinery. The princes and prelates liked what they saw and spread the fool's fame. One of the audience, the cardinal nephew Ludovico Ludovisi, who had labored with his papal uncle to achieve the canonization of Loyola and Xavier, was impressed enough to insure that Grassi would have the place of principal architect of the church he commissioned at the Collegio Romano. By no coincidence, the façade of Grassi's San Ignazio, perhaps the greatest church built in Rome during the seventeenth century, resembles his sketch for the stage set for the operatic apotheosis of the Jesuit saints. Both of these curiously associated works can still be enjoyed, San Ignazio in Rome and the opera on a CD issued in 2003.[135]

The lecture on the comet that Grassi published early in 1619 was eventempered, competent, useful, and, perhaps, a little brave, as it asserted against Aristotle, straightforwardly and nonhypothetically, that comets are celestial objects. In the playful, strained rhetoric admired by the Society, Grassi remarked that the comet of 1618 was the first news from the sky since the celestial messenger's, without, however, mentioning the messenger's name.[136] Instead, he gestured toward Tycho by suggesting that the comet

might occupy an orbit between the moon and sun. To support these conclusions he observed that if it were the sort of burning sublunar exhalation
specified by Aristotle, it would have to consume fuel in incredible quantities
to maintain the size indicated by its vanishing parallax. Furthermore, like
the fixed stars it showed little or no magnification through the telescope, an
effect, Grassi said, of distance. He thought this argument important and controversial enough to add that people who did not accept it had given insufficient attention to optical principles. Again, no mention of Galileo, who,
however, centering everyone's orbit on himself, would read the remark as
an insinuation that he did not know how his telescope worked.[137] Otherwise
Grassi's undogmatic lecture had nothing in it to arouse Galileo's ire.[138]

Galileo was not in a good mood when he received a copy. He had been
sick all winter, as he had been the winter before; a trip to and an ex-voto
left at the ex-home of the Virgin at Loreto during the intervening summer
did not return him to health.[139] Also, he had then recently received a set-
back in his cherished project of making the Medici stars the pilots of the
seas. He had realized as he perfected their schedule of eclipses that he might
have in it a solution to the problem of finding longitude on shipboard. The
solution required knowledge of the time at some fixed location. The Jovian
system could serve as the clock. Observing the positions of the moons at
local time t_1 and finding from the table the time t_2 of the same configuration
as calculated for Florence, the navigator immediately had his longitude L in
degrees from the meridian through Florence as $L = 15 (t_2 - t_1)$, the times being
measured in hours. Thus the theory. In practice the tables had to be accurate
to say 2 minutes of time (giving an error of half a degree at sea) and, what
was much more difficult, the observer had to be able to measure to the same
accuracy through a Galilean telescope from the deck of a pitching ship.

While in Rome in 1616, Galileo had discussed his invention with Spanish
officials and offered to go to Spain to show it to King Philip III. The terms were
stiff: 3000 scudi for the trip and necessary instruments, a royalty of 4,000
scudi per year for himself and 2000 per year to his heirs after his death, and
appointment to the royal order of San Iago. In return, Galileo would update
the tables every year and provide a method to steady the gaze of the navigator. This was headgear (*celatone*) with a single telescopic sight, leaving one eye
free to find Jupiter and the other fortified to track the moons. Castelli tested
the apparatus at sea, a labor of love as he suffered from seasickness. During
1617 Spanish interest cooled and Galileo preserved his record of never ven-

turing out of Northern Italy.[140] A pity. A stint in Spain might have kept him out of serious trouble.

When Grassi's lecture arrived, Galileo was desultorily sorting his papers on motion with the help of Guiducci, who was looking for a topic for lectures he was to deliver as consul of the Accademia Fiorentina. People were asking for an opinion about the comet, some of whom, like Archduke Leopold V of Austria-Tyrol, could not be put off. Eureka! The archduke could have his answer, Guiducci his subject, and Galileo an outlet for his black bile if he wrote a lecture on comets for delivery and publication by Guiducci.[141] As Viviani almost said, this ill-advised Discourse on comets "gave rise to all the controversies and injustice that Galileo suffered from then on." The judgment was correct, the application wrong. Viviani ascribed the offense to Grassi's rejoinder, not, as was the case, to Galileo's provocation.[142]

"Guiducci" opened the campaign by repairing Grassi's omission of the name of that "noble and sublime intellect who adorns the present age." This anonymous ornament had to suffer persecution, and also robbery, by inflated ignoramuses, who puffed themselves up with ideas they did not understand, "pretending to be Apelleses, when with poorly colored and worse designed pictures they have aspired to be artists, though they could not compare in skill with even the most mediocre painters." Like his fellow Jesuit Scheiner, Grassi was a false Apelles. His performance was "suspicious," his reasoning weak, his optics "idle." The mathematicians of the Roman College who approved his lecture could be no better than he. "Let those others be assured that we shall justly pass upon them the judgment which they have wrongly made against us."[143] What judgment? The extravagance of the accusation suggests that Galileo's main motive in going after Grassi was to repay Jesuit mathematicians for deserting him in 1615.

Galileo–Guiducci proceed to judgment. Aristotle suffers as usual for his errors about comets, fire, exhalations, meteors, orbs, and orbits, though they do not figure in Grassi's lecture. Tycho, the great Tycho, who first persuasively applied parallax measurements to comets, is an idiot, "inexcusably" negligent, "astonishingly wrong," "[lacking] even ordinary intelligence." Grassi is a swindler as well as an idiot. He deduced the place of the comet by parallax and its orbit by projection, and confirmed the finding by physical arguments; but we must reject them all, root and branch, because he did not establish that "all other possibilities [are] vain and fallacious."[144] This was to erect against Grassi the same impossible standard of proof that Barberini had urged on Galileo and Bellarmine had employed against Copernicanism.

Galileo–Guiducci suggested a model that certainly would not pass the Barberini–Bellarmine test. It portrayed the comet as an exhalation from the earth that mounts in a straight line toward the zenith of its source, moving at a constant speed, and, when high enough, shining in the night sky by reflecting solar rays. This new application of Galileo's old wrong theory of the nova of 1604 had the advantages of raising doubts about standard theories and of capturing comets in the forceless physics with which he was comfortable. The comet's irregular motion became merely apparent and its observed distancing from the sun simply geometric (Figure 6.2). Since on this theory some of the trajectory must lie beneath the moon, why do they never show parallax? Well, if comets moved with the observer, like rainbows and moon rings, they would be no fitter for parallax observations than Grassi. He knows

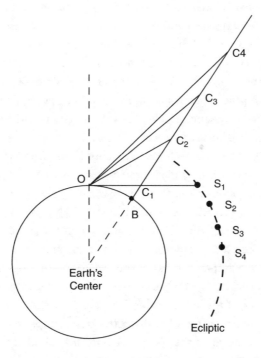

FIG. 6.2 A cometary path according to Galileo–Guiducci. B is the origin of the exhalation, O the observer, C the comet, S the sun. The subscripts indicate positions at the same time on successive sidereal days.

nothing of optics! He thinks that bodies enlarge the less the greater their distance from the observer. How can that be, Galileo–Guiducci ask, referring, tellingly, to terrestrial observations at distances effectively "infinite" (with the telescope adjusted to viewing the moon). They add, captiously, that a telescope lengthened for closer viewing (and so giving greater magnification) is not the same instrument as the same telescope shortened for astronomical use, and, erroneously, that without magnification stars invisible to the naked eye cannot become visible through the telescope.[145] A better answer, given by Grassi, is that the objective lens gathers more light than the unaided pupil does and so can make an otherwise invisible star perceptible. Galileo's instrument could not magnify stars.

Lecturing by proxy to a literary academy gave Galileo a platform from which to insinuate Copernican ideas and destroy competitors without asserting anything himself or endangering his mouthpiece. He has Guiducci teach that comets form a material bond between the terrestrial and celestial region; that all true planetary motions must be in the same direction (true for Tycho and Copernicus but not for Ptolemy); that Venus encircles the sun; and that spin and revolution (motions that do not alter relations among a body's parts) are tantamount to rest. And, with inspired dissimulation, Guiducci admitted that his otherwise very plausible, almost certain theory of comets could not be brought into agreement with observation without an additional cause he feigned not to know.[146] Grassi would understand the admission as a veiled hint at the earth's motion.

The intensity of Galileo–Guiducci's rebuttal of Grassi's assertion that the telescope does not magnify stars suggests that something more than optics and *amour propre* was at stake. Here is a possible explanation. A principal argument against the Copernican system, which Tycho had regarded as conclusive, derived from its implications about the size of stars. Tycho's instruments could detect angular distances of around one arc-minute but could not find any annual parallax in the fixed stars. On Copernicus' theory, the maximum parallax χ of a star a distance s from the sun is $= d/s$, where d is the diameter of the earth's orbit. The apparent size α of a star at this distance is c/d, where c is the star's diameter. Pretelescopic astronomers supposed that α for a star of the second or third magnitude was about 1 arc-minute. Assume that one of these stars has the just undetectable parallax of around $1'$. Then $\chi = \alpha$ and $c = d$. If Copernicus was right, such a star rolled onto the ecliptic plane would just fill the entire orbit of the earth. That was too monstrous to credit. Both Scheiner and Ingoli urged this objection as fatal.[147]

Galileo had pointed out in *Sidereus nuncius* that stars shrink when viewed through the telescope owing to loss of the irradiation or scintillation with which the naked eye endows them. However, according to Galileo, the telescope also magnifies them, stripping them perhaps to invisibility but then restoring them to greater brilliance by enlarging their naked disks. (For the planet Venus, Galileo deduced a true magnification after removal of scintillations of about 2.5 through a 20-power telescope.) Thus although he knew that the naked stellar disk had an angular width of well under a minute, he accepted that it was perceptible and guessed that it amounted to 5″ for a star of the first magnitude and to 5″/6 for one of the sixth. In the latter case, α = $\chi/72$ and c = terrestrial radii (t.r.) on the usual assumption that d = 2400 t.r., which did not seem incredible. In this way, Galileo thought that he could answer Tycho's challenge to Copernicus and retain the idea of finite stellar disks.[148] Why did he not escape the problem altogether by agreeing with Grassi that the stars had no perceptible size when viewed through the telescope? It was not Galileo's style to accept corrections from others.

In his influential *Astronomia pars optica* (1604), Kepler had suggested that comets move not in a Tychonic circle but in a straight line, the apparent curvature in their paths being an effect of the earth's rotation.[149] Galileo–Guiducci had adopted this idea with some modifications, though, of course, they could not say so; but, to a knowledgeable astronomer like Grassi, the connection between the unspecified additional cause that would bring their cometary theory into line with observation and Kepler's model was obvious.[150] In 1619 Kepler reaffirmed and refined the model, from which, as he put it, he could derive as many arguments for the annual motion there were comets in the sky.[151] That left Galileo–Guiducci badly exposed.

"The Jesuits are much offended by [your discourse]." Thus Ciampoli informed Galileo of the reaction he might have desired. That was in July 1619. In December the good fathers were heard to mutter "annihilate" when Galileo's name came up.[152] The patient Grienberger was shocked at the harsh treatment of Apelles, Grassi, and the mathematicians, nay, the mathematics, of the Roman College. They had not attacked Galileo (nor, for that matter, did Biancani in the *Sphaera* being printed when Guiducci addressed the Accademia Fiorentina), but rather treated him well and openly. "I cannot imagine what was in his mind that he could think himself attacked and prefer to suspect than to excuse if perhaps something was asserted not entirely agreeable to his opinions."[153] Grassi too was flabbergasted: "I do not know what reason he has for vilifying the good name of this Collegio Romano" and for writing in so "exasperated and angry a spirit."[154] But he understood perfectly that Galileo opposed him because he followed Tycho. Galileo had no room for Tycho's awkward compromise, quite apart from the question of cometary orbits. Those who took refuge in the Tychonic asylum would reject Galileo's staging of the war of the worlds as a fight between progressive modern science and outmoded ancient ignorance. The Jesuits' adoption of the Tychonic system, a false system invented by a false Christian, was a strategic blow to Galileo. His reaction to Grassi's lecture on comets betrayed his realization that the new respectability of Tycho's system made his mission—if he should have a chance to resume it—much more difficult.[155]

According to these considerations, Galileo's vehemence against Grassi's innocuous lecture betrayed irritation at an imagined insult and fear that the Jesuits might succeed in placing Tycho on the pedestal from which he, Galileo, had just toppled Ptolemy.[156] Two further considerations will help to explain the quixotic behavior that placed Galileo on the path to the windmill. One is psychological. He answered so harshly (so he wrote in his copy of Grassi's reply to Guiducci) for the same reason that he was so eager to renew the fray: "for the truth."[157] He could not defend the truth, the Copernican system, openly. An unhorsed Quijote, he had to depend on a Sancho Panza like Guiducci. He was sick, muzzled, frustrated, melancholic, irrational. Never one to nuance character, Galileo put competent Jesuit astronomers in the same box with the Horkys and delle Colombes. As Aristotle had observed, and Cesi had echoed in a recent statement of the aims of his academy, "By nature all men desire to know." Applying the observation to Jesuit mathematicians and philosophers, Galileo inferred that they thwarted the higher as well as the lower dictates of human nature. They sought knowledge and yet obeyed the orders of ignorant superiors, as if they were dead weights forcibly prevented from seeking the center of the earth. The true natural inquirer, freed, as Cesi hoped he had made the lynxes, from artificial constraints to fulfilling their natures, "swore in the words of no master."[158]

The second consideration is that Galileo's behavior was not irrational but carefully calculated. He responded not as a philosopher, world builder, or frustrated prophet, but as a "competent courtier." To cut the Jesuits with their many scattered mathematicians down to his size, Galileo devised a theory that eliminated one of their competitive advantages. If comets were immune from parallax, Grassi would obtain no advantage from the Society's far-flung observers. Throwing out aggressive challenges to individual combat knocked Grassi off balance. He would fall for the ploy and try to answer in kind, since he too was caught in a patronage network. But he had no chance in literary combat, or exchange of insults, with Galileo.[159] Galileo's worry that he might lose prestige with his patrons could have augmented his already robust concern for his reputation. The ingenious notion that he put forward his cometary theory specifically to destroy the value of the Jesuits' parallax observations, and not because he thought it probable, or because it agreed with Copernican theory as construed by Kepler, may be more Machiavellian than Galileian.

There was one more round in the exchange between Grassi and Guiducci before Galileo delivered the knock-down rhetorical blow in 1623. Grassi proposed to weigh Guiducci–Galileo in *An astronomical and philosophical balance*. To shield himself and the Jesuits from the inevitable return thrust, he pretended to write as his student "Lotario Sarsi Sigensano." Guiducci–Galileo easily discerned "Oratio Grassi Salonensi" in this transparent but imperfect anagram; for Grassi was not from Salona but from Savona, and Salona is precariously close, orthographically speaking, to Salone, an ancient place famous for its pigs. Rhetorical play on "Grassi" and pigs was hard to resist; but after a trial in draft Guiducci decided to spare Grassi the insult and reply to Sarsi as if he existed.[160]

Sarsi made three weighings. The first set aside Guiducci as a front, a mere Sarsi, as indeed he was. Galileo sent out Guiducci's book as if it were his own, and Archduke Leopold and other recipients of it replied to Galileo as its author.[161] The first weighing also assayed Galileo's manners and methods. As to manners: "[you] much preferred to lose a friend than an argument." Touché. Turning to methods, Sarsi neatly observed that the parallax argument, bringing mathematics against Aristotle, should have pleased Galileo; that the telescope did not operate on the stars as Galileo thought it did; and that his pose of ignorance about the missing "cause" of the apparent motions of comets was a dissimulation. Then, hitting Galileo on the wound inflicted by the Holy Office, Sarsi recalled that saving the phenomena in the manner of Kepler, by means of a moving earth, is "in no way permitted to us Catholics."[162] The second weighing found Galileo's Aristotelian notion of an exhalation, his assimilation of comets to rainbows, and his referral of the twist in comet tails to refraction, wanting on optical grounds and physical principles. Grassi's line of reasoning enabled him again to drop his balance on the open wound. "If the earth is not moved, [Galileo's] straight motion does not agree with the observations of the comet; but it is certain that among Catholics the earth is not moved." This was not to accuse Galileo of disobedience or unfaithfulness, but of something far worse: incoherence, equivocation, and incompetence in logical argument.[163]

The third weighing evaluated arguments against Aristotelian physics discharged *en passant* by Galileo–Guiducci. "My whole desire here is nothing less than to champion the conclusions of Aristotle"—and nothing more. "I shall not delay over the question whether the remarks of that great man are true or false." The main conclusion at issue derived from the assumption that the concave surface of the sphere of the moon (not the moon itself, but the hypothetical orb peripatetics had invented to bear it) could drag around the fire and air immediately under it in its rotation. Neither Grassi nor Galileo believed in the existence of the orb about which their masked avatars disputed. For the sake of argument, however, they supposed that its effect on fire could be simulated by that of a rotating body on a candle flame or a suspended leaf placed near or in it. Would the flame or leaf tremble? Galileo: no, or very feebly, if the body's surface is rough. Grassi: yes, and a little roughness can be supposed in the non-existent orb since, as Galileo had discovered, the moon itself is covered with prickles. Also, Galileo's explanation of the sunspots as cloud-like material rotating with the sun assumed just the sort of drag that Aristotle had supposed; and further, if we were allowed to mention it, Galileo's concept of the non-existent diurnal motion implies that the earth conveys its atmosphere with it. In any case, Sarsi observed, the experiments required patience. The transfer to enclosed air of motion from its rotating container would take much longer than a similar transfer to water. Ciampoli witnessed these experiments and reported their outcomes to Galileo.[164]

Aristotle taught that air friction could heat bodies to incandescence. Galileo denied it. Grassi made the tactical error of adducing the evidence of ancient poets and philosophers, and "other men also of great authority and trustworthiness." From the Byzantine lexicographer Suidas he gathered the quaint example that Babylonian soldiers used to cook eggs by whirling them rapidly in a sling. Seneca had reported the melting of lead pellets shot

1 The Abbey of Vallombrosa, where Galileo studied as a boy and took steps toward becoming a monk. Giovanni Stradano, 1580s.

2   Dante's Inferno as proposed by Antonio Manetti, calculated by Galileo, and drawn
by Giovanni Stradano ca. 1590. The topmost circle is the river Acheron; the small area
above the short black cylinder at bottom center is Malebolge, the eighth circle (the
Deceitful).

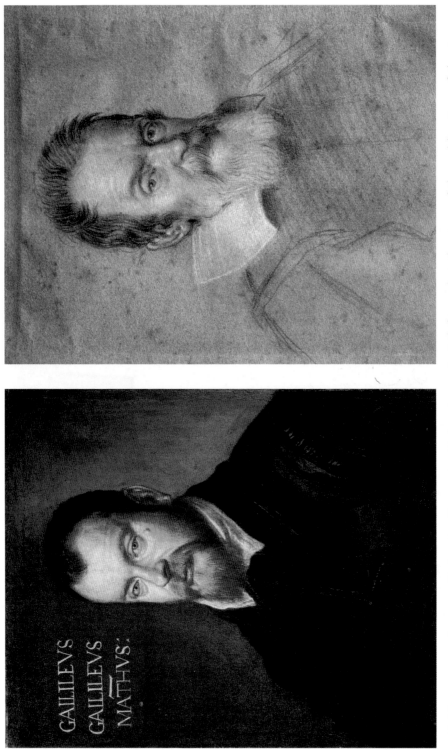

3 Galileo as a professor at Padua, by Domenico Robusti, ca. 1606 (left), and at the time of the publication of the *Assayer*, by Ottavio Leoni, 1624 (right). Note the asymmetry of the eyes in the Leoni portrait.

4   Galileo at 75, holding a telescope lens in one hand and showing his lynx ring on the
other. Commissioned from Justus Sustermans by Ferdinando II ca. 1640.

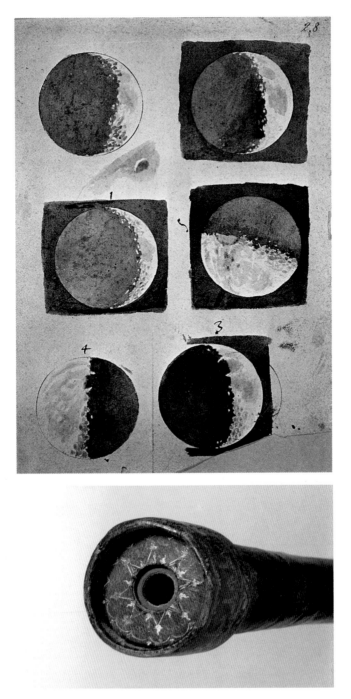

5 The objective end of an early Galilean telescope, 92.7 cm long, probably a gift to Cosimo II, ca. 1610; and Galileo's composite wash drawings of the lunar landscape, made with the help of a similar instrument, 1609.

6 From upper left, clockwise, Giambattista della Porta, playwright and natural magician; Christoph Clavius, Jesuit mathematician; Giovanni Camillo Glorioso, Galileo's successor at Padua; and Jacopo Mazzoni, concordist Pisan philosopher. Crasso, *Elogii* (1666).

7 From upper left, clockwise, Giovanni Ciampoli, glib and genial opportunist; Fortunio Liceti, fluent Paduan philosopher; Saint Robert Bellarmine, Jesuit theologian; and Cesare Cremonini, friend of Galileo and bête noire of the Inquisition. Crasso, *Elogii* (1666).

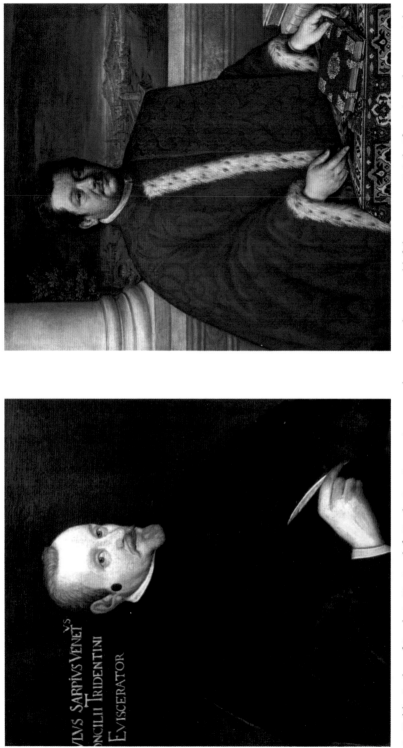

8 Galileo's close friends in Venice: left, Paolo Sarpi, wearing a patch to cover the wound left by an assassin's knife, artist unknown; right, Gianfrancesco Sagredo, wearing the robes of office of a Venetian patrician, by Leandro and Girolamo Bassano, 1618/19.

9 Galileo's daughter Virginia, Suor Maria Celeste, and his closest disciple and surrogate son, the Bendictine monk Benedetto Castelli, both by unknown artists.

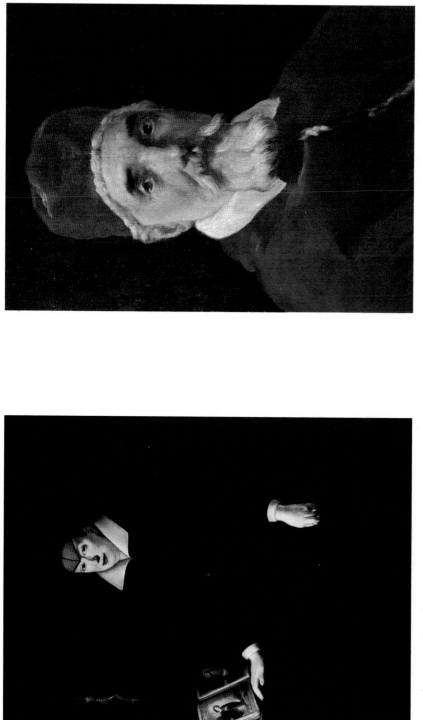

10 Religion and politics represented by the devout and domineering Christina of Lorraine, Grand Duchess of Tuscany, by Tiberio Titi, after 1609; and the willful and poetic Pope Urban VIII, by Gian Lorenzo Bernini, 1631/2.

11　The founder of the Accademia dei Lincei, Federico Cesi, by Pietro Facchetti, 1612 (left), and his cousin, Virginio Cesarini, poet and editor of the *Assayer*, by Anthony van Dyke, early 1620s (right).

12 Self-characterization of Dominicans as hounds of God. The relevant action, to the right of the cathedral (the Florentine Duomo with Buonaiuti's solution to its then incomplete dome), shows St Dominic unleashing dogs against wolves that would devour the lambs of the Lord. St Thomas helps by scaring the faithful with one of his big books. Detail of a fresco by Andrea Bonaiuti, 1365–67, in the Spanish Chapel of Santa Maria Novella, Florence.

13 Self-conception of Urban VIII: Divine Providence brings the papal tiara and keys to crown an open escutcheon containing the Barberini bees. Detail of a ceiling painting by Pietro da Cortona, 1630s, in the Palazzo Barberini, Rome.

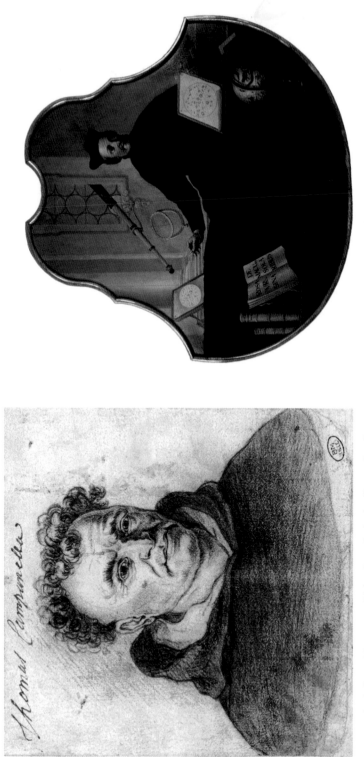

14 On the left, Galileo's would-be friend, the brilliant, flamboyant Dominican Tommaso Campanella, Francesco Cozza, date unknown; on the right, Galileo's whipping-boy, the Jesuit mathematician Christoph Scheiner, by Thomas Scheffler, 18th century.

15 Galileo and other illustrious Tuscans from a wall painting by Cecco Bravo, 1636, commissioned by Michelangelo Buonarroti the Younger to adorn his house in Florence. The architect Antonio Manetti stands at the far left holding his and Galileo's model of Hell.

16 Above, Galileo overlooks the door of Viviani's house in Florence and reliefs alluding to his contributions to navigation and gunnery; the text under D.O.M. addresses the "passerby of right and generous mind." The statue, right, depicting Galileo as an Old Testament prophet, by Aristodemo Costolli, 1851, stands in the Loggia of the Uffizi Gallery, Florence.

through the air. Aristotle himself attested to the heating of flying arrow tips. A modern historian had described the disintegration of lead linings of careering cannonballs. Sarsi: "Who believes that [such] men…would wish egregiously and impudently to lie?" Coming still closer to Aristotle's method of kindling comets, Sarsi resourcefully invoked the eerie ignition of vapors rising over cemeteries on hot days.[165]

Finally, against Galileo's denial that bodies can be seen through flame, and, consequently, that comets, whose tails do not occult stars, cannot be burning matter, Father Grassi allowed himself a little fun. The man who prided himself on his mastery of experiment, on his "facility in the explanation of very difficult matters by very simple demonstrations," had overreached. "I have found each experiment entirely false—may Galileo spare me for speaking the truth." Grassi pointed out that printed type close behind a candle flame could be read through it and that the visibility of other objects depended on their distance from the candle and the eye, on their nature, and on other sources of illumination present.[166] Perhaps to ignite Galileo, Grassi illustrated the transparency of flame by a story from scripture. Bad King Nebuchadnezzar of Babylon, having ordered his people to worship a golden statue, did not take kindly to the report that the Jews Shadrach, Meshach, and Abednego, who held high office under him, had not bowed down as instructed. Nebuchadnezzar commanded them to do so with the inducement that, if they refused, he would throw them into a burning fiery furnace. They refused. The infuriated king had the furnace heated seven times hotter than ever before. It fried the executioners but not the doomed men. "Lo!" said Nebuchadnezzar, "I see four men loose, walking in the midst of the fire, and they have no hurt: and the form of the fourth is like the Son of God."[167]

Guiducci did not try to answer Grassi's physical arguments. Instead he wrote a letter of complaint to a prominent professor of rhetoric at the Roman College, Tarquinio Galluzzi, who may well have been his teacher when he was trained there "with incredible and paternal love." The letter's main interest lies in Guiducci's claim to authorship of the *Discorso* and in inflammatory *ad hominem* arguments against Grassi (immodest, biting, cutting, at best a copier of Galilean style) and Scheiner (plodding, inartistic, thieving, at best a copier of Galilean discoveries). Guiducci compared himself to a Clavius, who made a creative composition largely from the work of others, and to an Andrea del Sarto, who did not disdain to copy a picture by Raphael. By following Galileo, Guiducci exceeded Grassi by the same margin that

Michelangelo outperformed Apelles.[168] For the rest, Guiducci downgraded Galileo's comet theory to a possibility introduced merely for discussion, disposed of Grassi's experiments as "very fallacious and not free from suspicion of fraud," and neatly bettered Grassi in exegesis.[169] In a verse in Daniel omitted by Grassi (and also not present in the King James version), an angel put out the fire when Shadrach and his friends entered the furnace. Guiducci genuflected ironically in the direction of Trent. "I do not mean to place my interpretation upon this, and I defer the matter completely and entirely to the declarations and expositions of holy doctors and professors of divinity; let them judge… [whether] King Nebuchadnezzar saw the holy ones amid the flame…and let them say also whether or not it is praiseworthy to cite Holy Scripture in this manner."[170] Checkmate.

Since Guiducci's letter did not answer Sarsi's main optical and physical arguments, and since the Jesuits advertised Grassi's *Balance* as a defeat for Galileo, his friends, notably the lynxes Cesi, Cesarini, Ciampoli, and Stellutti, urged him to respond. This represented a change in strategy. Cesi and Stellutti had advised Galileo to avoid confrontation in general, and they regretted that he had chosen to offend the Jesuits and close a door through which his science might have entered the schools. Cesi had worked to keep the door open by sending Bellarmine an erudite concordance of modern astronomy and the utterances of the Fathers. Bellarmine had responded with praise and the friendly advice that Cesi find something better to do. The best would be to behave so as to end up in heaven, "where and when everything immediately will be clear."[171]

While awaiting this enlightenment, Cesi pressed Galileo to answer Grassi's *Balance*, quickly, lest the ignorant think the Roman College had defeated the great lynx and destroyed the pretensions of the others. Ciampoli was the most exigent, anticipating a devastating and novel riposte, "it being understood everywhere that nothing comes from your hand but precious jewels entirely unknown to others."[172] The literary lynxes met to discuss the mode of the reply—whether through Guiducci or another disciple, or in the form of a letter to a third party. Although Galileo took his time (some two years where Grassi had taken six weeks), he was not reluctant to pursue a course that would bring defense of his *amour propre* into line with the desires of his friends and patrons. He followed their advice in going after Sarsi rather than the Roman College and in choosing Cesarini as correspondent; but he could not bring himself to avoid the "pungency" that the lyncean editors could not hope to remove. We are reminded again how provincial or peninsular Galileo's purview was: his subject may have been the universe, but his audience was a few dozen highly-placed literate Italians whose good opinion he prized.[173]

Unfair balance

The *Assayer* is a heroic poem in prose, a "Sarsiad," a protracted tale of right against wrong, good against evil, innocence against deceit. It is also a mocking or mockery of a scholastic analysis of an Aristotelian text, divided into 53 snippets of the original each followed by an extensive commentary in the lively style of the staged debates with which Galileo amused *letterati* in Florence and Rome. Where Sarsi was content to peck around "like a blind chicken" to find passages to weigh in his crude balance, Galileo would behave as a "man ought to do" in assaying the work of others, leaving no word unturned.[174] "Now read this, your Lordship," Galileo says of each tidbit, as if he were exhibiting an idiocy.[175] Why bother with such stuff? Quoting a line from Ariosto, Galileo intimated that he was condescending to argue over truths he already possessed. *Noblesse oblige.* The line referred to a fight between Orlando and Mandricardo over Orlando's sword. "Mine it is by right, let us stage a chivalrous duel for it."[176]

The *captatio benevolentiae*—Galileo's opening intended to obtain the good will of the reader—has nothing to do with chivalry or the matter in hand. It is the old complaint, increasingly tinged with paranoia, that people had either stolen his discoveries or, if too weak for plagiarism, deprecated and dismissed them.[177] This strident refrain follows awkwardly on three prefatory hymns of praise to Galileo from two lynxes and, of all people, the book's censor, the Dominican Niccolò Riccardi, who thus entered a story in which he would play a major part. He had made his reputation in Spain, where he cut so grand a figure as a preacher that the king himself, Philip III, came to hear, and dubbed him "Father Monster," some say because of the size of his belly and memory, others because of a head hard enough to crack walnuts. In weighing the *Assayer*, the monster reckoned himself "fortunate to be born when the gold of truth is no longer weighed in bulk and with the steelyard, but is assayed with so delicate a balance." You may have had predecessors and emulators, O Galileo (thus Johann Faber, hymnist, fellow lynx, and pontifical pharmacist), "but you surpass the others by as great a distance as the

celestial stars are separated from the earth." The faithful and laborious Stellutti added an appreciative ode in 24 verses, which concluded with the curiously non-Copernican image, "You from whom [Heaven] much received /
shall, so long as it may turn / shine as bright as the stars that burn."[178]

As an assayer Galileo seldom gives honest weight. He quibbles, bamboo-
zles, misleads, and misrepresents with such brilliance and invention as to
pass himself off as the injured party and, moreover, modest, unassuming,
and inevitably right.[179] The incongruous opening, in which Galileo settles scores with the thieves and belittlers among the imaginary legions of
his persecutors, is a good example of his even-handedness. After a generic
accusation he goes after Simon Mayr with lance and mace, fire and sword,
that Simon Mayr who, Galileo was convinced, had put up Capra to his infamous plagiarism and, in his *Mundus Jovialis* (1614), had redoubled his offence
by claiming independence of Galileo in spying on the satellites of Jupiter.
The crime was the more foul because Kepler had preferred Mayr's values
of their periods to Galileo's, from which, to be sure, they differed but little,
and because—the thief's impudence knew no bounds!—Mayr had corrected
Galileo's finding that the satellites' orbits lie in planes parallel to the ecliptic.
Galileo struggled to avoid the fact by converting it to an effect of perspective;
but, as he said elsewhere, nature prefers its way to ours. Fortunately he was
spared witnessing Mayr's names for the satellites—Io, Europa, Ganymede,
Callisto—supersede "Medici stars." In accusing Mayr ("my old adversary,
and invidious enemy not only of me, but of the entire human race"), Galileo neglected that Mayr reported many things that he could not have stolen
from the starry messenger.[180]

In his old fight with Capra, Galileo had condemned the theft of intellectual
property as worse than murder. With your life you lose interest in everything
else, while the victim of the theft of intellectual property feels constantly the
loss of "the honor, fame, and merited glory, which he obtained not by inheritance or from nature, fate, or chance, but from [his] own studies, efforts,
and long vigils."[181] Galileo had not hidden his views about plagiarism. How
then could he not have foreseen Scheiner's likely reaction at thinking himself
classed among the thieves, or the Roman College's anger at seeing one of its
professors staged as too perfect an imbecile for this imperfect world? The
Assayer amused Galileo's friends, multiplied his enemies, and brought him
new readers who could appreciate the brilliance of the style and the asides

that have made excerpts from it chestnuts in the history of science and in the teaching of Italian literature.

Perhaps the most celebrated of these famous asides is the admonition that philosophy is not fiction, as, for polemical purposes, Galileo has Sarsi think, not a romance like the *Iliad* or *Orlando Furioso*, "books in which the least important thing is whether what is written in them is true."[182] Oh, no, Sig. Sarsi, "Philosophy is written in this grand book—I mean the universe— which stands continually open to our gaze. It is written in the language of mathematics, and its characters are triangles, circles, and other geometrical figures, without which it is humanly impossible to understand a single word of it." This famous dictum too is a fiction, to which Galileo clung and on which he built: triangles and circles do not take you far in the analysis of nature. Historians have pointed to Plato's creation story *Timaeus* as a probable source for Galileo's reduction of nature to elementary geometrical figures, although Timaeus had in mind that the particles of which he supposed the elements composed had the shape of the regular (or Platonic) solids. Since the faces of these five solids are either triangles or polygons reducible to triangles, Timaeus managed a metaphorical triangulation of the universe.[183] But it is more likely that Galileo had in mind the matter in hand, which was a criticism of Sarsi–Grassi for slavishly following Tycho, who, in Galileo's version of the facts, could not tell a triangle from an icosahedron.

To annihilate Tycho's reputation completely, to expose his "foolish fabrication" and "fantasies," Galileo derided the demonstration indicated in (Figure 6.3). A and B represent Copenhagen and Prague, D a fixed star. Tycho

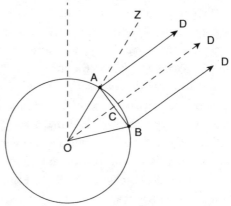

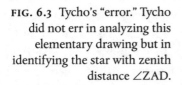

FIG. 6.3 Tycho's "error." Tycho did not err in analyzing this elementary drawing but in identifying the star with zenith distance ∠ZAD.

made DAB a right angle, and so it is, very near, since DCB is a right angle
and the chord AB virtually zero in comparison to the distance OD. Since the
difference in latitude between A and B is 6°, $\angle AOD = \angle ZAD = 3°$. The zenith
distance of the star D at Copenhagen therefore had to be 3°. Tycho erroneously wrote that D fell in the constellation Aquila, the Eagle, whose zenith
distance when on the meridian at Copenhagen is over 48°. Hence, eagle-eyed
Galileo deduced, Tycho did not know even the rudiments of geometry, and
Sarsi, whose supposed errors in mathematics Galileo delighted to catalogue,
was a fool to follow him. "Sig. Sarsi, philosophy is not fiction."[184] That was
too much for Kepler. The Imperial Mathematician countered that in his haste
Tycho had made use of a diagram he had drawn for one purpose for another,
and had written down the wrong star. But there was nothing wrong with his
geometry.[185]

"To return to the subject," as Galileo often had to say to the virtually
present Cesarini, a second famous aside from the *Assayer* tells the tale of a
hermit who knew musical sounds only from bird calls. One day he decided
to go back to civilization. He soon discovered musical instruments, singing crickets, and many other sources of pleasing sounds. "Thereupon his
knowledge was reduced to such diffidence that when asked how sounds
are generated he used to reply tolerantly that although he knew some of
the ways, he was certain that many more existed which were unknown and
unimaginable."[186] Galileo accepted that humans could not attain to a Theory
of Everything, but not the implication of modesty in pursuit of what might
be knowable. True philosophers find truth without fanfare. Like eagles, they
fly alone and are seldom seen. The crowd of philosophers know no truth and
flock like starlings: "[they] fill the sky with shrieks and cries wherever they
settle, and befoul the earth beneath them." Kepler was an eagle, a strong and
independent intellect, but very different in his philosophy from Galileo. That
was to be expected. Eagles are loners. Thence arises a significant problem.
Being an eagle will become increasingly challenging as previous eagles leave
ever fewer discoverable truths. "Hence the less attractive [science] will be,
and the smaller will be the number of its followers." The number of disgusting chattering starlings will grow, however, since magnificent claims, fallacies, frauds, and chimeras will always be more popular than true knowledge
and the number of fools is infinite.[187] The vast domain of the unknowable
will remain the domain of the Super Eagle, God Himself, to whom alone all
things are known.[188]

"Well, Your Excellency, let us turn to substantive matters." Sarsi had tried to reply to Guiducci's quibble that a telescope lengthened to see things up close is not the same instrument it was when shortened to view distant objects. What was Sarsi's reply? A trombone is the same instrument whatever the position of the slide! What a sophism! Would Sarsi say that a short thick organ pipe that sounds a low note is the same as a long thin one made from it that sounds a higher one? Who knows. Would Galileo say that the lute he is playing multiplied into different instruments as he stopped the strings at different frets? As we know, this silly argument covered an embarrassment. Although Galileo understood the justice of Sarsi's explanation of the light-gathering capacity of telescopes, he could not admit that it was the effective cause of the promotion of stars from invisible to perceptible status. He may have worried that conceding this point would have called into question not only his omniscience but also the reliability of discoveries interpreted through his understanding of the working of his telescopes.[189] His dismissal of poor Sarsi is pure bombast. After remarking, correctly, that the telescope gathers light and magnifies simultaneously, Galileo wrongly concludes that if one of these processes gives a perceptible effect, so must the other. "In order to hold a different opinion, [Sarsi] would be obliged to show that the telescope sometimes unites the rays [gathers light] without enlarging the angle [magnifying], but that this happens only when the fixed stars are being observed. This he will never show to the end of time, for it is a most foolish fiction, or, to be blunt, a falsification."[190] Thus Galileo cleverly made his mathematical error into his adversary's moral failing. Cleverly? It may be that Galileo truly did not understand how his telescope made invisible stars visible.[191]

In a related error, Galileo berated earlier astronomers for not noticing that the angular diameters they assigned the planets were much too large. As usual, he converted error into sin, "hardly excusable," nay "inexcusable." Kepler replied that as the old star gazers lacked telescopes to identify and remove the irradiation that had misled them, they had no reason or way to attack the problem.[192] With the telescope, Galileo had determined that Venus' apparent size varies by a factor of 40 and that of Mars by a factor of 60 owing to changes in distance and phase, and, to emphasize the importance of the finding, he had advertised that it destroyed the Ptolemaic system. Kepler countered that all the competing systems predicted changes in the size of planetary disks and that phases, not sizes, cooked the Ptolemaic goose.[193]

With Ptolemy dethroned and Copernicus prohibited, Tycho would seem to have no competition in Italy. Galileo tried to defeat this logic by removing Tycho from the competition: his was not a serious system, but a sketch and a caricature of one. Kepler observed that it had the same status as its competitors, as it was an amalgam of the two.[194] Here Galileo opened himself to the same ridicule to which he subjected Sarsi. Be sure, Sig. Sarsi, when attacking a position to make it as strong as you can; otherwise its supporters will deny they held it as you describe it and slip away, leaving you stunned and empty-handed, like Ruggiero trying to grasp the invisible Angelica.[195]

There remained the charge of Copernicanism that Sarsi insinuated lay in or behind Galileo's cometary theory. To repudiate it, Galileo insisted that since he had admitted that he did not know the cause that might save the theory he presented as a mere possibility, he did not understand how Sarsi could know that that cause was motion. The decree of 1616 had foreclosed this possibility: it remained only to demolish Kepler's claim that every comet is a proof of the Copernican system.[196] Well Sig. Sarsi? Can you do it? If you cannot, "If the movement attributed to the earth (which I, as a Catholic and pious person, consider to be most false and vain) lends itself to yielding explanations for so many widely diverse appearances observed in the heavenly bodies, then I should not be sure that so false a thing might not deceptively correspond with the appearances of the comet."[197] Some unknown person perceived this cloudy statement as a defense of Copernicus and referred it to the Holy Office. The task of evaluation ended in the hands of Giovanni di Guevara, a learned man close to the Barberinis and the lynxes. The matter went no further.[198]

Galileo's response to the third weighing, in which Sarsi fielded experiments in defense of Aristotle's cometary theory, yielded two golden asides dear to Galileists. One concerns the stories Sarsi yanked from antiquity about the melting of arrow points and the cooking of eggs. If you want to know how flying arrows heat up, Galileo advised Sarsi, you must read your Ariosto. Observe the clash of Mandricardo and Ruggiero. So violent was their collision that the splinters of their lances burst into flames—not because of the friction of the air, but because the impact carried the debris into the sphere of fire, from which it fell back incandescent, "[as the historian] Turpin writes, truthfully at this point."[199] After this deft exploitation of Ariosto's irony, Galileo applied an equally light logic to egg cookery. Sig. Sarsi, he said, I've tried the experiment and it failed. Rather than doubt ancient authority, however, which would be presumptuous, we must assume that our experiments omitted something present in theirs. "Now we do not lack eggs, or slings, or sturdy fellows to whirl them...And since nothing is lacking to us but being Babylonians, then being Babylonians is the cause of the eggs hardening."[200]

"Now, Your Excellency, to return to the matter at hand..." The second famous aside digresses from a long argument against Aristotle's claim that motion creates heat. Balderdash, screeches the eagle, the dissolution of solid bodies releases the very fine particles of fire they contain, and these, on penetrating animal bodies, produce within them the sensation of heat. When the dissolution occurs via friction, it might be said, though improperly, that motion causes heat. In every case, however, part of some body must be consumed, and a fine emanation released, for warming to occur. This was all Galileo needed to argue that a comet cannot be a dry exhalation ignited by motion through the air, since in the process no solid body disintegrates.[201] He did not stop here, however, but applied his insight into heat to consume the foundations of Aristotelian physics. "Your Excellency..., I must give some consideration to what we call 'heat', for I suspect in general people have a conception of this which is very remote from the truth." They think, with Aristotle, that it is a "real attribute, property, and quality which actually resides in the material by which we feel ourselves warmed."[202]

In fact, fire particles possess no qualities but shape, motion, and size. We manufacture the sensation of heat when they pervade our bodies, a pleasant warmth or a disagreeable pain depending on their speed and numbers. It is exactly the same with odor and taste, only the particles have sizes and shapes different from those of fire. Sound is merely our ears' response to vibrations of the air. As to vision, it is caused by the ultimate atoms of fire.[203] In the special vocabulary applied to Galileo's distinctions, motion, shape, and size are "primary qualities" and the sensations they produce in our minds through our senses, "secondary qualities." The distinction goes back to the ancient atomists whose theory Aristotle had rejected for a fistful of good, though not unanswerable arguments. Philosophers began to flirt seriously with atomism again in the sixteenth century. The Catholic church did not like it, as it smacked of materialism and had no obvious place for an immortal soul. As will appear, Galileo's enemies were not slow to capitalize on this latest of his hostages to fortune, with which they would associate the particulate structure of water in *On floating bodies*.

As a practiced propagandist, Galileo knew the value of repetition. Through-out the *Assayer* he uses words like "puerile" and "childish" to characterize Sarsi's arguments; accuses him, sometimes justly, of misrepresenting Guiducci; and constantly refers to Sarsi's errors, "repeated error[s]."[204] Of these the one hardest hit is, of course, Galileo's own mistake about the angular width of a naked star. Galileo ends his five weighings with an inspired account of the effectiveness of the telescope in removing irradiation as a function of the size of the enlarged image. For planets the removal is complete. For even the grandest star, the dog star, a little fulgor remains. Do you not agree then, Sarsi, that invisible stars become visible by enlargement, just as the planets are magnified? How can you not agree? You must agree, Sarsi. "Therefore yield, and be silent."[205] That is the way a knight does science.

Grassi rejoined nonetheless, but Galileo did not again deign to contest the sword of Orlando with him. Very pugnacious marginal comments in Galileo's copy of Grassi's *Ratio* (1626) show, however, that Galileo condescended to wrestle with him in private. In public Galileo had no need to take further action, since the lynxes judged that he had saved their honor and his. Cesi wrote, "I think that everyone knows full well that your Lordship has no need to joust, or obligation to enter any arena...Mons. Ciampoli agrees, as do other courtiers and literati who love and esteem [your] work."[206]

7

—∞∞∞—

Vainglory

Pope Gregory XV died on 8 July 1623. Eleven days later, having heard Ciampoli deliver the sermon traditionally given to prepare them for the task, 54 cardinals undertook to determine whom among them God wanted to lead them all. The conclave, protracted for 18 days in sweltering heat, was a trial of the fittest. Eight cardinals did not survive the ordeal. Finally the Spanish faction yielded and the Francophile Christian humanist Maffeo Barberini emerged as Pope Urban VIII. Ciampoli had not known his own strength. "My oration had…a better result than I could have hoped for."[1]

7.1 THE POPE

Any competent astrologer who knew Barberini's geniture could have computed his success. The sun, the lord of his horoscope, stood just beyond the midheaven, regarding at trine the rise of the solar constellation Leo and happily aspected by Jupiter and the moon. When the conclave began, Barberini's sun dawdled in Cancer while the cardinal nephews of the previous two popes maneuvered Borromeo, del Monte, and other aging favorites, and ignored their healthy young colleague Barberini. As the sun moved into Leo and conjunction with Jupiter and Saturn, and malaria thinned the electorate, the heavens spoke and so did the cardinals. *Habemus papam!* The new pope, who took the name Urban in reference to his and Rome's urbanity, left the conclave more certain than ever of his solar lineage. He had an image of the sun painted on a wall of his bedroom in the Vatican to collect solar rays and remind him of his power.[2]

Literary Rome was ecstatic at the rise of the poetical pope. The "marvelous conjuncture" (as Galileo put it, in reference to the happy event and the astrological signs) returned to the Holy See a knowledgeable and generous, indeed, a spendthrift patron of the arts. Urban's combined income as head of the Roman Catholic Church, and, as he liked to be styled, "King of Rome," was around 2.5 million scudi annually. The principal beneficiaries were his brothers and nephews, three of whom he transformed into cardinals, others into dukes, princes, and governors. Altogether they consumed around 12 million scudi (25 per cent!) of papal income during the two decades of Urban's reign. And this was perhaps no more than a tenth of the total proceeds from benefices, bribes, fees, retainers, and the pope's pocket. The most powerful of Urban's relatives, the cardinal nephew Francesco Barberini, was an ideal alter ego of the egoistic pope: an intelligent, reliable, and resourceful executor of his uncle's wishes, he was a humanist in his own right, an excellent scholar in Latin and Greek, a patron of the arts and sciences, and the owner of the largest library in Rome after the Vatican.[3] In acknowledgement of Francesco Barberini's achievements and the hope of his protection, Cesi invited him to join the lynxes; in confirmation of the Barberinis' closeness to the academy and the new regime's respect for learning, he graciously accepted. Further to togetherness, the cardinale nipote took lynx Cassiano dal Pozzo, one of the editors of the *Assayer*, as his private secretary, and since dal Pozzo was a knowledgeable botanist, his chief gardener.[4]

Friends who came to congratulate Urban did not go away empty-handed. Although he claimed to have little to bestow, the deaths of the cardinals martyred in the recent conclave freed substantial assets for redistribution.[5] Among the old friends thus rewarded were Giovan Battista Strozzi, who derived a double benefit from his visit to the Vatican. The pope gave him a pension of 300 scudi and Strozzi took away a like amount from his protégé Ciampoli, whose lavish lifestyle under the new regime he did not approve. Ciampoli objected to this retrenchment not because he needed the money but because he was dependent entirely on Urban and the few hundred scudi he received as a canon of Saint Peters. And Urban was not 100 percent reliable.[6]

Another old friend who came to pay his respects and collect his reward was Buonarroti, who, like Strozzi, judged the new court guilty of excess. "Music always and always poetry / music and poetry morning and night / music in every time and season." "I would rather hear frogs sing."[7] Many of Urban's own poems figured in this musak, in settings by Johannes Kapsburger, the lutanist who wrote the music for Grassi's opera.[8] The pope grew excessively fond of his verses. Buonarroti again: "He raised me up [from the obligatory genuflection] and recited to me a hundred of his poems, paraphrases or translations of the psalms." Buonarroti requested copies. "He referred me to Monsignor Ciampoli, who distributed them as if they were orders on the treasury."[9] At Urban's court, according to the poet Agostino Mascardi, "cultivating literature is not a matter of decorum but of necessity." Not all poetry, of course. Marino had hurried from Paris to Rome after Urban's election with his new *Adonis* in hand. Its earthiness did not suit the pious classicism of the new Parnassus, however, and he soon left for his native Naples. *L'Adone* earned the honor of a place on the Index and on Urban's list of exemplary bad poems. The pope did not want a rival as strong as Marino in Rome.[10] Urban's vanity about his poems, like his belief in astrology, opened him to damaging manipulation.

The extent of the Barberinis' promotion of the decorative arts may be gauged from the swarm of painted and sculptured bees in Rome, over 10,000 by a contemporary estimate, and by such masterworks as Bernini's baldachino over the main altar of St. Peters and that immense monument to nepotism, the Palazzo Barberini, now the Galleria Nazionale d'Arte Antica. The main public room in this palace boasts an immense painted ceiling depicting Divine Providence in the act of arranging Urban's bees, keys, and tiara (Plate 13). The pope rated the ceiling, completed in 1638 after six years of work, as the equal of Raphaël's *stanze* in the Vatican.[11]

To the heavy expenses of peace Urban soon added the grievous burdens of war. He strengthened Castel Sant'Angelo and built or rebuilt frontier fortresses. They functioned more as staging for theft than as strongholds for defense. In his need for money Urban did not disdain to rob his neighbors. The first to suffer was the Grand Duke of Tuscany, not Cosimo, alas, who had died in 1620, but in effect his mother Christina, who would dominate the government until the heir, Ferdinando II, came of age in 1628.[12] Although Ferdinando did not care to oppose his bigoted grandmother in matters of religion, he attempted to block Urban's theft of the Duchy of Urbino, which he claimed through his betrothed child-wife Vittoria della Rovere and other dynastic ties. Christina's acquiescence in the pope's rejection of her grandson's claim dashed all protests. Urban tried to repeat the adventure, with disastrous financial consequences, in Castro and Mantua.[13]

"Whereas Clement VIII was usually found with the works of Saint Bernard, and Paul V with the writings of the blessed Justinian of Venice, Urban VIII had the latest poems or plans for fortifications on his work table." Thus Leopold Ranke characterized the difference between Barberini and his immediate predecessors, juxtaposing his humanism and his hardness, his devotion to the arts and his ambition for power. Obsessed with being, and showing, himself a great prince, he declined advice, made decisions capriciously, and suffered from mood swings.[14] A poet who took 500 pages to summarize Urban's virtues ends with an unwittingly apt figure: "With sweet urbanity he plans to rule the world /.../ And thus intending, he takes the name of Urban / From the far west to the Indian Ocean / The world is urban, and the heavens echo "Urbano.""[15] Since "urbano" also signified "of the city," urbanizing the world meant imposing papal Rome upon it. The powerful Spanish colony in Rome led by the belligerent Cardinal Borgia did not care for this vision of empire under a Francophile pope. No more did it please the Society of Jesus, whose saintly founder, several of its generals, and many of its members were Spaniards.[16] Their quarrel with Urban chanced to come to a head during Galileo's final bout with the Holy Office.

Galileo had shared the enthusiasm of the literati and lynxes at the dawn of the new age and hastened to fulfill Urban's expectation of a congratulatory visit. Perhaps he did not know that one of Urban's first acts had been to pressure the Venetians into foregoing a great monument to Paolo Sarpi, who died in January 1623. It was not regard for Sarpi, however, that delayed Galileo's trip for almost a year but, as usual, bad health and bad weather.[17] The lynxes kept his person and projects in Urban's mind by making the pope the dedicatee of the *Assayer* of "our Galileo, the Florentine discoverer not of new lands but of hitherto unseen portions of the heavens, containing investigations of those celestial splendors that usually attend great wonders." Ciampoli read Urban choice passages from the *Assayer* during dinner. The pope enjoyed the hits at the Jesuits, read the rest of the book himself, and expressed his admiration and affection for its author.[18]

Galileo interrupted his journey to Rome at Acquasparta to rest and plan with Cesi. During his stay there he received the news that Cesarini, who had written of his ardent desire to see Galileo again, had at last succumbed to his illness.[19] Cesarini's sad and early death squeezed from his inseparable companion Ciampoli a rare expresson of real emotion. "I had hoped [he wrote Cardinal Archbishop Borromeo] that our feeling for one another would become an example at court and would be known throughout Italy, showing the envious and quarrelsome that collaboration in literature is a bond of love, not a cause of discord." Cesarini's death was also a great blow to the cause of the *Assayer*.[20] It brought to light hairline cracks in the bonds among the lynxes and between them and the Barberini. Cesarini had asked to be buried in Jesuit garb in a Jesuit church. Unknown to his friends, he had dickered schizophrenically with the Jesuit General Vitelleschi for privileged entry into the order while negotiating the imprimatur for Galileo's malicious attack on it. A Jesuit lynx was, by Lincean ordinance, an oxymoron. Thus the internal crack. Externally, Cesi wanted to implement his cousin's request for a modest, if not Jesuit burial, and Urban wanted to erect a grand monument advertising his favorite's rare literary accomplishments. Cesi objected: since the society of lynxes ranked high above the literary academies of Rome in subject and aspiration, it would not do to give them equal billing on Cesarini's tombstone. And since many of these literary academicians were courtiers, and ran "a great peril of falling into the despicable role of parasites or buffoons," they were not fit associates for a lynx dead or alive. The prince and the pope compromised by burying Cesarini in a dignified non-Jesuit setting (the Sala dei Capitani in the Palazzo dei Conservatori), under a marble that does not mention academies.[21] As is often the case, compromise created coolness between the compromisers.[22]

7.2 THE KNIGHT

Third Roman campaign

Urban treated Galileo papally, gave him six private interviews, two medals, a promise of a pension for his son Vincenzo, and, at parting, a jumble of gifts of diverse value. These included a picture of unknown nature, two medals, a sackful of *Agnus Dei* specially blessed by Urban, and an ornate letter, written by Ciampoli, to Grand Duke Ferdinand II praising Galileo's genius and piety.[23] He did not obtain what he wanted, however, which was permission to reopen the Copernican question.[24] He tried to get a reading from several cardinals. One of them, Frederick Eitel von Zollern, sounded out Urban.

He reported that the pope believed that the church had not and would not condemn the Copernican theory as heretical. The pontifical reasoning as rendered by the cardinal showed no relaxation from the position Barberini had taken in 1616: there being no truth in the way astronomy saves the accidents we see in the heavens, it made no sense to say that any of its systems was wrong, let alone heretical.[25]

In February of 1624, Cesarini had written Galileo that Father Monster Riccardi desired the honor of his acquaintance. They met in Cardinal von Zollern's rooms in late May. Lynx Faber too was present. Riccardi agreed with the pope that Copernicanism was not a matter of faith and that astronomical systems could not be right or wrong. He advised taking no advantage of these epistemological truths, however, since a discussion of Copernicus would only "rekindle a debate that has died down." As a censor who preferred peace to precision, he recommended leaving the motion and arrangement of the planets to the angels.[26] Riccardi advised further that if Galileo could not keep his peace, he should write out his ideas in such a way as to give no opening to his enemies. Father Monster did not know Galileo very well.

Whenever Galileo visited Rome he displayed wonders: improvements on Archimedes in 1586, telescopic marvels in 1611, show-off disputations in 1615/16, and now, in 1624, gigantic insects. He had with him a compound microscope that he had perfected, again, as with the telescope, proceeding from a Dutch original. Faber looked through it before Galileo gave it to Cardinal von Zollern for delivery to Maximilian of Bavaria. "I examined a fly that Galileo showed me [Faber wrote Cesi]; astonished, I said to him that it was another creator since it brought things into view whose existence previously was unknown." In the fall of 1624, Galileo sent Cesi a better instrument and an indication of its merits. "I've beheld a great many tiny animals with infinite surprise. Among them the flea is most horrible, the mosquito and the moth very beautiful; and with great pleasure I've observed how flies and other little animals walk on mirrors..."[27]

The lynxes soon turned their fortified eyes to the little animals of greatest interest in Barberini Rome. Combining their interest in natural sciences, a reference to their most famous member, and an obeisance to their great patron, they published a beautifully engraved broadside, *Melissographia*, for the Jubilee of 1625 featuring magnified images of a bee observed by Stelluti. The dedication mentions "the divine discovery of the new art," the genius of the *lincei*, and, of course, the bees in Barberini's bonnet. The lynxes could not leave them alone. Soon after the broadsheet they brought out a Latin poem entitled *Apes Dianiae* (Diana's bees) illustrated with engravings of bee-bearing ancient coins. The bond between the insect and the huntress was chastity, bees being supposed sexless and Diana doubling as goddess of continence. Both therefore projected the "chaste and virginal model" that Urban cultivated. The poem admirably instances the control of literary form prized by the men among whom Galileo made his career. A third, mighty variant on the bee theme, a gigantic broadsheet measuring 167 by 70 cm, the *Apiarium* of 1626, decorates a treasury of literary, artistic, and scientific information about bees with arcane references to classical learning and outrageous compliments to Urban. The pope could not look at his escutcheon without being reminded of lynxes and, as they put it in their *Melissographia*, "the power of polished glass."[28]

Back in Florence in manic mood, Galileo decided that his discussions in Rome did indicate that an essay or two, written in his unprovocative way, might gain a new hearing for Copernicus. Two trial balloons resulted. One was an important improvement in the theory of the tides, which removed the obvious objection that it had no place for the moon and fashioned it into the capstone of the great work on cosmology Galileo hoped to publish under the catchy title, *On the tides*. The second balloon was an answer to Francesco Ingoli, the author of the corrections to Copernicus issued in 1620.

On the tides is a dialogue between the resurrected (Salviati and Sagredo) and the unredeemable (Simplicio). After rehearsing the tidal theory we know, Salviati allows that it cannot be the whole story as it does not explain the monthly variation, spring tides at full and new moon, neap tides at the quarters. How does the moon influence the cause of the tides, the diurnal and annual motions of the earth? "[H]ow many hours, how many days, and how many more nights I spent on these reflections; and how often, despairing of ever understanding it, I tried to console myself by being convinced, like the unhappy Orlando, that that could not be true which had nevertheless been brought before my very eyes!"[29]

The reference is to Orlando's discovery of abundant evidence of Angelica's perfidy, which he tried to explain away, "deceiving himself with far-fetched notions," "searching in his mind a thousand ways for not believing what he could not help believe."[30] Orlando could not square his discovery of the facts with his theory of Angelica and went mad. Sagredo: "Thank God for not letting you [go the way of] Orlando," or of Aristotle either, who, according to

unreliable testimony, jumped into the sea from frustration at not being able
to understand its motions.[31] Aristotle would not have drowned himself had
he known the pendulum clock. Attend to it and the problem of the monthly
tide evaporates. A pendulum beats faster the shorter its length; a planet goes
slower the greater its orbit. Let the earth-moon system be the bob of a pendulum. When the moon lies between the earth and the sun (at conjunction) the
effective pendulum is shorter than when the moon is in opposition. Hence
the pair moves more quickly around new than around full moon. Salviati:
"From this it may be clear that the annual movement of the earth in its orbit
along the ecliptic is not uniform, and that its irregularity derives from the
moon and has its periods and restorations monthly."[32] Sagredo: "You, Salviati, have guided me step by step so gently that I am astonished to find I have
arrived with so little effort at a height which I believed impossible to obtain."
Against this singular solution was the little difficulty raised by Sagredo that
astronomers had not noticed any consequences of the supposed monthly
variation in the annual motion. Salviati: they have not looked and the effect
may be small.[33]

Two of Galileo's prevailing misjudgments as a natural philosopher come
into view again here. Neglecting physical cause, he advanced his pendulum
analogy, which was no more than a metaphor, as an explanation. What is it
that binds the earth and moon so strongly together that they act as a single
pendulum bob? Galileo liked the analogy all the more for this weakness. In
the paradoxical way he loved, it gave the moon a role in the drama of the tides
"without [its] having anything to do with oceans and with waters."[34] It also
allowed him to sidestep the hidden connection between the lunar motions
and the *diurnal* tides, and to rap Kepler, who, "though he had at his finger-
tips the motions attributed to the earth…has nevertheless leant his assent
to the moon's dominion over the waters."[35] The second of Galileo's enduring misjudgments also concerned Kepler. The arguments from which Kepler
deduced the elliptical form of the earth's orbit ruled out Galileo's monthly
variation as a cause of the inequalities of the earth's motion.[36]

Galileo did not intend his definitive account of the tides as just another
bizzarria. As if in dialogue with Urban, Salviati asserted the impossibility of
any other explanation while conceding that the difference in simultaneous
accelerations of the two ends of a seabed, on which the explanation rested,
was most "remarkable." He also faced up to the uncomfortable question
whether the diurnal motion alone could produce tides. To obtain the desired

answer, Galileo made the gratuitous assumption that without a considerable annual velocity the oceans would accommodate themselves to the spin and cease to slosh. In a final application of the theory, Salviati beautifully explains an imaginary annual variation in the strength of the tides arising from the alignments of the diurnal and annual motions at the solstices and equinoxes.[37] No more can be known about the "fixed and constant" causes of the tides except that theories invoking forces acting directly on the water are vain imaginings:

> These are so far from being actual or possible causes of the tides that the very contrary is true. The tides are the cause of them; that is, make them occur to mentalities better equipped for loquacity and ostentation than for reflections upon and investigations into the most hidden works of nature. Rather than being reduced to offering those wise, clever and modest words, "I do not know," they hasten to wag their tongues and even their pens in the wildest absurdities.[38]

This was the biggest bluff of Galileo's career.

Having thus made his theory of the tides sufficient (Copernican motions are enough), necessary (no other will do), and offensive, Galileo could turn his attention to countering objections to a moving earth. A convenient target existed in an essay composed by Ingoli as a follow-up to their dispute in Rome in the winter of 1615/16. Since then Ingoli had made a career harrowing Copernican astronomy. He had become a small hero for sustaining a counterattack by Kepler, who dismissed his theological arguments and referred him to the *Epitome of Copernican astronomy*, which Kepler had just published, for answers to his mathematical and physical objections. Ingoli responded by engineering the condemnation of the *Epitome* by the Congregation of the Index. Thus he continued to prosper. When Galileo decided to tackle him in 1624, he held the conspicuous position of secretary to the Congregation for the Propagation of the Faith set up by Pope Gregory XV in 1622.[39] A client of the Ludovisi, the benefactors of the Jesuits, Ingoli made a good target for a Copernican who did not fear the consequences.

Galileo motivated his attack as a defense of the reputations of Italian astronomy and Catholic theology. By showing that Ingoli's physical arguments had no force, Galileo would demonstrate to heretics that the Holy Office had not condemned Copernicus from a misunderstanding of the natural world. It had made its decision solely on the basis of scripture. "We

Catholics continue to be certain of the old truth taught us by the sacred authors, not for lack of scientific understanding, or for not having studied so many arguments, experiments, observations, and demonstrations as they have, but rather because of the reverence we have toward the writings of our Fathers and because of our zeal in religion and faith."[40] This pious pretence recurs in the preface to the *Dialogue*.

The first of Ingoli's arguments will suggest the level of most of the rest. It attempted to prove that Copernicus' theory required that the sun be closer to the earth than the moon—a silly argument based on word play and a bizarre definition of parallax. Galileo opposed a lengthy, crystal-clear account of parallax (yet again!).[41] A quibble by Ingoli over heaviness and its natural place produced something more interesting: all the planets, including the sun, have heavy parts tending to their centers; the question was, which center coincided with the center of the system. Galileo knew: the sun. "I have other evidences not previously observed by anyone [how would he know that?], which are necessarily convincing, about the certainty of the Copernican opinion." Mad Orlando had returned. And immediately returned to his senses: "'necessarily convincing' [Galileo added] as long as we remain within the limits of human and scientific inquiry."[42]

Ingoli had brought up the standard physical objections to a spinning earth and claimed in proof the malleable experiment of the rock let go from the mast of a moving ship. Galileo countered with the doctrine that everything on the earth participates in its motion, towers, loose balls, falling rocks, clouds, and birds; the experiment of the rock-and-mast did not destroy but confirmed the doctrine of shared motion. Then Galileo proposed a thought experiment worthy of Einstein. Enclose yourself, a friend, a jar dripping water vertically into a pot, a tank of fish, and a flock of small winged creatures in the cabin of a docked ship. Play catch, jump back and forth, and observe that the fish and flying things move easily in all directions. Now suppose the ship under sail at constant speed. Everything will remain as before: the jumping, ball throwing, flying, and swimming still take place as easily toward the prow as toward the stern, and the water continues to drip vertically.[43] Galileo imagines Ingoli's objecting that, as no wind exists to drive the heavy earth through the celestial spaces, the thought experiment is irrelevant. "Ah Sig. Ingoli," Galileo replies to his straw man, "you are incapable of stripping yourself of the old ideas impressed in your mind, and so you confuse heaven and earth and utter great inanities."[44] Circular motion is the "congenital, proper, and completely natural behavior" of spherical bodies, their "natural, primary and eternal inclination." A big heavy sphere like the earth should enjoy trouble-free constant spin and revolution until it pleases God to end its existence.[45]

The campaign aborted

A few straws in the Roman wind in 1624 showed that it had not veered in Galileo's direction. In the autumn, in the well-attended ceremony for the reopening of classes at the Roman College, the main speaker compared freelance philosophers and mathematicians who placed themselves above tradition to the builders of Babel, spreaders of confusion, men who put the satisfaction of their vanity above the interests of Christian solidarity. And did so when Catholicism once again confronted heresy in the Germanies![46] On 21 December a great sower of confusion and discord, Archbishop Marc Antonio de Dominis, was burnt in Rome together with his books. Fortunately for him, he had been dead for three months. He had also been a mathematician, a pal of Paolo Sarpi, and an apostate. He had fled to England, embraced reformed religion, and received, as a sinner saved, the high post of Dean of Windsor. There he made history by helping to publish Sarpi's anti-Roman *History of the Council of Trent*. Troubled by the weather and perhaps his conscience, de Dominis returned to Rome in 1622, recanted, and again received the benefits of the saved sinner. For a short time he prospered. He resumed his archbishopric and published a book on, of all things, the tides, for which Grassi served as censor. A persistent interviewer from the Holy Office teased from him that he did not hold entirely to the Tridentine decrees and thought that a reunion of the Christian churches might be possible. Incarceration in the Castel Sant'Angelo rewarded his frankness. He died while awaiting trial. On papal authority, and in response to an uncharitable suspicion of poisoning, lynx Faber conducted an autopsy. What he left of de Dominis was burnt in the Campo dei Fiori.[47]

A third straw was Grassi's furious reaction to the *Assayer*. Stellutti had witnessed it, in a bookstore. Grassi was a dangerous enemy. He had guile, ingenuity, and courage, all displayed in the ancient Roman setting of his opera on the Jesuit saints and the original design of his church in the Roman College. Today the most interesting feature of San Ignazio to most visitors is the persuasive fake cupola painted on the inside of its flat roof. Grassi had not designed a canvas *trompe l'oeil* to finish off his masterwork, however, but a huge dome that would have blocked the morning light from the library of the Dominicans and outdo the huge headpiece of the Theatines' church of San Andrea della Valle.[48] That was to take on too many powerful enemies at once. Would he answer the *Assayer*? Galileo asked Guiducci to find out. Grassi turned the tables by calling on Guiducci, declaring a wish to end hostilities, and extracting the information that Galileo was more attached than ever to his theory of the tides. In return, Grassi confided the disinformation that he would reply to the *Assayer* without raising any new questions. Galileo felt himself outmaneuvered. Ciampoli advised against delivering the inflammatory letter to Ingoli. The Roman *lincei*, who now included the belletrist Guiducci, put Galilean matters on hold.[49]

Grassi's response brought something new and worrisome.[50] It suggested that Galileo's discussion of primary and secondary qualities committed him to atomism and hence to a theory incompatible with Tridentine canons concerning the sacred sacrament of the eucharist. The canons read, in an appropriately baroque translation:

> *Canon 1.* If anyone denieth that, in the sacrament of the most holy Eucharist are contained truly, really, and substantially the body and blood together with the soul and divinity of our Lord Jesus Christ, and consequently the whole Christ; but saith that He is only therein as in a sign, or in figure, or virtue; let him be anathema.
>
> *Canon 2.* If anyone…denieth that wonderful and singular conversion of the whole substance of the bread into the Body, and of the whole substance of wine into the Blood—the species only of the bread and wine remaining—which conversion the Catholic Church most aptly calls Transubstantiation; let him be anathema.[51]

Grassi capitalized on the menace of the tone of the whole and the mystery of the phrase, "the species only of the bread and wine remaining."

The species signifies the properties of the bread and wine supposed to persist after the priest's incantation, *Hoc est corpus meum*. The Aristotelian system possesses technical terms in which the miracle can be dressed up to appear less magical. A piece of bread unites certain "substantial qualities" that characterize it as bread with "accidental qualities" like size, color, place, and savor that distinguish it from all other pieces. Transubstantiation preserves the accidents while transforming the substance of the bread. What could be clearer? If, however, there are no such entities as real accidents (size, color, savor) but only the substantial primary qualities of size, motion, and arrangement claimed by the atomists, canon 2 cannot be fulfilled and the atomist must be anathema. An anonymous denouncer pointed out to the Holy Office the ominous bearing of Galileo's primary–secondary distinction on the Tridentine teaching. Pietro Redondi, the historian who discovered this document, attributed it to Grassi, who, however, had no reason to play the sneak since he had published the problem through his alter ego Sarsi. The accusation did not go far at the time, probably owing to the intervention of Francesco Barberini. Nonetheless Sarsi's insinuation that Galileo's teachings subverted a fundamental Tridentine decree alarmed the lynxes.[52] Galileo did not reply. He continued to work on *De motu* and to tend his garden and his family.

Placing Virginia and Livia with the Poor Clares of San Matteo in Arcetri did not relieve Galileo of responsibility for their care. They called upon him frequently for protection against the hunger and cold that stalked their "prison," as Maria Celeste called their refuge, usually for themselves but also for other nuns on the edge of starvation or bankruptcy.[53] Galileo always met these requests, which directed his charity toward the monastic life that had attracted him as a boy. During the 1620s and early 1630s, for which Maria Celeste's moving letters to her father are extant, the plight of his daughters kept the dark and light of Catholic experience constantly before his eyes. Maria Celeste stoically describes her cold and hunger, and her distressingly frequent illnesses, the loss of all her teeth, her misery in "this wretched world"; and joyfully expresses her satisfaction in serving Galileo and the convent in small things, and in contemplating "the reward that awaits us, after the brevity and darkness of the winter of the present life, when at last we will enter the clarity and happiness of the eternal spring of Heaven."[54]

Meanwhile there was this life and its winters to get through. One nun tried to shorten her sufferings. Maria Celeste was there to tend and comfort the failed suicide. Sister Arcangela suffered severe depression. Maria Celeste asked Galileo for special food and wine for her, and, because she "often finds interaction with others unbearable," moved out of the tiny cell they shared "in order to be able to live in the kind of peace and unity befitting the intense love we bear each other." This left Maria Celeste on the charity of another nun with more space until one of the best rooms in the convent became available. Galileo gave the considerable sum required to buy it.[55] Maria Celeste dealt with infirmities of the body as well as of the soul. She ran the convent drug store, made potions, and gave medical advice. Galileo received both pills and counsel. Her refrain: work less in the garden, especially on cold spring days, and drink less wine on all days.[56] She sugarcoated this advice with candied fruits, of which Galileo was particularly fond. So poor, however, was she that often she asked for the fruit and the sugar, and for plates and vessels to transport them and the cordials she prepared in her laboratory.[57]

During these years—he was now in his sixties—Galileo did not live in Florence but in the hills south of the Arno at Bellosguardo, about an hour's ride by mule from Maria Celeste's convent of San Matteo. He devoted much of his time to cultivating his garden, fruit trees, and vineyard, and looking after his chickens, nephews, and grandchildren.[58] His brother Michelangelo came to visit in 1627. It was not a happy time. Galileo was ill and crotchety. Michelangelo's wife Chiara also was ill. Trying to preserve some independence, Michelangelo declined Galileo's offer to establish him in Florence and returned to Munich, leaving Chiara and their younger children with Galileo. Two daughters went back with Michelangelo. One of them, Mechilde, was so pretty and promising that the convent in which she studied Latin had waived her fees.[59] That helped. Michelangelo was always in need of money to feed, house, clothe, and educate his family, and to keep himself in the wine he regarded as a necessity. The brothers had something more than the lute in common.

What would happen to Michelangelo's needy young family if their ultimate support died of his frequent ailments? Michelangelo allowed himself to raise the question. "If you were to die (God forbid) without having arranged your affairs... the misery of my situation would be inexpressible." Galileo did not like this reminder of his mortality. Michelangelo had to apologize. "I am very weak-minded, as you know." But he was not too weak to remove Mechilde from her convent when he discovered that the nuns had little Latin to teach but many ways to exploit girls on fee waivers, or to call his family back to Munich when Galileo complained that his young nieces and nephews disturbed his philosophical repose. Their Bavarian reunion did not last long. Poor, overburdened, overshadowed, weak-minded Michelangelo died in January 1630. Chiara and the children lived precariously on a pension provided by the Duke of Bavaria and, probably, on help from Galileo.[60]

Galileo's son Vincenzo married in 1629 and soon had a son who stayed with Galileo while Vincenzo withdrew to avoid the plague then menacing Florence.[61] For once a Galilei married well. Vincenzo's wife, Sestilia, had been educated properly in a convent. Her brother, Geri Bocchineri, a secretary to Ferdinando II, an open and generous man, became a good friend to Galileo as well as to Vincenzo. When Vincenzo returned to Florence, he was able to purchase a house adjacent to the small one he already owned with financial help from Bocchineri and Galileo. This improved accommodation, in the Costa San Giorgio upriver from the Pitti palace, lay on the road to the village of Arcetri, where Galileo moved in the summer of 1631 at the urging of Vincenzo and Maria Celeste.[62] The family thus reassembled could eat together provided Galileo brought the food to the nuns' guests' dining table.[63]

Galileo confided in Maria Celeste some of the hopes he pinned on Rome. She energized her sisters to work for their fulfillment in the only way they could. "We shall not fail to pray the Lord…to bless you by letting you achieve all that you desire, so long as [she knew her father] that be for the best." The nuns hoped for something in return, for alms and the man, for a small income and a proper confessor. Their current spiritual guide was "more suited to hunting rabbits than guiding souls." Poverty enticed danger. Priests unpaid for their services tended to recoup in kind, by coming uninvited to dinner and becoming overfamiliar with the flightier nuns. Galileo seems to have succeeded in replacing the confessor but not in augmenting the alms.[64]

Galileo's contributions to the economy of San Matteo—food from his garden, wine from his cellar, delicacies as required, wool and linen for clothes and bedding, allowances and pocket money on request, larger sums when needed (Maria Celeste's cell, Arcangela's overdraft for food for the convent), an occasional loan to a needy nun—did not cause him to stint on himself.[65] However, this outlay plus expenses for the rest of his family and for the upkeep of Bellosguardo or Arcetri with its two servants probably left him with little extra. He did not live extravagantly. He sold the produce of his garden and vineyard when the yield exceeded his needs.[66] Maria Celeste thought him very generous. "We are overwhelmingly committed to you, not only as daughters, but as the abandoned orphans we would be, Sire, if not for you." "I…confess myself indebted for an almost infinite multitude of blessings conferred by you."[67] She felt that she could ask him for any small gift or service: relics from Rome, cloth for cuffs, music for a defective organ, repair of the convent clock, though that was work "more suited to a carpenter than to a philosopher."[68]

Dangerous dialogues

Despite the prolonged visits of his family, Galileo was relatively free from illness, or unusually quiet about it, during the later 1620s. He complained to Maria Celeste about his usual spring ailments in 1627 and 1628, and occasional short indispositions, but they abated enough that he could complete his treatise *On the tides* in good time when he returned vigorously to it in September 1629.[69] When in good form he made do with fewer than seven hours of sleep a day. Exploiting this capacity and earlier texts like the letter to Ingoli, Galileo swiftly created three "days" of philosophical comedy to preface the already written fourth day's dialogue on the tides.[70] Day 1 demolishes Aristotelian physics; Day 2, arguments against the diurnal motion; and Day 3, arguments against the annual motion.

Day 1 of Destruction takes on Aristotle's distinction between the ungenerable and incorruptible heavenly bodies, rotating in their spherical perfection around the center of the universe, and the sublunar elements and their combinations, perpetually exposed to the likelihood of transforming into their opposites. Salviati approaches the distinction as Aristotle did in *De caelo* by inquiring why the world has three and only three dimensions. Simplicio replies that three is complete, perfect, etc., and that there are exactly three natural motions, toward, away from, and around the center. Salviati prefers the experimental route. Harking back to a Ptolemaic argument that Galileo could have learned from his teacher Buonamici, Salviati declares that, since no more than three mutually perpendicular lines can be made to intersect at a point, there can be no more than three dimensions. The business has nothing to do with three natural motions. Indeed, there is but one such if by natural motion we mean one that can go on forever. Motion toward or away from the center occurs only briefly in order to convey out-of-place material to its proper distance from the center. Rocks can enjoy perpetuity of motion as well as planets; for a rock is indifferent to motion by which it neither approaches nor recedes from the earth, that is, a motion that seems horizontal locally but in fact is an arc of a circle concentric with the earth. To prove which Salviati tantalizes his friends with a glimpse of Galileo's theory of descent along inclined planes.[71]

Simplicio objects that, as his Aristotelian texts taught, the natural motions toward, away from, and around the center of the universe are manifest to the senses. Poor Simplicio! He has fallen for a *petitio principii*: it is just the existence of this center that is in question. Perhaps the universe has none. "If any center may be assigned to the universe," however, says Salviati, more boldly than Bellarmine would have liked, "we shall find the sun to be placed there." Then the motions up and down, *sursum et deorsum*, can refer only to the earth's center and can scarcely ground a universe. It is the same with the moon; out-of-order lunar rocks fall to the lunar center, and similarly with other planets. There is but one natural motion and no quintessence. Simplicio: *Negantes principia non disputandum est!* What? Yes, you cannot argue with imbeciles who reject the principles underlying the disputation. There are no such things as moon rocks or planetary parts. Those bodies are incorruptible, unbreakable, and perfectly spherical, encased in a firmament harder than adamant.[72] That, anyway, is what Galileo made Simplicio believe, although by then most peripatetics had replaced the hard spheres, which were the work of Arabic commentators on Aristotle, with the softer heavens of Tycho Brahe.[73]

The discussion had not advanced. Sagredo thought he knew why. Like most philosophers, Simplicio feared freedom. "Who would there be to settle our controversies if Aristotle were to be deposed? What other author should we follow in the schools, the academies, the universities?" Abhorrence of a philosophical vacuum no doubt is an important force in the world. "It is wrong [said Aristotle, rightly] to remove the foundations of a science unless you can replace them with others more convincing." Replying 2,000 years later, Sagredo likened stubborn peripatetics to fine gentlemen who, having invested heavily in their palaces, in pictures, mosaics, marbles, murals, refuse to admit that the entire structure needed replacement, but try to shore it up with jerry-built expedients, chains, props, and wedges. To which Salviati added that to introduce a new philosophy, a reformer had to do much more than refute one philosopher or another. "It is necessary first to teach the reform of the human mind and to render it capable of distinguishing truth from falsehood, which only God [and his prophets] can do."[74]

Salviati and Sagredo attack in the obvious way. What do you say, Simplicio, about novas, comets, sunspots, moon mountains? Simplicio appeals to delle Colombe's crystal moon and Scheiner's solar starlets, theories then widely discredited and, in Scheiner's case, discarded by its inventor. Anyone who does not see that the mountains and the spots lie on the surfaces of the luminaries suffers from "the rankest ignorance of perspective."[75] Salviati and Sagredo endeavor to teach Simplicio the true reason for the secondary light of the moon, but it is difficult going because he keeps relapsing into incorrupt-ibility and cannot bring himself to assimilate the moon to the earth.[76] They all reject the possibility of human life on the moon, a dangerous concept that Galileo had always repudiated.[77] This unusual concord breaks to allow the gratuitous annihilation of Scheiner for referring moonglow to sunlight dif-fusing through the lunar body and for plagiarism, deceit, and stupidity.[78]

There were other familiar dragons to slay, too, delle Colombes, Chiara-montis, and their like, who degraded the human mind by enslaving it to the authority of old Greeks who lacked telescopes.[79] To be sure, Salviati concedes, we cannot know God's creation extensively, in all its details; but (replying to Barberini?) "with regard to those few [things] which the human intellect does understand, I believe that its knowledge equals the Divine in objective certainty, for here it succeeds in understanding necessity." Simplicio: "This speech strikes me as very bold and daring." And so it was. Sagredo: Let's go for a ride in a gondola. "Tomorrow I shall expect you both so that we may continue the discussions now begun."[80]

Day 2. They reassembled to hear Salviati present physical demonstrations of the earth's diurnal rotation and destroy arguments opposing it. He turned out to be too persuasive for his own good. As he admitted, he enjoyed show-ing off and often had to run for cover. "[I] am impartial…and masquerade as Copernicus only as an actor in these plays of ours." "Be guided not by what I say when we are in the heat of acting out our play, but after I have put off the costume, for perhaps then you shall find me different from what you saw of me on stage." "It should be almost as if we had met to tell stories."[81] Under this armor Salviati and Sagredo blaspheme freely against peripatetics. "Oh, the inexpressible baseness of abject minds," Sagredo sighs; "[nothing] more revolting," echoes Salviati.[82] Having thus aroused the goodwill of his adversaries, Salviati advances the main arguments pro-spin: it is more eco-nomical for the earth to revolve than the entire starry firmament, especially a firmament as solid as steel; and also more symmetrical, since if the earth revolves and the heavens stand, celestial rotations fall into order, growing progressively slower from Mercury to the stars.[83]

Simplicio brings up the heavy artillery of common sense. If the earth spun, birds and clouds would fall behind. Salviati replies that all objects in or near the earth participate in its rotation. A motion common to all is as good as nonexistent. Not so, cries Simplicio: has not Aristotle said, and countless peripatetics after him, that a weight dropped from the mast of a ship in motion would fall astern and even to sea? Not so, rejoins Salviati, claiming to have performed many times on this tiresome floating laboratory. Had he, or rather Galileo, in fact done the experiment in question? Almost. On his way to Rome in 1624, Galileo had enjoyed a boat ride with Cesi and Stellutti on a lake near Cesi's villa. The discussion turned to Aristotle's claim. Galileo asked for something heavy. Stellutti offered his keys. Galileo threw them vertically upward. Would they return to the hand that sped them or fall into the lake? Salviati amplified this quasi test with the thought experiment about the birds, balls, and fish that Galileo had set out for Ingoli.[84]

Meanwhile Sagredo was pondering the consequence that, since the common motion has no effect on the behavior of the bodies that share it, the weight from the mast must fall to the deck in the same time however great the speed of the boat. What would the fall look like to an observer on the shore? In a beautiful insight, Sagredo saw that the released weight is to the stationary observer what a cannon ball fired point blank is to the artilleryman, the horizontal velocity imparted by the gun being analogous to the common motion of the boat and all things on it. From the analogy it followed that cannon balls shot horizontally over level ground hit the ground at the same time irrespective of their range. "[That] seems a marvelous thing." Salviati: "This reflection is very beautiful by reason of its novelty...I have no doubt about its correctness."[85] There is no need for experiment in a well-constructed natural philosophy.

Unwilling to be outdone, Salviati disclosed two beautiful reflections that temporarily silenced Sagredo: "I cannot find words to express the admiration they cause in me." The first considers the trajectory in space of a rock dropped from a tower on the supposition that it can attain the earth's center. Assume that it gets there along a semicircle that keeps it always in touch with the tower AD (Figure 7.1) until it hits the ground at F. In a time t the earth turns through an angle $\alpha = \omega t$ and the stone falls a distance BX. Since $\angle AYX \approx 2\alpha$, the rock's speed along the tangent at X, v_x, is to its speed before release at A, v_A, as $2\omega t \cdot YX$ is to $\omega t \cdot AO$. But $XY = AY = AO/2$. So $v_x : v_A = 1$. In falling, the rock moves from one circle to another without changing speed and, if

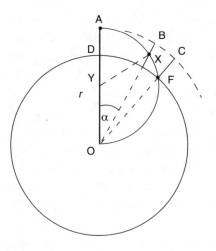

FIG. 7.1 Galileo's *bizzarria*. A weight released from the tower at A reaches the earth's center by the semi-circle AXFO.

allowed to do so, would reach the earth's center in six hours independent of the height of the tower. Hence, as Salviati argued in Day 1, separated rocks return to the earth along circular paths, conserving their tangential velocity and allowing Galileo, as was his custom, to avoid bringing causes or forces into play.

This first reflection is the *bizzarria* or oddity mentioned earlier as a playful marvel in Galileo's philosophical comedy. If taken seriously, it results in a velocity of free fall proportional to the time elapsed, as in Galileo's rule, but far off quantitatively.[86] Salviati professed a great interest in obtaining a value of the constant of acceleration in free fall, a "problem," he said, previously unknown to "any philosopher or mathematician whatsoever." To solve it, Salviati gave another glimpse at Galileo's theory of motion—the spaces traversed in equal intervals of time are as the odd numbers beginning with one. Sagredo: "This is a remarkable thing…Is there a mathematical proof?" Salviati: "Most purely mathematical," one of many such in the science "our friend" has founded in place of a thousand worthless volumes. "Not a single one of the infinite admirable conclusions within this science had been observed and understood by anyone before [him]." After many experiments, Galileo had decided that a falling body covers 100 yards in five seconds.[87]

The second reflection, another extraordinary piece of kinematics, is even less plausible than the first. Ptolemy had argued that objects not fastened to the earth would be thrown off if the globe spun. Galileo countered that irrespective of the speed of rotation, no such extrusion can occur. The argument invokes a theorem of Euclid's: the tangent (HG in Figure 7.2) is a mean proportional between the secant FG and its external part GE, that is, GE:HG = HG:FG. Suppose that in a unit of time the rock would reach G if it did not fall towards E; and let the entire fall occur from G as if the tangential velocity v vanished there. Suppose further that t and therefore the angle α are very small; FG then approaches the constant value $2a$ and $HG^2 \approx 2a GE$. Thus, since GE goes to zero much faster than HG no matter how large v is, for sufficiently short times GE can always be small enough that the rock can regain the earth or, better, never leave it. "Take note, Simplicio, just how far one may go without geometry and philosophize well about gravity." Or, rather, just how absurd a conclusion the theoretical physicist (*filosofo geometra*) can reach by subliming to mathematics. Salviati tacitly takes HG, which measures the velocity of extrusion in time t, as related to GE, the distance fallen toward earth in the same time. But GE is determined by g, the acceleration under gravity, HG by the velocity of rotation, two entirely independent quantities. The velocity v can be as large as required but g cannot be changed. The usually shrewd Sagredo allowed himself to be hoodwinked. "The argument is truly very subtle, but nonetheless convincing, and it must be admitted that trying to deal with physical problems without geometry is attempting the impossible."[88]

Galileo's striking error here combines three typical misconceptions. First, he could not think himself away from the earth anymore than Simplicio could imagine himself on a planet. Consequently, just as he conceived the "horizontal" on which a ball rests or moves indifferently to be a spherical shell concentric with the earth, so he tied the velocity of extrusion functionally to the

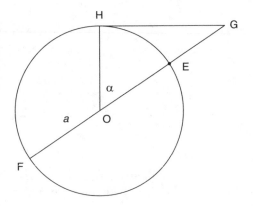

FIG. 7.2 Diagram to show that an unattached body cannot be extruded from the earth no matter how fast the earth spins.

earth's gravity. Second, he swindled himself again by confusing space and time, as appears from Figure 7.3, where HI, IJ, JK represent equal intervals of *time* and IL, JM, KN the *velocities* gained under gravity. He then interpreted the diagram as it appears to be, a drawing of *spatial* relationships, taking IX, JY... as the equivalent of the subtangent GE in Figure 7.2 and HI, HJ...as the associated tangent HG. Since the subtangent goes to zero faster than the tangent, no extrusion can occur.[89] Third, he overreached. He needed only to demonstrate, against Ptolemy and Tycho, that extrusion can be prevented by gravity, not that, irrespective of the magnitude of the spin, it can never occur at all. The idiosyncrasy of Galileo's black-and-white approach appears when comparing his treatment of extrusion with Sarpi's. Fra Paolo understood that two independent quantities were at play, the velocity of spin and the acceleration under gravity, and that extrusion does not occur for us because on earth the latter overcomes the former. If, however, the velocity of rotation increased indefinitely while the gravity remained the same, eventually we would be thrown off our earth like so much mud from a wagon wheel.[90]

When not talking philosophy or praising one another, Salviati and Sagredo treat Simplicio as the dolt in a Socratic dialogue, drawing from him the demonstrations they put in him, patronizing him when he succeeds ("[you] have shown yourself half a geometer") and insulting him when he fails ("if you had paid attention...you would not have thought up such a silly idea").[91] The poor man is hopelessly ill-equipped. "If Aristotle is to be abandoned...Suppose you name some other author?" An excellent question: without texts

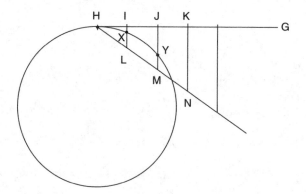

FIG. 7.3 A confusion of space and time: the horizontal segments represent time or velocity, the vertical ones space or acceleration.

where does the investigator begin? Salviati has no answer. Instead he offers
a false dichotomy. "It is impossible that one of two contradictory propositions should not be true and the other false." Simplicio—he is simple—
acquiesces.[92] Now, when Salviati proves that Ptolemy and Aristotle are
wrong, and opposed to heliocentrism, it will follow that Copernicus must
be right.

The second day concludes repetitiously by responding to three anti-
Copernicans whose arguments Simplicio reads out one by one. These old
sinners are Chiaramonti, Scheiner, and Scheiner's student Johann Georg
Locher. Scheiner and Locher had remarked that if stones falling onto the
earth from the sphere of the moon also participate in the earth's hypothetical spin they must execute spirals of varying complexity depending on their
angular distance from the equator. The same thing must be true of birds. In
order to stay suspended over their nests or to swoop down on their prey,

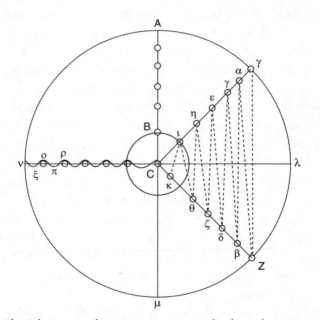

FIG. 7.4 The Scheiner–Locher argument against the diurnal motion. All points
above the earth's center C rotate around the axis νλ in 24 hours. A ball let fall
from the pole ν would reach the earth along a straight line; from A, on a spiral
in the plane of the equator; from Y, along a spiral on the surface of a cone. After
Koyré, APS, *Trans.*, 45:4 (1955), 332, a simplification of the diagram in Scheiner–
Locher, *Disquisitiones mathematicae* (1614), 30.

they must perform acrobatics of staggering intricacy. An imposing drawing, on which Sagredo and Salviati exercise their sarcasm, depicts everyday space traffic—birds, balls, and a snail (bird food?) going down, rockets, fireworks, cannon shots going up (Figure 7.4), all improbably managing apparently vertical motion by intricate unseen dances. How do they do it? Is the principle intrinsic or extrinsic, natural or violent, compound or simple, angelic or aerial, accidental or substantial? In every case neither prong of the dichotomy will do if the earth turns.[93]

Here is another conundrum, aimed at Copernicus' teaching that separated parts of the earth return not to the center of the universe but to the great body of rocky matter. Would a stone at the center of a spherical cavity concentric with the earth rise to the shell or stay put? Salviati: the cavity would collapse. Again, according to Scheiner and Locher, the earth is far too small for the tasks that Copernicus imposed on it. For in a year it would roll out only 365 times its circumference. But the suppositious annual circuit is 1,200 times the earth's circumference. Copernicus requires almost four years where one is seen to do. Simplicio likes the argument. What do you think of that? Not much, says Sagredo, and threatens to go off in a gondola.[94]

It is Chiaramonti's turn. He holds that Copernican theory would ruin philosophy. All philosophers agree that knowledge begins with sense impressions. Copernicans deny the testimony of their senses. If we follow them, "we must necessarily suspect our own senses as wholly fallible or stupid…Then what truth can we hope for?" None, to be sure. Unfortunately for him, Chiaramonti descends from this secure position to disallow compound motion. Wherefore? Because nature gives joints to bodies that have to perform more than one motion simultaneously and the earth has no joints. Simplicio, who has been reading out this nonsense, thinks it ingenious. Salviati: "Are you serious?" Simplicio: "I am giving you the very best that is in me." After disposing of a jointed earth, Salviati points out so many trivial errors of Chiaramonti about the Copernican system that even Simplicio blushes for his author.[95] Jointless or not, Salviati protests, everybody picks on the earth. It may be dark, but so are the planets. Earth is no more the sink of corruption than the moon or Jupiter. And if it were the muck pile of the universe, why should it be placed at the center? Sagredo: that would be like placing the lazarhouse in the middle of the city. Salviati: "Sagredo, you are too caustic and sarcastic."[96]

Sagredo calls it quits for the day. There would be no gondola ride. Simplicio had to study Scheiner's arguments against the annual motion, Salviati

Chiaramonti's against the parallaxes of novas. As we know, Sagredo had other ways to amuse himself at night. "Tomorrow we shall take up the discourse again…hoping to hear great new things."[97]

Day 3. While awaiting the arrival of Simplicio, Salviati and Sagredo amuse themselves at the expense of philosophers who dishonor the human race by slavishly following some author. "Their company may be not only unpleasant but dangerous." They except Simplicio from the condemnation "as a man of great ingenuity and entirely without malice." A pity he knows no geometry. Lo! The man without malice or geometry enters, out of breath, having been detained in a narrow canal where his gondola ran aground. He had used the time productively by watching the tide ebb and flow, the tiny residual trickles turning into rivulets without an instant of rest between fall and rise. Salviati undertakes to explain the effect until reminded of his assignment to report on Chiaramonti's arguments against the parallactic measurements that placed the novas in the heavens.[98]

Chiaramonti's main argument, a continuation of his *Antitycho* (1621), turned on comparing the locations of the nova of 1572 as determined by a dozen observers in different places and presented and analyzed by Tycho. Taking the observations in pairs and calculating the implied parallaxes, Chiaramonti found that some pairs put the nova under the moon, others among the fixed stars, and still others beyond the firmament. There is no truth in mathematics! To bring the matter to Simplicio's level, Salviati drew Figure 7.5, in which ABE is the earth centered at D, A and B are the two observers stationed, for ease of representation, on the same meridian of longitude, DE is the intersection of the equator with the plane of the paper. The observers measure the zenith distances α and β of the star S. The angular distance λ between their stations as seen from D is the measurable difference in their latitudes; and so the parallax angle, $\angle ASB = \gamma$, may be determined. The method is simple: the sum of the angles in the quadrilateral DBSA must be 360°; two of the angles are the supplements of α and β; hence $\gamma = (\alpha + \beta) - \lambda$. If S is very far away, observers at A and B see it in the same direction, $\gamma \approx 0$, and $\alpha + \beta = \lambda$. If $\lambda > (\alpha + \beta)$, S has no place, or is beyond the firmament; if $\gamma > \gamma_0$, the lunar parallax, S is below the moon. I know all that, says Simplicio, as I've read in Aristotle that the sum of the angles in a triangle is two right angles. Thank the Lord, Salviati replies, in his annoying condescending way. "I was worried about not being able to

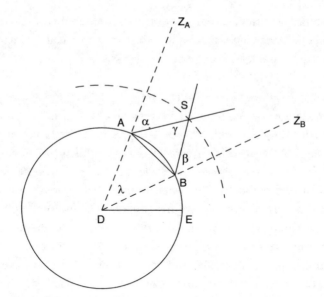

FIG. 7.5 The parallax γ of an object S as determined by observers at A and B for whom it has zenith distances α and β.

explain it in such a way that a pure philosopher and Peripatetic would get a firm grip on it."[99]

Now the work begins, computing the distance to S in terrestrial radii from the known angles and the law of sines. The computations, which baffled Simplicio, will be obvious to trigonometers and of no interest to anyone else. Salviati exhibits them in detail, making many computational errors in the process. He premises that all the observations erred to some extent. The job of the astronomer is not to select the pairs that put the star wherever he wants but to calculate whether with minimal corrections most of the observations can be brought into agreement. Salviati's shrewd and commonsensical discussion concerns the manipulation of data rather than an analysis of instrumental error. It comes to this: five of ten measurements place the nova beyond the moon, and another five do so also with a total correction of 10¼ minutes of arc; to obtain ten measurements that place it halfway between the earth and the moon required repairs totaling 756 minutes. Simplicio: "the ineffectiveness of this author's proofs thus seems to me to have been very clearly exposed."[100] However, they occupy only a small part of Chiaramonti's big book. Perhaps his other proofs are better? They must be worse,

Salviati assures Simplicio, though he has not read them, since they suppose observations more difficult than those already discussed. In contrast, other observations—that the nova never changed its distance from any fixed star and that every observer found its zenith distances at upper and lower continuation to be equal—confirm the true picture.[101]

Returning to Socratic mode, Salviati directs Simplicio to draw the positions of the luminaries and the planets. Put the earth and the sun wherever you want. Now, where do the planets go? Simplicio dutifully puts the inferior planets around the sun because of their limited elongation and Venus' phases. And where does Mars go? Evidently around the earth, since we can see it at any angular distance from the sun; and also around the sun, since it does not run into or occult Venus. Jupiter and Saturn have similar orbits enclosing that of Mars. Well done! "So far you have comported yourself uncommonly well." And now comes the moment of truth. Which is at rest in the suppositions center, the earth or the sun?[102] While Simplicio hesitates to embrace Salviati's view, his cleverer friends go back to philosopher-bashing. Sagredo: why have people not admitted the Pythagorean scheme that Simplicio almost drew spontaneously? Salviati: it is better to wonder how this world of fools and idiots can have generated a few minds so superior that "through the sheer force of intellect [they have] done such violence to their own senses as to prefer what reason told them over what sensible experience plainly showed them." Only "superhuman souls" can perform such feats.[103]

Simplicio admits to being one of those simpletons who cannot understand how a heavy body like the earth can run around in space. The wise men ignore the implied question. Instead Salviati tells Simplicio about the telescope, "[which] it has pleased God to concede to human ingenuity." This instrument has resolved puzzles that perplexed even Copernicus. It shows the correct sizes of celestial objects by stripping them of their irradiation, provides the missing phases of Venus, and annuls earth's uniqueness in possessing a moon. Still, no one needs a telescope to persuade himself of the sun's rest and the earth's motion. A naked eye connected to an unprejudiced mind can infer the truth from the retrogradations of the superior planets. What Salviati had in mind appears from Figure 7.6. From time to time a superior planet slows in its easterly motion, stops, moves backwards ("retrogrades") to the west, stops again, and resumes its easterly journey; in mid-retrogradation, the planet, being in opposition to the sun and closest to the

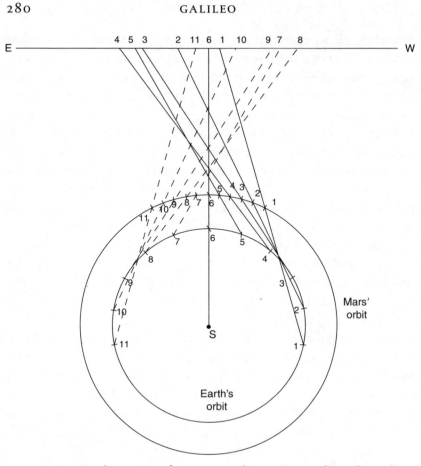

FIG. 7.6 Retrograde motion of a superior planet. As seen from the moving earth between positions 1 and 4, Mars appears to move "directly," from west to east; between 4 to 8, including opposition (6), "in retrograde," from east to west; and between 8 and 11, again directly.

earth, is at its brightest.[104] Everything falls into place if the earth occupies the third heaven.

Salviati proceeds to a second argument for the annual motion, a new one, unprecedented, based on the motion of sunspots, and discovered, like "all other novelties in the sky," by Galileo. In a passage of sustained bluff and deceit, Salviati steals Scheiner's observations of the trajectories of the spots and the solar theory based on them and makes both serve the Copernican cosmology. With the help of a network of Jesuit observers, Scheiner had found that the trajectories curved upwards for half the year and downward for the other half and appeared straight only at the transition. He remarked that if the axis around which the spots turned inclined toward the ecliptic, observation would agree with theory; and, after many careful measurements, he determined the inclination to be 7° or 7°30′, a very good approximation to the modern value (7°15′). Scheiner worried that Galileo would learn about it before his long-winded masterpiece *Rosa ursina* appeared, and he asked his correspondents to maintain it in confidence lest the Censor (as he called the Assayer of Jesuit astronomy) steal it.[105] And so he did, although *Rosa* was published two years before the *Dialogue*.[106]

On Day 2, Salviati had made the solar axis perpendicular to the ecliptic and the spot trajectories parallel to it, as in Galileo's *Letters on sunspots* (1613). On Day 3, after claiming the discovery of the spots for Galileo and dismissing Apelles as "vain and foolish," Salviati admitted that the trajectories curve, and that he had known it all along. In proof whereof he told the following cock-and-bull story. One day, when staying with Salviati at Le Selve, Galileo saw a fat spot traversing a conspicuously curved path. He grasped at once that the axis must be tilted. And if the axis is tilted? "Filippo," quoth Galileo, "if the axis around which the sun revolves is not perpendicular to the plane of the ecliptic...then we shall have a more solid and convincing theory of the sun and earth than has ever yet been offered by anybody."[107]

Galileo needed such a story because, even before laying eyes on *Rosa*, he was trumpeting that it could contain nothing new. "I am sure that if he says anything different from what I said in my *Letters on Sunspots*, it will all be vanity and lies." That was to dig himself into a hole from which he could only lie his way out. He was encouraged by Castelli, who had sniffed the *Rosa* and found it "stinking." Together with other delicate Galileists, Castelli suggested that someone complain to the Jesuit General about the nausea the work caused them.[108] What would be the charge? Enticing Galileo to trample the truth? Using a Keplerian rather than a Galileian telescope? Publishing a book so lavish in layout, so full of excellent plates, so prolific with depictions of instruments, that Galileo could never hope to equal it?[109] Scheiner complained that Galileo had laid "violent hands" on his life's work. That was not to do violence to human rights. To Galileo Scheiner was an animal, a pig, an ass.[110]

Returning to the sublime, Salviati explained that if the sun stands in the center with its direction of spin fixed in space (Figure 7.7), everything observed takes place without special additional hypotheses. Straight trajectories do

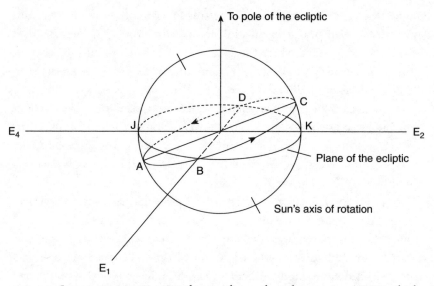

FIG. 7.7 Sunspot trajectories. Let the spot lie on the solar equator ABCD, which meets the plane of the ecliptic at B and D. Seen from E_1 (E_1 B is perpendicular to the plane of the paper), the spot moves along the straight line AC; seen from E_2, along a path curving up toward C and down toward D; seen from E_3 (not shown) beyond D on the line E_1BD, along CA; and seen from E_4, along a path curving down toward A and up toward B.

occur, but only when the line of sight from the moving observer falls perpendicularly on the solar axis. Simplicio brought himself to ask why the same phenomenon would not follow if the sun moved. Salviati replied that the sun then would have to move simultaneously in three distinct ways: once a year around the earth and once a month around its axis; and, while this is going on, the axis would have to revolve around the perpendicular to the ecliptic. Scheiner had introduced this third motion to keep the axis of the moving sun pointing continually in the same direction. He borrowed the device from Copernicus, who had ascribed such a revolution to the earth's axis of spin to keep it always parallel to itself. On the assumption that the annual revolution takes place as if the earth rotated like a weight on a stick (Figure 7.8), an axis EA fixed in it would be carried from E_1A_1 to E_2A_3 during half a year; to keep it parallel to itself, to bring it to E_2A_2, it would have to rotate 180° around the moving perpendicular to the ecliptic EB, keeping the angle of inclination BEA constant. As appears from consideration of a quarter turn, Copernicus'

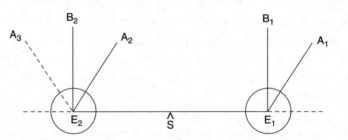

FIG. 7.8 Copernicus' third motion. If the earth went around the sun like a rock in a sling, its axis would change direction, as from E_1A_1 to E_2A_3. The axis would have to rotate back through 180° to E_2A_2 to remain parallel to itself.

third rotation has the additional awkwardness of opposing the sense of the annual revolution.[111]

If the sun rested at the center it would not require the retrogression of the axis to save the phenomena of the sunspots. That was enough to ravish Sagredo. "I have not, among the many profundities I have ever heard, met with anything which is more wonderful to my intellect or has more decisively captured my mind…than these two conjectures, one of which is taken from the stopping and retrograde motion of the five planets, and the other from the peculiarities of movement of the sunspots." Salviati met this enthusiasm with dissimulation: he has been giving the best arguments he can, leaving the decision to others. The decision came down to a binary choice, Salviati went on, with another of his preposterous paralogisms, "since one of the arrangements must be true and the other false."[112] Simplicio must have looked surprised at this violation of sound logic since Salviati returned to the question whether the Copernican system gave the simpler explanation of the motions of the spots. If the earth has three motions, why not the sun? The answer gave another glimpse at Galileo's kinematics. There is no need for the third Copernican motion, Salviati said; the earth will naturally conserve its direction of spin. In its annual motion it does not revolve like a rock in a sling but like a float in a tub of water turning around a vertical axis through its center. As the tub revolves clockwise, the float, say a wooden disk, appears to rotate counterclockwise so that a needle lying on it always points in the same direction. That is the way things are. "Any suspended and balanced body you please [will behave that way], and without requiring any cause of motion."[113] Here Galileo's kinematic intuition led him to a useful result, and also a correct one, since the conservation of angular momentum, a principle unknown to him, annuls Copernicus' third motion.

But—to return to Simplicio's question—would the argument not also apply to a moving sun? If it did, the claim that the Copernican theory afforded a simpler explanation of spot trajectories would fail. Salviati held his ground, muddily, by asserting that a third motion would be necessary if the sun did not rest. In this he was right, since if, during its diurnal revolution around the earth, its axis maintained the same orientation in space, the spots would appear to spin around it in a day. This consequence jumps out from Figure 7.9, where OQ is perpendicular to the ecliptic, OP is the sun's axis of spin, and X is a sunspot as seen on the sun's disk by an observer on earth. Let ETO be perpendicular to SOU at sunrise: X will appear on the sun's eastern limb. At midday, the line of sight EO passes through S; the great circle QUS and therefore X lie in the plane of the meridian, and X appears to have traveled halfway across the disk. At nightfall, EO passes through V and X appears on the sun's western limb. To correct for this effect the sun would require a retrogression similar to Copernicus' third motion. The argument is not decisive but does have an appeal to non-Simplicios who like simplicity.[114]

In partial extenuation of the *Dialogue*'s putdowns of peripatetics and magnifications of Galileo, it may be placed in the genre of epideictic rhetoric, a

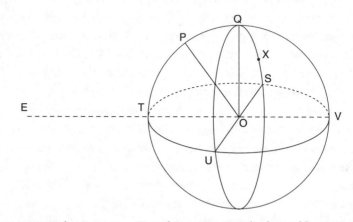

FIG. 7.9 A spot that appears at X on the sun's eastern limb would appear near its center at noon (when SXQU is in the meridian) and on its western limb at sunset if the sun circled the earth and its axis remained parallel to itself.

favorite Renaissance style of praise-and-blame. Often the cardboard characters in these exercises wore masks to give them greater leeway in throwing bouquets and brickbats. Galileo's mastery of the style extended to the postils, or marginal notes, in the *Dialogue*. They introduce another unseen party, another voice in the praising and blaming. The postil writer is pungent and pithy: "a philosopher's ridiculous answer"; "very puerile conclusions"; "paralogism of this peripatetic." Another standard ingredient in epideictic rhetoric was the sort of inversion in which Galileo excelled: "you are not only wrong, sir, but the truth is the exact opposite of what you say."[115] If Galileo consciously adopted this rhetorical mode in the expectation that his victims would grin and bear it, he was mistaken.

7.3 THE WINDMILL

Urban's mood swings made maneuvering at his court unusually hazardous. He could drop highly placed officials without a moment's notice. Ciampoli thus fell in 1632 after a decade's service as secretary, special agent, and entertainer.[116] He nursed his wounds in a bitter diagnosis of life at Urban's court in the form of advice to a would-be courtier. Here is a summary. Seek patronage among functionaries and chamberlains, for they, and not the court cardinals, have the power; but do not aim too high. Confide in no one; believe in no one; do not meet with other courtiers in your room, lest someone spread rumors of a plot. Avoid ostentation; do not talk about the prince or about court scandal, do not talk cleverly, try not to talk at all. Never criticize priests or monks publicly, show no preference for any religious order; appear to be religious, pious, and zealous, for hypocrites do well; "truth in all courts, and especially in Rome, is a great impediment to advancement." Are you spied on? Honor the spy; "simulation is the life of the court." Do you want to destroy a rival? Expose his love of women or money. But watch out for the sharpers, bad mouthers, and double dealers who would do the same to you. Then, if you serve your patron as if he were a god, keep your mouth closed and ears open, avoid appearing clever, and remember that patience is to the courtier what chastity, poverty, and obedience are to the monk, you might do well before you fall.[117]

Fall was inevitable, if only because the next pope would replace his predecessor's clients and creatures with his own. Meanwhile the courtier around the Curia had to speak and write in a dead language, negotiate the pitfalls of

an all-male society, square profane art with sacred scripture, escape tramp-
ling in the scuffles among the religious orders, avoid damaging political
allegiances, and curry favor with cardinals likely to be pope.[118] Needless to
say, the person most exposed to the plots and vagaries of the Vatican and the
external events that drove them was His Urbanity, the Pontiff himself.

Dominicans

A world-class acrobat among high-wire courtiers was Galileo's acquaintance
and would-be correspondent Tommaso Campanella. During his 27 years
of confinement in Naples before his release in 1626, Campanella had written many compromising things, for example, a forecast of an imminent
schism in the papacy and the dissolution of the College of Cardinals, news
he forwarded to Paul V, "because...I thought Your Beatitude would like to
know."[119] Campanella had obtained this information through "God's benevolence" and his own reading of the stars. He had picked up astrology from
della Porta and made the fatal mistake of calculating the success of his plot
against the Spanish government of Naples from his own distinguished nativity. In prison he had leisure to refine his procedures to the point that, despite
his incarceration as an enemy of the state, distinguished citizens and prelates came to his cell for astrological consultations. He considered *Sidereus*
nuncius a confirmation of his prediction that a universal restructuring of the
sciences, if not of the Neapolitan regime, was in the offing. He proposed to
reciprocate the discoveries of the starry messenger by prescribing for Galileo's ailments, at a distance, from knowledge of his birth chart. In declining
to furnish it, Galileo expressed reservations about the art. Campanella, who
knew dissimulation like the back of his hand, spied hypocrisy. I know you
believe in astrology, he wrote, from your references to Jupiter in your dedication of the *Starry message* to Cosimo. "It would not have been correct to make
use of opinions believed only by common people."[120] Campanella's mastery
of astrology would enable him to leapfrog over the other amphibians grop-
ing for Urban's favor.

    Campanella's arrival in Rome in 1626 followed that of his *Apologia*, or
Defense, of Galileo, written in 1616, published in 1622, and banned in 1623.
The occasion of its composition, according to its preface, was a request to
Campanella by Cardinal Caetani for a theological evaluation of Galileo's
Letter to Castelli. It may be that Campanella had completed his *Defense* before

the March decree of the Holy Office banning Copernican works, and therefore did not disobey it; but whether the Cardinal solicited the opinion of a jailbird and, as Campanella claimed, urged him to publish it, may be questioned.[121] In any case, Campanella's ingenious defense of Galileo's position as more consonant with scripture than the philosophico-theological interdiscipline that opposed it could not stand in 1623. Campanella had built on two pillars. The Fathers disagreed about whether earth, sun, or firmament stands or moves; Moses spoke the language of common people and neglected astronomy; standard theological sources therefore left room for Galileo's interpretation, indeed, rightly considered, favored it.[122] The second pillar, which Campanella erected with many spiteful grunts, crushed the authority of Aristotle and the "potbellied theologians who locate the limits of human genius in [his] writings." Those who hold this opinion were either crazy, invincibly ignorant, or "embarrassed to become students now that they are called teachers." In hitching theology to Aristotelian cosmology these ignoramuses failed to notice that it is both inadequate, as it cannot even explain the stations and retrogradations of the planets, and impious, as it teaches the eternity of the world. "[I]t is astonishing how many more vistas, in which God reveals his wisdom, power, and love, have been discovered by Galileo." Down with Aristotle! Down with his plague of heresies, his eternal world, his mortal soul, and his frightful God, indifferent to our prayers and a slave to necessity![123]

More closely considered, Galileo's astronomical ideas, far from being heretical, derive from Moses.[124] Yes, assuredly. The Copernican theory goes back to Pythagoras, and Pythagoras was either a Jew or a vehicle for Jewish thought. That evil genius Aristotle had no time for Moses or Pythagoras. So Galileo came to rescue "the sacred philosophy of Moses from the insults of the pagans by using the most discriminating arguments and instruments."[125] Let us therefore be Jews, or, rather, Christians, not Aristotelians, and admit with Galileo that the moon has mountains and that most philosophers are fools. Galileo must not be condemned. Foreigners would laugh, theologians distort, philosophers strut. "[Galileo] has not called on us to do anything wrong but rather to search for the truth."[126]

The Roman Inquisition jailed Campanella in June 1626, a month after the Neapolitan state had released him. He extricated himself by praising Urban's poetry. He is greater than Dante, Plato, and Moses; nay, a new David, a new Orpheus. His poems excel in meat and meter, grammar and philosophy, and, unexpectedly, freedom. (The liberty relates to Aristotelian rules, which Campanella had urged abandoning many years earlier.[127]) Proceeding in this vein, "with demure seriousness and prolix erudition, as if he dealt with a great masterpiece," Campanella appealed to a vanity inflated "beyond the limits of good taste and even of modest common sense." Still, he managed to go too far. Campanella pointed to Urban's sublime verses praising Galileo's astronomical discoveries as harbingers of the new world of humane letters and religious harmony that Christianity was about to enjoy. The pope's words already radiated from the center just as the rays of the sun do, "according to Copernicus." Had not Urban, when only a cardinal, made sure that "the opinion of great Copernicus, purged of error, could be read hypothetically, to the advantage of philosophers and the well being of the state"?[128] The brilliant poet pope, "from whose house a new day in the protection, promotion, and reformation of the divine muses begins to dawn," deserved something special. Should we not name those unemployed blobs that Galileo had found around Saturn after him?[129]

Urban took the bait. In 1628, after a quick withdrawal of his eccentric commendation of the pope's Copernican commitments, Campanella exchanged his cell in the Inquisition's prison for one in the convent of Santa Maria sopra Minerva. A few months later, in January 1629, he obtained his freedom.[130] Urban had directed this act of clemency out of fear as well as vanity. He thought he needed Campanella's expertise to counter the predictions of the most reliable astrologers money could buy. For months they had been busy circulating predictions of Urban's imminent death by comet or eclipse. They had an eager audience in courtiers hoping for preferment under the next pope and in Spanish and Austrian agents praying for regime change. During his imprisonment by the Holy Office, Campanella had received a visit from one hopeful successor to Urban, no less a person than the Master of the Sacred Palace, the Dominican Niccolò Ridolfi, who had with him calculations that forecast Urban's death in September 1628. Campanella found an error in the reckoning but sweetened the news by predicting Ridolfi's elevation to cardinal in June 1629. That was almost right; in that month Ridolfi rose to the post of Dominican General. The astrologers, among whom another Dominican, Raffaelo Visconti, was a leader, refined their calculations: Urban would die in February 1630 and be succeeded seriatim by Dominicans, first Ridolfi and next Father Monster. Either they or the Angel of Death missed the mark. Urban's brother Carlo, the propagator of

VAINGLORY   289

the Barberini line, died that February. The calculators postponed the due
date for Maffeo to the summer. Borgia summoned the Spanish cardinals to
Rome, and the French and Germans followed, as if the vacancy they longed
for already existed.[131]

In the summer of 1628 the pope had summoned Campanella to his rooms
on the Quirinale to work astral magic against the malign effects of eclipses.
Campanella had a theory apt for the occasion. Since, according to a famous
dictum of Saint Thomas, the stars work only on our bodies, a person able
to resist his appetites can thwart them. To which Campanella added, from
bitter experience, that resistance to astral forces was easy for anyone who
had not broken under torture.[132] To counter the influence of an eclipse (a
great danger for the son of the sun), it was only necessary to close yourself in a dark chamber and there recreate the normal state of the heavens,
with lamps representing the planets and luminaries, and incense, aromatic
herbs, and music to indicate the harmonies of the universe. The star-sick
pope and his astral doctor performed these rites together. Urban emerged
more robust than ever. In October 1629, at Castel Gandolfo, he displayed his
strength in a wrestling match with Ciampoli and his good sense in not trying
a fall with the Monster. Campanella dismissed the dire predictions regarding
the summer of 1630 for neglecting the beneficial effects of Mercury.[133] He
was right again. Campanella's enemies worried that Urban would make him
a consultor of the Holy Office or even a cardinal.

To clip the wings of the former jailbird, the cardinal nephew and the hopeful Dominicans Ridolfi and Riccardi hatched a plot to which Campanella's
manipulations had exposed him. Eager to publish his store of manuscripts,
he had cultivated Ridolfi and Riccardi, both of whom, in sequence, held the
highest censorial post in Rome. Among the items Campanella had in press in
1629 was a treatise on astrology in six books. Around the same time, he gave
Riccardi a copy of a seventh book, "On avoiding the fate of the stars." Ridolfi
and Riccardi decided to publish it. They obtained the manuscript of the first
six books from the printer in France and brought out the lot in Rome, with
a fake imprint and no imprimatur.[134] A disgraceful proceeding for chief censors! Since the volume had no license and contained the silly prophylaxis
against eclipses that Campanella and Urban had performed at the Quirinale,
and since all Rome knew that the pope had consulted a "Dominican...who
professes astrology and also necromancy" (so the Venetian ambassador
wrote home), the publication embarrassed Urban.[135] He was very angry.

Campanella denied authorship, blamed Ridolfi and Riccardi, and issued the news that Urban, "as a very wise pope, detested astrology." That saved his skin but ended his rise.[136]

The center for astrological computations in Rome was the Vallombrosan convent of Santa Prassede, presided over by Galileo's schoolboy friend Orazio Morandi. The two had remained in contact despite Morandi's service as astrological advisor to Don Giovanni de' Medici. In May 1630, when in Rome to try to push the *Dialogue* through the censorship, Galileo dined with Morandi in his astrological center. Soon thereafter Urban ordered Morandi's arrest for presuming to calculate the chances of the cardinals likely to succeed him.[137] The case against Morandi turned on his manuscripts and library. To protect him, his monks burned the papers and hid the books. Eventually, however, their location came out and so did the truth. Morandi read very good books—Aristotle, Plato, the classical poets, the best mathematicians—but also terrible ones, banned books on magic, demonology, and history, notably the works of Paolo Sarpi. Important people compromised by entanglement in his circle found his continuing existence inconvenient. Therefore he died in prison, of fever said the doctor, of poison rumbled rumor.[138] Although Galileo had no part in Morandi's research into the death order of the high and mighty, or in the circulation of the forecasts of Urban's death, he took care to clear himself, with Buonarroti's help, of every shred of suspicion.[139] The episode showed Urban that astrology was too dangerous to be allowed to anyone but himself. In 1631 he issued a bull, *Inscrutabilis*, against divination, especially against foretelling the deaths of popes or members of their families.[140] The order stigmatized the art as the product of pride and ambition, and included a statement in harmony with Urban's voluntarism and unfriendly to cosmologists as well as to astrologers:

> The inscrutable profundity of the judgments of God does not allow the human intellect, confined to the dark prison of the body, to rise beyond the stars. Yet not only does it dare explore with impious curiosity the mysteries buried in the depth of the divine and unknown even to the saints, it also presumes, with arrogant and dangerous example, to circulate these mysteries as certainties, with contempt for God, disturbance of the state, and danger to princes.[141]

Verbum sapienti!

Jesuits

Among Galileo's sparring partners who turned up in Rome in the 1620s was Scheiner, who came on business of his college in Neisse and stayed to publish his masterwork on sunspots. Ignorance of Italian had spared him from feeling the full sting of the references to him in Galileo's letters to Welser of 1613 and the implication of plagiarism in Guiducci's attack on Grassi. Up to the wrangle over the comet, Scheiner had regarded Galileo as a comrade in arms, sometimes rash but often right. In 1615, having swallowed lunar mountains and Galileo's proof of Venus' circumsolar orbit, he defended Galileo's priority over Mayr in the discovery of Jupiter's satellites. He presented Galileo with several of his books dealing with telescopes, vision, the shape of the sun, and the flaws of heliocentrism as elaborated by Locher.[142] But his efforts to open a correspondence with "the distinguished and brilliant Italian mathematician...in the hope that from it greater light be shed on the truth" met with the same silence as his presents and concessions.[143]

By the time the distinguished brilliant mathematician deigned to reply, in the *Assayer*, Scheiner had learned enough Italian to read Galileo's insults with full understanding and proportional irritation. He was incensed at the insinuation that he had plagiarized Galileo's work on sunspots ("Apelles...accused of infamous thievery!") and aggrieved at Galileo's offhand dismissal of one of his better discoveries as trivial and mistaken ("[Scheiner]...exposed to mockery and childish disdain!")[144] He had announced this discovery, that atmospheric refraction squashes the sun into an ovoid shape when it nears the horizon, in one of the books, *Sol ellipticus* (1615), he had given Galileo in unreciprocated collegiality.[145] Galileo did not like it. It showed that he had missed something important in his solar surveys. He thanked Scheiner in the *Assayer* by reducing the effect to a trivial case of foreshortening owing to the displacement of the observer's eye from the center of our atmosphere. "The entire mystery requires no doctrine more profound than to understand why a circle seen straight on looks round to us, while looked at in foreshortening it appears oval."[146] The effect of atmospheric refraction on astronomical observations was then an agitated and fundamental problem implicated, as Scheiner remarked, in optics and meteorology as well as in astronomy.[147] Again Galileo ran from physics to geometry and disposed of the baby with the bathwater. And again his inability to appreciate the difficult position of a Jesuit struggling to introduce new knowledge caused him to make an enemy of a potential ally.

Scheiner's ongoing exposition of sunspots and marginally freethinking about their cosmic implications had placed him in a position deserving more sympathy than censure.[148] General Acquaviva had ordered him to avoid novelties.[149] Acquaviva's successor Muzio Vitelleschi frequently admonished Scheiner to diminish his expenditure on instruments, to remember his vow of poverty, and to be more charitable to his better-disciplined colleagues.[150] Had Galileo known these orders, they would not have mollified him. He picked on Scheiner, as he did on Grassi, as representatives of the Jesuit mathematicians too cowardly to stand up for him in 1616.[151]

Another matter on which Vitelleschi felt obliged to instruct Scheiner was politics. Avoid them, he ordered, or, rather, avoid appearing political when using your position as confessor to Archduke Karl I, and friend to Archduke Leopold, to influence their brother Ferdinand II, the Holy Roman Emperor, "for the benefit of themselves and of all Christendom."[152] The brothers were at the heart of the Counter Reformation beyond the Alps. Ferdinand's policies prompted the action that set off the Thirty Years' War, and Karl and Leopold strongly supported the papal shock troops, the Jesuits. Leopold built them a church in Ingolstadt, Karl a college in Neisse. Scheiner supervised the construction of both and, perhaps, Leopold's study of astronomy, in which the archduke took an informed interest. When he visited his sister the Grand Duchess of Tuscany in 1618, Leopold formed a tie to Galileo, who subsequently sent him copies of his treatise on the tides ("read it as fiction") and his *Letters on sunspots* (read it as a corrective to Scheiner). Galileo and Guiducci dedicated their attack on the Roman Jesuits to Leopold. None of this drove a wedge between Leopold and Scheiner. In 1624 the archduke sent his protégé to Rome to obtain Urban's approval for the Neisse college and its finances. Scheiner found the going tough, since Urban had little sympathy for Jesuits or Austrians; but it proved easier to find support for the college than for the Ingolstadt church, which, to Scheiner's infinite chagrin, collapsed as it climbed into the sky.[153]

While Urban procrastinated, Scheiner tried to print his 784-page, in-folio page-turner on sunspots. Being four years in press, it made a perfect occupation for a waiting supplicant.[154] The odd title of the book, *Rosa ursina*, paid tribute to the dukes of Orsini, whose emblem was a rose and whose cash paid for the printing. This represented a serious and symbolic shift of patronage, since the Orsini had been friendly to Galileo.[155] When printing began, Scheiner expressed his fear that Galileo would get wind of it and somehow make trouble. And no doubt Galileo would have protested had he known that the first of *Rosa's* four books would be dedicated to him. In it Scheiner itemized their respective contributions by date, claimed priority for what was his, and lovingly chronicled Galileo's errors, some two dozen in all. He would pay for his lèse-majesté.[156] Scheiner followed this chronicle with explanations and depictions of the instruments he had used. His accumulated observations occupy no fewer than seventy plates. The final section of *Rosa*, some 310 pages, considers the theory of sunspots, the tilt and spin of the sun, the nature of flares and prominences, and the bearing of it all on scripture. The spin of the sun by no means implied that of the earth, which rests while the heavens turn, "as the holy fathers require."[157]

The book did not impress the Galileists. We know Castelli's reaction: "Revolting…I am disgusted by the atrocious and poisonous rage…arrogance…ignorance, conceit, hatred." He added from the vantage point of Urban's entourage that he could not abide Scheiner's parade of intimacy with all the archdukes in the universe.[158] The ponderous *Rosa* proved a greater weight to Scheiner than to his enemy. The book's first patron, Cardinal Alessandro Orsini, died before *Rosa* appeared. Paolo Giordano II Orsini, once and then perhaps still a friend of Galileo's, picked up the tab. He considered his investment an advance against sales. It turned out, however, that few people cared enough about sunspots and sun spats to buy 800 pages of Latin on the subject. *Rosa* failed commercially and Scheiner suffered under debt for many years. After a decade of nagging by General Vitelleschi, he discharged much of it by selling his instruments.[159] Vitelleschi's successor Vincenzo Caraffa had no patience with poor Scheiner. Relieve yourself of your belongings! Burn the letters you have written on the shortcomings of others![160] Obey! The recipient of these demeaning directives, a man 77 years of age, had been a distinguished member of his order for over fifty years.

During Scheiner's time in Rome, the big man at the Roman College was Grassi, whose persecution by Galileo Scheiner knew from its beginning, since Archduke Leopold, the dedicatee of Galileo–Guiducci's *Discourse*, sent him a copy. Grassi in turn followed the sun spat and doubtless sympathized with Scheiner, whose contributions he knew only too well, as he read (or said he read) the entire *Rosa* for the Jesuit censorship.[161] Unlike Scheiner, Grassi had too many obligations during the late 1620s to squander time nursing the wounds inflicted by the *Assayer*. But the fool, dunce, beast, buffalo, scorpion, and snake found enough leisure while serving simultaneously as professor, vice-rector, and librarian of the Roman College, and architect and builder of its church, to publish a response to the *Assayer* in Paris in 1626, without the usual permissions.[162] As we know, it played the high theological card of the eucharist. Grassi admitted that Galileo had said too little about atomism to allow a definitive deduction of his opinion about transubstantiation. "Yet I cannot avoid giving vent to certain scruples." They came to this: atomism deserves a stronger sanction than Copernican astronomy. "What has not been granted for the opinion on the earth's motion, although its immobility is not considered among the fundamental points of our faith, will be even less permissible, if I am not mistaken, for that which constitutes the essential point of faith or contains all other essential points."[163]

Galileo regarded the charges about the eucharist as equally shameful and impotent. The highest authorities had granted the imprimatur for the *Assayer*; Grassi's response did not have the permission even of local authorities. The Roman Jesuits repaired that omission by issuing a new edition with the required permissions. Galileo asked Castelli, whose service to the Barberini had brought him close to Ciampoli and other high officials, to ask Riccardi whether Grassi's charges now carried weight. Riccardi replied that he did not think so and that he would always stand up for Galileo—but not in every little matter that might come before the Holy Office. He was feeling pressure from his Dominican brothers for his leniency toward novelty. He and the Roman lynxes advised Galileo to ignore Grassi on this round and return to the *Dialogue*.[164] Galileo went back to sticking pins in Simplicio. He finished the first draft of the great work in 1629. That was also the year in which Melchior Inchofer, the third Jesuit conspicuous in Galileo's case, turned up in Rome and began his rapid climb to favor.

Inchofer was perhaps the quirkiest character in Galileo's quixotic adventures. Born into a Lutheran family in Hungary, he heard the message of the Jesuit missionaries, converted to Catholicism and, in 1607, at the age of 22, began his novitiate. After completing his studies he was assigned to the Jesuit College at Messina in Sicily, where he taught all subjects from logic to theology, held responsible administrative posts, and interested himself in the correspondence of the Virgin Mary. According to a legend kept green in Messina although condemned as apocryphal by the Inquisition, the Mother of God had instructed the Messinians in the true faith by a letter then still extant. Inchofer demonstrated its genuineness in a large folio published in 1629 that attracted the rapt attention of the Congregation of the Index. Closer to home, the book outraged the Archbishop of Palermo, Giovanni Doria, who regarded it as an aggravation of the contest between the center of Spanish influence in Palermo and the College of Jesuits in Messina for cultural control of the island. General Vitelleschi offered Inchofer every support and ordered him to Rome to justify his book there. It could not escape censure in Sicily, where the Spanish Inquisition held sway.[165]

Did Inchofer believe that Mary of Galilee was in correspondence with the citizens of Messina? The freethinking Gabriel Naudé, who arrived in Rome almost simultaneously with Inchofer to become librarian to a French cardinal, Jean-François de Bagny, thought not. Naudé put it to Inchofer that the citizens had forged the letter. "He replied that he knew all the reasons to think so as well as I did, and that he had written his book to please and obey his superiors...and that he did not believe at all what was in the letter." With the logic he had learned from study with Cremonini, Naudé deduced that Inchofer was a "sly and crafty fellow."[166] That he was willing to do others' dirty work he made clear in one of the many pseudonyms he invented to cloak his later writings: Benno Durkundurkhus S[k]lavus, perhaps German-Latin pidgin for "Benno, through and through a slave (or Slav)."

Durkundurkhus must have had considerable charm. He obtained a decision from the qualificators of the Holy Office and from Father Monster that the earlier decree regarding the Virgin's letter applied only to assertions of its genuineness; "but if the author only proposes [his argument] as probable, or not impossible...he can publish a second, corrected edition of his book." And so he did, with "conjecture" in place of "truth vindicated" in his title, using the same palette of cosmetics that Ingoli had employed in improving Copernicus' De revolutionibus. Inchofer became a great friend of Riccardi and a political advisor to Francesco Barberini. In this position, which he held from 1637 to 1647, Inchofer would have had opportunities to disclose his character to Naudé, who became Barberini's librarian in 1641.[167] It was through Riccardi, who had misjudged his character, that Inchofer, having just worked his way clear from charges against his own work, had a place on a committee to decide whether to charge Galileo. Through him Grassi and Scheiner could have a say in the proceedings.

7.4 THE TILT

Suiting up

"Oh what a pleasant and precious gift you give me this Christmas by telling me that your Dialogue is now happily concluded!" Thus Ciampoli, who had been urging Galileo to finish his masterpiece, reacted to the news that after thirty years on and off the job, Galileo had managed to bring it to fruition.[168] The Christmas in question was that of 1629. The following spring, as Galileo finished his polishing, Castelli and Buonarroti, the mathematician and the poet, were smoothing its way to Rome. Since Cesi and the lynxes were to see it through the press as they had the *Assayer*, Galileo needed permission to print in Rome. Castelli reported that Riccardi, then recently named Master of the Sacred Palace, was well disposed and that Ciampoli would undertake to secure Urban's assent. The protective cardinal nephew, Francesco Barberini, had reservations, however, about the argument from the tides. Castelli reassured him. The argument was entirely hypothetical: if the earth moved with the diurnal and annual motions, tides would occur necessarily. The cardinal objected that if the earth so moved, it would be a star, which seemed to him contrary to scripture. "To this I replied that you would have...proved that the earth is not a star, a thing very easy to show, since it is easy to prove that the moon is the moon and not the earth, Mars is Mars and not the moon or Venus, and so on." To which the insightful cardinal replied that Galileo had better supply this easy proof if the rest of his book was to pass muster.[169]

Cardinal Francesco correctly divined that Galileo did not intend to bring the earth down from the Copernican heaven. Puffing the *Dialogue* just before finishing it, Galileo wrote to Elia Diodati, a Parisian lawyer who had come to Florence especially to meet him, that it contained besides material on the tides "many other problems and a very ample confirmation of the Copernican system [accomplished] by showing the emptiness of everything that Tycho and others have brought against it."[170] However, the tides alone were more than enough. "If the earth is immobile, tides cannot occur; if it moves with the motions already assigned to it [by Copernicus], they follow necessarily." To the Bolognese count Cesare Marsili, then soon to be elected to the lynxes, Galileo confided that he could not separate his book from heliocentrism. "I'm dragging forward my *Dialogo del flusso e riflusso*, which drags along with it the Copernican system".[171] But on 3 May 1630, when Galileo arrived in Rome with his long manuscript, no one other than he knew exactly what was buried in it, although the local gossips chattered that it refuted Jesuit opinions and that its author cast horoscopes. Galileo received his usual cordial welcome from the Barberini.[172]

On 17 or 18 May Galileo had a long private interview with Urban. Although the gist of their conversation is not known, it eases understanding of subsequent events to suppose that by 1630 Urban had come to see that a demonstration by Galileo of the superiority of the Copernican hypothesis to its competitors might be useful to the church. He had liked the preamble to Galileo's unsent *Letter to Ingoli*, as read to him by Ciampoli; which, it may be remembered, justified presenting the advantages of the heliocentric hypothesis as a demonstration that the Roman Catholic Church had understood the astronomical and physical arguments when it condemned the Copernican system as opposed to scripture.[173] The May interview might have run as follows:

Gal. You know, Your Holiness, that knowledgeable people here and abroad laugh at the policy toward Copernicus decided in 1616. I want to counter their mockery by showing that Catholic astronomers know at least as much about the cosmos as Protestants do.

Urb. And you, Your Lordship, know that consideration of God's omnipotence makes asserting the absolute truth of any astronomical system very rash and perhaps heretical. If we allowed you to proceed, how would you accomplish both your purpose and ours: to show the world that Italians understand astronomical hypotheses and their uncurably hypothetical nature?

Gal. I would write—in fact I have written—a Dialogue in which persons representing the old and new systems and a man of aggressive common sense discuss the pros and cons; and I have given the Copernican spokesman the strongest arguments to show that we know them. It is a duel in masquerade, like my *Assayer*, which I believe Your Sanctity was gracious enough to like, and does not ascend above the level of hypothesis.

Urb. We could develop some enthusiasm about this project if we trusted you to stay strictly within the realm of hypothesis. Then you could frame your masquerade with a preface similar to the one you wrote in your *Letter to Ingoli* and a postscript declaring that however cogent the arguments given, neither system can ever be shown to be true. If we give our permission, you must put the preface and postscript in your own voice. This message is much too important to be imparted by masked men. Coming from you, it would establish the proper status of natural science and help free the church from the unhealthy coupling of theology and Aristotelian philosophy on which the Dominicans and the Jesuits have built their empires.

Gal. Agreed. The stronger the case made for Copernicus, the more effective will be the denial that we can ever know it to be true.

Urb. To be sure. And to be sure that you do not get carried away, we shall assign your friends Ciampoli and Riccardi to work with you on the preface and postscript. Your *Dialogue* will be as a shield against those who would derive absolute truth from a source other than Holy Church. I put my trust in you—policed by Ciampoli and Riccardi.[174]

A week after his long interview with Urban, Galileo dined, at Morandi's invitation, with Raffaelo Visconti and Ludovico Corbusio, whose positions suggest the purpose of the get-together. Corbusio, onetime Inquisitor of Florence, had become a consultor to the Holy Office. Visconti, whom Galileo had met through Morandi in 1624, had become Riccardi's lieutenant and would read the *Dialogue* for the Roman censorship. On 17 June, Urban spoke to Riccardi about the book.[175] Perhaps the pope then gave the three orders that Riccardi repeated a year later in Urban's name to the Florentine inquisitor Clemente Egidi. First, the title must not refer to the tides but to mathematical considerations of the Copernican view. Second, "with the intention of proving that without God's Revelation and sacred doctrine the phenomena could be saved on this view," Galileo should explode all the objections against the Copernican hypothesis suggested by experience and peripatetic philosophy. Third, "it must be made clear that the purpose of the work is solely to demonstrate that the Catholic Church knows all the reasons that bear on the question, and that it was not for not knowing them that the Copernican opinion was banned in Rome."[176]

In July 1631, Riccardi sent Egidi a fourth requirement, the administration of a "medicine of the end" to balance the draft preface he also supplied. This medicine, which we will call "Urban's Simple," must be compounded of the "arguments about Divine Omnipotence explained to [Galileo] by Our Lord, which should quiet the mind though there might seem no way through the Pythagorean arguments." Riccardi further instructed Egidi that the postscript (*fine*) containing the medicine "must concern the same argument" as the preface, that is, return to the argument of the *Letter to Ingoli*.[177] Why should the Master of the Sacred Palace instruct the Florentine Inquisitor about a book to be published in Rome? Therein lies a story.

After hurriedly reading the text that Galileo had with him in Rome in 1630, neither Riccardi nor Visconti raised any serious objection, merely small points that Galileo could correct before coming back to Rome for final approval and printing of the text. Following a final friendly meeting with the pope and the cardinal nephew, Galileo left for Florence "having completed his business to his entire satisfaction." Thus ambassador Niccolini, who went on, no doubt to the grand duke's satisfaction, "he is esteemed and honored by the entire [Papal] Court, as he deserves to be."[178] He was virtually a member of it, on two counts. For one, Urban assigned him a pension, or, rather, reassigned to him the pension he had promised Vincenzo, who had refused the qualifying condition of wearing a tonsure. That did not put off Galileo, who became a cleric in the summer of 1630 to the extent of wearing a tonsure and reciting a daily office. He did not realize much from this investment. He wore his hair clipped for several years before the pension, drawn from a canonry of the Cathedral of Brescia, started, and then it too was clipped, to 40 scudi from the 100 that Urban had promised.[179]

Secondly, Galileo figured on an official list of savants whose presence in Rome between 1630 and 1633 distinguished Urban's reign. The original draft of this list of *apes urbanae* ("Urban's bees") praised Galileo too fulsomely to survive his condemnation. It contained, besides the titles of his published works, the verses in which Urban had praised his discoveries, mention of unpublished studies on mechanics that would "illuminate all of philosophy," and the information that Diodati had come all the way from France just to talk with him. The author of the *Apes*, Leone Allacci, reader in Greek at the Vatican Library, also volunteered, without a hint of censure, that he had read a tract by Galileo on the motion of the earth. None of this material apart from the mention of Diodati survived into the published *Apes urbanae*.[180] (That mention would have been cancelled too if Allacci had known that Diodati had promoted French editions of Sarpi's condemned histories.) Allacci's hive included many important actors in Galileo's story, for example

Ciampoli, Grassi, Riccardi, Inchofer, and Scheiner, who was still in Rome in 1630, packing the 14 crates of books, instruments, relics (10 saintly skeletons), and Agnus dei (12,000 in all) that he had acquired over the previous six years.[181]

Soon after, Galileo, now back in Florence, began intensive revisions of his manuscript, the aspects that had looked so favorable in Rome began to dim. The metaphor applies only too well. The Morandi case and the bull *Inscrutabilis* against astrologers made it prudent to avoid everything astrological. The crackdown did not hit Galileo directly, but his enemies took comfort in his reputation as an astrologer and his friendship with Morandi and Visconti, whereas Scheiner had caught the crest of the new wave in Rome by condemning astrology in *Rosa* as "laughable" and "worthless."[182] Galileo thought it useful to insert a jibe at astrologers in the *Dialogue*. It is surprisingly mild and awkwardly placed. Sagredo and Salviati have just had a good laugh at philosophers who claimed that Aristotle's writings "contained" anticipations of modern inventions like the telescope. Of course, Sagredo jeered, Aristotle contains these anticipations, but only in the sense that a block of marble does a statue, or as heathen oracles do the future, that is, after the event. Salviati: "And why do you leave out the prophesies of the astrologers, which are so clearly seen in horoscopes…after their fulfillment?"[183]

It did not take an astrologer to see that the death of the Prince of the Lynxes on 1 August 1630 would compound the difficulties of publishing the *Dialogue*. It proved catastrophic. Cesi's death effectively ended the lynxes. The obvious successor, the cardinal nephew, declined the honor; no one else had the position, dedication, and wherewithal.[184] Still, since Galileo had almost obtained permission to print from Riccardi, and he had friendly editors near the Barberini in the persons of Riccardi, Ciampoli, Visconti, and Castelli, it made best sense to complete the original plan and publish in Rome. Without Cesi, however, Galileo would have to negotiate directly with Monster Riccardi and oversee the printing. That would require a long stay in Rome. "A plague on it," he might have said, and, lo, there was plague, carried into Italy by a Holy-Imperial army come to do battle against Urban and the French in the subwar of the Mantuan succession, a disastrous consequence of the pope's irrepressible rapacity. By the summer of 1630 plague was in Florence. One of Galileo's rustic servants died of it. "[U]se every possible precaution to protect yourself from the danger," Maria Celeste wrote, prescribing as "the best remedy of all…a thorough contrition and penitence." Even with this medicine Galileo

could not have traveled easily to Rome, which established elaborate precautions to keep the plague at bay. These barred or delayed entry to people
from plague-ridden places and required fumigation or confiscation of items,
including manuscripts, that might carry contagion.[185] Galileo asked to print
in Florence.

After consulting Visconti, who would soon be banished from Rome over
the Morandi affair, Castelli reported that publication in Florence would not
be a problem and urged that Galileo obtain permission from the censors
there quickly, as Urban's favor was fickle. With astonishing speed—on 12
September, some two weeks after receiving Castelli's letter—Galileo had the
desired imprimatur.[186] The business was highly irregular as no censor had
approved the text. Riccardi had expected Galileo to return to Rome to iron
out the remaining wrinkles; but because of the plague he agreed that most of
the book could be reviewed in Florence, reserving for himself the beginning
and the end. A letter to Galileo from the wife of the Tuscan ambassador to
Rome, Caterina Riccardi Niccolini, set forth the arrangement. Caterina Niccolini had helped pressure her cousin the Monster into separating responsibility for the body of the book (which was assigned to the Dominican
theologian Giacinto Stefani in Florence) from the head and tail; and, when
the Monster, buffeted by shifting political winds that made piloting people
who sailed close to reefs hazardous, continued to procrastinate, she obtained
copies of these *disjecta membra* and insisted that he approve them promptly.[187]
In the summer of 1631, having sent his preface and instructions to Egidi, Riccardi washed his hands of the problem. If you follow the instructions faithfully, he wrote, " the book will not encounter any obstacle here in Rome and
Your Reverence will be able to please the author and serve His Serene Highness [Grand Duke Ferdinand II], who is very anxious about the matter." The
matter seemed to be in hand. Stefani required only a few small changes.[188]

The *Dialogue* went to press around 1 June 1631, before Galileo had Riccardi's
corrections and draft preface in hand. The printer finished the first three
days at the end of the year, the entire book on 21 February 1632. The next day
Galileo presented a copy to its dedicatee, Grand Duke Ferdinando.[189] Galileo regretted that he had had to drop all reference to the tides from his title,
since the publisher had counted on the word, with its hint to those who knew
Galileo's arguments, to boost his sales. The change of title may well explain
why the *Dialogue* begins without a proper mise-en-scène; perhaps the original
opening concerned the tides, and survives in Simplicio's contemplation of

their turning when his gondola beached en route to the third day's discussion.[190] The preface, "To the discerning reader," followed Riccardi's draft. His purpose, Galileo wrote, was to defend the church and Italian science from the charge of ignorance in banning heliocentrism. "Upon hearing such carping insolence, my zeal could not be contained." He had been in Rome in 1616 and had experienced for himself the competence with which the Congregations of the Index and the Inquisition drew up their "salutary edict" against Copernicus. To right the balance, Galileo took the Copernican side in the *Dialogue*, "proceeding as with a pure mathematical hypothesis and striving by every artifice to represent it as superior to supposing the earth motionless."[191]

Riccardi had not specified a place for the administration of the medicine of the end. Galileo chose to put it in the mouth of Simplicio. The nincompoop allowed that although he esteemed Salviati's ideas about the tides he could not admit them as true and conclusive. "Indeed, keeping always before my mind's eye a most solid doctrine that I once heard from a most eminent and learned person, and before which one must fall silent, I know that if asked whether God in his infinite power and wisdom could have conferred upon the watery element its observed reciprocating motion using some other means than moving its containing vessels, both of you would reply that he could have, and that he would have known how to do this in many ways which are unthinkable to our minds." Salviati: "An admirable and angelic doctrine, and well in accord with another one, also Divine...that we cannot discover the work of His hands." Sagredo: "Let us go and enjoy an hour's refreshment in the gondola that awaits us."[192]

Galileo knew from first hand how strongly, even emotionally, Urban clung to his almost unanswerable argument.[193] Why then did he not give Urban's Simple to Salviati or Sagredo? That is because both of them, and Simplicio too (if he was a distillate of Borro, Buonamici, and Cremonini) opposed mixing theology with physics. The out-of-character argument would spoil the masquerade. That was Urban's view too. He wanted Galileo to say, in his own voice, something like this:

> The comedy has ended. Italian honor and the reputation of Holy Church are saved from the calumniators who would destroy our faith. But do not deduce from this happy outcome, gentle reader, that you are free to assert the absolute truth of a physical system even if the arguments in its favor seem unanswerable. For as I learned long

ago from His Holiness, any such assertion would derogate from the Omnipotence of God, who in His wisdom and power can do or make anything that does not involve Him in a contradiction. The argument of His Holiness is in fact and logic unanswerable. A true Christian must bend the knee and fall silent before it.

Had Galileo written such words, philosophers, mathematicians, and astrologers whose theories implicitly limited God's past and future actions would find themselves opposed to the greatest mathematician in Italy, and perhaps the world. But Galileo could not speak Urban's words. That would have been entirely out of his character. It would have amounted to a denial of his mission.[194]

Galileo did not, and probably could not, anticipate the effect of allowing Simplicio to express Urban's Simple. Throughout his trial and after it he protested that he had done nothing wrong, that he had honored Urban's instructions by administering the prescribed medicine of and at the end.[195] In his black-and-white world, the medicine was either present or absent; and if present and taken, it did not matter much who administered it. He reckoned that the least implausible dispenser among the dialogists was Simplicio. In giving the words to him, Galileo had performed as required while remaining true to himself and the two progressive characters in his comedy. To Urban's subtler mind, the medicine was not fit for purpose unless swallowed by the man most in need of it, his assertive old friend Galileo.

Riding forth

Galileo's masked comedy staged in Venice received high marks there if Micanzio's response was representative. "I do not mean to flatter, but I tell you sincerely that *Non est factum tale opus in universa terra* ... But my God, how well you have brought to life that worthy man Sig. Sagredo! God save me, I think that I hear him speak." On the road, however, in Rome, the faithful resurrection of Sagredo did not count.[196] Eight copies reached the Holy City in late May 1632, in the baggage of Filippo Magalotti, a friend of Galileo and Guiducci and a close relative of the Barberini. Galileo directed that the first copy go to the cardinal nephew and the others to Riccardi, Niccolini, Ciampoli, Campanella, and two representatives of the enemy, Ludovico Serristori (a consultor to the Holy Office) and Leone Santi (a professor at the Roman

College). Magalotti was to retain the eighth.[197] Castelli, still near the supreme power, read the cardinal's copy, with as much absorption, he told Galileo, reaching for the perfect compliment, as if it were the *Furioso*. Interested Jesuits could read Santi's copy. Soon word and rumor about the book's content reached the ears of most people in Rome interested in such things. Who first undertook to inform Urban about the treatment of his cherished principle may have been blown up in the explosion that followed. The pope ordered through Riccardi that the publisher cease distribution and that all copies already sold be recalled or confiscated. That was in July 1632.[198]

The order ended with an instruction to the Florentine Inquisitor to look into the meaning of the three dolphins biting one another's midriffs that appeared on the title page of the *Dialogue*. That may indicate the pope's hair-trigger sensitivity at the time the "Galileo affair" began. Were the dolphins a cabalistic insult to Urban and his two brothers? Or to the Barberini cardinals? Did they signify a hope or prediction of a replacement of the Barberini bees by a family of fish? "In Rome they pardon atheists, sodomites, libertines and other sorts of rascals, but they never pardon those who bad mouth the Pope or his court, or who seem to question papal power."[199] No doubt Urban had some cause for his heightened sensitivity. He sensed betrayal all around him. He had had to drop his long-time friend and agent Ciampoli for presuming to know the papal mind and for flirting with the Spanish, who had paid him a small pension (or retainer) for years; but after the great set-to in the consistory of 8 March 1632, when Cardinal Borgia and Urban's younger nephew Antonio Barberini almost came to blows over the pope's foreign policy, the time for a purge of fellow-travelers in the Spanish interest had arrived.[200] Worried about assassination, Urban took care to have his food tasted, withdrew to Castel Gandolfo under high security, and ordered armed patrols of the road to Rome. This worry was better founded than his suspicion about the dolphins, which turned out to be the printer's trademark.[201]

The fall of Ciampoli had several elements in common with Galileo's. Both men had enjoyed the friendship and admiration of Maffeo Barberini and received substantial tokens of his favor when he became pope. He did not want to act against either of them. But both made the mistake of taking him for granted, of ignoring one or another of his pet opinions, and of behaving in ways that made protecting them increasingly costly. Ciampoli's lavish lifestyle, occasional orgies, and curtness when not toadying paralleled in their awkwardness Galileo's incurable badgering and ridicule of respectable savants and clerics who disagreed with him.[202] And after their fall, both of Urban's former friends made the same misjudgment in expecting him to relent. Ciampoli left Rome quietly in October 1632 for a governorship of a territory in the Marches, from which he tried desperately to return. Urban remained firm; Ciampoli's banishment was perpetual. So, in a harder degree, would be Galileo's.[203]

It was rumored that the Jesuits had a hand in informing Urban about the violations of the edict of 1616 and of Galileo's misuse of the medicine of the end.[204] They had regained influence at the Vatican as Urban was forced to favor the Habsburgs. In 1631, in their Good Friday oration in the Sistine Chapel, they had pointed out the destruction of their missions and colleges by "the mob of furious pillagers" roaming Catholic Austria. The orator was Orazio Grassi and the dedicatee of his printed text, the Italian head of the Spanish faction in the curia, was his patron Cardinal Ludovisi. The Jesuits sweetened their message by bringing out a new and sumptuous edition of Urban's poems with designs by Urban's favorite architect, Bernini.[205] Whether Grassi used his connections to help bring Galileo to book may be doubted. He said that he had nothing to do with it and regretted the outcome. "He ruined himself by being so fond of his own ideas and by not appreciating others. It is no wonder that everyone conspired against him."[206]

However it came about, Urban knew in July 1632 that Galileo had not honored his commitment to deal equally and mathematically, that is, hypothetically, with the world systems, and that the powerful talisman of God's omnipotence had turned to dust in Simplicio's mouth. When Urban came to examine the book, his papal hackles rose before he reached the text. Something fishier than the dolphins of the title page appeared on its verso: three imprimaturs, one, undated, from Riccardi as Master of the Sacred Palace and the others, dated 11 September 1632, from Pietro Niccolini and Clemente Egidi, respectively the Archbishop and Chief Inquisitor of Florence. Since Riccardi had no authority outside Rome, Urban's suspicious nose smelled a rat, or rather two. Ciampoli and Riccardi had misled him to think that the manuscript of 1630 contained nothing not easily repaired; Riccardi granted the imprimatur on Urban's voicing no fundamental objection (which, believing the work to be sound, he had no cause to do); and then, having at last read the manuscript or some of it, Riccardi tried to be true to his friendship with Galileo and his obligations as a censor by shifting the burden of approval to a jurisdiction outside Rome. Riccardi may have been as surprised as the pope to see his imprimatur alongside those of the Florentine officials. It appeared that Galileo had divided and conquered the censorship by translating a provisional approval obtained in Rome into a license to print, without strict prior review, in Florence. This was probably close to what did happen. Urban demanded an explanation. Riccardi blamed the business on the Florentines and on Ciampoli, who had told him, he claimed, that Urban had no objection to Galileo's general argument.[207] Urban accepted that Riccardi too was a victim of deceit and that Ciampoli had choreographed the comedy of the licenses.[208]

Much of the information about Ciampoli's involvement comes from a report by Niccolini of unpleasant interviews he had with Urban on 4 and 11 September 1632. In the first interview, following the Grand Duke's orders, Niccolini inquired why any suspicion should attach to a book so well fortified with licenses as Galileo's *Dialogue*. That was not the right approach. The pope railed at Ciampoli and Riccardi. Niccolini then asked that Galileo be allowed to come to Rome to justify himself. Urban grew angrier: the Holy Office (to which the matter had not been referred) does not negotiate. And then: Galileo's book "involv[es] great harm to religion (indeed, the worst ever conceived)"; he has meddled with "the most perverse subject one could ever come across"; his book is "extremely perverse," "troublesome and dangerous," "pernicious."[209] A few clues to what this rhetoric, which seems more appropriate to plague than astronomy, applied occur in other phrases of Niccolini's reports. Urban said that Galileo knew full well where the problems lay, "since we have discussed them with him and he has heard them from ourselves." Galileo had "dared [to enter] when he should not have, into the most serious and dangerous subjects which could be stirred up at this time."[210]

What was so perverse and deadly, so menacing to religion ("indeed, the worst [menace] ever conceived"), what so troublesome at that very moment? No doubt undermining the papacy by misleading a trusting pope during war and plague would qualify as perverse and pernicious. Still, this would not make Galileo's book so dreadfully harmful to religion. When Urban gave vent to these extravagant accusations, a special panel, which he had appointed in August to determine whether Galileo should be brought before the Holy Office, was hard at work. The unusual procedure of a preliminary review indicated that the pope, as Niccolini reported on 5 September, then still regarded Galileo as a friend. The panel's findings convinced Urban that Galileo valued neither his friendship nor his advice. It had obtained a document from the Inquisition's archives that, as Riccardi volunteered to Niccolini, "is alone sufficient to ruin Mr Galilei completely." This was the injunction or precept read in Bellarmine's house in 1616. We know that the precept forbade Galileo to "hold, teach, or defend [Copernicanism] in any way whatever, orally or in writing," under threat of further proceedings by the Holy Office. Now certain of Galileo's bad faith, Urban informed the ambassador that he had no alternative but to allow the Inquisition to proceed.[211] Still, the charges remained at the level of bad behavior. The discovery of the injunction might even have done Galileo some good, since it deflected attention from Riccardi's negligence and opened the possibility of disposing of the case without entering into dangerous subjects.[212]

Urban's special panel consisted of Master Riccardi, who had his monstrous hide to save; the papal theologian Agostino Oreggi, Urban's Bellarmine; and Riccardi's friend, the Jesuit Melchior Inchofer, expected to be sympathetic to people hounded by the censorship.[213] Inchofer turned out to be hostile to Copernican ideas and a champion of Scheiner, the target, as they both saw it, of most of the vituperation in the *Dialogue*. It is likely that the panel assigned Inchofer the job of reviewing the anonymous accusation, based on the *Assayer*, that Galileo believed in an atomism irreconcilable with Tridentine teaching. An unsigned statement in Inchofer's hand, which appears to be a rephrasing and considered judgment of the accusation, survives among the papers of the Congregation of the Index. Assuming that Inchofer drew up this judgment for and during the deliberations of the special panel, we have a second and more powerful reason for Urban's wrath and alarm. Inchofer's statement would have reinforced a connection in his mind among personal betrayal, disdain for his voluntarism, and menace to dogma.[214] We are coming closer to pernicious doctrine.

Pushed to extremes, Urban's basis for evaluating crimes against the faith could convict Galileo of denying such divine attributes as freedom of action, omniscience, and omnipotence. These denials truly would have been pernicious and subversive. Many collateral matters could be bound up with Galileo's lèse-majesté, for example judicial astrology, which Urban had then recently anathematized on voluntaristic and other grounds.[215] The problem of salvation and grace, which had brought the Jesuits and Dominicans into unresolvable conflict, involved the same sorts of questions of divine and human freedom, or so Oreggi insinuated by inserting his report of the conversation between Barberini and Galileo in 1616 into a discussion of just these topics.[216] And there is a logical parallel between Urban's reasoning, which took as fundamental the possibility of saving the appearances while transforming their causes, and the Tridentine doctrine of the eucharist, whereas Galileo's atomism might make transubstantiation a logical impossibility by requiring an atom to appear as bread and not-bread simultaneously. Although Galileo's trial was not about atomism, astrology, freewill, salvation, grace, or divine attributes, many or all of them probably were in Urban's mind when he castigated Galileo's doctrine as "the worst [menace] ever perceived."[217] Had Galileo only lived up to his bargain, had he not spoiled, but fortified, Urban's medicine!

The panel's unsigned report of September 1632 twists the tale slightly to exonerate Riccardi from any oversight or wrongdoing. In order to allow Galileo to talk with publishers, the report declared, Riccardi had given a provisional imprimatur for printing in Rome on the understanding that he would review the sheets as they came from the press. When Galileo insisted on printing in Florence, Riccardi told the Florentine Inquisitor what needed to be done, furnished the core of a proper preface, and let go of the business. Against the instructions sent by Riccardi and issued by Bellarmine, the report continued, Galileo argued cosmology absolutely and not hypothetically. Specifically, he used the Roman imprimatur without permission; printed the preface in distinctive type as if it did not belong to the book; put the "medicine of the end in the mouth of a fool"; mistreated opposing authors (here may be an echo of Scheiner's complaint) "and those most used by the Holy Church"; declared that he understood geometrical theorems as well as God did; and attributed the tides to nonexistent motions of the earth. All true and all, in the opinion of the commission, amendable, "if the book were judged to have some utility which would warrant such a favor."[218]

Knocked down

In late September 1632, soon after the special panel had given its judgment, the cardinal nephew instructed the Florentine Inquisitor to direct Galileo to make himself available to the Holy Office in Rome during October. Galileo tried to decline this invitation on the grounds of age and infirmity. Through his friend and admirer ambassador Niccolini he wrote to the cardinal nephew asking to be allowed to defend himself in writing or, if a trial was necessary, that it be held in Florence. The journey, hard enough at any time, would be the death of him during a plague. Galileo added a most mysterious reference to a wise and saintly man, and a most exact and economical theologian, who had encouraged him to write the *Dialogue*. "[A]lmost like an echo of the Holy Spirit, [I heard] a very short but admirable and most holy assertion suddenly come out of the mouth of a person who is most eminent in learning and venerable for the holiness of his life; that is, an assertion which, in no more than ten words cleverly and beautifully combined, summarizes what one can gather from the lengthy discussions scattered in the books of the sacred doctors." Galileo did not disclose the name of the holy man who had reduced St Thomas to the length of an Ave Maria.[219] Niccolini advised against delivering the letter as the Inquisition would demand to know the identity of the taciturn theologian. On further instruction, however, he passed it to the cardinal who showed it to the pope. They did not fall for any of Galileo's stories. No more did the copies of his *Letter to Christina*, which he had sent to Riccardi, help his case.[220]

Niccolini tried to move Urban "to pity poor Mr. Galileo, who is now so old and whom I love and adore." Urban pitied, and allowed the culprit to come in a litter with all comforts, but come he must. Urban added the papal prayer that God would forgive Galileo's error of falling into "an intrigue like this after His Holiness himself (when he was cardinal) had delivered him from it."[221] From which it appears that Urban believed that he had kept Galileo's name and writings out of the condemnation of Copernican works issued by the Index in 1616. Galileo not only had not recognized the author or favor of his earlier deliverance, but had rejected both by repeating his original sin. Niccolini returned to his observation that Galileo had published with the approval of Riccardi. That did not help. Riccardi was in trouble over endorsing the *Dialogue*, "which he should never have done, as the general of the Dominicans and everyone else says." Urban may not have known that Niccolini had some responsibility for the endorsement, since he had appealed to friendship with Galileo, and his wife to allegiance to the Riccardi, to persuade the vacillating Monster to sign off on the manuscript of the *Dialogue*.[222]

As pressure mounted to push him to Rome, Galileo fell into depression. Buonarroti saw him on 10 October and reported to the cardinal nephew that their friend was plunged into deep melancholy.[223] Galileo had himself examined by three physicians who found him a physical wreck. His pulse was intermittent, from which they inferred a general debilitation of "the vital faculty." He complained of frequent dizziness, hypochondriacal melancholy,

weakness of the stomach, sleeplessness, pain throughout the body, and
other ailments, to which, the doctors said, others could attest. They had
no trouble locating his heavy fleshy hernia and ruptured peritoneum. "All
of which is serious and evidently any small external cause could place his
life in danger." The Holy Office did not believe the doctors' certificate and
directed Inquisitor Egidi to inquire further into Galileo's health: should he
be malingering, he should be sent to Rome in chains. The malefactor took
the hint and left Florence on his own, on 20 January 1633, for quarantine in
Siena.[224] His departure came between two other events most gratifying to
the pope. In November 1632 the invincible Protestant champion, Gustavus
Adolphus, whose alliance with France had placed Urban in an impossible
diplomatic position, fell in battle. And in February 1633 the Roman people
officially thanked him for his "paternal providence and exquisite diligence"
in keeping the plague from the city.[225]

   Galileo arrived at Niccolini's home in Rome on 14 February in good health
and spirits, and immediately started chatting up officials of the Holy Office.
The Barberinis had not expected such boldness. The cardinal nephew agreed
that Galileo could remain with Niccolini but only if he ceased his visits and
avoided all socializing. Niccolini gathered from this relative kindness that
things might not go badly "despite the fact that His Holiness feels so negative about this business." Confined to quarters, knowing nothing about the
top-secret deliberations of the Holy Office, Galileo could do nothing in his
own defense. In this unfamiliar situation he received from Niccolini some
unfamiliar counsel. "I have advised him to show himself always ready to
obey and submit to whatever he is ordered to do, because this is the way
to mitigate the fervor of those who are fiercely excited and treat this trial as
a personal thing."[226] By the end of February Niccolini could report that the
main issue seemed to be the personal injunction Galileo received in 1616.
The pope said as much in an interview on the 26th and added that Galileo's
doctrine was very bad, without, however, specifying the evil. The cardinal
nephew was less reticent. The problem involved "introducing some imaginary dogma into the world, particularly in Florence, where...intellects are
very subtle and curious." Niccolini: perhaps the Copernican is the stronger
side and Galileo could not help himself in seeming to prefer it? The Lord
Cardinal: "[Galileo] knew how to express exquisitely and to justify wonderfully whatever he wanted [to say]." Here the cardinal spoke the truth.[227] The
celebrated clarity of Galileo's prose left him no hiding place.

In his last important interview with the pope before Galileo went to the Holy Office for examination, Niccolini obtained valuable information— valuable to the historian, not to Galileo—about Urban's thinking. The pope said that in dealing with new doctrines and Holy Scripture it is best to follow the common opinion. May God forgive Galileo and help his fellow plotter Ciampoli! "God is omnipotent and can do anything; but if He is omnipotent, why do we want to bind him?" Niccolini replied that, as far as he knew, that was Galileo's opinion too. "However, [the pope] got upset and told me that one must not impose necessity on the blessed God; seeing that he was losing his temper, I did not want to continue discussing what I did not understand." Niccolini had good cause to worry. "I do not like His Holiness's attitude, which is not at all mollified."[228]

Meanwhile Urban recalled the special panel (replacing Riccardi by a Theatine theologian, Zaccaria Pasqualigo) to consider whether Galileo had violated the injunction "not to hold, teach, or defend [Copernicanism] in any way whatever, orally or in writing." All members of the committee agreed that Galileo taught and defended Copernicanism in the *Dialogue*. Oreggi did not offer an opinion about whether Galileo also held it. Pasqualigo also did not pronounce definitively, though he entertained the "strong opinion" that Galileo held the condemned system.[229] Inchofer had no doubt. "I am of the opinion that Galileo...is vehemently suspected of firmly adhering to [the Copernican] opinion, and indeed that he holds it." Inchofer adduced some 27 passages to support this conclusion giving particular emphasis to those in which Galileo treated "despicably" those who held the contrary opinion. "What Catholic ever conducted such a bitter dispute against heretics, even regarding a truth of faith, as Galileo does against those who maintain the earth's immobility?" Galileo had dismissed fixed-earth philosophers as less than human, ridiculous, small-minded, half-witted, idiotic, and praised Copernicans as superior intellects. This was not the way to dispute, to joust hypothetically, to exercise the mind; it was war, "arrogant war," against defenders of traditional philosophy.[230] The reports of the special commissioners made it difficult to restrict Galileo's infraction to mere disobedience.[231]

In so far as Inchofer's report set the tone for discussion at the Holy Office, the Jesuits helped to hasten and deepen Galileo's fall. Inchofer's expression "vehemently suspected," which, as will appear, had a technical meaning, reappeared in the formal charge eventually brought against Galileo. Several well-placed contemporary observers, including Galileo, attributed his troubles to the machinations of the Jesuits. About the time that Inchofer was finishing his report, the informed gossip Naudé wrote that Scheiner and other Jesuits, "who want to ruin [Galileo]," were behind his troubles.[232] Certainly Scheiner was outraged by Galileo's "shameless attacks" on him and felt, as Inchofer faithfully transmitted, that the *Dialogue* was a diatribe against his and Locher's *Disquisitiones mathematicae*. And there may be an echo of Grassi in Inchofer's dwelling on Galileo's megalomania.[233]

It is unlikely, however, that the Jesuit order bothered itself about Galileo in the spring of 1633. General Vitelleschi had his hands full trying to salvage colleges in the Germanies pillaged by Protestants, to protect and direct missions in the war zone, and, what might have been even more difficult, to reconcile Francophile and Habsburg factions within the Society.[234] Apart from Scheiner, Inchofer, and, perhaps, Grassi and their immediate circle, the Jesuits probably contributed no more than indifference to Galileo's troubles. Father Grienberger said as much to a mutual friend, and Galileo reported his saying, perhaps with some touching up, to Diodati. "If Galileo had known how to keep on friendly terms with the fathers of [the Roman] College, he would be enjoying fame in the world, he would not have had any misfortunes, and he would be able to write freely about anything, even the motion of the earth." To which Galileo added the characteristic self-deceiving gloss, "Thus you see that it is not this or that opinion which has caused the past and present warfare against me, but rather it is my being held in disfavor by the Jesuits."[235] It was rather his opinions, and his insistence on forcing them on others, that interested the Holy Office.

The Inquisition at last called Galileo to an interview in mid-April 1633. He was kindly received and lodged, exceptionally, in the prosecutor's rooms at the Minerva. Galileo had his own servant to wait upon him and food directly from Niccolini's kitchen. Nonetheless he considered the treatment harsh. On 30 April, on the intervention of the commissary and the authority of Francesco Barberini, Galileo returned to Niccolini's to "recover from the discomforts and from his usual indispositions."[236] He immediately regained his health. On 22 May Niccolini reported that the affair would probably be completed by the end of the month and offered a guess at the outcome. The book would be prohibited and Galileo would be given some salutary penance to wash away his disobedience to the injunction of 1616.[237] The final act did not take place until June. On the 19th Urban, in good spirits, gave Niccolini an indication of the decision. The book certainly would be condemned and Galileo detained for some time in a convent like Santa Croce in Florence. Urban wanted it known, to discourage other free thinkers, that the leniency of Galileo's treatment was owing neither to the man nor to his physical condition. "Every punishment has been mitigated out of regard for our master the Most Serene Grand Duke, and this is the real and only reason why all possible accommodations have been and will be made." Urban did allow Galileo creature comforts during his trial in Rome that no one else in similar circumstances, whatever his civil or ecclesiastical rank, had ever enjoyed.[238]

The Holy Office had interrogated Galileo four times officially between 12 April and 21 June and once irregularly by the Commissary General, the Dominican in charge of the case, Vincenzo Maculano, just the man for the job, a combination of engineer (he was one of the pope's favorite military architects) and inquisitor.[239] Maculano had relied on two accusations: Galileo had violated the injunction of 1616 and had obtained the double imprimatur of 1632 deceptively. Against the first accusation Galileo displayed the letter he had solicited from Bellarmine to stop the rumors that he had been made to abjure. No doubt the document surprised and disoriented Maculano, as it not only countered the rumors against Galileo but stated that "only the declaration made by the Holy Father and published by the sacred Congregation of the Index has been revealed to him." This declaration prohibited defending or holding the Copernican motions. The further injunction or precept given to Galileo at the same session in which Bellarmine read him the decree of the Index was much more restrictive. Bellarmine's letter expressly limited the "revelation" to Galileo to the milder text of the general prohibition published by the Index.[240] But perhaps Bellarmine had not meant that no one had spoken further to Galileo at the meeting but only that he had not?

Maculano: who else was present at the session and did any of them issue a precept or injunction against you? Galileo: There were some Dominicans in attendance. "It may be that I was given an injunction not to hold or defend the said opinion, but I do not recall it." Maculano presented a copy of the injunction. Galileo could not remember receiving it. "I do remember that the injunction was that I could not hold or defend, and maybe that I *could not teach*. I do not recall, further, that there was the phrase *in any way whatever*, but maybe there was." In any case he had not bothered to keep in mind anything other than Bellarmine's certificate, "which I relied upon and kept as a reminder." Maculano turned to the imprimatur. Had Galileo told Riccardi about the injunction? Galileo made a bad slip in replying. He should have said, "no, I could not have done so even if I had thought it appropriate, since by then I had long forgotten any such injunction, if, indeed, I had ever received one." Instead he said: "I did not say anything to the Father Master of the Sacred Palace about the above-mentioned injunction because I did not judge it necessary to tell it to him, having no scruples since with the said book I had neither held nor defended the opinion of the earth's motion and the sun's stability; on the contrary, in the said book I show the contrary of Copernicus' opinion and show that Copernicus' reasons are invalid and inconclusive."[241]

With this fib the examination of 12 April ended. Galileo, sworn to secrecy, went to his comfortable rooms and Maculano retired to consider what remained of his case.[242] The judgments of Oreggi and his fellow panelists, submitted on the 17th and accepted by the Congregation of the Holy Office on the 21st, strengthened his hand. So did Galileo's cries of pain induced by one or more of his ailments, which made a speedy settlement desirable for all parties. Maculano asked for authorization to deal "extrajudicially with Galileo to make him understand his error and, having recognized it, to bring him to confess it." The cardinals of the congregation at first thought the proceeding too bold (so Maculano wrote the cardinal nephew); but they authorized it on learning the "basis" on which he would operate. What was the basis? Perhaps a plea bargain? Galileo would confess to having gone too far, the Holy Office would accept that his error was inadvertent, and some mild penalty would be imposed. It would not be necessary to reconcile Bellarmine's certificate and the injunction.[243] There is not much evidence for this gentlemen's agreement.

Maculano's letter to Cardinal Barberini motivated the extrajudicial proceeding as a way of saving Galileo from himself. If he were to maintain the fib with which the first interrogation ended, "it would become necessary to use greater rigor in the administration of justice and less regard for all the ramifications of this business." The "basis" of Maculano's new strategy was fear. Maculano might well have told Galileo that if he insisted on a battle of wits and documents, the Holy Office would treat him as it did others who waited their chance with the tribunal. He would exchange his comfortable rooms for a jail cell for an indefinite period. People died in these unpleasant places; witness the late abbot–astrologer Morandi and poor Archbishop de Dominis. Furthermore, Maculano might have added, the Holy Father is still very angry and there is no way that you can escape punishment. Would you like it gentle or harsh? The choice was not difficult. Galileo agreed to confess that he recognized that he had gone too far and repented doing so. This result, Maculano hoped, would please the pope and everyone else. "The Tribune will maintain its reputation [for always getting its man]; the culprit will be treated with benignity; and, whatever the final outcome, he will know the favor done to him." The inquisitors might be merciful. "[H]e could be granted imprisonment in his own house."[244]

On 30 April Galileo made his confession. He had reread his *Dialogue*, he said, and examining it minutely, as if the work of someone else, he found places in it where a reader ignorant of his intention might gather incorrectly from the force of the reasons given that the Copernican view was true. The arguments from sunspots and tides might be especially difficult to answer, "[although] I inwardly and truly did and do hold them to be inconclusive and refutable." Then comes something true: "[I yielded to] the natural gratification everyone feels for his own subtleties and for showing himself to be cleverer than the average man…To use Cicero's words, 'I am more desirous of glory than is suitable'…My error then was, and I confess it, one of vain ambition, pure ignorance, and inadvertence." And then a perfect piece of Galilean impudence: to show his good faith and rhetorical skills, Galileo proposed to add one or two more Days to the *Dialogue*, in which he would confute the condemned opinion "in the most effective way that the blessed God will enable me. So I beg this holy Tribunal to cooperate with me in this good resolution, by granting me the permission to put it into practice."[245] He felt in fine spirits, having had, he wrote Maria Celeste, "triumphant successes" owing to her prayers and the Niccolinis' care. Maria Celeste perceived something else at work. "I implore you not to confuse yourself with drink."[246]

Ten days after this confession Galileo was allowed to make a statement in his defense. It centered on the effacement of the memory of the injunction's fatal phrase ("in any way whatsoever") by Bellarmine's protective certificate. Since Galileo did not think himself under an order stricter than the published edict of the Index, whatever excesses he committed in the *Dialogue* were not evidence of "the cunning of an insincere intention." They arose rather through vain ambition, a fault, no doubt, but not a deceit. Please also bear in mind (so Galileo ended his appeal to "the clemency and kindness of heart" of his judges) my advanced age and pitiable state of health, and the constant "slanders of those who hate me." Indeed it was to silence such slanders that he had obtained his certificate from the Most Eminent Lord Cardinal Bellarmine.[247]

The Congregation reported to Urban in May or early June. Their unfair summary of the process included besides Galileo's depositions, the certificate, and the injunction, a résumé of Caccini's accusations, a few points from the *Letter to Christina*, and a reference to the Copernican dicta in the *Letters on sunspots*. This gave the impression that Galileo had been pushing a doctrine contrary to scripture for two decades. In order to save the implausible argument of a plea bargain, this summary report has to be attributed to dishonest persons eager to see Galileo punished harshly; had it been drawn up fairly, the argument continues, Urban would have been better advised and the affair would have terminated more happily.[248] But Urban had no need of a summary, biased or fair; he knew the salient facts and the necessary conclusions. On 16 June he ordered that Galileo be interrogated again as to his motives and that if he answered satisfactorily he was to abjure *ex vehementi* before the Congregation of the Holy Office, and thereafter be imprisoned at the Inquisition's pleasure. And, of course, the *Dialogue* would be banned.[249]

Abjuration *ex vehementi*, a process seldom undergone by laymen, was required to clear the culprit from "vehement suspicion of heresy." Being so suspected was itself a crime. The "trial" preceding abjuration sought to persuade the accused to confess to the crime his judges had decided he had committed. The procedure amounted to a public display of confession, contrition, and absolution. There was the sticky question, however, what heresy Galileo may have held that had raised vehement suspicion in inquisitorial minds. The process *ex vehementi* was not the only or the obvious punishment for Galileo's offences. Indeed, it may have been the harshest punishment in form (though not in its execution) to which Galileo could have been subjected. Its purpose was to humiliate him.[250]

On 21 June, Galileo gave the necessary assurances. "I do not hold the Copernican opinion and have not held it after being ordered by injunction to abandon it." On the following day he heard his sentence before seven of the ten cardinal-inquisitors. Among the missing was Francesco Barberini; among those present, his brother Antonio and Guido Bentivoglio, a former student of Galileo's who had remained well disposed toward him. The sentence recapitulates the case as written up in the summary report, to which it added that the Bellarmine certificate aggravated rather than ameliorated Galileo's situation since, "while it says that the said opinion is contrary to Holy Scripture, yet you dared to treat of it, defend it, and show it as probable; nor are you helped by the license you artfully and cunningly extorted since you did not mention the injunction you were under." By these acts Galileo had made himself "vehemently suspected of heresy, namely of having held and believed a doctrine that is false and contrary to the divine and Holy Scripture." According to the canons and in the names of Jesus Christ and His Ever Virgin Mother the cardinal-inquisitors were willing to absolve Galileo from his pernicious error and transgressions if he first renounced them with a sincere heart and unfeigned faith. To complete the business, they prohibited the *Dialogue*, condemned Galileo to "formal imprisonment in this Holy Office at our pleasure," and imposed on him the obligation to recite the seven penitential psalms once a week for the next three years.[251]

It was time to recant. The ailing old man knelt down painfully before his judges and twenty witnesses and read out, lighted candle in hand, the statement prepared for him in the format for removing vehement suspicions:

> I, Galileo Galilei…kneeling before you Most Eminent and Most Reverend Cardinals Inquisitors-General against heretical depravity in all of Christendom, having before my eyes and touching with my hands the Holy Gospels, swear that I have always believed, and believe now, and with God's help will believe in the future all that the Holy and Apostolic Church holds, preaches, and teaches…[D]esiring to remove from the minds of Your Eminences and every faithful Christian, this vehement suspicion, rightly conceived against me, with a sincere heart and unfeigned faith I abjure, curse, and detest the [Copernican] errors and heresies, and in general each and every other error, heresy, and sect contrary to Holy Church; and I swear that in the future I will never again say or assert, orally or in writing, anything that might cause a similar suspicion about me…So help me God and these Holy Gospels.[252]

Galileo rose without muttering *eppur si muove* ("still it moves") and returned, shattered, to the Medici palace.

8

—∞∞∞—

End Games

8.1 DRAMATIS PERSONAE

The minor characters

Several protagonists in the events of 1632/3 carried on the vendetta while others retired to nurse their grievances. Chief among the ongoing duelists was Inchofer, who toward the end of 1633 published a little book purporting to prove that a stationary earth and a mobile sun were matters of faith.[1] Its dedicatee was God Almighty; its authority, scripture and the Fathers; its method, a jumble of formal logic, theological argot, and mathematical hocus pocus. For example, "it is a matter of faith that the heavens are up and the earth is down," from which it follows that the sun cannot be at the center or the earth above Venus. For if they were so placed, as Copernicans require, how are we to understand how Christ descended into Hell and rose to Heaven? However, false is not useless. The motions supposed by Copernicus can be employed in calculations, and might even be useful to the faith if mathematicians emphasized their falsity along with their utility. Here Inchofer had in mind the minor truth later rediscovered by Karl Popper: "mathematicians [should]...work more and more toward trying to falsify theories rather than to defend them." To this anticipation of modern epistemology Inchofer added a pinch of ancient wisdom, Urban's Simple in the words of the Preacher: "no man can find out the work that God maketh from beginning to end."[2]

Then comes nastiness and nonsense. Mathematicians who play with philosophy can become "wild and ridiculous," even heretical, if they claim the
truth of their hypothesis from its agreement with observation. Epicycles,
eccentrics, syzygies, zodiacal arcs, are just artifices, which become absurdi-
ties, indeed, monstrosities if applied on the assumption of a moving earth.
"If the translation of the earth were the sole cause of the motion of the sun
to the senses, and if the parallaxes are still the same when the earth is taken
to be at rest, then it follows in the given case that the earth both stands still
and moves..."[3] With even more powerful arguments Inchofer annihilated
the defenses that Copernican theory could not be heretical because Paul III
had sanctioned it and Pythagoras had received his version of it from Moses.
"False and disproven," "empty and foolish," "an imposture." Tycho's system
was just as good. "There is no need to invent any other theories that upset
the system of the universe, especially when they agree neither with true philosophy nor with the sacred scriptures."[4]

The Franciscan consultor who read Inchofer's manuscript for the Roman
censorship liked the message and the method. "[It] shows rightly that mathematics and the other human sciences should be subordinated to the rule
of sacred scripture, lest in our day there occur a dangerous detour into an
excessive freedom..." Scheiner read it for the Jesuit censorship. He did not
like the message at all, since, in his opinion, the sun's motion and the earth's
rest were not matters of faith.[5] He soon had in hand his own anti-Copernican
exercise in response to a request from Urban for a short book against astrology and the new astronomy.[6] He worked at the agreeable task from 1632
until his death in 1650, nourishing his hatred for the "conceited and cunning
braggart" who had so grievously insulted him.[7] The book, which appeared
in 1651, charged Copernicus with dismissing the basis of sound philosophy
(sense perception) and the unanimous opinion of the Fathers, and Galileo
with starting errors and insults too numerous to itemize.[8]

Like Scheiner, Grassi left Rome in 1633, driven out, according to a good
story, because of his open advocacy of the cause of the Austrian Habsburgs.
It appears rather that he returned home to Savona to attend to his dying
father's affairs.[9] During his time in Liguria, he designed churches and worked
on optics while the Jesuit generals—Vitelleschi's successors Vincenzo Caraffa and Francesco Piccolomini—became more and more insistent on strict
adherence to Aristotle and St Thomas. That was a great blow to Grassi and
other mathematicians in the Society who recognized that some modern

discoveries made much of what they had to teach untenable. The Jesuit hierarchy shut up Grassi more effectively than Urban silenced Galileo. "I see that
my work on colors can not be published because of the rigorous measures
of the last general congregation." The heavy-handed revisionists had prohibited some opinions on which Grassi's treatise depended, "not because they
think them bad or false, but because they are new and out of the ordinary."
He would have to do what Galileo could not countenance. He would have to
sacrifice his book to Obedience, "by which no doubt I will gain more than I
would by bringing it out."[10]

Riccardi managed to remain as Master of the Sacred Palace despite Maculano's opinion or hope that he would be disciplined for his "inadvertence
and neglect in approving the [Dialogue]." He regained Urban's confidence by
continuing to portray himself as a victim of Ciampoli and Galileo; and perhaps also by undertaking to refute Sarpi's repellent history of the Council
of Trent.[11] He had been preternaturally ingenuous for a monster.[12] But he
had recovered enough to be in line for a cardinal's hat when, in 1639, he was
called to a higher life, to the regret of many, including the Galileists, who did
not regard his actions in saving his skin as traitorous.[13] Maculano succeeded
to the mastership and, in 1641, to the cardinalate. Oreggi received his reward
more quickly. Urban made his Bellarmine a cardinal shortly after the conclusion of the trial.

Inchofer gave the funeral oration for Riccardi. The deceased Master, according to the self-styled Sklavus, was as sensitive as a babe and as erudite as
Solomon. He wept like Job on any day he could not perform a scared rite. He
had no equal for knowledge of sciences, arts, languages, and literature, or for
strength of mind or tenacity of memory. "And for these accomplishments he
was so much admired that he bore, as an honor, the title of Monster."[14] This
generous oration marked the height of Inchofer's prosperity. Soon his risky
writing and sneaky methods, as well as his closeness to the Vatican, gave wings
to Scheiner's negative opinion of him among his brethren. In return Inchofer
grew disenchanted with the Society. In January 1648 it brought him to trial for
contributing to an anti-Jesuit tract. In a bizarre replay of Galileo's ordeal he
confessed and in exchange received salutary penance and a sentence of indefinite imprisonment. The term became definite nine months later when, on 28
September, Inchofer died in detention.[15]

Although Urban had more important business than the aftermath of the
Galileo affair, he could not escape pestering and criticism arising from his

involvement in it. Both Galileo and Ciampoli continued to petition directly and through others for relaxation of their sentences. Nonetheless, Ciampoli stayed marooned in the Siberian Marche and Galileo in detention near Florence. Both turned to natural philosophy to pass the time.[16] The flamboyant Ciampoli projected a work in thirty volumes, or maybe sixty, providentially cut short by his death in 1643. He left a prodigious quantity of manuscripts, including four volumes on the history of Poland written for his patron King Ladislav IV, and a number of essays uneasily combining Galileo's method of reaching truth—sensible experience and necessary demonstration—with Urban's teaching that the only accessible truth lies in revelation. From these principles Ciampoli demonstrated in several ways that the marriage of Aristotle with the Bride of Christ, that fatal interdisciplinarity to which, among other things, Galileo owed his fall, could produce only monsters.[17] In pursuing his new physics, Ciampoli had the help of a student of Castelli's, who served as his secretary for several years between 1637 and 1641. Thus Evangelista Torricelli became the complete Galileist, combining, through his teachers and patrons, the mathematical and literary sides of the master.[18]

Another reason—other than appeals for Galileo's release—that kept his case before Urban was the Holy Office's incompetence. Perhaps even it did not know what Galileo's condemnation had decided. Was Copernicanism a heresy or not? Did the proscription against teaching it apply only to Galileo? The uncertainty suggests a difference of opinion among Galileo's judges over the nature and severity of his crime. A recent retelling makes the trial a struggle between a lenient party (Francesco Barberini, Maculano, Bentivoglio) and a severe one (Urban, Oreggi, Inchofer). They compromised on the charge of "vehement suspicion of heresy" without specifying the heresy suspected.[19] Whether or not this was the origin of the awkward charge, the consequent confusion became manifest when the papal nuncios, obedient to Urban's orders, distributed copies of Galileo's sentence and abjuration to mathematicians within their jurisdictions. Some fifty of these dangerous calculators, as many as the Inquisition could round up, assembled in Florence to hear the official documents read to them.[20] Some of the nuncios did their duty zealously, others more languidly, but none could say for certain what the documents, which, as handwritten copies, differed somewhat among themselves, signified.[21]

The decisions of 1616, which framed the "trial" of 1633, had not been clear either. As we know, they consisted of the secret finding by the Holy Office that a moving sun was "formally heretical because contrary to scripture" and the published prohibition by the Index of Copernican books without mention of heresy. Galileo's semi-private interview with Bellarmine, whether or not it ended in the injunction or precept from Seghizzi, went no further: he learned that his view of the universe was *contra scripturam*, not that it had been declared heretical. As the omission of "formally heretical" from the Index's version shows, a finding *contra scripturam*, which could be made by the tribunal, did not mean "heretical," a determination that could be made only by a pope or general council.[22] What then was the heresy of which Galileo was vehemently suspected?

In his abjuration, Galileo itemized the acts that had troubled his judges: disobeying the precept, defending a false doctrine already condemned as contrary to scripture, and writing and publishing a book giving strong reasons in its favor without refuting them. "Therefore I have been vehemently suspected of heresy, namely, of having held and believed that the sun is the center of the world and motionless…" The tribunal used the same phrase in its sentence of Galileo; but even this formulation seems only to say that the suspicion arose because he clung to a doctrine he had been told contradicted scripture, not to a proposition that a competent authority had declared heretical. Continuing his recantation in the standard formula for abjuration *ex vehementi*, Galileo had to "abjure, curse, and detest *the above-mentioned* errors and *heresies*, and, in general, each and every other error, heresy, and sect contrary to the Holy Church." It appears that the first official public identification of Copernicanism as a heresy came in the curious and irregular form of a pronouncement not by a pope or a council but by a layman! Since Galileo's recantation had the enthusiastic but silent support of Urban, it has been construed as sufficient to make condemnation of Copernicanism a matter of faith. On this interpretation, Galileo would have announced a newly defined heresy and sworn that the Inquisitors erred in thinking he held it in the same breath. It seems safer to say that Galileo's abjuration followed the standard form where inappropriate, and Copernicanism became a heresy not by papal edict but by poor editing.[23]

It was not hard to fall under vehement suspicion of heresy without being supposed to hold one. The Inquisition needed only to notice that a suspect held the beliefs, and behaved in ways, that in its experience often led to heresy. Sample offences were hindering the work of the Inquisition; favoring, defending, advising, or receiving heretics; and denying openly

well-known tenets of faith. Galileo's violation of the admonition and precept, his sleight-of-hand with the imprimatur, and his vigorous defense of a
condemned opinion would qualify him as a suspect under this head. There
is another way that he might have been deemed suspect although the belief
that gave rise to the suspicion was not a heresy. Inquisitors seem sometimes
to have applied (or misapplied) the term "heretical" to rebellious, licentious,
or impious talk and acts that showed an inclination to one heresy or another.[24] Galileo's disobedience might have qualified under this generous definition. We remember Sarpi's observation that the Roman hierarchy made a
habit of crying "heretical" against ideas and behavior it did not like.

In this confusion the nuncios could agree how to threaten but not how to
advise. Those who thought they could read Urban's mind inferred that he had
determined Copernicanism to be a heresy. Others could suppose that with
progress in physics and exegesis the condemned view might be safely held;
the finding *contra scripturam*, contrary to one of heresy, was amendable. The
nuncio to Venice allowed himself to characterize both Galileo's views and
those of Holy Writ as opinions.[25] We know that neither Grassi nor Scheiner
believed that heliocentrism was, or had been declared, a heresy. The initial
perplexity of Descartes and his manner of resolving it may be representative
of temperate Catholics outside the Roman hothouse. At first he thought to
burn or hide the manuscript in which he had developed his version of the
banned system. It then occurred to him that since neither pope nor council
had condemned Copernicanism, it was not a heresy or a matter of faith. "It
might turn out as it did for [belief in] the antipodes, which once was condemned in almost the same way..."[26] Mathematicians within the reach of
Rome censored themselves and characterized the Copernican system as a
hypothesis or supposition if they discussed it at all. Only Galileo suffered for
teaching and defending it.

The circumstances of Galileo's challenge to the church—his zeal in pushing a doctrine declared to be contrary to scripture, his invention of a hermeneutics to undermine constituted authority in biblical interpretation, his
outspoken disdain for people who disagreed with him, and his treatment of
Urban's voluntarism—would seem sufficient reasons for his downfall. Some
historians add Spanish pressure on Urban, which, they suppose, he thought
he could reduce by demonstrating his vigilance against Galileo's "heresy."
They are off by several orders of magnitude. The Spanish faction demanded
that Urban intervene vigorously, that is, financially, on the Habsburg side in

the ongoing war and that he reverse his reversal of the policy of three generations of his predecessors, who had encouraged the growth of Spanish influence in Rome. Galileo had no value as a trophy in this high-stakes political
game. His condemnation made no difference to Urban's expressions of partiality for the French, which occasionally precipitated armed conflict between
Frenchmen and Spaniards in the streets of Rome. When Urban fell seriously
ill in 1637, Spanish troops mobilized on the Neapolitan border to insure a
friendlier successor. Knowing that the sacrifice of Galileo could not appease
Spain, the Barberinis barricaded themselves in Castel Sant' Angelo to weather
the interregnum they feared, having first stripped the Vatican of the furnishings they fancied.[27]

Urban lived to continue to favor France and tether Galileo. Several influential French Catholics pleaded for Galileo's release, notably Fabri de Peiresc,
who squandered in the cause the entire capital of good will he had acquired
by publishing Urban's poems.[28] In a routine case he would have succeeded.
Customarily people who had abjured *ex vehementi* stayed for a time in jail
and then in a monastery, after which, on application, they obtained their
freedom. Campanella underwent exactly that course of rehabilitation in the
1590s. "Perpetual incarceration" in inquisitorial practice meant detention for
three to eight years.[29] The Holy Office maintained its clampdown on Galileo,
which included monitoring his visitors, because Urban continued to regard
him as a clear and present danger. That was as great a compliment to the
power of Galileo's words as to the perniciousness of his philosophy.

Castelli did not compromise himself during Galileo's trial. He spent the
time in Brescia, sent there, some say, by a caring pope, to keep him from
acting rashly, or, others say, drawn there to liberate his brother from jail with
the help of the sympathetic cardinal nephew. The Barberinis' solicitude did
not extend to allowing Castelli to return to the chair of mathematics at the
University of Pisa offered him after Aggiunti's death in 1635. That would
have brought him too close to Galileo. Instead, Castelli received the largely
pro-forma charge of a Benedictine abbey in the Veneto to help him meet his
expenses in Rome, which included outlays for good wine and the tobacco he
recommended to Galileo as a cure for everything.[30]

Castelli's ongoing service as "mathematician to Pope Urban VIII and professor at the University of Rome" dealt with such enduring problems as
controlling the River Reno and draining the Pontine Marshes. His advice
concerning a proposal to straighten a part of the river Bisenzio, a tributary

of the Arno, is worth recalling. Castelli, Guiducci, and other Galileists familiar with Galileo's unpublished ideas about motion on inclined planes argued
that the cut would not improve the flow because change in elevation, not
length of channel, determined change in velocity. The projectors replied that
the longer and sharper the meanders, the greater the friction and the slower
the flow. When consulted, Galileo affirmed that a flowing river behaves like a
cascade of balls down an inclined plane (water particles do not cohere!); since
the length of the river bed, like that of the plane, makes no difference, the
proposed cut would have no benefit. This analysis omits the most important
factor in the calculation: Guiducci owned land along the Bisenzio threatened
by the engineering works.[31] The episode displays to advantage the solidarity,
rather than the science, of Galileo's disciples. Throughout his employment
by the Barberini, Castelli remained fiercely loyal to Galileo.[32]

The main actor

Picking himself up from his humiliating posture before the cardinals and
the gospels, Galileo returned to Niccolini's residence while awaiting Urban's
decision about his place of detention. On 30 June 1633, after an appeal to his
former friend, he received permission to stay within the palace of the archbishop of Siena, Ascanio Piccolomini, in anticipation of a return to Florence.
This last step, conceded by the pope in December, brought Galileo back home
to Arcetri after an absence of over a year. The main condition placed upon
him apart from not leaving was to keep at bay people eager to discuss world
systems. Above all, he could not make his house a meeting place for a society
of philosophers and mathematicians.[33]

The six months that Galileo spent in Siena at Piccolomini's house and
table revived his spirits. He started a new work on mechanics—"full of many
curious and useful ideas"—and enjoyed the conversation of the archbishop.
Among the subjects discussed was astrology. Piccolomini's elder brother
Ottavio had asked him to supply a certain birth chart. The archbishop
could not obtain it from Sienese astrologers, whom Urban's bull had driven
underground. Fortunately, so Ascanio replied to Ottavio, he had an accomplished astrologer as a guest. "I have largely lost credence [in the art] since
learning that Messr Galileo, famed as an Astrologer...derides it entirely, and
makes fun of it as a profession founded on the most uncertain, if not false,
foundations."[34] That was not what Ottavio wanted to hear.

Ottavio Piccolomini was the unscrupulous lieutenant of the imperial general Albrecht von Wallenstein, who went nowhere without his astrologers. For a time beginning in 1628 this baggage included Kepler, driven from Prague by imperial stinginess and religious persecution. Kepler knew the details of Wallenstein's geniture peculiarly well. He had drawn it up in 1608, and, at the general's request, had updated it in 1624 with the prediction that a horrible tragedy would strike Wallenstein in February 1634. Kepler did not have the satisfaction of seeing his prediction realized. Not because it did not happen as he had foreseen, but because he died in 1630. Piccolomini, who knew about it, and knew further that Wallenstein believed it, led a plot against his fearsome leader. Naturally it unfolded in February 1634, when the general had fallen from favor and the stars assured success. Piccolomini did well from his collaboration with fate and ended the Thirty Years War in command of the imperial forces.[35]

The report of the archbishop to the general probably accurately reproduced Galileo's final considered opinion about astrology. He had placed some credence in it when he cast birth charts for himself, his daughters, and his students, and sought to reassure astrologers that the discovery of the moons of Jupiter did not make their art any less reliable than it had been.[36] But experience had exposed its shortcomings (his calculation of Virginia's character suited Livia's better, and vice versa), the Morandi affair its dangers, and the Bull *Inscrutabilis* its illegality. "It would be very remarkable if [anyone] managed to place astrology at the top of human sciences," so Galileo wrote Diodati as he set forth for Rome in 1633. Or rather, as he put the point to Archbishop Piccolomini later in the year, astrology was almost certainly false.[37] Another possible reason for Galileo's cooling toward the art is that it did not fit with his tidal theory. According to a strong traditional plausibility argument, planets exercise physical influences here below just as the sun has jurisdiction most evidently over the seasons and the moon less evidently over the waters. Galileo denied the middle term, the lunar influence, as part of his kinematical clean-up of the Aristotelian heavens and the spooky qualitative cosmic magnetic forces invoked by Kepler.

The free conversation at the archbishop's table gave rise to the accusation that he allowed his infamous guest to introduce dangerous ideas into susceptible minds. The local inquisitors learned that Galileo spread "opinions hardly Catholic" with the support of Archbishop Piccolomini, "who has told many here that Galileo was unjustly treated by the Sacred Congregation and that he could not and should not reprove philosophical opinions that he had sustained with true and invincible mathematical reasons, and that he is the greatest man in the world and will live forever in his writings, and that he is followed by all the best modern minds."[38] All of which was brave and true. The fine story that when rising from his knees before the inquisition Galileo muttered, "still it moves," is associated with the Piccolomini. A portrait, representing the scene of Galileo's recantation, perhaps by Murillo and perhaps commissioned by General Ottavio, displays the slogan, *eppur si muove*.[39]

Despite the silk hangings and rich furnishings of his apartment, Galileo longed to return to Florence and, as he fancied, the grand duke's service. He urged Cioli to ask Ferdinando II to ask the pope for his release, "and, to give the petition more force, he might adduce the loss of my services over so long a time, exaggerating their value somewhat over their true worth." The ambassador, the knowledgeable Niccolini, judged that Urban was still too angry to approach.[40] Galileo had therefore to continue to operate in Florence at a distance, primarily through Maria Celeste, who with the help of Guiducci and Bocchineri had attended to Galileo's business during his confinement in Rome. It took some effort to run Arcetri with its servants, dovecote, orange trees, vegetable garden, grape vines, and "most original mule."[41] Maria Celeste managed well, let the house, sold the surplus crops, and made a profit that she shared with her sister. She managed even to sell a year's supply of spoiling wine. Meanwhile Galileo was enjoying premium wine at the archbishop's table. Maria Celeste: "I pray that you continue [in good health] by governing yourself well particularly with regard to the drinking that is so hurtful to you, for I fear that...your social obligations to your host afford you ample opportunity for indulging."[42] But in Christian charity she responded to Galileo's worry that Arcetri would be dry on his arrival. The good angel ordered in a goodly supply, three mule loads at one go; and, in her last recorded service to her father, could announce, on 10 December, that "the casks for the white wine are all in order."[43]

Maria Celeste did not neglect the spiritual comfort of her psychologically labile father. After learning his sentence she wrote, "Now is the time to avail yourself more than ever of the prudence [!] which the Lord God has granted you, bearing these blows with that strength of spirit which your religion, your profession, and your age require. And since you, by virtue of your vast experience, can lay claim to full cognizance of the fallacy and instability of everything in this miserable world, you must not make too much of these storms." Galileo smiled and made the best of things, to his daughter anyway, by allowing, no doubt with some reservations, that his travail had ended "to the satisfaction of both [himself] and [his] adversaries."[44] Maria Celeste proposed to free her father by mobilizing women—good Mrs Niccolini and Mrs Barberini, the mother of the cardinal nephew, and all the nuns of San Matteo, who would pray to God to enlighten his Vicar. The method did not procure Galileo's release but may have been responsible for his unusual run of good health.[45] More practically, the innocent nun allowed Bocchineri and the professor of mathematics at Pisa, Niccolò Aggiunti, to remove all the manuscripts from Galileo's house in which the Inquisition might find anything compromising.[46]

Despite her incarceration, Maria Celeste kept Galileo's family together. She persuaded him to help enlarge her brother Vincenzo's house in Costa San Giorgio between Florence and Arcetri, which would be a refuge for him when ill; she succored the ingrate Vincenzo Landucci, left in a pitiable state with two children when the plague took his wife; and she urged Galileo to other good works, like helping the relatives of his mother.[47] Perhaps Galileo took the many services his daughter rendered too much for granted. In reply to his complaint that she had neglected her weekly letter, she flared out: "If only you were able to fathom my soul and its longing the way you penetrate the Heavens, Sire, I feel certain you would not complain of me." Galileo loved Maria Celeste as much as he could anyone. He was devastated when, three months after he returned to Arcetri, her longing soul fled the prison to which he had consigned it.[48]

Maria Celeste's death coincided with a hardening of the conditions of Galileo's detention. He described them to Diodati:[49]

> I am restricted to this little villa a mile from Florence with the strictest prohibition against going there and against having meetings with many friends together...So I live quietly frequently visiting a nearby convent where I had two daughters whom I much loved, especially the elder, a woman of fine mind and singular goodness and most affectionate to me...Returning home [the day before she died] I found a deputy of the Inquisitor come to tell me of an order from the Holy Office in Rome...that I must stop asking for permission to return to Florence or they will make me return to a real prison of the Inquisition. From this and other incidents...you see that the rage of my most powerful persecutors is continually aggravated.

He could do nothing but fulfill at last the purpose for which his friends thought him uniquely suited. "God willing, I want to publish my books on motion and other works, all entirely new..."[50]

8.2 WINDING UP

When on his knees before the Inquisition, Galileo had sworn not to say or write anything about a moving earth or stationary sun, "et contra," lest he fall again under suspicion of heresy. Presumably "et contra" meant that he could not write against Copernicanism either, a measure directed, perhaps, against his proposal to add two more days to the *Dialogue* to refute the arguments of the previous four. He was not required to remain silent about other matters, however, and he began to plan his long-delayed work on mechanics while still enjoying Piccolomini's hospitality in Siena. He continued with this work, encouraged by friends and notwithstanding the loss of some of his manuscripts during their precautionary removal from his house by Bocchineri and Aggiunti. He planned to publish the resulting dialogues with appropriate license in Venice through the good offices of his old friend Fulgenzio Micanzio.[51] The plan snagged on Micanzio's discovery of a miserable example of Roman malignity. He asked the Inquisitor of Venice for a license to republish Galileo's *Dialogue on floating bodies*. The inquisitor refused. He had received an order from Rome. Micanzio replied that the order had to do with the Copernican system. "No," he replied, "it is a general prohibition *de editis omnibus et edendis*," against everything Galileo had published and wanted to publish. Micanzio was outraged at this effort to deprive posterity of "the greatest progress in philosophy that had been made in two thousand years." It would be a "crime against humanity."[52]

Micanzio could estimate what posterity would lose since he had received drafts of Galileo's new work before visiting the inquisitor. Micanzio expressed his admiration in the style of a character in a Galilean dialogue. "It is inexplicable how in everyday trivial things known to everyone you observe the effects of nature and rise to the deepest thoughts... The novelty of the things, the reasoning and demonstrations put me into a new world..." "I cannot admire enough how the book of nature is so fully open to your mind that you find the most profound and undetected marvels in everything." "Our good old master Fra Paolo [was right in saying] that God and nature had made your intellect unique in all time for understanding motion, and that what did not occur to you [about it] was beyond the reach of human thought."[53] As Micanzio fretted how to print the great work in Italy, Diodati was busy beyond the Alps sponsoring a Latin edition of the *Dialogue*. He chose a Lutheran professor of history as translator. This was Kepler's great friend Matthias Bernegger, who had rendered Galileo's booklet on the compass into Latin. The demanding new task allowed him simultaneously to enlighten posterity and embarrass Rome. "You see [Bernegger wrote a fellow Lutheran and Keplerian] to what stupidity those purple-coated priests have come. Let's not allow them to deprive the public of such a book." The Elsevier Press in Strasbourg delivered Bernegger's version of the *Dialogue*, entitled *Systema cosmicum*, in 1635. The volume also contained Foscarini's condemned hermeneutics, and, almost, Galileo's then still unpublished letter to Christina, which had to be issued separately because Diodati's Latin translation missed the printing deadline.[54]

The published *Letter to Christina* carried a preface defending Galileo assembled by Diodati from excerpts from Galileo's letters. It blamed everything on rabid, relentless, and, except for the Jesuits, anonymous enemies. What had hurt Galileo most, so he had written Fabri de Peiresc and Diodati, and so Diodati wrote in his preface, was the slander that he was not a good Catholic. "Many might have been able to behave and speak in a much more learned manner, but no one, not even among the Holy Fathers, would have been able to do so with more piety or with a greater zeal for the Holy Church, or ultimately with a holier intention than mine…In reading all my works, no one will find even the smallest shadow of anything straying from piety and reverence toward the Holy Church." Perhaps a strict interpreter might spy a lapse of piety or zeal in Galileo's collaboration in the publication of the *Letter to Christina*, since he supplied the manuscript and also commented on Diodati's draft preface. He may have felt the risk of further dealings with the Holy Office worth running not only because publication of the preface and the *Letter* would put his case before the Republic of Letters, but also because he believed that his essay in theology might convince Rome to relax the conditions of his detention.[55]

The obvious publisher for Galileo's "Dialogues on motion," as he referred in correspondence to the "fruits of my studies I value the most," was Elsevier.[56] The route was circuitous. Galileo had sent copies of the manuscript, completed in 1636, to several sets of friends. One came into the hands of Giovanni Pieroni, who served the Holy Roman Emperor and saw no problem in publishing in Prague. Jesuit influence ruled out Vienna and, as Pieroni discovered, Prague too. He therefore proposed to acquire a press to print Galileo's work, "which is equally beautiful and novel, as marvelous as it is certain." The Cardinal Bishop of Olmütz had a press, and seemed willing to print; the figures were almost finished; but, in the end, Pieroni could not overcome the obstacles and had to return the manuscript. "What an unhappy place we live in, where there reigns a determined resolution to exterminate all novelties, especially in science, as if we already knew everything knowable!"[57] Thus Galileo, growing impatient and fearful that in the end his enemies would succeed in silencing him.

The obvious solution then appeared in the form of Louis Elsevier, who obtained a copy of the manuscript through Mincanzio during an extended visit to Venice. Another copy went north directly in the baggage of the French ambassador to Rome, the Comte de Noailles, who had been Galileo's student in Padua in 1603. Having finished his embassy in Rome, Noailles obtained from Urban, who could not refuse a high French diplomat a reasonable request, permission for Galileo to meet him along his route back to France in 1636.[58] The manuscript he then obtained also reached the Elseviers, who issued Galileo's favorite work from their Leyden press in 1638 as *Discourses on two new sciences*. The title diminished Galileo's satisfaction in the book. He thought it plebian, vulgar, and unimaginative in comparison with "Dialogues on motion."[59]

Galileo dedicated *Two new sciences* to Noailles with an open appeal for protection. "If I may be permitted to say so, you are now obliged to defend my reputation against anyone who attacks it, you having entered me in the lists against all adversaries." Orlando had found his Charlemagne, and a more promising field for the celebration of his deeds, in the vast open tracts beyond the Alps. The narrow arena defined by Florence and Rome, in which he had tried universal truths against parochial interests, now appeared in its true measure. Failure among the priests and prelates of Italy might be a recommendation elsewhere. Regarded from Paris, Noailles wrote, "you are the greatest mind in Italy."[60] Galileo thought so too. In the preface to *Two new sciences*, written by him but signed by the printer, we read that "wise antiquity" honored its great inventors and light-bringers, "even to the point of making them gods." Galileo is such a light-bringer, the greatest in our time; in the present work, dear reader, you will find abundant evidence of "the grace conceded to this man by God and nature."[61]

Days 1 and 2

The *Two new sciences* develop in conversation among Salviati, Sagredo, and Simplicio, who have moved from Sagredo's airy palace and world systems to the Venetian Arsenal and down-to-earth machinery. The discussion opens with a complaint by Sagredo: despite his best efforts, he has not been able to make workmen understand that if a wooden beam of given length and cross-section can support a weight w, a similar beam with dimensions increased ten-fold should be able to support a weight 10w. The workman was right, Salviati replies, and claims to be able to show geometrically the relative weakness of bigger structures scaled up from smaller ones. Consider only that a child can sustain a drop that would break the bones of an adult; that a cat can fall further than the child without injury; that a cricket might survive a plunge from a tower and an ant one from the moon. A giant sixty feet tall could scarcely stand if his bones were only ten times thicker than those of ordinary men.[62]

What then causes resistance to breakage? In a wooden beam the case is clear: fibers, as in a rope. What about stone or metal? Salviati remarks that two flat, highly polished pieces of marble will adhere when their smooth surfaces are pressed together, and asks whether nature's reluctance to admit a void (the cause offered for the adherence of the marbles) can explain the coherence of bodies? To settle the question, Salviati adduces Galileo's demonstration that water has no force of aggregation among its parts. If it were only possible to make a rope of water and hang a weight from it! Nothing easier, says Sagredo, the experiment takes place every day: suction pumps can raise water only thirty feet or so before the water column breaks under its own weight. Since by hypothesis water contains no glue, a column of it thirty feet high, independent of cross-section, must measure nature's abhorrence of a vacuum. Now a stone pillar heavier than a thirty-foot water column of the same cross-section can sustain itself. It appears that something other than fear of the vacuum holds the world together.[63]

Sagredo suggests and Salviati agrees that the conclusion is precipitous. Perhaps the force at play is not resistance to a measurable void, like that supposedly created in the instant of separation of polished marble slabs, but resistance to exposing the infinity of infinitesimal voids that might exist between the infinity of infinitesimal particles that might constitute the surface of the stones. Simplicio interjects that the idea smacks of the teaching of atomists. Sagredo: "At least you did not add, 'who denied Divine Providence,' as in a similar instance a certain antagonist of our Academician very inappropriately did add." Even Simplicio has no time for this antagonist (Sarsi). "Indeed I perceived, not without disgust, the hatred in that malicious opponent...I know how far such ideas are from the temperate and orderly mind of such a man as you [Salviati], who are not only religious and pious, but Catholic and devout." With this thin protection, the friends go for a random walk through the vacuum, through "many intricate labyrinths," where the paradoxes of the infinite and the infinitesimal greet them at every turn. From all of which it followed that the rim of a disk is made up of an infinite number of matter bits and infinitely divided nothingness.[64]

This bizarre picture resulted from an analysis of a famous problem in Aristotle's *Mechanica*. A rolling wheel advances along the ground a distance equal to its circumference during each rotation. Its hub during the same rotation lays down a smaller circumference along a line parallel to the ground (Figure 8.1). Yet they return to the same configuration. Thus the successive points C appear to skip forward (the successive points A being at rest). Or, if the rotation occurs so that C is instantaneously at rest (CA being free to move in a ditch), the successive points A would appear to skip backward. The skipping arises when material points come opposite infinitesimal voids. It was this argument that had unleashed Micanzio's excitement over *Two new sciences* since it revealed the mechanism of contraction and dilation (without removal or addition of material) that he, Sarpi, and Galileo had puzzled over in the old Paduan days.[65] Sarpi's position as preserved in his notebooks does indicate puzzlement. "Rarefaction and condensation will be a sort of configuration of the plenum and vacuum, condensation by expelling the *alieno*, rarefaction by admitting it." Fra Paolo had not identified the "alien" in question.[66] Galileo kept at the puzzle until one morning at breakfast at the real Salviati's he saw how nature could condense and expand bodies without requiring void space or bodily interpenetration. The theory as sketched in the *Dialogue* exploits the infinity of infinitesimal voids scattered among the infinity of matter points that together formed bodies. A moving together of the points produced condensation, the opposite motion expansion. In keeping with his kinematical approach, neither here nor in his account of the strength of materials did Galileo specify forces that could push mass points in and out of their microvoids. Simplicio noticed that there was too much geometry and too little physics in Salviati's philosophy, but had no idea how to repair it.[67]

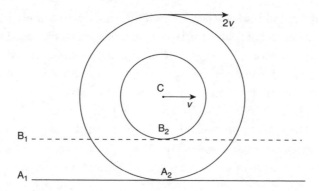

FIG. 8.1 Aristotle's wheel, which presents several puzzles of the infinite.

Salviati turned to exploding Aristotle's theory of the void with the arguments Galileo had developed by 1610: velocity of fall independent of weight, Archimedean buoyancy, descent along inclined planes, and pendulum swings strictly isochronous irrespective of amplitude. From pendular vibrations Salviati sublimed to music, to Vincenzo Galilei's experiments on monochords, once more avoiding dynamics in favor of arithmetic. Since a string's pitch depends on its thickness and tension as well as on its length, none of these physical qualities can be "the direct and immediate reason behind the forms of musical intervals"; the reason must be sought, rather, in "the ratio of the numbers of vibrations and impacts of airwaves that go to strike our eardrum." Sagredo explains why the "direct and immediate" cause, 5:3, gives us such pleasure before noticing that the day has gone in digressions.[68] Still, it had been worthwhile to hear Salviati's "delicate and wonderful," "subtle...novel...remarkable...ingenious," speculations. Salviati promised to return on the morrow, "to serve and please you."[69]

When the friends meet again they are all business. Salviati has cast his ideas about the strength of materials into propositions and corollaries couched in geometry. His basic idea is that the equilibrium of a horizontal wooden beam may be likened to the operation of a bent lever. Neglect at first the weight of the beam and ascribe its resistance to a large number of "threads" parallel to its axis each under tension T. We can suppose (incorrectly as it happens) that these fibers all act at the center of the base AC (Figure 8.2), so that the total power of resistance R to the pull of W is proportional to a^2T. To keep the beam from breaking, the clamped portion must pull to the left on the lever arm CA with the moment $Ra/2$. The law of the lever then

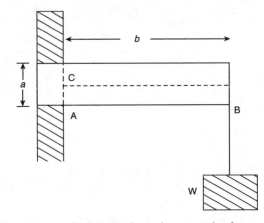

FIG. 8.2 Galileo's beam, which assimilates the strength of materials to the law of the lever.

gives Proposition 1, $R:W = 2b:a$. Propositions 2–4 bring nothing new except a catchy corollary: "the moments of resistances of prisms and cylinders of equal length are as the three-halves power of their volumes."[70]

Then come a few propositions bearing on the original problem of scaling. Salviati compares two geometrically similar beams made of the same material each bearing only its own weight. Let the dimensions be a,b and c,d, where $a:b = c:d$, and let the larger weigh k times the smaller. Then $c = k^{1/3}a$, $d = k^{1/3}b$. The area of the base and hence its resistance R increase as $k^{2/3}$. Therefore the weight, which increases by k, goes up as $R^{3/2}$, or, as Salviati puts it, with moment M for weight, $M_1:M_2 = (R_1:R_2)^{3/2}$.[71] Simplicio: "This proposition strikes me as not only new but surprising, and at first glance very remote from the judgment I had originally formed." For it showed that the resistance of a beam to fracture does not increase linearly with its weight, but at a higher power. Sagredo: "This demonstrates the proposition which, as I said at the beginning of our discussion [after Salviati corrected me], seemed then to reveal itself to me through shadows."[72]

Salviati shows that for a given ratio $a:b$ there is only one prism at the edge of breaking under its own weight; a longer one would fail, a shorter one could carry an additional load. Two nice departures from this line of reasoning mark the end of the day. Sagredo observes that material might probably be removed from a prismatic beam without lessening its resistance or shortening its length. "Without lessening resistance" signifies that the carved beam is no more likely

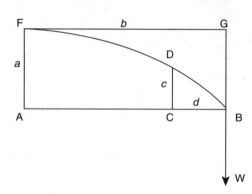

FIG. 8.3 Statics of a parabolic beam.

to break transversely at one point than at another. Assume that the beam weighs nothing and that the shape required is the unknown curve in Figure 8.3. The moment bW balances the moment of resistance at FA, which is proportional to a^3. Can it also be made to balance the resistance proportional to c^2a exercised by the base CD? (The dimension of both bases perpendicular to the plane FABG is a.) That would require $a^2/b = c^2/d$. A curve satisfying this property for all points D along it is a parabola. Salviati shows off by demonstrating that its volume is only two-thirds that of the prism ABGF. "A beautiful and ingenious demonstration" (Sagredo), and quite a saving in weightless wood![73]

The second refreshing departure explains the cleverness of birds and plants in employing hollow cylinders to strengthen their bones and stalks with little expenditure in mass. A hollow tube made of the same amount of material as a solid one can be considerably stronger. Let them have the same length and external diameters a and b. Then since the base of the tube and of the rod anchor the same number of fibers (they contain equal matter) their moments of resistance are as their lever arms, that is, as $a:b$.[74] (In Salviati's approximations, the average lever arm of the base of the tube is $b/2$.) With this contribution to bird economy the second day ends, without as much as a *Nunc dimittis*, perhaps because Galileo had intended to continue it before the Elseviers concluded the discussion by publishing it.

Day 3

Without a word of greeting, Salviati begins the third day's conversation by reading aloud in Latin from Galileo's perfected treatise on motion. It is a very fine book, filled with clever theorems, a basic text for students of mechanics, but also tedious, repetitive, and, because Galileo stuck with the ancient style of geo-metrical presentation, inefficient. What life it has comes from interruptions in Italian when Sagredo or Simplicio requires fuller explanations. Sagredo is the first to interject. He does not like Galileo's definition of uniformly accelerated motion as "add[ing] on to itself equal momenta of swiftness in equal times." He offers two objections: since time is indefinitely divisible, the falling body would have to move through every degree of slowness before attaining any discerna-ble speed; and the proposition that velocity increases in proportion to distance fallen would do as well or better than Galileo's definition. Salviati replies that he too had trouble with these obstacles as had the Academician himself, but that with a little attention they disappear. The falling body does go through all degrees of tardiness, but spends no finite time in any of them; and an object whose velocity increases with distance would have to cover any space assigned in the same time. Whence Salviati made a sociological deduction: since Gali-leo's demonstrations are so clear and effective, his adversaries must understand them too, and suppress them "merely to keep down the reputations of other men in the estimation of the common herd of little understanding."[75]

Salviati resumes his reading. The material is familiar: velocity acquired in descending along an inclined plane depends on the vertical drop, not the slope; successive intervals traversed in free fall from rest increase as successive odd numbers and, consequently, total distances as the squares of the time elapsed. Simplicio breaks in to declare a preference for the arithmetic formulation (the odd-number rule), which he fully understands, over Galileo's geometrical one (areas of triangles), which gives him trouble; and he demands some experi-mental evidence to anchor the theory before the reading continues. Salviati obliges with a detailed description of trials made by the Academician in his presence. They had taken a smooth, straight board, rabbeted out a groove or channel down its length, lined the channel with smooth vellum, rolled a hard bronze ball down it from different places, and timed the rolls with a water clock. In a hundred trials, so Salviati affirmed, the rule of fall—distance rolled proportional to the square of time elapsed—held "to within a pulse beat."[76]

The ground being thus prepared, Salviati reads off a series of propositions about broken journeys along inclined planes, all requiring the proposition that acceleration along an incline of height a and length b is weaker than acceleration in free fall by the factor a/b. Apparently Galileo decided that this capital proposition required stronger arguments than Salviati's assertion and in 1638, too late for publication then, he added a strong argument included in later editions. This begins with the observation that the tendency of motion down an inclined plane declines with its slope, becoming zero on a horizontal surface, where it has no tendency to move up or down. The argument continues, in a gratuitous but precious deduction irrelevant to the proof, that motion in the horizontal, "which here means a surface equidistant from the center [of the earth]," would, if once started, continue indefinitely. It appears, therefore, that in his last formal analysis of motion, Galileo saw no point in considering a moving body not constrained by a center.

The problem under consideration concerns planes of finite slope. Galileo resolves it by an application of the law of the lever. In Figure 8.4, the weights being in equilibrium, $A<B$, the downward moment of A is equilibrated by that of B, only part of which acts vertically. This portion is (a/b) since, if A sinks by an amount x, then B, constrained by the plane and the string, will rise only by y, which geometry requires to equal $(a/b)x$: as usual in machines, the smaller weight moves through a greater distance to raise a larger weight a smaller distance. The law of the lever tells us that the weights are in inverse proportion to the distances: $B = (b/a)A$; or, taking the pulley as the fulcrum, that the moment of B along the plane is $(a/b)B$. From which it follows— Salviati here returns to the text of 1636—that the times of descent along planes of unequal length but the same height are as the lengths. Sagredo remarks that Salviati need not have troubled to prove it, since he had already proved (or rather assumed) that the velocity acquired in such descents depends only on the drop.[77]

The rest of the third day's "dialogue" rings changes in Galileo's very beautiful theorem that descent along all chords beginning at the top or ending at the bottom of a vertical circle takes place in the same time. This truth prompts a "very beautiful...reflection" from Sagredo. Let a great number of balls be released at the same time along planes spreading downward in all directions from a single point. The balls lie on a circle that grows in

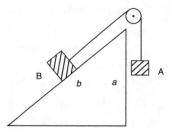

FIG. 8.4 Galileo's final demonstration that the weight of a body on an inclined plane of height a and length b is (a/b) times its weight when suspended freely. Cf. Fig. 2.10.

time. Another way of producing a widening circle of motion is to drop a pebble into a still pond; but whereas in the first case the expanding circles stay tangent to one another at the point of launch, in the second they spread outward from the pebble's splash as center. These considerations elicit from Simplicio his most original contribution. "Since we can assign as the site of such emanations the lowest center as well as the highest spherical surface, I believe that some great mystery may perhaps be contained in these true and admirable conclusions—I mean a mystery that relates to the creation of the universe...and perhaps to the residence of the first cause." Had he a Pisan Drop in mind? Salviati does not object but refuses to be drawn. "Such profound contemplations belong to doctrines much higher than ours."[78]

From this sublimity Salviati falls to reading and proving increasingly tedious theorems and artificial cases, of which Proposition XII is the finest specimen. "If a vertical and a plane however inclined intersect between given horizontal lines, and mean proportionals are taken between [each of] these and its part contained between the intersection and the upper horizontal, the time of movement in the vertical line will have, to the time of movement made in the upper part of the vertical and then in the lower part of the cutting plane, the same ratio as...."[79] Who cares? Among the thirty such propositions that display Galileo's surpassing playfulness and cleverness in plane geometry there is one that for its beauty and simplicity deserves notice. He asks for the ratio of the time of fall along the vertical diameter AB = d (Figure 8.5) to the combined times of the journeys along chords AE + EB. Since the velocity at E is the same whether the body passes along AE or GE, the transit time t_{EB} will be the same for the broken journey or the straight shot GB. Now $t_{EB} = t_{GB} - t_{GE}$, and the answer sought therefore is (since $t_{AB} = t_{AE}$) "$t_{AE}{:}(t_{AE} + t_{GB} - t_{GE})$." Now comes the geometry. Galileo represents t_{AE} by AE and t_{GE} by GE, which works because times of descent along planes of different slopes between the same parallels are proportional to their lengths. He must now find a line along GB to represent t_{GB} on the same scale. Let this line be GF = xGE, where $x = (t_{GB}{:}t_{GE})$ is the required scaling factor. So $x = (GB{:}AB)(t_{AB}{:}t_{GE}) = (GB{:}AB)(AE{:}GE) = AG{:}GE$. (The last step follows from consideration of the similar triangles AGB, EGA.) We have GF = AG. Since GE represents t_{GE}, EF represents t_{EB} and the ratio required, $t_{AB}{:}(t_{AE}+t_{EB}) = AE{:}(AE + EF)$, the simple form that Galileo gives.[80]

It remained to praise. Sagredo: "[W]ith what ease and clarity, from a simple postulate, he deduces the demonstrations of so many propositions!" Salviati: "[O]nly now has the door been opened to a new contemplation,

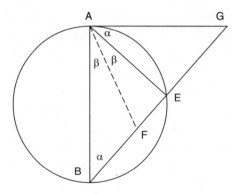

FIG. 8.5 Diagram to relate the time of fall along the broken path AE + EB to that for fall along the vertical AB.

full of admirable conclusions, infinite in number." Sagredo: "[W]hen these [demonstrations] have passed into the hands of others of a speculative turn of mind, [they] will become the path to many others, still more marvelous." Descartes: "[Galileo's] way of writing in dialogues with three persons who do nothing but praise and exalt his inventions in turn certainly makes the most of his wares."[81]

Day 4

Simplicio arrives, Salviati starts to read. They approach what Galileo considered to be his greatest discovery. The subject is the trajectory of cannon balls compounded from a uniform rectilinear motion arising from their projection and a constantly acting downward acceleration exercised by the earth. Although a shot at pointblank would, on Galileo's physics, continue its "horizontal" motion in a circle concentric with the earth, he treats it as linear because the distances under consideration are small relative to the earth's circumference. Thus Proposition 1: "When a projectile is carried in motion compounded from equable horizontal and from naturally accelerated downward [motion], it describes a semi-parabolic line."[82] Hold on, Salviati, Sagredo interrupts; you will have to tell us about parabolas if you want Simplicio and me to follow you. Salviati graciously excerpts some theorems of Apollonius to the purpose. They came to this: the points on a parabola satisfy the condition that the ordinates squared are proportional to their abcissae. A drawing (Figure 8.6) is worth a thousand words: $p^2 : q^2 = OP : OQ$. In the case of point blank, the ball shot from C (Figure 8.7) will proceed horizontally covering equal spaces in equal times while it drops through spaces increasing as the square of the

FIG. 8.6 Salviati's definition of a parabola.

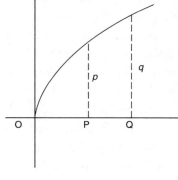

FIG. 8.7 Generation of a parabola
from uniform horizontal motion
and accelerated natural fall.

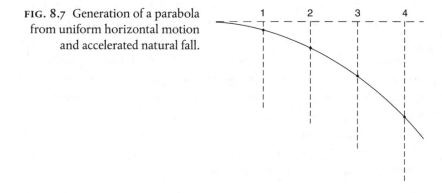

times. The abcissae grow linearly, the associated ordinates quadratically, the trajectory is a parabola.[83]

The dialogue ends with the reading of an entertaining table containing 315 numbers indicating the ranges and heights to which cannons aimed at elevations from 1° to 90° would send their shots. There are certain conditions: the gunners use identical balls fired with identical charges; the guns and their targets stand on the same horizontal surface; and everybody neglects the resistance of the air. To this last restriction Simplicio rightly objects; Salviati counters, wrongly, by adducing the almost simultaneous fall of bodies of different weights and substances, and the almost isochronous beats of pendulums, as evidence that the atmosphere does not offer much friction to objects moving with the speed of cannon balls. In any case, Salviati declares, eager to get on to the calculations whether realistic or not, "to deal with such [messy] matters scientifically, it is necessary to abstract from them. We must find and demonstrate conclusions abstracted from the impediments..."[84]

The calculations toward which Salviati pushes the others are exemplary for their form, power, and unnecessary difficulty. To help appreciate their nature, a sketch of a simpler account, conceptually available to Galileo, will be helpful. In Figure 8.8, α is the elevation, GD = R/2 half the range, BD = h the height of the trajectory, A the intersection of the tangent of the orbit at G with the axis AD. At B the upward velocity of projection, $v\sin\alpha$, equals the downward velocity gt acquired in the time t since firing; in this time the horizontal velocity $v\cos\alpha$ has carried the shot through the half-range GD; hence $R = 2v\cos\alpha \cdot (v\sin\alpha)/g = (v^2/g)\sin 2\alpha$, or, as Galileo would have expressed it in proportion, $R{:}R_0 = \sin 2\alpha{:}\sin 2\alpha_0$, where α_0 is an arbitrary reference point. The obvious choice, which Galileo makes, is $\alpha_0 = 45°$, so that $R{:}R_0 = \sin 2\alpha$. Evidently R is greatest when $\alpha = 45°$. The properties of $\sin 2\alpha$ make the ranges at elevations equidistant from 45°, like 40° and 50°, equal. The height h may be obtained most instructively by supposing that the shot initially fell from rest through AB during time t and, after reaching B, fell again from rest through h in an equal time t. Therefore AD = 2h, and, since $\tan\alpha = 4h/R$, we have $h = (v^2/g) \cdot \sin^2\alpha$, or, old style, $h{:}h_0 = \sin^2\alpha$. Galileo's tables did not require much calculation by him: R is a standard list of sines, which he could have found in Copernicus, h of sines squared.[85]

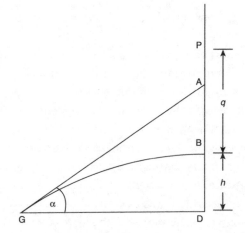

FIG. 8.8 The tangent to the trajectory of a shot at $\alpha°$ above the horizon cuts the vertical at twice the height h to which the shot rises.

Galileo goes about gunnery backwards. He begins not from the gun G but from the summit B, a relic of his experiments with inclined planes. To proceed geometrically as much as possible, he replaces w, the horizontal velocity at B, with the distance, $q = PB$, which a body would have to fall to attain w on arriving at B: from the law of free fall, $w \propto \sqrt{q}$. Now let the body move horizontally from B with velocity w and simultaneously begin to fall again from rest. By the time it arrives at G its vertical velocity will be proportional to $\sqrt{h}$. Hence at G the square of the total velocity v goes as $q + h$ and its direction or elevation is given by $\tan\alpha = \sqrt{(h/q)}$. The half-range $R/2 = wt = \sqrt{(2gq)} \cdot \sqrt{(2h/g)} = 2\sqrt{(hq)}$. The discussion centers at first on trajectories that have a constant range and, consequently, different total velocities proportional to $\sqrt{(q+h)}$ on arriving at G. It is useful to know that the minimum value of v, and so of v^2, for which the shot descending from B has the half-range $R/2$ occurs at $45°$. There $\tan\alpha = 1$, $q = h = R/2$, and $q + h = R$. If $\tan\alpha > 1$, $h > q$; if $\tan\alpha < 1$, $h < q$; in either case $q + h > R$. Galileo takes from this analysis that a shot propelled in the reverse direction, from G, would require the least charge for a given range if aimed at an elevation of $45°$. But gunners cared more to know how ranges differed by elevation if they employed standard charges and balls. Taking v or v^2 as a measure of charge, Galileo geometrized their problem by making charge proportional to $q + h$ or $\sqrt{(q + h)}$. The restriction to constant charge came to requiring $q + h$ to be a constant for all trajectories.[86]

We may reasonably call this God's view of gunnery since Galileo explicitly related it to the Pisan drop of the *Dialogue*: the idea that the Creator placed the planets in their orbits by dropping them all from the same point and converting their vertical motions into horizontal ones when He was satisfied with their performance. As we know, Galileo had a special fondness for this brave conjecture. Here he put it into Sagredo's mouth "in the guise of a true story." To check it out required knowledge of planetary speeds "and the distances from the center about which they turn," information, Sagredo continued, now available through "very competent astronomical doctrines." To this whiff of Copernicanism Salviati replied that the Academician had told him once that he had made the necessary computation, "and also that he found it to answer very closely to observations." Here he drew back. The Academician had not wanted to talk more about it, "judging that he had [already] discovered too many novelties that have provoked the anger of many. So let us get on with our material."[87]

The problem now is to calculate the range for each degree of elevation under the relations $\tan\alpha = \sqrt{(h/q)}$, $h + q =$ constant, $R = 2\sqrt{(hq)}$. This formulation displays immediately the maximum at 45°: it comes to asking for the rectangle hq with the greatest area of all rectangles with the same perimeter ($h + q =$ const.). Everyone knows the answer: the square $h = q = R/2$. It also indicates that for values of α equidistant from 45° the range is the same. For whether h exceeds or falls short of $R/2$ by an amount x, q must do exactly the reverse, and $R = 2\sqrt{[(R^2/4) - x^2]}$ for both cases.[88] So we have for the relative ranges $R(\alpha):R(45°)$ for balls shot with the same charge,

$$R(\alpha): R(45°) = \sqrt{(qh)}: R/2 = h\,ctn\alpha: R/2 = \sin2\alpha.$$

The last two steps come from $h = q\tan^2\alpha$ and $h + q = R$.

As a reward for slogging through these demonstrations, Salviati offered his friends a little "wonder and delight." Know then that a chain suspended from two points on a horizontal line closer together than its length will take the form of a parabola. Or almost.[89] Simplicio was not delighted with this news or with the geometry he had had to bear to learn a little physics. A glance at what he might have to endure—Galileo's old Archimedean propositions that the Academician had added to the book from which Salviati had been reading—put him off completely. He did not participate in the fifth day's discussion on percussion, which Galileo drew up after the publication of *Two new sciences*. This was not farewell. Galileo loved Simplicio too much to part with him and, as will appear, conjured him up for a last talk at his deathbed.[90]

8.3 LAST DAYS

Of the mind

Galileo's analysis of his fall, as he presented it to correspondents outside Italy, was less penetrating than his analysis of the flight of cannon balls. It all began with "calumnies, frauds, and conspiracies deployed 18 years ago [*recte*, 19] to befuddle the authorities." Should they ever become public, "my entirely religious and holy intentions [in campaigning for Copernicus] would appear all the clearer."[91] Why then the plots? "The principal, rather the unique and only cause of my downfall [was] having discovered many fallacies in received ideas, some of which I published... [This] excited in the minds of those who only wanted to be esteemed learned such disdain that, as they are very clever and powerful, they managed to suppress what I had found and published and to block the publication of what remains to me to bring to light." Everything is prohibited, past and future.[92] Galileo shut his eyes to the evidence that Urban's inflexibility had nothing to do with academic jealousy, or, by 1636, with a sense of personal insult. When Noailles undertook to persuade the pope that Galileo had not intended a slight by allowing Simplicio to describe Urban's medicine, Urban replied, "we know it, we know it," but added, as Galileo relayed the report to Micanzio, "reading my *Dialogue* was most pernicious to Christianity."[93]

Urban carried his pastoral care in this matter to the extraordinary degree of reserving to himself the authority to grant exemptions to read the *Dialogue*. He did not give permission easily.[94] Typically bishops and inquisitors dealt with applications for exemptions on a need-to-read basis. Urban's insistence on attending to so small an administrative matter while still deeply engaged in war and nepotism measures his fear that the program of reading the book of nature without restrictions might shred the wholesome doctrine with which he hoped to protect the church against every challenge based on natural science or unaided reason. Urban's vigilance gave Galileo an opportunity to show once again how narrowly he too viewed the great contest. To Micanzio, who wrote with delight that the notoriety of the edicts of 1616 and 1633 were converting mathematicians to Copernicus everywhere, the Prisoner of Arcetri replied that he did not like the news because it might stir up the inquisitors. "Since licensing reading of the *Dialogue* is now reserved to the pope alone, I can reasonably fear that finally even the memory of it will be destroyed."[95] This extravagant judgment suggests that Galileo could no more remove his struggle over the freedom to philosophize from the narrow confines of Rome and Florence than he could think himself free from the earth when reasoning about the unforced motion of bodies. The prohibited *Dialogue* was then so sought after in Italy that it sold for as much as six scudi a copy.[96] That was a month's salary for a junior mathematician.

Despite the severe constraints on him, Galileo did not intend to leave the field to small-minded adversaries. As Ariosto wrote, "Everyone runs to gather wood from the tree the wind blows down." Against these scroungers Galileo would publish the inflammatory remarks he had seared into the margins of his copies of their books—an enterprise luckily unrealized.[97] A more promising project, the republication of his works apart from the *Dialogue* in Latin, he tried to sell to the Elseviers, with such inducements as translations done at his expense and in his prison by a holy mathematician, and an offer to buy a hundred copies for himself. The first Latin *Opera* did not appear in Galileo's lifetime or from the Elsevier press.[98] As stop gaps Galileo urged the flooding of Italy with copies of Bernegger's translations of the innocuous manual on the military compass and the provocative *Letter to Christina*, "to the confusion of my enemies."[99]

While completing the dialogue that became *Two new sciences*, Galileo tried again to find a buyer for his method of finding longitude at sea. Negotiations with the Spanish government had sputtered out in 1620. They resumed briefly around 1630, when Galileo had the help of Ippolito Francini, an accomplished lens maker employed in the Medici glass factory. Although Galileo offered to send Francini and Vincenzo Galilei to Spain to repair broken lenses and negotiate licenses, once again nothing came of it.[100] In 1635, in keeping with the shift of maritime power and his own business to the North, Galileo turned to the Dutch, who had offered a substantial prize to the inventor of a reliable method of finding longitude at sea. Galileo had better tables and techniques on offer than before and a "Jovilab," an instrument that represented the motion of Jupiter's satellites in a relatively easy, if not an adequately exact, manner.[101] And this time he could offer an irresistible incentive. Encouraged by his dealings with Elsevier and now interested more in gaining credit than income from his invention, Galileo proposed to give it to the Dutch government, "knowing it to be more able than all other powers to put it to use since the confederation has a great many ships and, what is more important, abounds with experts in astronomy." One of them, Martinus Hortensius (Maarten van den Hove), professor of mathematics at the Athenaeum in Amsterdam and, later, at the University of Leyden, would be Galileo's main technical contact with the Dutch authorities. In corresponding with him directly or through Diodati, Galileo ran a risk, since Hortensius was a Copernican as well as a heretic. He willingly ran it, however, trusting in Hortensius's discretion, honesty, and acuteness of mind: "Since you are very intelligent, I am certain that you will know that there is no other method available to determine longitude than the wonderful properties of the stars around Jupiter."[102]

The States General of the United Provinces accepted Galileo's gift and, as the Republic of Venice had done with the telescope, gave him a present in return—a gold chain worth some 200 ducats. They also granted Hortensius a sum to procure instruments and an observatory to check observations and calculations and to find a way to reduce the proffered method to the capacity of navigators.[103] He appealed to Galileo for theoretical underpinnings (the parameters of the satellites' orbits) and instrumental help (telescope lenses). But Galileo did not have the parameters or the time or sight to recalculate them from "a mess of thousands of observations" and offered instead their reification in the Jovilab. As for the lenses, he would try to engage Francini, and, as a stopgap, would send the best lenses that had ever been constructed, the telescope "with which I have discovered all the celestial marvels." He could no longer use this companion of his night vigils, "this discoverer of so many novelties in the heavens, the agent of the greatest advancement of the noble science of astronomy."[104] He was going blind.

In the summer of 1638, as Galileo awaited a visit from Hortensius and the arrival of the gold chain, the vigilant new Inquisitor of Florence, Giovanni Muzzarelli, got wind of the affair. Should agents of the United Provinces be allowed to visit Galileo? Muzzarelli wrote Cardinal Antonio Barberini for instructions. The answer: allow the visit only if the representatives are Catholics from a Catholic country, and then under the restriction that there be no discussion *de motu terrae et stabilitate caeli*. Galileo took the subtle hint and rejected the visit and the gift, in place of which he had a word of commendation from the cardinal. The inquisitor could not work out whether Galileo behaved so well from fear of the danger of violating this latest order or because he had not yet perfected the invention. In the inquisitor's opinion he was unlikely ever to succeed, "being totally blind and with his head more in the grave than his mind on mathematics."[105] Galileo asked Diodati to tell Hortensius not to bother to come, "for even if he should manage to find me alive (which I doubt), he would find me altogether unable to give him the least satisfaction." "Such is the malignity of my fate...However I acquiesce in so much adversity, since it would be vain temerity to wish to fight against the necessity of destiny."[106]

Galileo calculated correctly that the universe would not support his way of navigating on earth, but he had not reckoned rightly the ways of fate. It was not his death, but those of the commissioners assigned to evaluate his method, Hortensius and the Italian-speaking Admiral Lorenzo Reael, that effectively closed the business. Galileo tried to keep it alive by commissioning the new professor of mathematics at Pisa, Vincenzio Renieri, "a young, vigorous man with the most acute vision," to observe the moons and calculate ephemerides, and by urging the Dutch to continue discussions through their ambassador to Venice. Micanzio took the matter in hand with his usual optimism. "I beg you to dispose yourself to enjoy while still living the glory of so marvelous an invention. Remember, it is the misfortune of mankind when it does not receive the rare things and recondite inventions made by able minds inspired by God and nature. Do not do this wrong to humanity."[107] Galileo died before he could act on his old friend's admonition to live.

Galileo's method required an accurate method of telling time between noon solar sightings. He boasted that he had a suitable clock, of such accuracy that if four or six of them were set going together, they would agree to within a second in a month. "These clocks are really admirable for observers of motion and celestial phenomena," Galileo wrote the Dutch States General, "and their construction is very simple." The idea behind this characteristic bluff centered on a timekeeper regulated by a pendulum, of which Galileo then had neither design nor model. The idea was not far to seek—Leonardo and others had proposed it—but no one seems to have implemented it until Galileo devised (in theory) a novel escapement that permitted a pendulum to control the fall of a weight. He elaborated the theory in discussions with his son Vincenzo and his disciple Viviani but had no prototype in hand when he died. Vincenzo constructed one eventually, from which Viviani made a sketch by which it became known in northern Europe.[108] By then, 1655 or 1656, Christiaan Huygens had hit independently on a better design, not only in construction but also in concept. Knowing that only if the bob traveled along a cycloid would its period be independent of its swing, Huygens constrained the pendulum to wrap around cycloidal cheeks.[109]

While trying to perfect navigation here below by tracking moons on high, Galileo pushed ahead on several other problems that might seem beyond the reach of an aged blind valetudinarian. Some dealt with astronomical observations, others with mechanics, and still others with fundamental questions. Examples from each category will indicate how Galileo kept up, or rather could not stop, the play of his "restless mind" during his long twilight.[110] "I do not stop with my speculations, although with considerable damage to my health, since along with my other troubles they deprive me of sleep, which increases my melancholy at night."[111]

Immediately after this sad disclosure, Galileo told his correspondent, Micanzio, that he had made a "most admirable discovery in the face of the moon." Observers over the centuries always found that the moon always presents precisely the same face to us. "I find it is not true." It turns slightly in all possible ways, left and right, up and down, and sideways, around the line joining its center with the earth's. Astonishingly, these oscillations had different periods, daily, monthly, and yearly. "Now what will your Reverence say when you compare these three lunar periods with the diurnal, monthly, and annual motions of the sea, over which everyone agrees the moon is the arbiter and overseer?" These guesses perplexed Micanzio if he remembered that in the *Dialogue* Salviati had already pointed out the daily period of lunar "libration" and referred it, correctly, to parallax: an earth-bound observer looks at slightly different limbs of the moon when it rises and sets, and only if and when it crossed his zenith could he see exactly what the ideal astronomer placed at the earth's center would see. In 1632 Galileo had offered the phenomenon as a way to show that the moon's orbit centered on the earth.[112] But now he wanted to use it to give his condemned theory of the tides a greater plausibility. The moon does indeed librate in some such manner as Galileo guessed, although he did not see or describe the motions he intuited.[113]

"The sky, the world, and the universe that my marvelous observations and clear demonstrations enlarged a hundred or a thousand times beyond what all the sages of all the ages before me had seen, are now diminished and restricted for me to no more than the space my person occupies."[114] That did not prevent Galileo from improving the technique of observations he could not make. In the fall of 1637 he proposed determining the period of a reference pendulum by counting its swings between successive meridianal passages of the same star. To obtain the length of a pendulum that beats seconds, so Galileo advised Baliani, just compute the period of the reference pendulum from the 24-hour count of its swings and employ the "golden rule" that periods of pendulums go as the square root of their lengths.[115] This was a stratagem, not a clock, since the pendulum did not regulate a machine. But once again, the blind seer pointed to the future. The isochronous pendulum clock had become an essential instrument of astronomy within a quarter century of Galileo's death. So much for time. As for space, Galileo recommended determining small angular separations by refining the game with strings he had developed to "measure" stellar diameters. Substitute a rectangular upright beam on a distant mountain for the string and look through a telescope for a star that the beam just

blocked from view. A little geometry would then produce the diameter of
the star, which Galileo still believed his telescope magnified. This suggestion did not have a future. As the pendulum clock came in, so did microm-
eter wires, which cannot be fitted to Galilean telescopes. Already by his
death in 1642 the Keplerian form of the telescope, which would easily
accommodate a measuring grid at the common focal planes of its objective
and eyepiece, had superseded the type with which Galileo had made his
discoveries.[116] It was said that a Frenchman in Turin, operating with such a
telescope 38 palms long, had seen waves on the *maria* of the moon.[117]

As for mechanics, during 1638 Galileo dictated a Day 5 on percussion to
add to the already published four-days' work in *Two new sciences*. Galileo's
former student, Father Paolo Aproino, who had died that March, joined Salviati and Sagredo as a substitute for Simplicio. Aproini easily outdid Simplicio in knowledge of mechanics and admiration of "the greatest man who
has ever lived."[118] Day 5 is entirely qualitative and, for Galileo, weak. He had
spent thousands of hours (he said) trying to understand why a large static
weight cannot move a nail easily sent home by the blows of a relatively light
hammer. His favorite technique of reducing mechanical problems to the
action of a lever suggested an analogy between, on the one hand, the lifting
of a great weight through a short distance by the displacement of a small
force through a great distance and, on the other hand, the slow progress of
the nail under wide swings of the hammer. But that did not take him far and
the Fifth Day finished a fragment.[119]

Torricelli tried to finish it using the quintessential Galilean technique
of replacing the mechanics of percussion by the kinematics of free fall. A
descending weight adds at each instant a degree of *momento* to all the other
momenti previously acquired in its descent; the "power" of percussion lies
in the accumulation, which is to the static "force" of the same weight as the
number of *momenti* is to unity. Since the time of fall can be divided into an
infinite number of indivisible instants, the power of percussion, "which wins
the crown of the greatest of all marvels," was wonderful indeed, being in
principle infinite. Neither the master's kinematics of natural acceleration nor
his construction of matter and motion from "infinite" indivisible infinitesi-
mals equipped the disciple to develop a successful dynamics.[120]

Galileo returned to basics in response to Fortunio Liceti (Plate 7), a Paduan
Aristotelian in Cremonini's mould but without his style. In 1639 Liceti published a book on the phosphorescent Bologna stone, which he put forward

as a model of the secondary light of the moon. On this analogy, the moon's surface imbibes light from the sun and releases it in the dark just like (though slower than) the phosphor from Bologna. It followed that Galileo's explanation of moonglow was wrong. Galileo replied with a few of his typical compliments, which Liceti took in good part and published, with Galileo's permission, in a new edition of his book. The encounter with Liceti caused Galileo to review his attitude toward Aristotle. The analysis led to a startling conclusion. "I am impugned as an impugner of the Peripatetic doctrine, whereas I claim (and surely believe) that I observe more religiously the Peripatetic or I should rather say the Aristotelian teachings than do many who wrongfully put me down as averse [to them]."[121]

"That's news to me," Liceti replied, "I'd rather gathered the contrary from your writings." Galileo returned his reasons for admiring Aristotle. First of all, his logic, his rules for right reasoning, his *demonstratio potissima*. From his study of mathematical demonstrations, Galileo had mastered this logic; consequently and happily, "very rarely have I fallen into mistakes in my argumentation. Thus far, therefore, I am a Peripatetic." Secondly, Aristotle put sense experience before reasoning. "[W]e should deny authority to ourselves wherever we find that sense shows us the contrary." People who prefer Aristotle's teaching to their own sense experience, and distort his philosophy to cover their errors, are not good Aristotelians. "I am sure that if Aristotle should return to earth he would accept me among his followers on account of my few but conclusive contradictions [of him]." Aristotle had taught that the Heavens do not alter because he had detected no change in them. If he could see what Galileo had found, he would grant that they are not immutable because everyone sees them change.[122]

Liceti wrote faster than ordinary men read, about a book a week it was said. One of the three he sent Galileo late in 1640 proved that the earth occupies the center of the universe. That, snapped Galileo, "is among the least worthy of considerations in all astronomy." We do not know the shape of the universe or whether it has a center. A more sensible question is whether the earth is at the center of the planetary system. It permits a decisive answer: "no." The appearances of Mars show plainly enough that we are far from the center of its motions. "One place that could almost be put as the center for all planets but the moon would be the sun..." Give up Aristotle on this point, Liceti! He was just plain, incontestably wrong to put the earth at the center of the universe.[123] In writing this Galileo too was wrong, since it violated his solemn oath not to teach Copernican theory. But no harm could come of it. On this point his correspondent was not teachable.

The most basic of texts for Galileo was Euclid's *Elements*. At the very end of his life he began a new dialogue to clarify for Sagredo and Simplicio the Euclidean theory of proportion on which his geometrical physics relied. They had no difficulty with proportions containing only whole numbers, as, say 3:7 = 51:119. It holds because the first number when increased by a certain factor equals the third, and the second increased by the same factor (in this case 17) equals the fourth. No ambiguity resides in the innocent word "equal." But what if the proportion relates continuous magnitudes, like line segments? Galileo had labored to show in his first proposition on motion in *Two new sciences* that if a body moves with a constant speed, the times required to pass through different spaces will be as the spaces: representing times by the lines AB, CD, and the corresponding distances by the lines WX, YZ, AB:CD = WX:YZ. But since the geometrician cannot in general assign whole numbers to lines, what are we to make of "equal"? Galileo appealed to a Euclidean definition. That did not satisfy Sagredo. "[I've] long entertained doubts about that definition." Simplicio: "I never encountered a more serious obstacle than this, in the little of geometry which I studied in school." Their reasons for puzzlement may be clear from the offending definition: "Magnitudes are said to be in the same ratio, the first to the second and the third to the fourth, when, if any equimultiples whatever be taken of the first and third, and any equimultiples whatever of the second and fourth, the former equimultiples alike exceed, are alike equal to, or alike fall short of, the latter equimultiples respectively taken in corresponding order." People have cracked their heads for 23 centuries over the meaning and utility of this formulation.[124]

Salviati clarifies the definition by introducing the concept of "somewhat greater [or less] than necessary." Thus, if a is just a little larger than needed to make $a:b = c:d$, then $xa:xb > yc:yd$ and $xa:yb > xc:yd$; which, with a little *reductio ad absurdum*, Salviati makes equivalent to Euclid's definition. Since the argument satisfied Simplicio, it should be obvious to the reader.[125] Galileo dictated this fragment to Torricelli, who arrived in Arcetri in October 1641, fresh from the hands of "our intrepid Maecenas," the "most learned and very famous Monsignor Ciampoli." With Torricelli's arrival, Galileo had the fortune usually enjoyed only by royalty of having two generations of his successors living with him.[126] Torricelli became mathematician to Ferdinando II soon after Galileo died. On Torricelli's untimely death in 1647, at the age of 39, Viviani came into his birthright and held the post into the eighteenth century.

Of the body

In 1634 Spanish troops brought the plague to Munich. Ten thousand people died, among them Galileo's sister-in-law and four of her children. Michelangelo's once numerous family was reduced to Galileo's nephews, Vincenzo and Albert(in)o. Galileo sent them money through Micanzio and invited them to Arcetri. Before they received his invitation Albertino wrote describing their plight. "It made me weep to read it."[127] Galileo's losses in 1634—his dearest daughter and tenderest friend Maria Celeste, three nieces and a nephew, his sister-in-law, and his freedom—would have drawn a tear from Torquemada. To lighten his darkening days there remained of his nuclear family his son Vincenzo and Vincenzo's wife and children, and his estranged daughter, Suor Angelica. Vincenzo shared his father's interests in mechanical devices and poetry. We already know one consequence of this alliance, the first semi-practical pendulum clock. Another was a meeting between Galileo and Milton. The English poet had attended a literary society in Florence to which Vincenzo belonged. With Vincenzo as intermediary, Milton satisfied his wish to see the famous man who could not see, "the starry Galileo with his woes," "the famous Galileo, grown old a prisoner of the Inquisition, for thinking in astronomy otherwise than the Franciscan and Dominican licensers thought."[128]

Although a prisoner, Galileo could have visitors, even heretics provided they were not mathematicians, and over-night and long-staying guests. Viviani stayed almost three years, Marco Ambrogetti, the Latin translator, 20 months, Torricelli three months; and Piarist monks, Clemente Settimi and his colleagues, put at Galileo's disposal by direct order of their general, remained with him when he needed them. The Olivetan friar Vincenzio Renieri, who updated the calculations for Galileo's longitude method, frequently came to talk and compute.[129] The poet Giovanni Carlo Coppola amused Galileo in January 1637 by reading his play *Nozze degli dei*, performed later in the year at the wedding of Ferdinando II.[130] Castelli stopped by a few times while on Benedictine or Barberini business, but only after special pleading as he was too close intellectually and emotionally to Galileo for Urban's comfort.

Toward the very end, in 1641, Urban permitted the brief visit during which Castelli proposed Torricelli as an assistant. Later that year, under tight restrictions, Castelli returned to Arcetri to help Galileo prepare for death. The authorities at first prescribed three interviews in the presence of witnesses. But the old reprobate's soul could not be prepared for its upcoming journey so easily and Castelli appealed for indefinite access. Urban agreed, under threat of automatic excommunication should their discussion of the destinations of deceased Christians touch on "the opinion about the earth's motion condemned by the Supreme and Universal Inquisition."[131] Apart from Vincenzo Galilei, Torricelli, and Viviani, most of those who brought solace and diversion to Galileo during his later years at Arcetri were clerics, even friars, though, of course, no Jesuits or Dominicans. Galileo could retain the friendship of monks in minor or less assertive orders, like the Olivetans, Piarists, Servites, and Benedictines, but not of members of powerful ones.

After 1637 Galileo carried on his still imposing correspondence through amanuenses. He exchanged letters with Micanzio, Diodati, the Dutch longitude hunters, Castelli, Florentine gentlemen, and local courtiers up to the grand duke, without reprimand or reprisal. A particularly interesting correspondence took place with Francesco Rinuccini, the Tuscan agent in Venice, whose mathematical interests had qualified him to attend the famous reading of Galileo's sentence and abjuration at inquisitorial headquarters in Florence in 1633. Rinuccini asked about Ariosto and Tasso. Unfortunately, Galileo replied, since he had lost the book in which he had recorded his analysis and could not repair the loss by reading, he could only supply a few parallels from memory. They were all quite exact.[132] In return, Rinuccini sent Galileo news to rejoice a Copernican. It appeared that Pieroni had detected stellar parallax! Galileo replied as if Urban were at his elbow: "The falsity of the Copernican system must not be doubted, especially by us Catholics, who have the irrefragable authority of Holy Scripture interpreted by the greatest masters in theology...The conjectures of Copernicus and his followers offered to the contrary are all removed by that most sound argument, taken from the Omnipotence of God," that is, Urban's Simple. After stating it, Galileo rejected the implication that nothing can be known for certain by the light of reason. "Just as I deem inadequate the Copernican observations and conjectures, so I judge equally, and more, fallacious and erroneous those of Ptolemy, Aristotle, and their followers, when [even] without going beyond the bounds of human reasoning their inconclusiveness can be very easily discovered." As for Pieroni's argument, no doubt if correct it would displace the earth from the center, but as it depended on delicate measurements within the likely margin of error, it could not carry the day. A later reader shocked at this opinion scratched out Galileo's signature from the letter transmitting it. Apparently he or she did not understand Galileo's need or mastery of dissimulation.[133]

Despite the attention of friends and the distraction of work, Galileo suffered grievously during his last four years at Arcetri. Blindness came over him during 1637. From the symptoms he described—fluxions from one eye and then the other, progressive decline with occasional improvements— doctors have diagnosed his problem retrospectively and variously. The most recent and authoritative assessment points to glaucoma probably unrelated to Galileo's chronic illnesses.[134] Naturally he complained of "melancholy."[135] "Horsù, Sig.r Galileo, caro, caro, allegramente," "now, dear dear Galileo, be more cheerful," wrote Castelli, in an effort to reconcile his teacher to perpetual darkness. God tests those he loves (Plate 4). Castelli sent the same cheer to Dino Peri, the moribund professor of mathematics at Pisa. "I rejoice with you [Sig.r Peri] as I see you visited with such trials by the benign and loving hand of God...I beg you to tell our dear Sig.r Galileo about my joy so that he can mix this feeling with his tribulations and joyfully bear them for the love of God... [W]hat he now *seminat in lacrimis in exultatione metet, et si ad vesperum demoratur fletus, ad matutinum erit laetitia.*"[136]

Castelli advised Galileo to seek spiritual help and, when that did not suffice, to request permission from Rome to go to Vincenzo's house for easier access to medical care. Galileo made the request using Castelli's draft. The Holy Office sent Inquisitor Muzzarelli to check that Galileo, known to it as a hypochondriac, was not malingering. The inquisitor turned up with a doctor and without an appointment. "I found him totally blind," Muzzarelli reported. "Moreover, he has a very serious hernia, constant pain, and insomnia, which he and his servants say prevent his sleeping more than one hour in 24; for the rest, he looks more like a cadaver than a living person." Muzzarelli recommended that Galileo go to his son's for medical help. In approving this grace, Urban directed that Galileo not go into Florence and that he not have visitors with whom he could discuss "his damned opinion of the motion of the earth." As the cardinal nephew informed the Inquisitor of Florence, "His Holiness [orders] particularly that he be prohibited under the severest penalties from discussing such material with any one at all." A further concession granted Galileo permission to attend mass in a nearby church on holidays, "provided that there is no crowd."[137] Vincenzo agreed to enforce these conditions, "as he is most grateful...for the favor" and (we deal here with a cynic by profession) "it is in his own interest that his father behaves properly and lives a long time, since with [Galileo's] death the thousand scudi he receives annually from the grand duke will stop."[138]

In the fall of 1638 Galileo added to his usual complaints colic, lack of appetite, and delirium. "I find everything disgusting, wine absolutely bad for my head and eyes, water for the pain in my side...my appetite is gone, nothing appeals to me and if anything should appeal [the doctors] will prohibit it. These, my friend [Diodati] are great trials for me. But much worse are the afflictions of my mind and imagination ...."[139] In September 1639 he cries out to Castelli in arthritic pain, "I can no longer stand it." In February 1640, he describes his situation as a Hell on earth, or rather, still the exact geographer of the inferno, "a hell on the earth's surface."[140] His last illness began in November 1641. A visitor found him suffering from a fever that had confined him to bed for two weeks. Still, he was not done. "He told me that he had the greatest satisfaction in the new mathematician Torricelli and that he had derived great pleasure from listening to him and Viviani argue over some new demonstrations."[141]

Galileo died on the evening of 8 January 1642 in the presence of his son and his mathematical protégés. His corpse was deposited at his wish in Santa Croce, but not, as he had wanted, in the tomb of his ancestors. Not wishing to challenge the Roman authorities by commemorating the death of a man once vehemently suspected of heresy, the family deposited Galileo's mortal remains in an obscure chamber under the bell tower. Grand Duke Ferdinando planned a more fitting monument, a precise counterpart of Michelangelo's, in the great nave. When Urban learned of the plan he summoned Ambassador Niccolini to an informal chat. "He wanted to tell me that it would not be a good example to the world for [us to erect such a monument], as that man had been here before the Holy Office for a very false and erroneous opinion, which he had also impressed upon many others..." The cardinal nephew delivered the same message to Ferdinando through the Florentine Inquisitor. "[I]t is not good to build mausoleums to [such men]...because the good people might be scandalized and prejudiced with regard to Holy Authority."[142] The different routes by which uncle and nephew reasoned to the same conclusion underscore Urban's irrationality in the affair. To him Galileo's doctrine remained dangerous and contagious; to his executive secretary, it was a matter of preserving authority. Either way, Urban denied Galileo his tomb in Florence as he had done Sarpi his in Venice, uniting in limbo the two old friends who had had the courage to mobilize natural science and ecclesiastical history against the established order.

Viviani devoted much of his life and fortune to bringing a suitable memorial into being. He had a bust sculptured from Galileo's death mask, designed an appropriate monument, and developed the pious fiction that Galileo came into the world as his soulmate Michelangelo left it. The fiction supported a wealth of analogies that promoted Galileo to equality with the "master...sent by God as an example of what an artist could be." At the same time Viviani painted a picture of his master as a true Catholic willing to put aside his opinions in full and unfeigned obedience to the determinations of Holy Church. Ever vigilant to remove the slightest impediment to rehabilitation, he tried to suppress the printing of Galileo's letters to Sarpi in an edition of the excommunicated monk's works. Despite these moves neither the Medici nor the Inquisition wanted to reconsider the monument.[143] Viviani responded by making the façade of his house into the memorial mentioned earlier. Erected to coincide roughly with the 50th anniversary of Galileo's death, it may still be consulted in the Via Sant'Antonino by patient lynx-eyed Latinists arrested and flattered by the salutation, "O passerby of right and generous mind! [Plate 16]"[144]

Reading on, the right-minded wayfarer would learn the accomplishments and discoveries, the biographical facts, and the sterling character of the dedicatee; the "ornament of Italy, light of Tuscany, delight of cultivated Europe, beacon of a world philosophizing in darkness, garrison and guide to nature, Oedipus of the wisdom of mortals." The notice concludes with the qualities of the master that inspired the disciple to follow the way of truth. They include, first of all, "the example of a life [lived] in the odor of sanctity." Galileo's strict adherence to God's commandments, deep scientific knowledge, and command of geometry enabled him to bring to light "a few truths from among the infinite and eternal mysteries." He recognized that knowledge of these truths could help humans draw closer to their God. Dealing with men was more difficult and less certain, but Galileo knew how to do that too. A true Galileist "comes firmly to the defense of truth and justice; flees like the plague all lying, adulation, and hypocrisy; avoids leisure and laziness; writes of good deeds in bronze and of bad ones in air; rewards, at least with gratitude, those who help him; fulfills promises scrupulously and honors his word; rejects sordid avarice and ill-gotten gains; gives wisely and employs honestly wealth honestly acquired..."[145] It appears that Viviani confused aspirations with achievements, ideals with realizations, and Galileo with the Man of Galilee.

8.4 THE END OF THE AFFAIR?

Off the Index

The uncertainty concerning the heretical status of Copernicanism took time to settle—some 200 years. A routine text by a professor of mathematics at the University of Rome, Giuseppe Settele, precipitated the dénouement. Settele submitted a book on modern, heliocentric astronomy to the Roman censorship in 1820. The Master of the Sacred Palace, Filippo Anfossi, believing it a heresy, refused the license to print. With the help of people eager to see this view of the matter exploded, Settele appealed to the pope, Pius VII, in whose person the Church had suffered the humiliation of incarceration by Napoleon. The pope referred the complaint to the Congregation of the Index, which granted the imprimatur, and to the Holy Office, which instructed the Master that by "contrary to scripture" the old inquisitors had meant not "contrary to faith" but "opposed to the traditional reading of scripture." Concerned more for his soul than for his job, Anfossi protested the findings of the congregations and refused the imprimatur until Pius ordered him to grant it. That settled the question whether the heresy of which Galileo had been suspected was Copernicanism. The church silently removed all books from the Index that, like Galileo's, had a place there merely for their advocacy of Copernican theory. That was in 1835, in the first new edition of the Index after the Settele affair.[146]

In justification of its verdict, the Holy Office observed, without saying so publicly, that "[n]othing is opposed to defending Copernicus' opinion about the motion of the earth in the manner in which it customarily is now held by Catholic authors." This was to acknowledge that the decree of 1616 had not been enforced in Italy for some time, not, in practice, since the middle of the

eighteenth century, when the censorship allowed or condoned the reprint-
ing of the condemned *Dialogue*. The earliest instance occurred in 1744, in an
edition of Galileo's *Opere* edited by Giuseppe Toaldo, later professor at the
University of Padua. To protect the innocent, Toaldo removed a few postils in
which Galileo had slipped into stating that the earth truly moves, and added
the further antiseptic of an introductory essay by a theologian on the theme
of humankind's incorrigible ignorance. "It seems that God, being jealous, as it
were, of the beauty and magnificence of his work, has reserved to himself the
perfect understanding of its structure, and the secrets of its motions."[147]

The censorship licensed Toaldo's edition of the *Dialogue* because by then,
1744, Catholic astronomers already customarily employed the Copernican
system in their work. Good administrators had turned a blind eye to the
practice as it had its advantages and posed no threat to anything but an
authority they were powerless to exercise. Pope Benedict XIV was a good
administrator. He not only allowed Toaldo's edition but also struck the
general prohibition against Copernican works from the Index. He had not
thought it worth the candle, however, to do battle with the Anfossis of his
time over the *Dialogue*, which remained indexed.[148]

With a little fudging, the customary handling of Copernicus by Italian
astronomers between the defeat of Galileo and the victory of Settele can
be divided into four successive phases. During the first, from 1633 to 1670,
the old interdiscipline of Aristotelian physics and Thomistic theology broke
down under the weight of Galileo's falling bodies and Descartes' corpuscu-
lar philosophy. By 1670 most informed people recognized that no physical
argument of the sort developed by Tycho—a spinning earth would leave
clouds and birds behind, and so on—could be sustained. The break can be
pinpointed. In 1651, Giovambattista Riccioli, SJ, published a very valuable
compendium of the astronomy of his day, an *Almagestum novum*, or updating
of Ptolemy, which contains among ten thousand other things 126 arguments
philosophical, mathematical, and theological for and against Copernicanism
(49 pro, 77 contra).[149] Among the physical arguments, Riccioli judged those
aimed against a stationary sun equivocal, but those against a moving earth
decisive. Galileists were too demoralized in 1651 to protest effectively. When
Riccioli repeated the old arguments in 1665 several had become bold enough
to do so, and in 1669 Riccioli conceded that none of the physical or mathematical arguments he had marshaled in his *Almagest* decided the question.
Copernicanism could not be judged false in philosophy. There remained in

proof of the moving sun and stationary earth only the decree of the church. "I firmly hold, infallibly believe, and openly confess [the earth-centered cosmos]…solely at the command of the faith, by the authority of Scripture, and by the direction of the Roman See."[150]

Riccioli was clear that he rejected Copernicanism in obedience to Rome, not because the Catholic faith required him to do so. It was not a heresy. In the *Almagestum novum* he had declared, with the permission of his superiors, that the Holy Office on its own could not proclaim a heresy or an article of faith. Only a pope or council approved by a pope could so bind the church. "It is not a matter of faith that the sun moves and the earth stands still on the strength of the decree of the congregation; but only at the most by force of the Holy Scripture on those to whom it is morally evident that this is God's revelation." Still, prudence and obedience obliged Catholics to observe the decree, "or at least to teach nothing contrary to it."[151] And yet, "the deeper one dips into the Copernican hypothesis, the more ingenuity and precious subtlety one may unearth." Had only Copernicus presented his theory *ex suppositione*! The same went for Galileo, "a mathematician of immense power wonderfully skilled in astronomy: he would have been greater still if he had put forward the opinion of Copernicus as a mere hypothesis."[152]

During the second phase, 1670–1710, Catholic astronomers gained the right to teach and even develop Copernican theory if they designated it expressly and repeatedly as an hypothesis. Bolder writers, like Honoré Fabri, SJ, picking up on Riccioli's demonstrations that Copernicanism was not heretical, supposed that the decrees against it arose from prudence and surprise, and that as knowledge progressed they would be amended.[153] The censorship allowed the fig leaf of fictionalism. In 1685, the Master of the Sacred Palace reported to the cardinals of the Holy Office that he had required the addition of the words "erroneous hypothesis" to the title page of a book on the Copernican system and to the text the phrase, "Since the Church has declared that the Holy Scripture expressly teaches the contrary, this system cannot be defended in any way." The cardinals complimented him on his vigilance. But he and they allowed the book's publication.[154] Apparently the effective administrative ruling in 1685 was to tolerate Copernicanism as an ineradicable evil after warning the faithful against it, as modern societies allow the sale of cigarettes bearing a notice of their harmfulness.

That was the practice of good administrators. When zealots took over, as during the reign of Innocent XII (1691–1700), who believed in discipline and suffered from literalness, the church could turn its machinery against people who lusted after novelties. Thus Innocent targeted some lawyers and doctors in Naples who followed Copernicus and Descartes. But his intervention ended in the ejection of his agents.[155] His successor, Clement XI (1700–21), whose court intellectual Francesco Bianchini introduced Newtonian ideas into Italy and looked for evidences of the earth's motion, seems not to have cared much about a question of no significance compared with the problems presented to him by the War of the Spanish Succession.[156] His inquisitors turned a blind eye to the unauthorized republication of the *Dialogue* in 1710 by a Neapolitan printer who enjoyed bringing out prohibited works.

Anyone willing to take the trouble to turn philosophy into poetry might teach up-to-date cosmology without the protective shield "ex suppositione" otherwise required. Thus in 1704, in the first of several editions of his *Philosophia nova-antiqua*, Tommaso Ceva, SJ, versified Descartes and Newton, accepted Galileo's theory of motion (but not his cosmology), rejected most of Aristotle, detailed the system of Copernicus and ascribed Protestant preference for it to opposition to the pope. This mongrel mixture was used for many years in Jesuit schools as a text in astronomy and an exemplar of Latin poetry.[157] A Sicilian poet, Tommaso Campailla, then set Cartesian philosophy to poetry in an extravaganza on the education of Adam by the angel Raphael, whose divine physics was nothing other than the vortices of "l'immortale Renato e de la Carte." The Royal Society of London, amazed to discover so bright a light in Sicily, sent Campailla a copy of Newton's *Principia*, perhaps in the hope that he would versify it too. The desired poetical counterpart to Descartes was completed only in 1752. The author, Benedict Stay, SJ, required 24,000 lines for the job, about two-thirds of a *Furioso*. It took forty years to print.[158]

The third phase in the rehabilitation of Copernicus and Galileo ran from 1710 to 1760, from the requirement of explicit fictionalism (except in poetry!) to pro-forma expressions of it. In the 1720s and 1730s, older astronomers practiced the self-censorship they had interiorized earlier. Angelo Marchetti, professor of logic, and the son of the professor of mathematics, at the University of Pisa, provides a good example. The second edition of his *Brief introduction to cosmology* (1738) invites its readers to decide which among the systems of Ptolemy, Copernicus, and Tycho is "more like the truth and more in conformance with the sacred dogmas of our Catholic faith." It offers some guidance: Ptolemy's system is wrong; Copernicus', confirmed by "the great Galileo," meets every reasonable test, but Catholics suspect it because "it does not conform well with Divine Scripture and [because of] the sentence against Galileo"; Tycho's, though neither simple nor coherent, is preferred by Catholics obedient to the Sacred Congregation of Cardinals. And what does the author prefer? "For my part I say... that I do not intend to approve the Copernican theory; but in everything and for everything, I defer to those competent to judge the matter, contenting myself with having explained it in the form in which its inventors proposed and approved it."[159]

The more relaxed approach of younger men toward 1760 is represented by a *Disputation on the diurnal motion of the earth* (1756) by a successor of Marchetti, the bold Barnabite Paolo Frisi. Modern astronomy, mechanics, and physics were so many confirmations of the "most elegant and most celebrated opinion of the great Galileo." New discoveries—Frisi mentioned the aberration of light and the nutation of the stars—are decisive. "[They] can be explained easily and wonderfully in the [Copernican] system... Is this not a kind of certainty and demonstration?" Whereas on the hypothesis of a stationary earth, neither the aberration nor the nutation can be explained, "or to explain them intricate and arbitrary hypotheses, which are implausible when judged by the light of reason, must be adopted."[160]

Frisi grew even bolder as documents pertaining to Galileo's trial became public. Galileo had not only discovered fundamental truths but had also worked to win for everyone the right to seek them until the Jesuits fiddled the system to bring him down. The Jesuits did not reply to this provocation or to the documents, published in 1775, supporting it. They could not. In 1773, under pressure from heads of state impatient with their intrigues, Pope Clement XIV suppressed their order. This act, unimaginable in Galileo's time, drove ex-Jesuits who wished to maintain the shreds of their previous life to Austria or Russia. They formed a nucleus from which Pius VII reestablished their order during the conservative reaction following the fall of Napoleon. A few years later he began the official rehabilitation of Galileo by removing the *Dialogue* from the Index.[161]

On the rota?

Galileo resides high on the short list of founders of modern physics and sufferers for the freedom of thought. And he is still making progress, though his momentum is slow, up the register of servants of the church. The first milestone in this advance was the encyclical *Providentissimus deus*, issued in 1893 by

Pope Leo XIII, who some years earlier had made Thomism the official theology and philosophy of the Church. Leo turned to Galileo's hermeneutics to disarm attacks on scripture mounted by the higher critics, who treated the bible as literature and exposed its many errors in history and science. Leo removed scriptural statements about the natural world from the field of battle by the device of accommodation. But whereas Galileo used accommodation to gain space for natural science, Leo required it to save space for scripture. They agreed that God did not intend to teach physics through Moses. They relied on the same arguments from Augustine to sustain their positions. But Leo did not mention Galileo.[162] A century later Pope John Paul II, who ranked Galileo better at theology than the congregation of cardinals that condemned him, repaired Leo's omission.

The 300th anniversary of Galileo's death brought a second milestone. To demonstrate the church's openness to science, Pope Pius XII, "filled with the spirit of Galileo," commissioned an unrestricted biography of Urban's bugbear.[163] The commission went to Monsignor Pio Paschini, an historian known for his balance. This virtue worried some of Pius' advisors, who had the satisfaction of being proved right. Paschini took Galileo's part, declared his condemnation an error, and blamed it on the Jesuits. The Jesuits objected and Paschini's two-volume manuscript disappeared in the review mechanism. In consolation, he was made a bishop. The next conspicuous opportunity to display eagerness to rehabilitate Galileo was the 400th anniversary of his birth. That fell in 1964, during the Second Vatican Council. As a sign of their program for the peaceful coexistence of science and religion, the popes (John XXIII and Paul VI) authorized the printing of the defunct Paschini's biography, which had been improved by the keeper of the Jesuits' archives. John Paul II pointed to it as an example of openness in 1979 when he inaugurated a multidisciplinary inquiry into the Galileo Affair. On this occasion he endorsed Galileo's principles of hermeneutics. Perhaps he did not know that in effect Paschini's biography had been indexed *donec corrigatur* and corrected more severely than Copernicus' *De revolutionibus*.[164]

Although John Paul's committee sponsored several useful publications, it did not resolve fundamental questions of responsibility. It was adrift in lethargy and dullness when, at the pope's direction, the President of the Pontifical Council for Culture, Cardinal Paul Poupard, who had served on the committee since its inception, pulled its findings to a conclusion. The world learned the result in 1992, 350 years after Galileo's death. Poupard's conclusion adopted the analysis developed under the committee's aegis by the German Jesuit Walter Brandmüller. It surprised unsophisticated people. Brandmüller and Poupard reduced the collision between the cardinals and Galileo to a no-fault accident. Galileo had proceeded correctly along the difficult road of scriptural exegesis; the cardinals had negotiated with equal expertise the equally difficult road of epistemology. Yes, the theologians knew more about the nature of science than Galileo, who mistakenly thought he had proved Copernican theory, and Galileo knew more about the business of theology than the theologians, who mistakenly took scripture as a guide to natural science. This counterintuitive and even comic formulation raised Galileo's hermeneutics almost to the level of Urban's medicine. The pope accepted it.[165]

John Paul might have been disappointed with the result that on the playing field of theology and epistemology the cardinals and Galileo had managed a draw. In his speech of 1979, from which the committee drew its terms of reference, he had remarked that Galileo "had to suffer a great deal at the hands of men and organisms of the church." The pope deplored this suffering: Galileo was a good Catholic who believed as the pope did that there could be no fundamental opposition between God's words and His works, between scripture and science. Galileo did Catholics a singular service at great personal cost by showing that the church must not oppose scripture to proved or provable assertions about the natural world.[166] This judgment echoed in the more careful words of a pope the opinion expressed by a theologian a quarter century earlier: Galileo's suffering before the Inquisition was the fulfillment of his "providential purpose."[167] God had appointed him to instruct misguided theologians about the relationship between words and works. The man who believed that human beings can reach truth by the light of reason had beaten the pope who denied that humans can recognize truth unaided by revelation.

Language like this describes saints and martyrs. In the heat of battle over the status of Copernicus in 1616, Galileo represented his struggle as that of a saint against devils.[168] His disciples warmed to the theme. Immortal, divine, "his fame will last as long as the universe"; "the bringer of light to the human mind," "one of those rare men destined by Providence to honor the sublime works of His hands," he is the "father of all true students of nature." "His mission was to restore human reason to the dignity it had lost for centuries."[169] He came to teach the church "a severe lesson in humility."[170]

"For most people, Galileo's greatest claim to glory is to have offered himself as a sacrifice for a cause that he had made his own. He entered into it with all the enthusiasm, ardor, and passion of a man who believes himself called from on high to the mission of revealing to the world the truth about the motion of the earth—just as Providence had given him the grace to be the first to see in the heavens new stars previously unknown to mortals."[171] Suffering, sacrifice, humility, mission, providential calling: elements for a case for canonization begin to accumulate.

We already have the relics. Besides Galileo's instruments and a telescope lens set in a reliquary, bits of his bones, now resident in museums, are available for higher purposes. Like many a true saint, some of his body parts exist in greater quantities after death than in life. During his reburial in Santa Croce in 1737, the curious or worshipful removed index fingers, a thumb, and a vertebra. For a time there were two competing right index fingers, one of which is displayed at the Museo Galileo in Florence along with two other digits and a molar.[172] Enough relics exist to stock the churches in Venice, Padua, Florence, and Rome at which Galileo customarily worshipped. There is also the beginning of a liturgy in a sonnet inspired by contemplation of "the great Galileo's finger."[173]

It might be objected that Galileo performed no miracles. What then were the miracles of Thomas Aquinas? In fact, Galileo performed a stupendous miracle. He obliterated the ancient distinction between the celestial and terrestrial realms, raised the earth to the heavens, made the planets so many earths, and revealed that our moon is not unique in the universe. Not since the creation had there been such a refashioning. Then there was the miracle of himself, a rare combination of talents and personalities, who, despite mania and depression, arthritis, gout, hernias, blindness, and overindulgence in wine and wit, lived to write three books—the *Messenger*, the *Dialogue*, and the *Discourse*—any one of which would have given him enduring fame.

According to Galileo's mechanics, the slightest force can move the greatest weight given sufficient time. The direction of motion is clear. Who can doubt that within another 400 years the church will recognize Galileo's divine gifts, atone for his sufferings, ignore his arrogance, and make him a saint?

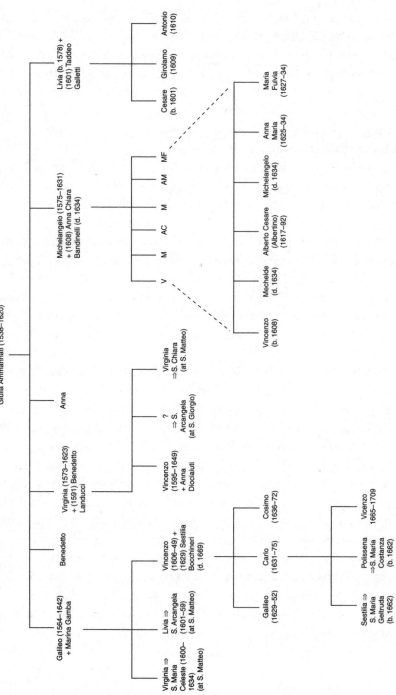

Vincenzo Galilei (1520–91) + (1562)
Giulia Ammannati (1538–1620)

Galileo (1564–1642)
+ Marina Gamba

Benedetto

Virginia (1573–1623)
+ (1591) Benedetto
Landucci

Anna

Michelangelo (1575–1631)
+ (1608) Anna Chiara
Bandinelli (d. 1634)

Livia (b. 1578) +
(1601) Taddeo
Galletti

Virginia ⇒
S. Maria
Celeste (1600–
1634)
(at S. Matteo)

Livia ⇒
S. Arcangela
(1601–59)
(at S. Matteo)

Vincenzo
(1606–49) +
(1629) Sestilia
Bocchineri
(d. 1669)

Vincenzo
(1595–1649)
+ Anna
Diociaiuti

?
⇒ S.
Arcangela
(at S. Giorgio)

Virginia
⇒ S. Chiara
(at S. Matteo)

Cesare
(b. 1601)

Girolamo
(1609)

Antonio
(1610)

Galileo
(1629–52)

Carlo
(1631–75)

Cosimo
(1636–72)

V   M   AC   M   M   AM   MF

Sestilia ⇒
S. Maria
Geltruda
(b. 1662)

Polissena
⇒S. Maria
Costanza
(b. 1662)

Vicenzo
1665–1709

Vincenzo
(b. 1608)

Mechelde
(d. 1634)

Alberto Cesare
(Albertino)
(1617–92)

Michelangelo
(d. 1634)

Anna
Maria
(1625–34)

Maria
Fulvia
(1627–34)

GLOSSARY OF NAMES

The Glossary contains names of Galileo's contemporaries mentioned in the text, of people real and imaginary whose writings or escapades engaged him, and of principal actors in the resolution of the "Galileo Affair." Where birth and/or death dates do not appear, they eluded the author, who does not guarantee all those he did find.

Real people

ACERENZA, Cosimo d' (?–1601), Neapolitan patrician and bibliophile. whose library formed the basis of the Ambrosiana Library in Milan.

ACQUAPENDENTE, Fabrizio d' (1533–1619), Paduan surgeon and anatomist, discovered the valves of the veins.

ACQUASPARTA, Federico, Marquese di Monticelli (1562–1630), father of Federico Cesi, founder of the Accademia dei Lincei.

ACQUAVIVA, Claudius (1543–1615), fifth general of the Society of Jesus, who solidified its position as the leading teaching order in Europe.

AGGIUNTI, Niccolò (1600–1635), student of Castelli, professor of mathematics at Pisa.

ALAMANNI, Luigi di Piero (1558–1609), Florentine poet, certified the originality of Galileo's Archimedean theorems.

ALDOBRANDINI, Pietro, Cardinal from 1593 (1571–1621), nephew of Pope Clement VIII, patron of Torquato Tasso.

ALLACCI, Leone (1586–1669), teacher of Greek, librarian, close to the Barberini.

ALTOBELLO, Ilario (1560–1637), Veronese, Minorite monk, poet, mathematician, astronomer.

AMBROGETTI, Marco, Florentine priest, translated Galileo's shorter works into Latin.

ANFOSSI, Filippo (?–1825), Dominican, Master of the Sacred Palace in 1820, central figure in the Settele affair.

ANTONINI, Daniele (1588–1616), military man, Galileo's student at Padua, killed in Venetian campaigns against Austria.

APELLES. *See* Scheiner.

APROINO, Paolo (c.1584–1638), canon of the cathedral at Treviso, Galileo's student at Padua, interlocutor in the fifth day of *Two new sciences*.

ARCHIMEDES (c.287–212 BCE), Greek mathematician, celebrated for his inventions both pure and applied.

ARIAS MONTANA, Benito (1527–1598), head of the Escorial Library, editor of the polyglot bible sponsored by Philip II of Spain.

ARIOSTO, Ludovico (1474–1533), Italian poet, author of *Orlando furioso* (1516, 1532).

ARISTOTLE (383–322 BCE), "The philosopher," still "The master of those who know."

ATTAVANTI, Giannozzo (c.1582–1657), cleric, accused by Tommaso Caccini of heretical views allegedly learned from Galileo.

BADOVERE, Giacomo (Jacques) (c.1580–c.1620), student of Galileo who supplied information about the Dutch forerunner of the telescope.

BALIANI, Giovanni Battista (1582–1666), official of the Republic of Genoa, mathematician, correspondent of Galileo and Grassi.

BANDINI, Ottavio, Cardinal from 1596 (1558–1629), uncle of Galileo's informant in Rome, Pietro Dini.

BARBERINI, Antonio (the elder), Cardinal from 1624 (1569–1646), younger brother of Urban VIII, member of the Order of Capuchins.

BARBERINI, Antonio (the younger), Cardinal from 1627 (1607–1671), nephew of Urban VIII, younger brother of Francesco Barberini.

BARBERINI, Carlo, Duca di Monterotondo (1562–1630), older brother of Urban VIII.

BARBERINI, Francesco, Cardinal from 1623 (1597–1679), cardinal nephew of Urban VIII, Vatican Secretary of State, lynx, sometime protector of Galileo.

BARBERINI, Maffeo. *See* Urban VIII.

BARDI, Giovanni de', Conte di Vernio (1534–1612), Florentine, composer, patron of Vincenzo Galilei.

BARDINELLI, Baccio (1488–1560), Florentine painter and sculptor, worked on the choir and altar of the cathedral of Florence.

BARONIO, Cesare, Cardinal from 1596 (1538–1607), became Superior of the Oratorians in 1593, author of the official church history, *Annales ecclesiastici* (1588-1607).

BARTOLI, Giovanni, Florentine agent in Venice when Galileo invented the telescope.

BELLARMINE, Robert (Roberto Bellarmino), Cardinal from 1599 (1542–1621), S.J., chief theologian of the Holy Office, canonized in 1930.

BELLONI, Camillo (?–c.1633), professor of moral philosophy at Padua, founding member of the Accademia dei Ricovrati in 1599.

BENEDETTI, Giovanni Battista (1530–1590), mathematician to the Duke of Savoy.

BENEDICT XIV, Prospero Lambertini, Pope from 1740 (1675–1758).

BENIVIENI, Girolamo (1453–1542), Florentine man of letters.

BENTIVOGLIO, Guido, Cardinal from 1621 (1579–1644), Galileo's student at Padua, one of the ten cardinal-inquisitors who tried Galileo in 1633.

BERNEGGER, Matthias (1582–1640), Lutheran professor of history at Strasbourg and a friend of Kepler, translator of Galileo's *Dialogue* into Latin.

BERNI, Francesco (1497–1536), Florentine poet who wrote mocking or burlesque poems.

BERNINI, Gian Lorenzo (1598–1680), Urban VIII's favorite architect and sculptor.

BIANCANI, Giuseppe (1566–1624), Jesuit mathematician and astronomer.

BIANCHINI, Francesco (1662–1729), astronomer, antiquary, cultural advisor of Pope Clement XI.

BOCCHINERI, Sestilia di Carlo, wife of Galileo's son Vincenzo.

BOCCHINERI, Geri (c.1590–1650), brother of Sestilia, secretary to Grand Duke Ferdinando II of Tuscany.

BOETHIUS, Ancius Manlius Severinus (c.480–524), translator and commentator of Greek philosophical works, whose *Consolations of philosophy* was widely read during the Renaissance.

BORGHESE CAFFARELLI, Scipione, Cardinal from 1605 (1576–1633), cardinal nephew of Paul V.

BORGHINI, Jacopo, Galileo's first teacher.

BORGIA, Gaspar de (Gaspar Borja y de Velasco), Cardinal from 1611 (1580–1645), Spanish ambassador to the Holy See, hostile to Urban VIII.

BORRO (or Borri), Girolamo (1512–1592), professor of philosophy at Pisa, taught Galileo physics.

BORROMEO, Federico, Cardinal from 1587 (1564–1631), a religious in the style of Baronio and Bellarmine, Archbishop of Milan in 1595, patron of Ciampoli.

BOSCAGLIA, Cosimo (c.1550–1621), professor of philosophy at Pisa, 1600–1621, poet, specialist in Greek literature.

BRAHE, Tycho (1546–1601), Danish astronomer.

BRANDMÜLLER, Walter (1929–), S.J., president (1998–2009) of the Pontifical Committee for Historical Sciences, founded in 1954 by Pope Pius XII.

BRENGGER, Johann Georg, physician from Augsburg, friend of Welser, challenged Galileo over the height of moon mountains.

BRENZONI, Ottavio (c.1575–1630), physician and astrologer from Verona.

BRUCE, Edmund, a Scot in Pinelli's circle, later in Kepler's.

BRUNELLESCHI, Filippo (1377–1446), Florentine architect, builder of the dome of the cathedral of Forence.

BRUNO, Giordano (1548–1600), Dominican, philosopher, mathematician and astronomer, condemned by the Inquisition as a heretic and burned at the stake in Rome.

BUONAMICI, Francesco (?–1603), professor of philosophy at Pisa, member of the Florentine Academy, taught Galileo physics.

BUONARROTI, Michelangelo (1475–1564), considered by some to be the greatest of Italian artists.

BUONARROTI, Michelangelo the younger (1568–1646), grand nephew of the artist, friend of Galileo and Maffeo Barberini.

BUONTALENTI, Bernardo (c.1536–1608), architect, stage designer, and military engineer, friend and collaborator of Vincenzo Galilei's patron Bardi.

BURTON, Robert (1577–1640), English scholar and clergyman, author of the *Anatomy of melancholy* (1621), a vast, witty, and learned compendium of madness.

CACCINI, Matteo (1573–1640), Florentine, brother of the Dominican preacher Tommaso.

CACCINI, Tommaso (Cosimo) (1574–1648), Florentine Dominican who preached against Galileo and denounced him to the Inquisition.

CAETANI, Bonifazio, Cardinal from 1606 (1567–1617), nephew of Enrico Caetani, helped moderate the decision of the Congregation of the Index against Copernican books.

CAETANI, Enrico, Cardinal from 1585 (1550–1599), supporter of Galileo's candidacy for a chair of mathematics in Bologa, where he was legate 1585–87.

CAMPAILLA, Tommaso (1668–1740), Sicilian poet who versified Cartesian philosophy.

CAMPANELLA, Tommaso (1568–1639), Dominican philosopher, theologian, astrologer and poet from Calabria, imprisoned from 1599 to 1626 for heresy and conspiracy against Spanish rule, subsequently astrological advisor to Urban VIII.

CAPPONI, Luigi, Cardinal from 1608 (1582–1659), Vatican librarian from 1649.

CAPRA, Baldassar (c.1580–1626), a student who plagiarized his teacher.

CARAFFA, Vincenzo (1585–1649), Neapolitan, succeeded Vittelleschi as Jesuit general in 1645.

CARDANO, Girolamo (1501–1576), astrologer, mathematician, naturalist, gambler.

CASTELLI, Benedetto (1578–1643), Benedictine from Brescia, Galileo's closest disciple, professor of mathematics in Pisa (1613), entered the service of the Barberinis in 1626.

CASTRAVILLA, Ridolfo (pseudonym), attacked Dante's *Divine comedy* for violating Aristotelian poetic norms.

CATALDI, Pietro Antonio (1548–1626), professor of mathematics at Bologna when Galileo applied for the junior professorship there.

CAVALIERI, Bonaventura (c.1598–1647), Jesuat (not Jesuit!) mathematician trained at Pisa by Castelli, elected professor of mathematics at Bologna (1629) with Ciampoli's help.

CECCO DI RONCHITTI, the name under which Galileo, Querenghi, Spinelli, and Castelli lampooned Lorenzini and Cremonini.

CERVANTES, Miguel de (1547–1616), creator of Don Quijote, an ardent reader of *Orlando furioso*.

CESALPINO, Andreas (1519–1603), professor of medicine at Pisa, philosopher with unusual ideas about motion.

CESARINI, Virginio (1596–1624), Federico Cesi's cousin, lynx, chamberlain to Urban VIII, editor and addressee of the *Assayer*.

CESI, Federico, Prince (1585–1630), founder of the Accademia dei Lincei, patron and publisher of Galileo.

CEVA, Tommaso (1648–1737), professor of mathematics at the Jesuit College in Milan, author of a physics text in the form of a Latin poem.

CHIARAMONTI, Scipione (1565–1652), professor of philosophy at Pisa, 1627–36, antagonist and target of Galileo.

CHRISTINA OF LORRAINE (Chrétienne de Lorraine), Grand Duchess of Tuscany (1565–1637), wife of Grand Duke Ferdinando I de' Medici, mother of Cosimo II, addressee of Galileo's "Letter to the Grand Duchess."

CIAMPOLI, Giovanni (c.1590–1643), childhood friend of Cosimo II de' Medici, poet, opportunist, lynx, devoted friend of Cesarini, Urban VIII's correspondence secretary.

CIGOLI, Lodovico Cardi, known as (1559–1613), Florentine painter of the late Mannerist/early Baroque style, close friend of Galileo.

CIOLI, Andrea (1573–1641), Tuscan Secretary of State.

CLAVIUS, Christoph (1537–1612), professor of mathematics at the Roman College from 1565, implementer of the Gregorian calendar reform, authoritative Ptolemaic astronomer.

CLEMENT VIII, Ippolito Aldobrandini, Pope from 1592 (1536–1605), austere, pious, strengthened the Inquisition and issued, in 1596, an enlarged and stricter Index of Prohibited Books.

CLEMENT XI, Giovanni Francesco Albani, Pope from 1700 (1649–1721).

CLEMENT XIV, Giovanni Vincenzo Antonio Ganganelli, Pope from 1769 (1705–1774), suppressed the Society of Jesus in 1773.

COIGNET (or Cognet), Michel (1549–1623), from Antwerp, mathematician and military engineer to Archduke Albert of Austria (1598–1621).

COMMANDINO, Federico (1509–1575), Italian humanist from Urbino, editor and translator of Archimedes and other Greek mathematicians.

CONTARINI, Giacomo (1536–1596), Venetian patron of the arts and sciences, superintendent of the Venice Arsenal.

CONTARINI, Niccolò (1553–1631), Venetian patron of Galileo, Doge 1630–31.

CONTARINI, Simone (1563–1633), Venetian ambassador to Rome in 1616.

CONTI, Carlo, Cardinal from 1604 (1556–1615), advised Galileo on theological aspects of Copernican ideas.

CONTI, Ingolfo de' (1565–1615), first lecturer in mathematics at the Accademia Delia in Padua, a military school, for which Galileo dew up a study-plan.

COPERNICUS, Nicholas (1473–1543), canon of the cathedral in Frombork, Poland, his *De Revolutionibus orbium coelstium* (1543), demonstrating the technical merits of a sun-centered universe, provoked the reform of astronomy and the crusade of Galileo.

COPPOLA, Giovanni Carlo (?–1652), poet who visited Galileo at Arcetri, author of a play performed at the wedding of Grand Duke Ferdinando II to Vittoria delle Rovere in 1637.

CORESIO, Giorgio (1554–1641), professor of Greek at Pisa, 1609–15.

CORNARO, Federico Balissera Bartolomeo, Cardinal from 1626 (1579–1653), Venetian patrician, founded the Accademia dei Ricovrati in Padua in 1599, Patriarch of Venice, 1632–44.

CORNARO, Giacomo Alvise (1539–1608), monsignore, Galileo's neighbor in Padua, testified on his behalf concerning Capra's plagiarsim.

CREMONINI, Cesare (1550–1631), popular professor of philosophy at Padua, friend of Galileo, constantly in trouble with the Inquisition for his faithful teaching of Aristotle.

DAL POZZO, Cassiano (1588–1657), lynx, an editor of the *Assayer*, botanist, collector, and antiquary, he served as private secretary and chief gardener to Cardinal Francesco Barberini.

DA MULA, Agostino (1561–c.1620), Venetian patrician and friend of Galileo, specialist in optics, member of Sarpi's group of telescopic observers.

DANTE Alighieri (1265–1321), Florentine poet, subject of a great debate in the 1580s to which Galileo contributed a correct map of the Inferno.

DE DOMINIS, Marco Antonio (1566–1624), Croatian mathematician, Dean of Windsor between stints as a Catholic archbishop, instrumental in publishing Sarpi's *History of the Council of Trent*.

DELLA PORTA, Giambattista (c.1535–1615), Neapolitan playwright and natural magician, founded the Accademia dei Segreti, frequently of interest to the Inquisition, lynx.

DELLA ROVERE, Franceso Maria II, Duke of Urbino (1549–1631), willed his duchy to the Papal States.

DELLA ROVERE, Vittoria (1612–1694), betrothed in infancy to the boy who became Grand Duke Ferdinando II, in the hope of obtaining the inheritance of Urbino for Tuscany.

DELLE COLOMBE, Ludovico (Lodovico) (1565–c.1615), Florentine philosopher, eponymous member of the "pigeon league" that opposed Galileo in Florence.

DELLE COLOMBE, Raffaelo (1563–1627), brother of Ludovico, Dominican preacher who thundered against Galileo from the pulpit in Florence.

DEL MONTE, Francesco Maria, Cardinal from 1588 (1549–1627), Venetian diplomat and connoisseur, brother of Guidobaldo, an important and steady patron of Galileo.

DEL MONTE, Guidobaldo (1545–1607), mathematician, philosopher and astronomer, a student of Commandino and friend of Tasso, he helped Galileo materially and intellectually.

DEMISIANI, John (?–1614), Greek, mathematician to the Duke of Gonzaga, proposed in 1611 the name "telescope" for Galileo's invention.

DETI, Giovanni Battista, Cardinal from 1599 (1580–1630) patron of an important Roman literary academy.

DIETRICHSTEIN, Franz Seraph von, Cardinal from 1599 (1570–1636).

DINI, Piero (c.1570–1625), Florentine, monsignore, who, when secretary to his cardinal uncle, Ottavio Bandini in the 1610s, acted as a confidential agent for Galileo in Rome.

DIOCIAIUTI, Anna di Cosimo (–1633), maintained in a convent by Galileo, whose nephew, Vincenzo Landucci, she later married.

DIODATI, Elia (1576–1661), Genevan, parliamentary lawyer in Paris, fan of Sarpi and Galileo, whose works he helped to publish outside Italy.

DONÀ, Leonardo (1536–1612), Doge during the Venetian interdict of 1606-07, received Galileo's gift of the telescope to the Venetian Senate in 1609.

DORIA, Giovanni, Archbishop of Palermo, Cardinal from 1623 (1573–1642).

DUODO, Pietro (1554–1610), Venetian patrician, ambassador, friend of Galileo, he founded the military school, the Accademia Delia, in Padua.

EGIDI DA MONTEFALCO, Clemente (1571–c.1639), Dominican preacher and theologian, Inquisitor General of Florence (1626–36), approved Galileo's *Dialogue* for publication.

EITEL VON ZOLLERN, Frederick (Eitel Friedrich von Hohenzollern-Sigmaringen), Cardinal from 1621 (1582–1625).

ELSEVIER, Louis (1604–1670), Leyden publisher of Galileo's work outside of Italy.

ERNEST of Bavaria, Elector-Archbishop of Cologne (1554–1612), recipient of one of Galileo's first telescopes.

FABER, Johann (Giovanni) (1574–1629), lynx, papal doctor, botanist, and art collector.

FABRI, Honoré (1607–1688), prominent Jesuit natural philosopher whose eclecticism had a place for Descartes.

FABRI DE PEIRESC, Nicolas-Claude (1580–1637), French polymath, studied in Padua, published Maffeo Barberini's poems, tried to gain Galileo's release from Arcetri.

FARNESE, Odoardo, Cardinal from 1591 (1573–1626), son of Alessandro Farnese, Duke of Parma, and Maria of Portugal, known for his patronage of the arts.

FAVARO, Antonio (1847–1922), professor at Padua, patron saint of Galileo studies.

FOSCARINI, Paolo Antonio (c.1565–1616), Venetian, Carmelite monk, theologian, attempted to bring scripture into line with the Copernican system.

FRANCINI, Ippolito (1593–1653), accomplished lens maker employed in the Medici glass factory in Florence.

FRANCO, Veronica (1546–1591), Venetian, famous *cortigiana onesta* and poet, whose clients included men Galileo knew.

FRISI, Paolo (1728–1784), Barnabite, professor of mathematics at Pisa, 1756–64.

GALEN, Claudius (c.130–c.201), Greek imperial physician whose writings were still the canon in medicine when Galileo studied them 1600 years after their composition.

GALILEI, Alberto Cesare (1617–1692), son of Galileo's brother, Michelangelo.

GALILEI, Galileo (1564–1642), the subject of this book.

GALILEI, Giulia, born Ammannati (1538–1620), Galileo's mother, who came from a family of cloth merchants in Pisa.

GALILEI, Livia I (1578–?), Galileo's sister, who married (in 1601) Taddeo di Cesare Galletti.

GALILEI, Livia II (Suor Arcangela from 1617) (1601–1659), Galileo's younger daughter, by Marina Gamba, lived her religious life in the Convent of the Poor Clares at San Matteo d'Arcetri near Florence.

GALILEI, Mechilde (?–1634), daughter of Michelangelo Galilei.

GALILEI, Michelangelo (1575–1631), Galileo's only brother to survive infancy, made his career as musician to the Elector of Bavaria, married Anna Bandinelli c. 1608.

GALILEI, Vincenzo I (1520–1591), Galileo's father, a professional musician (lutanist) and musical theorist, married Giulia di Cosimo Ammannati at Pisa in 1562.

GALILEI, Vincenzo II (1606–1649), Galileo's only son, by Marina Gamba, legitimized in 1619, studied at Pisa, married Sestilia di Carlo Bocchineri in 1629.

GALILEI, Vincenzo III (1608–?), son of Michelangelo Galilei.

GALILEI, Virginia I (1573–1623), Galileo's sister, married Benedetto di Luca Landucci in 1591.

GALILEI, Virginia II (Suor Maria Celeste from 1616) (1600–1634), Galileo's elder and favorite daughter, by Marina Gamba, lived her religious life in the Convent of the Poor Clares at San Matteo d'Arcetri near Florence.

GALLANZONI, Gallanzone, secretary to Cardinal François de Joyeuse.

GALLETTI, Taddeo di Cesare, husband of Galileo's younger sister Livia.

GALLUZZI, Tarquinio (1574–1649), professor of rhetoric at the Jesuit Roman College, teacher of Mario Guiducci.

GAMBA, Marina (c.1570–1612), Galileo's Venetian mistress and the mother of his three children.

GHETALDI, Marino (1568–1626), mathematician from Ragusa (Dubrovnik), student of Clavius, Coignet, and Viète.

GIACOMINI, Lorenzo (1552–1598), poet, uncle of Giambattista Ricasoli.

GILBERT, William (1544–1603), physician to Queen Elizabeth I of England and author of De magnete (1600).

GIAMBULLARI, Pier Francesco (1495–1555), Florentine literary man, wrote on the geography of Dante's Inferno.

GIOVANNA D'AUSTRIA, Grand Duchess of Tuscany (1547–1578), the first wife of Grand Duke Francesco I and an actor in Borro's dialogue on the tides.

GLORIOSO, Giovanni Camillo (1572–1643), Galileo's successor as professor of mathematics at Padua in 1613.

GONZAGA, Ferdinando, Duke of Mantua and Montferrat, Cardinal from 1605 (1587–1626), surrendered his cardinalate in 1612 to become a procreating duke.

GONZAGA, Vincenzo I, Duke of Mantua and Montferrat (1562–1612), a major patron of the arts, sciences, and theater, with whom Galileo negotiated for a position in 1603–4.

GRASSI, Orazio (c.1590–1654), professor of mathematics at the Jesuit Roman College, architect of St. Ignatius (church of the Roman College), and, through his quarrel with Galileo over comets, a main actor in Galileo's life.

GREGORY XV, Alessandro Ludovisi, Pope from 1621 (1554–1623), Bolognese, founded the Congregation for the Propagation of the Faith and canonized Teresa of Avila, Ignatius Loyola, Philip Neri, and Francis Xavier.

GRIENBERGER, Christoph (1561–1636), born in the Tyrol, Clavius' successor as professor of mathematics at the Roman College, an early supporter of Galileo.

GUALDO, Paolo (1553–1621), close to the Jesuits and also to Galileo, settled in Padua in 1591, became archpriest of San Antonio (Padua Cathedral) in 1609.

GUALTEROTTI, Raffaello (1548–1639), Florentine astronomer and astrologer.

GUEVARA, Giovanni di (1561–1641), Neapolian, mathematician, General of the Minor Regular Clerics, client of Francesco Cardinal Barberini and supporter of Galileo.

GUICCIARDINI, Piero (1560–1626), Tuscan ambassador to Rome from 1611, replacing Giovanni Niccolini.

GUIDUCCI, Mario (1585–1646), Florentine man of letters, student of Castelli, lynx, member of the Florentine academy, to which he presented Galileo's theory of comets.

GUSTAVUS ADOLPHUS, Gustav II Adolph Vasa (1594–1632), King of Sweden, scourge of Catholic forces during the Thirty Years War until his death at the Battle of Lützen.

HABSBURG, Karl I, Archduke of Austria (1590–1624), brother of Ferdinand II, Holy Roman Emperor, and of Maria Maddalena, Grand Duchess of Tuscany.

HABSBURG, Leopold V, Archduke of Austria-Tyrol (1586–1632), brother of Karl I Habsburg, correspondent of Galileo and patron of Scheiner.

HABSBURG, Rudolf II, Holy Roman Emperor from 1576 (1552–1612), a great patron of the arts and sciences, the more occult the better.

HARRIOT (HARIOT), Thomas (c.1560–1621), English mathematician and astronomer, who anticipated several of Galileo's telescopic discoveries.

HEECK, Johannes van (Johannes van Eck) (1579–?), unstable Dutch physician, alchemist and astronomer, one of the founding lynxes, expelled from the group in 1616.

HENRI III, King of France (1551–1589), a client of Veronica Franco's.

HENRI IV, King of France from 1589 (1553–1610), assassinated in 1610, desired that Galileo discover a star or two to name after him.

HOMBERG, Paul (1559/60–1634), music teacher at the Lutheran secondary school in Catholic Austria where Kepler also taught.

HORKY, Martin (c.1590–after 1650?), born in Bohemia, Magini's assistant in Bologna, friend of Capra, foe of Galileo.

HORTENSIUS, Martinus (Maarten van den Hove) (1605–1639), professor of mathematics in Holland, examined Galileo's means of determining longitude for the Dutch authorities.

HULSIUS, Levinus (1546–1606), mathematician, student of Galileo, instrument maker and printer, published books on instruments like Galileo's proportional compass.

INCHOFER, Melchior (c.1584–1648), Hungarian Jesuit, friend of Riccardi, played an important role in Galileo's trial of 1633.

INGOLI, Francesco (1578–1649), drew up "corrections" to Copernicus for the Congregation of the Index, first secretary of Gregory XV's Congregation for the Propagation of the Faith.

INNOCENT XII, Antonio Pignatelli, Pope from 1691 (1615–1700).

JAMES I, King of England and Ireland from 1603, and, as James VI, King of Scots from 1567 (1566–1625).

JOHN XXIII, Angelo Giuseppe Roncali, Pope from 1958 (1881–1963), convened the Second Vatican Council, 1962–65.

JOHN PAUL II, Karol Józef Wojtyla, Pope from 1978 (1920–2005), the first non-Italian pope in 450 years, inaugurated a multidisciplinary enquiry into the Galileo Affair in 1979.

JOYEUSE, François de, Cardinal from 1584 (1562–1615), Cardinal Protector of France in 1587, interested in astronomy, received one of Galileo's first telescopes.

KAPSBURGER, Johannes Hieronymus (Giovanni Geronimo) (c.1580–1651), Venetian, a favorite performer and composer in Rome, wrote the music for Grassi's Jesuit opera.

KEPLER, Johannes (1571–1630), a most original mathematician, astronomer to Emperor Rudolf II, abetted acceptance of Galileo's telescopic discoveries.

LADISLAV IV, King of Poland (1595–1648), King of the Polish-Lithuanian Commonwealth from 1632.

LANDUCCI, Benedetto di Luca, husband of Galileo's sister Virginia.

LANDUCCI, Vincenzo (1595–1649), son of Galileo's sister Virginia.

LANDUCCI, Virginia, Galileo's great-niece, who became Suor Olimpia in the Convent of San Giorgio in Florence.

LEMBO, Giovanni Paolo (1570–1618), S.J., a member of the Clavius group in Rome.

LEO XIII, Gioacchino Vincenzo Pecci, Pope from 1878 (1810–1903), a cautious modernizer.

LIBRI, Giulio (c.1550–1610), professor of philosophy at Pisa, and, from 1595 to 1600, at Padua.

LICETI, Fortunio (1577–1657), professor of philosophy at Padua, 1609–37, and of medical theory, 1645–57, after a period in Bologna, 1637–45.

LOCHER, Johann Georg, a student of Scheiner, whose dissertation of 1614 included original arguments against Galileo's version of Copernicanism.

LORENZINI, Antonio (c.1540–?), Aristotelian philosopher ridiculed by Cecco di Ronchitti.

LORINI, Niccolò (1544–c.1617), Florentine Dominican favored by Grand Duke Ferdinando I and Grand Duchess Christina, attached to the "pigeon league."

LOWER, Sir William (c.1570–1615), collaborator of Thomas Harriot.

LOYOLA, Ignatius (1491–1556), Spanish soldier and co-founder of the Society of Jesus (1534) with Francis Xavier, canonized in 1622 by Gregory XV.

LUDOVISI, Ludovico, Cardinal from 1621 (1595–1632), cardinal nephew of Pope Gregory XV, close to the Jesuits, leader of the Spanish faction in the curia.

MACULANO, Vincenzo, Cardinal from 1642 (1578–1667), Dominican, Commissary General of the Holy Office in charge of Galileo's trial in 1633.

MAELCOTE, Odo van (1572–1615), S.J., born in Brussels, professor of mathematics and Hebrew at the Roman College, gave an oration there in praise of Galileo.

MAGALOTTI, Filippo (1558–?), friend of Galileo and Mario Guiducci, relative of the Barberinis.

MAGINI, Giovanni Antonio (1555–1617), chosen over Galileo for a professorship of mathematics at Bologna, competent astrologer, astronomer, and instrument maker.

MANETTI, Antonio (1423–1497), Florentine mathematician and architect, biographer of Brunelleschi, devised a geography of Dante's *Inferno* later defended by Galileo.

MARCHETTI, Angelo, Galileo's successor (several times removed) in the chair of mathematics at Pisa.

MARIA MADDALENA OF AUSTRIA, Grand Duchess of Tuscany from 1609 (1589–1631), sister of Ferdinand II, Holy Roman Emperor, married Cosimo II of Tuscany in 1608.

MARINO, Giambattista (Giovan Battista) (1569–1625), Neapolitan poet, famous for his long epic, *L'Adone*, published in Paris in 1623 and soon banned in Rome.

MARSILI, Cesare (1592–1633), Bolognese patrician, natural philosopher, lynx (1625).

MARZI-MEDICI, Alessandro (1563–1630), Archbishop of Florence from 1605, a member of the "pigeon league."

MÄSTLIN, Michael (1550–1631), professor of mathematics and astronomy at Tübingen, Kepler's teacher, an early Copernican.

MAURI, Alimberto (a pseudonym, perhaps of Galileo), replied to Ludovico delle Colombe's *Discourse on the new star of 1604*.

MAXIMILIAN I, Duke/Elector of Bavaria (1573–1651), employer of Michelangelo Galilei.

MAYR, Simon (1573–1624), tutor of Baldassar Capra, from 1605 mathematician to the Margrave of Ansbach, whose name he conferred on Jupiter's satellites in 1614.

MAZZOLENI, Marcantonio (?–1632), coppersmith from the Venetian Arsenal, became Galileo's instrument maker in 1597.

MAZZONI, Jacopo (1548–1598), famous philosopher and literary man, defender of Dante, friend of Tasso, and, as professor at Pisa, Galileo's closest colleague.

MEDICI, Antonio de' (1576–1621), natural son of Grand Duke Francesco I and his mistress, Bianca Capello, took a great interest in Galileo's scientific work and dabbled with alchemy.

MEDICI, Cosimo I de', first Grand Duke of Tuscany from 1569 (1519–1574).

MEDICI, Cosimo II de', Grand Duke of Tuscany from 1609 (1590–1621), son of Ferdinando I and Christina of Lorraine, Galileo's tutee and employer.

MEDICI, Ferdinando I de', Grand Duke of Tuscany from 1587 (1549–1609), son of Cosimo I, resigned his cardinalate in 1589 to succeed his elder brother Francesco I.

MEDICI, Ferdinando II de', Grand Duke of Tuscany from 1621 (1610–1670), eldest son of Cosimo II and Maria Maddalena of Austria, whose reign began under the regency of his grandmother, Christina, and his mother, continued Cosimo II's generosity to Galileo.

MEDICI, Francesco I de', Grand Duke of Tuscany from 1574 (1541–1587), married Johanna of Austria in 1565, and, in 1578, after her death, Bianca Cappello.

MEDICI, Giovanni de', Don (1563 or 1567–1621), gifted natural son of Cosimo I, engineer, alchemist, philosopher, and bibliophile.

MEDICI, Giuliano de' (1574–1636), Florentine ambassador to Prague, 1608–18, later Archbishop of Pisa.

MEDICI, Jacopo de', one of Galileo's gambling associates.

MEDICI, Marie de', Queen of France, wife of Henri IV (1573–1642), daughter of Francesco I de' Medici.

MEI, Girolamo (1519–1594), Florentine patrician who taught Vincenzo Galilei enough about ancient music to write a dialogue about it.

MERCURIALE, Girolamo (1530–1606), professor of medicine at Padua and then at Pisa, physician to Grand Duke Ferdinando I.

MICANZIO, Fulgenzio (1570–1654), Servite monk, Sarpi's disciple, biographer, and successor as theologian to the Venetian Republic in 1623, loyal friend to Galileo.

MICHELANGELO. *See* Buonarroti

MILTON, John (1608–1674), English poet and pamphleteer, probably visited Galileo in Florence in 1638.

MOLETTI, Giuseppe (1531–1588), Galileo's predecessor at Padua.

MONTAIGNE, Michel Eyquem de (1533–1592), French essayist and lawyer, who knew the pleasures of Paris and Venice.

MORANDI, Orazio (c.1570–1630), Galileo's boyhood friend, Abbot-General of the Vallombrosan order, passionate astrologer, involved in predicting the death of Urban VIII.

MOROSINI, Andrea (1558–1618), Venetian patrician and historian, host with his brother Niccolò of a salon in Venice that brought politicians and intellectuals together.

MOROSINI, Niccolò, brother of Andrea Morosini.

MUTI, Tiberio, Cardinal from 1615 (1574–1636).

MUZZARELLI DA FANANO, Giovanni, Florentine Inquisitor General who oversaw, usually leniently, Galileo's detention at Arcetri.

NAUDÉ, Gabriel (1600–1653), Parisian, student at Padua under Cremonini, librarian of Cardinal Francesco Barberini (1641-42), then of Cardinals Richelieu and Mazarin.

NICCOLINI, Catarina Riccardi (1598–1676), wife of the Florentine ambassador to Rome, Francesco Niccolini, and cousin of Niccolò Riccardi ("Father Monster").

NICCOLINI, Francesco (1584–1650), Florentine ambassador to Rome, 1621–43, close friend of Galileo.

NICCOLINI, Pietro (?–1651), Archbishop of Florence.

NOAILLES, François de, Comte (1584–1645), studied with Galileo at Padua in 1603, served as French ambassador to Rome 1634-36, dedicatee of *Two new sciences*.

OREGGI, Agostino, Cardinal from 1633 (1577–1635), Tuscan, long-time theological advisor to Maffeo Barberini, played an important role in Galileo's trial in 1633.

ORSINI, Alessandro, Cardinal from 1615 (1593–1626), nephew of Ferdinando I and patron of Galileo, who dedicated his work on the tides to him (1616), later joined the Jesuits.

ORSINI, Paolo Giordano II, Duke of Bracciano (1591–1656), Roman collector and prominent patron of the arts, published Scheiner's *Rosa ursina* (1630), "the Orsini rose."

ORTELIUS, Abraham (1527–1598), Flemish geographer and cartographer, creator of the first modern atlas.

PAGNONI, Silvestro, disaffected employee of Galileo, denounced him to the Inquisition in Padua in 1604.

PAPAZZONI, Flaminio (c.1550–1614), Bolognese, appointed professor of philosophy at Pisa on Galileo's recommendation.

PARMIGIANINO, Girolamo Francesco Maria Mazzola, known as (1503–1540), prominent Mannerist painter and printmaker.

PASCHINI, Pio (1878–1962), historian whose biography of Galileo commissioned by Pope Pius XII was corrected by the Jesuits (whom it criticized) before publication.

PASQUALIGO, Zaccaria (1600–1664), Veronese, Theatine theologian who served on the special committee set up by Urban VIII to examine Galileo's *Dialogue*.

PAUL III, Alessandro Farnese, Pope from 1534 (1468–1549), convened the Council of Trent in 1545, dedicatee of Copernicus' *De revolutionibus orbium coelestium*.

PAUL V, Camillo Borghese, Pope from 1605 (1552–1621), Tuscan, precipitated the Venetian interdict and authorized the condemnation of Copernican books in 1616.

PAUL VI, Giovanni Battista Enrico Antonio Maria Montini, Pope from 1963 (1897–1978).

PAZZI, Maria Maddalena de', (1566–1607), visionary Carmelite nun from a noble Florentine family, sanctified in 1669.

PERETTI DI MONTALTO, Alessandro Damasceni, Cardinal from 1585 (1571–1623), cardinal nephew of Sixtus V.

382 GLOSSARY OF NAMES

PERI, Jacopo (1561–1633), composer of the first opera whose score has survived (*Erudice*, 1600).

PETRARCH, Francesco Petrarca (1304–74), son of a Florentine notary exiled with Dante, known for his *Rime* inspired by unrequited passion.

PHILIP III, King of Spain and Portugal (1578–1621).

PICCHENA, Curzio (1553–1626), Secretary of State to Grand Duke Cosimo II.

PICCOLOMINI, Ascanio II (c.1590–1671), Archbishop of Siena from 1628, in whose palace Galileo began to serve his sentence in the fall of 1633.

PICCOLOMINI, Francesco (1574–1651), theologian, Jesuit General, 1649–51.

PICCOLOMINI, Girolamo, professor of philosophy at the Roman College.

PICCOLOMINI, Ottavio (1599–1656), general in the armies of the Holy Roman Empire, elder brother of Ascanio Piccolomini, insured that Kepler's prediction of the death of Wallenstein in March 1634 came true.

PIERONI, Giovanni (1586–1653), Italian architect in the service of the Holy Roman Emperor, tried to find a publisher for *Two new sciences*.

PIERSANTI, Alessandro, Galileo's servant and godfather to Galileo's son.

PIGNORIA, Lorenzo, Paduan priest and archeologist who acted as a conduit for the care of Galileo's son.

PINELLI, Giovanni Vincenzo (1535–1601), Neapolitan owner of a rich library that functioned as the literary center of Padua around 1600.

PIUS VII, Luigi Barnabà Chiaramonte, Pope from 1800 (1742–1823), re-established the Jesuit Order following the defeat of Napoleon, presided over the Settele affair.

PIUS XII, Eugenio Maria Giuseppe Giovanni Pacelli, Pope from 1939 (1876–1958).

PLATO (c.428–c.348 BCE), Greek philosopher praised and criticized in Galileo's time for his mystical-mathematical abstractions as opposed to the common-sense, empirical philosophy of his student Aristotle.

POSSEVINO, Antonio (1534–1611), founder of Jesuit colleges in northern Europe, papal legate to Moscow, author of a guide to good books (*Bibliotheca selecta*) for Catholics.

POUPARD, Paul, Cardinal from 1985 (1930–), served on the Pontifical Council for Culture since its inception in 1982, prominent on Pope John Paul's committee on the Galileo affair.

PRIMI, Annibale, superintendent of the Medici villa in Rome when Galileo stayed there in 1615/16.

PULCARELLI, Costanzo (1568–1610), Neapolitan Jesuit who praised Galileo's *Sidereus nuncius*.

PTOLEMY, Claudius (c.90–168), Alexandrian mathematician whose geocentric model of the solar system dominated western astronomy until the time of Galileo.

QUERENGHI, Antonio (1546–1633), Paduan patrician, priest, patron, poet, diplomat, and admirer of Galileo.

REAEL, Lorenzo (Lorenzo Realio) Admiral (?–1637), Italian-speaking Dutch commissioner assigned to evaluate Galileo's method of determining longitude.

RENIER, Vincenzio (1606–1647), professor of mathematics at Pisa, whom Galileo asked to update and improve his astronomical tables of Jupiter's moons.

RICASOLI, Giovanni, the disputed heir to Giovanni Battista Ricasoli's estate.

RICASOLI BARONI, Giovanni Battista (?–1589), friend of Galileo, whose will was the subject of a law suit in which Galileo testified.

RICCARDI, Niccolò ("Father Monster") (1585–1639), Dominican, who acquired his nickname from Philip III of Spain for his size and his preaching, Master of the Sacred Palace (the chief censor of Rome) from 1629.

RICCI, Ostilio (1540–1603), Galileo's mathematics teacher, lecturer in mathematics at the Accademia del Disegno in Florence, mathematician to Grand Duke Ferdinando II.

RICCIOLI, Giovambattista (1598–1671), S.J., author of the best compendium of astronomy of his day, the *Almagestum novum* (1651).

RIDOLFI, Niccolò (1578–1650), Master of the Sacred Palace in 1622, Dominican General in 1629, consulted Morandi about astrological indications of the fate of Urban VIII.

RINUCCINI, Francesco (1603–1678), Tuscan agent in Venice from 1637, later Bishop of Pistoia and Prato (1652).

ROCCO, Antonio (1586–1652), libertine student of Cremonini, taught philosophy at the Benedictine monastery of San Giorgio Maggiore in Venice.

RUZZANTE, Angelo Beolco, known as (1502–1542), comic actor and playwright whose pieces in the Paduan dialect inspired the work of Cecco di Ronchitti.

SACROBOSCO, Johannes de (John of Hollywood, c. 1195–c. 1256), author of the basic introductory text of geocentric astronomy used in the Middle Ages and the Renaissance.

SAGREDO, Giovanfrancesco (Gianfrancesco) (1571–1620), Venetian nobleman almost a brother to Galileo, immortalized in Galileo's dialogues.

SALVIATI, Filippo (1582–1614), rich Florentine noble, natural philosopher, friend and patron of Galileo, immortalized as Galileo's alter ego in Galileo's dialogues.

SALVIATI, Giovanni, Cardinal from 1517 (1490–1553), nephew of Pope Leo X, friend of Macchiavelli.

SANTI, Leone (1585–c.1651/2), professor at the Roman College and successful dramatist.

SANTINI, Antonio, Venetian merchant who, in 1610, saw the Medici stars through his home-made telescope.

SANTORIO, Santorio (1561–1636), physician, close to the Morosini and Sarpi, professor of medical theory at Padua from 1611.

SARPI, Paolo (1552–1623), Venetian Servite monk and polymath, theological advisor to the Venetian Senate, advisor and sounding board to Galileo, author of the anti-Roman *History of the Council of Trent*.

SARROCCHI, Margherita (1560–1617), poet and mathematician, friend of Luca Valerio and Galileo.

SARSI, Lothario, anagram and pseudonym of Orazio Grassi.

SCAGLIA, Desiderio, Cardinal from 1621 (1567–1639), Dominican theologian and preacher, prominent member of the Congregation of the Inquisition that sentenced Galileo in 1633.

SCHEINER, Christoph (c.1573–1650), professor of mathematics at the Jesuit College in Ingolstadt, fought with Galileo over priority in the discovery of sunspots and other matters.

SCHRECK, Johann(es) (1576–1630), S.J., German astronomer and botanist, who resigned from the lynxes to join the Jesuits and go on mission to China.

SEGETH, Thomas (?–1627), a Scot from Edinburgh, Pinelli's librarian around 1600, subsequently a collaborator of Kepler.

SEGHIZZI, Michelangelo (1585–1625), Dominican, Commissary General of the Holy Office during its deliberations over Copernicanism in 1615–16.

SERRISTORI, Ludovico (?–1656), consultor to the Holy Office, Bishop of Cortona in 1624.

SETTELE, Giuseppe (?–1841), professor of mathematics at the University of Rome whose astronomy text precipitated the removal of books defending heliocentrism from the Index.

SETTIMI, Clemente (1612–?), Piarist (Scolopian) monk who sometimes served as Galileo's secretary after Galileo lost his sight.

SFRONDATI, Paolo Camillo, Cardinal from 1590 (1560–1618), nephew of Pope Gregory XIV, Dominican prefect of the Congregation of the Index.

SIXTUS V, Felice Peretti di Montalto, Pope from 1585 (1520–90), champion of papal prerogatives over princes.

SIZZI, Francesco (c.1585–1618), Florentine, client of Don Giovanni de' Medici, opponent of Galileo, executed for supporting Catherine de' Medici against Louis XIII.

SPINELLI, Girolamo (c.1580–1647), Benedictine monk in Galileo's circle, collaborator on the works of Cecco di Ronchitti.

STAY, Benedict (Benedetto) (1714–1801), S.J., author of a long poem on Newtonian physics.

STEFANI, Giacinto (1577–1633), Dominican theologian and preacher who read Galileo's *Dialogue* for the Florentine censorship.

STELLUTTI, Francesco (1577–1646), founding member of the Accademia dei Lincei and the most faithful of Cesi's associates.

STRADANO, Giovanni (Jan Van der Straet) (1523–1605), Flemish artist who made his career in Italy.

STROZZI, Giovan(ni) Battista (Giambattista), the younger (1551–1634), Florentine aristocrat, poet, patron of Galileo and Ciampoli.

STROZZI, Giulio (1583–1652), Venetian poet, man of letters, and opera librettist.

STROZZI, Piero (1550–1609), Florentine patrician and composer, member of Bardi's *Camerata* and friend of Galileo's father.

TASSO, Torquato (1544–1595), Italian poet, author of *La Gerusalemme liberata* (1580).

TAVERNA, Ferdinando, Cardinal from 1604 (1558–1619), Governor of Rome, 1599–1604.

TEDALDI, Muzio, an in-law of Vincenzo Galilei, he took care of Galileo in Pisa while Vincenzo was in Florence or elsewhere.

TENGNAGEL, Franz (1576–1622), student and son-in-law of Tycho Brahe.

TERRENTIUS. *See* Schreck, Johann

TOALDO, Giuseppe (1719–1797), editor of Galileo's *Opere* (1744), professor of astronomy at Padua.

TORRICELLI, Evangelista (1608–1647), mathematician trained at Pisa by Castelli, stayed with Galileo in 1641, succeeded him as mathematician to Ferdinando II.

URBAN VIII, Maffeo Barberini, Pope from 1623 (1568–1644).

VALERIO, Luca (1552–1618), the last great representative of the school of Commandino, close friend of Margherita Sarrochi, lynx.

VELLUTELLO, Alessandro (1473–?), a literary critic from Lucca, devised an alternative to the Florentine version of Hell.

VENIER, Domenico (1517–1582), Venetian patrician and poet, patron of Veronica Franco.

VENIER, Sebastiano, Venetian patrician, friend and patron of Galileo.

VIÈTE, François (1540–1603), lawyer and mathematician, renowned as an algebraist.

VINTA, Belisario (1542–1613), Tuscan State Secretary under Ferdinando I and Cosimo II.

VISCONTI, Raffaelo, Dominican, mathematician and astrologer, read Galileo's *Dialogue* for the Roman censorship, banished to Viterbo for involvement in the Orandi affair.

VITELLESCHI, Muzio (1563–1645), Venetian, Jesuit General, 1615–45.

VITRUVIUS POLLIO, Marcus (first century CE), author of the standard (and only extant) Roman treatise on architecture, which includes astronomy, gnomonics, and hygiene.

VIVIANI, Vincenzo (1622–1703), mathematician, student of Castelli, Galileo's assistant at Arcetri, last disciple, and first biographer.

WALLENSTEIN, Albrecht von, Duke of Friedland (1583–1634), supreme commander of the Habsburg armies under the Holy Roman Emperor, Ferdinand II.

WELSER, Mark (1558–1614), Augsburg banker close to the Jesuits, helped to circulate Galileo's ideas, made a lynx in 1612.

XAVIER, Francis (1506–1552), Spanish missionary, co-founder of the Society of Jesus (1534), canonized by Gregory XV in 1622.

XIMENES, Ferdinando (1580–1630), Dominican friar called to testify before the Inquisition in consequence of Caccini's denunciation of Galileo.

ZARLINO, Gioseffè (1517–1590), musical theorist and composer who contributed to the theory and practice of counterpoint, teacher of Vincenzo Galilei.

ZIECKMESSER (or ZUGMEISSER), Jan Eutel (c.1575–?), Dutch mathematician who studied at Padua and produced a compass similar to Galileo's, became mathematician to the Archbishop of Cologne.

ZUCCARI, Federico (1542/3–1609), mannerist painter and architect who worked on the Florentine cathedral.

Fictional characters

AGRAMANTE, King of Africa in OF, commander of the Saracen armies against Charlemagne.

ALCINA, sorceress in OF, seductress of Astolofo and Ruggiero.

ALEXANDER, Galileo's alter ego in his dialogue "De motu" (c.1590).

ANGELICA, daughter of the Emperor of Cathay in OF whose jilting of Orlando caused his madness.

ARGANTE, Circassian warrior in Egyptian cause in GL, scourge of Godfrey's Christians, killed in a dual by Tancredi.

ARMIDA, Syrian sorceress in GL, energetic seductress of crusaders, notably Rinaldo, whose eventual return to Charlemagne's service caused her to convert to Christianity.

ASTOLFO, son of King of England in OF, cousin of Orlando, another crusader ensnared by Alcina.

BRADAMANTE, Christian warrior maiden in OF, destined to marry Ruggiero and found the Este line.

CHARLEMAGNE (747–814), Emperor of the West, whose expedition against the Moors in Spain in 778 gave rise to the heroic literature played with in OF.

CHARON, the boatman who ferries dead souls across the river Styx in Dante's Inferno.

CLORINDA, Muslim warrior maiden in GL, who beguiled Tancredi, wreaked havoc among the crusaders, and was killed unwittingly by Tancredi.

DOMINICUS, Alexander's fellow student and foil in Galileo's dialogue "De motu" (c.1590).

DON QUIJOTE DE LA MANCHA, like Galileo an addicted reader of Orlando furioso.

DORALICE, daughter of King of Granada in OF, jilted Rodomonte for Mandricardo.

ERMINIA, pagan princess in GL, tended Tancredi's wounds to mutual benefit.

GABRINA, a perfect witch in OF, punished by Mandricardo, hanged by Odoric of Biscay, identified by Galileo with his mother Giulia.

GODFREY OF BOUILLON, commander-in-chief of the Christian armies besieging Jerusalem in GL.

MANDRICARDO, King of the Tartars in OF, constantly pursuing duels with Orlando and Rodomonte when not making love to Doralice, killed by Ruggiero.

MERLIN, wizard of King Arthur's court, an occasional character in OF.

ORLANDO, nephew and chief warrior of Charlemagne in OF, driven mad by Angelica, restored to his trade of Christian champion by Astolfo.

RINALDO, Christian champion in GL, like Ruggiero in OF an ancestor of the Este, detained in Armida's love nest until recalled to his duty to slaughter Saracens.

RODOMONTE, African King in OF, greatest of Agramante's champions, almost destroyed Paris single-handed, killed by Ruggiero.

RUGGIERO, Saracen champion in OF, converted to Christianity for love of Bradmante after adventures with Alcina and others.

SIMPLICIO, the gentle, innumerate, Aristotelian philosopher harried by Salviati and Sagredo in Galileo's masterworks.

TANCREDI, Norman prince in fact, and in GL besotted with Clorinda, almost killed by Argante, and nursed back to health by Erminia.

TIPHYS, navigator of the Argo on its voyage to pilfer the golden fleece.

TURPIN or Tilpinus (died c.794), French ecclesiastic, pseudepigraphic author of a version of the story of Roland written in the twelfth century by another French cleric.

VIRGIL, Publius Virgilius Maro (70-19 BCE), in life the author of the adventures of Aeneas and other enduring classics, in fiction Dante's guide through the underworld.

"Retombera-t-il?" (Will it fall back?), Varignon, *Conjectures* (1690).

NOTES

References to Galileo, *Opere* (see below) are given in the form "Op. x:y," or, if following a letter or document, "(x:y)," in both cases x being the volume and y the page(s). When no ambiguity results, the century dates are omitted, that is, 12 August 1616 appears as 12 Aug. 16. In addition, the following abbreviations are used:

Assayer   Galileo Galilei, *The assayer* [*Il saggiatore*, 1623]. In Stillman Drake and C.D. O'Malley (eds.), *The controversy on the comets of 1618*. Philadelphia: University of Pennsylvania Press, 1960, pp. 151–336.

BH, *Dial.*   Galileo Galilei, *Dialogo sopra i due massimi sistemi del mondo tolemaico e copernicano* [1632]. *Edizione critica e commento*, 2 vols., eds. Ottavio Besomi and Mario Helbing. Padua: Antenore, 1998.

BH, *Disc.*   Galileo Galilei and Mario Guiducci, *Discorso delle comete* [1619]. *Edizione critica e commento*, eds. Ottavio Besomi and Mario Helbing. Padua: Antenore, 2002.

BH, *Sagg.*   Galileo Galilei, *Il saggiatore* [1623]. *Edizione critica e commento*, eds. Ottavio Besomi and Mario Helbing. Padua: Antenore, 2005.

Camerota   Michele Camerota, *Galileo Galilei e la cultura scientifica nell'età della Controriforma*. Rome: Salerno, 2004.

Chiari   Galileo Galilei, *Scritti letterari*, ed. Alberto Chiari. Florence: Le Monnier, 1970.

DBI   *Dizionario biografico degli italiani.*

DSB   Charles C. Gillispie (gen. ed.), *Dictionary of scientific biography*, 16 vols. New York: Charles Scribner's Sons, 1970–80.

FA   Maurice A. Finocchiaro, *The Galileo affair*. Berkeley: University of California Press, 1989.

GG   Galileo Galilei.

GL   Torquato Tasso, *Gerusalemme liberata* [1581].

IMSS   [Museo Galileo] Istituto e Museo di Storia della Scienza, Florence.

JHA   *Journal for the history of astronomy* (1970–).

LCF   José Montesinos and Carlos Solís (eds.), *Largo campo de filosofare: Eurosymposium Galileo 2001*. La Orotava: Fundación canaria Orotava de historia de la ciencia, 2001.

NCCS   Paolo Galluzzi (ed.), *Novità celesti e crisi del sapere*. Florence: IMSS, 1983.

OF   Lodovico Ariosto, *Orlando furioso* [1516, 1532].

Op.   Galileo Galilei, *Opere*, 20 vols., ed. Antonio Favaro. Florence: Giunti-Barbera, 1890–1909.

OS   Eileen Reeves and Albert Van Helden, tr. *On Sunspots: Galileo Galilei and Christoph Scheiner*. Chicago: University of Chicago Press, 2010.

Pagano   Sergio Pagano (ed.), *I documenti vaticani del processo di Galileo Galilei (1611–1741)*, 2nd ed. Vatican City: Archivio segreto vaticano, 2009.

RG   Maurice A. Finocchiaro, *Retrying Galileo, 1633–1992*. Berkeley: University of California Press, 2005.

SN   Galileo Galilei, *Sidereus nuncius* [1610]. *The sideral messenger*, tr. Albert van Helden. Chicago: Univerity of Chicago Press, 1989.

TCWS   Galileo Galilei, *Dialogue concerning the two chief world systems*, tr. Stillman Drake. Berkeley: University of California Press, 1953.

TNS   Galileo Galilei, *Discourses and mathematical demonstrations concerning two new sciences pertaining to mechanics and local motions* [1638], tr. Stillman Drake. Madison: University of Wisconsin Press, 1974.

Preface

1. Vasari, *Lives* [1568], ed. Burroughs (1946), 258q; Viviani, *Vita* (2001), 74–6; Bentivoglio, *Memorie* (1648), 124 ("Archimede toscano").

2. According to Viviani, *Vita* (2001), 30.

Chapter 1: A Florentine Education

1. Lunardi and Sabbatini, *Il rimembrar* (2009), 50, 75, 89.

2. See Section 3.2 below.

3. Cf. Armour, in Aquilecchia et al., *Essays* (1971), 146, and Pino, in *Letteratura* (1955), 139–40, 143.

4. Cochrane, *Florence* (1973), 167; Galileo on GL, 4.4, 7–8 (Chiari, 544–6). Cf. Della Terza, *Forma* (1979), 216; Tongiorgio Tomasi, *Galilaeana*, 4 (2007), 20–1; Tannery, in McMullin, *Galileo* (1967), 167.

5. Op. 19:22. The ducat or scudo was a general money of account, equal to different numbers of lire in different places, 7 at Florence. An experienced artisan had an annual salary of about 50 scudi in Florence in 1590. Litchfield, *Emergence* (1986), xiii.

6. Favaro, *Scampoli* (1992), 1, 178–9; Caffarelli, *Galileo e Pisa* (2004), 11, 13–14; Rapp, in Galilei, *Fronimo* (1969), "Premessa."

7. Galilei, *Fronimo* (1969), 52; Rapp, in ibid., "Premessa."

8. From, resp., "Special discourse" (1588/9) and *Dialogi* (1581), quoted in Palisca, in Coelho, *Music* (1992), 143.

9. Favaro, *Studio* (1966), 2, 49–50; on the episode of the inquisitor of Florence, see Section 3.2 below; on the spying, Section 5.1 below.

10. See Section 3.2 below.

11. Muzio Tedaldi to V. Galilei, 13 Jan. 1574 (10:17).

12. Tosi, in Ciardi, *Vallombrosa* (1999), 259–60; Paschini, *Vita* (1965), 1, 56–7; Ercolani, *Riv. stor. bened.*, 2 (1907), 571; Brooks, *Graceful and true* (2003), 80–1 (Zuccari's sketches).

13. Ercolani, *Riv. stor. bened.*, 2 (1907), 572–4, 578–9, referring to abbot Dom Colombino d'Alfiano di Valdesa; Ernst, in Canone, *Bibliothecae* (1993), 21, 227–8; Section, 7:3 below.

14. Galilei, *Fronimo* (1568, 1584), 3.

15. Tedaldi to V. Galilei, 16 July 1578 (10:21).

16. Camerota, 33–7; Drake, *Galileo at work* (1978), 2–3.

17. Spini, in Maccagni, *Saggi*, 3:2 (1972), 426–7.

18. Camerota, 43–4; Drake, *Galileo at work* (1978), 3–6.

19. Zangheri, *Accademici* (2000), x–xi, xv, 27, 68, 142, 273; Waźbiński, *Accademia* (1987), 282–3, 494–5; Barzman, *Florentine Academy* (2000), 151–5, 158.

20. Viviani, *Vita* (2001), 30. Some of Galileo's sketches have survived; Bredekamp, in Flemming and Schütze, *Ars* (1996), 477–84.

21. See Section 4.2 below; Benedetti, *Diversarum speculationum liber* (1585).

22. Lattis, *Clavius* (1994), 24–6.

23. Meli, *Nuncius*, 7:1 (1992), 16–19; Mamiani, *Elogi* (1828), 47–8, 52–4; Sections 2.3 and 4.2 below.

24. Giacobbe, *Physis*, 18 (1976), 34–5.

25. *Prior analytics*, 25^{b}32–26^{a}2; McMullin, in Butts and Pitt, *Perspectives* (1978), 213–17. The strongest syllogisms are convertible, premise and conclusions being co-extensive; in Aristotle's example (*Posterior analytics*, 97^{a}35–97^{b}1–15), from the premise "all plants with broad leaves are deciduous" and the middle term, "vines have broad leaves," we can infer that vines are deciduous, whereas from the premise "all deciduous plants have broad leaves" and the middle term, "vines are deciduous," we may safely conclude that vines have broad leaves. Quite reassuring.

26. E.g., Pietro Catena, mathematics professor at Padua, 1547–76, in Giacobbe, *Physis*, 15 (1973), 181, 185–8, 191–2; Euclid, 1:32.

27. E.g., Benito Pereyra, SJ, professor of philosophy and theology at the Collegio Romano, in Giacobbe, *Physis*, 19 (1977), 53–6, 62, 68, 71–6.

28. Giacobbe, *Physis*, 14 (1972), 163, 168–9, 181, 186–92.

29. Giacobbe, *Physis*, 14 (1972), 357, 364–5, 372; Biancani, *De mathematicarum natura dissertatio* (1614), in Giacobbe, *Physis*, 18 (1976), 8, 17–30; Clavius, in Carugo and Crombie, IMSS, *Ann.*, 8:2 (1983), 21, 33–4.

30. Op. 19:77; Dupré, *Galilaeana*, 2 (2005), 150, citing Schiavo, *Benedictina*, 9 (1955), 44; Ciardi, *Vallombrosa* (1999), 265; Barzman, *Florentine Academy* (2000), 154. Cf. Favaro, *Scampoli* (1992), 1, 256–7; Del Lungo, *Patria* (1909), 1, 360, and Spini, *Galileo* (1996), 33–4.

31. Bernabeo, in Congresso, xx, *Atti* (1964), 200–1; Op. 1:183, 10:24–7; Section 2.2 below.

32. Palisca, in Galilei, *Dialogue* (2003), xxvii–xxxiii; Galilei, ibid., 13.

33. Palisca, in Galilei, *Dialogue* (2003), xxxiv–xxxxiii; Galilei, ibid., 4–5; Palisca, *Mei* (1960), 5–6, 12, 28–9, and *Camerata* (1989), 45–77. Mei knew virtually all the ancient musical treatises known today; Palisca, *Mei* (1960), 34–40.

34. Palisca, *Mei* (1960), 11, 13–15, 42–3, 46–7; Palisca prints Mei's most important letters to Galilei, ibid., *passim*, and that of 8 May 1572 in English in *Camerata* (1989), 56–75.

35. Galilei, *Fronimo* (1568), 13, and *Dialogo* (1582), cited by Tomlinson, *Music* (1993), 191–2.

36. Palisca, in Galilei, *Dialogue* (2003), liii–lix, quote on p. lv, and Palisca, *Camerata* (1989), 2–3, 10; Tofani, in Cropper, *Culture* (1992), 57, and figs. 16, 17.

37. Palisca, *Camerata* (1989), 154–8. Drake, in Coelho, *Music* (1992), 10–11, raises this finding to the first law of physics discovered by experiment.

38. Palisca, *Camerata* (1989), 162–3, 201–3, and Palisca, in Rhys, *Seventeenth century* (1961), 130–5; Drake, *Galileo studies* (1970), 55–61.

39. Palisca, in Coelho, *Music* (1992), 145–9, and *Camerata* (1989), 158, 162; Galilei, ibid., 185 (first quote), 181 (second).

40. Settle, *NCCS*, 229–33.

41. Palisca, *Mei* (1960), 23; *DBI*, 1:571 (Alamanni); Weinberg, *Giorn. stor. lett. ital.*, 131 (1954), 175–6, 192, and *Italica*, 31 (1954), 267; Cochrane, *Florence* (1973), 116–18; Op. 19:57.

42. All have since been published several times, e.g., in Op. 9; they are cited below from the convenient edition by Chiari, *Scritti* (1943).

43. The dating game begun by Favaro (10:13, 16, 21, 26), who assigned the "Considerazioni" to no later than 1609, declined to date the notes, guessed that the play was written in Padua, and left the rest to youthful indiscretion, is also played by Vaccalluzzo, *Gal. lett.* (1896), 18–30, Bigi, in Maccagni, *Saggi*, 3:2 (1972), 531, Wlassics, *Galilei* (1974), 17–25, 30, and Della Terza, *Forma* (1979), 198–9.

44. Weinberg, *Gior. stor. lett. ital.*, 131 (1954), 183–6, 189, 191, 193, and *Italica*, 31 (1954), 211–12; Palisca, in *New looks* (1968), 11–12, 20–3.

45. Brown, *Stud. seic.*, 12 (1971), 6, 12, 14.

46. Palisca, in *New looks* (1968), 12.

47. Op. 9:228–9.

48. Chiari, 13. The rhyme scheme is Galileo's, the lines mine and a foot shorter than his.

49. Graf, *Attraverso il cinquecento* (1926), 226.

50. Del Lungo (a collaborator of Favaro's), *Patria* (1909), 1, 352, 356–60, 387; Bettoni, *Scrittori* (1933), 102–3, has a similar view of Galileo's rhymes. A perhaps more perceptive critic, Colapietra, *Belfagor*, 11 (1956), 567, adduces the image of the ladies of the street as the best line in Galileo's poetry.

51. Aquila, in Giandomenico and Guaragnella, *Prosa* (2006), 249–51.

52. Anon., in Galileo, *Considerazioni* (1793), vi.

53. Mestica, *Scritti* (1889), v–vii, xiv; Del Lungo and Favaro, *Prosa* (1911); Del Lungo, *Patria* (1909), 1, 345q. The honorifica come from Olschki, in McMullin, *Galileo* (1967), 104, and Italo Calvino, quoted by Bolzoni, *Galilaeana*, 4 (2007), 157–8, resp.

54. Wilkinson, *Horace* (1968), 98–9, summarizes the accompanying technical advice: avoid excess, keep violence off stage, don't begin with the climax, suit words to the speaker, etc.

55. Weinberg, *History* (1961), 2, ch. 16.

56. Aristotle, *Poetics*, 1458^{a}8–30, 1495^{b}8–15, 1460^{a}4–10, 27–33qq.

57. Brunner, *Illustrierung* (1999), 10–11, 123, 127–8, 130; Sherberg, *Ren. quart.*, 56 (2003), 26–9.

58. Landolfi, *Stud. seic.*, 29 (1988), 125, 128–31, 135.

59. Serassi, *Vita* (1790), 1, 7–9, 19–22, 48 (quote), 49, 53.

60. Ibid., 76–8 (first quote), 89, 97 (second).

61. Musacchio and Pellegrino, *Mazzoni* (1982), 5–9.

62. Mazzoni, *Defence* (1983), 71–3, 77–8q; Musacchio and Pellegrini, *Mazzoni* (1982), 14–15, 60–1, 68–70.

63. Brunner, *Illustrierung* (1999), 77q, 79, 87; Vannucci, *Dante* (1965), 26–7, 59.

64. Although Galileo's name does not occur on the Academy's surviving lists, Favaro, *Galileo* (1968), 1, 26–7, and Vaccalluzzo, *Gal. lett.* (1896), 95, 99, give good reasons to believe that he was a member in 1587/8.

65. Scolari, *Scritti* (1937), 42–3, 53–9; Brown, *Stud. seic.*, 12 (1971), 14–15 (Camillo Pellegrino, *Il caraffa* (1584), was the attacker); Hempfer, *Lektüren* (1987), 58–9; Weinberg, *History* (1968), 2, 958.

66. OF (1974, 1998), a breezy rendition in excellent prose, will probably satisfy most English-speaking readers.

67. OF, 10.92–11.8, ed. Waldman, 107q.

68. OF, 32.65–110.

69. Ramat, *Critica* (1954); Brand, *Ariosto* (1974), 184–5, 188 (John Byrd's madrigals); Palisca, *Camerata* (1989), 11, and in *New looks* (1968), 35.

70. Spini, in Maccagni, *Saggi*, 3:2 (1972), 431.

71. Brand, *Tasso* (1965), 79–82, 87–92, 95–102, 106, 111–15.

72. Galileo on OF, 4.74, in Chiari, 611.

73. Galileo, on GL, 4.43 and 5.92 (Chiari, 554, 568–9). Cf. Wlassics, *Galilei* (1974), 51, and Varanini, *Galileo critico* (1967), 35.

74. Del Lungo, *Patria* (1909), 1, 375; Battistini, *Intro.* (1989), 8; Wlassics, *Galilei* (1974), 140, 160–1.

75. Brown, *Stud. seic.*, 11 (1971), 3–4, 14–15, 17–21; Weinberg, *Italica*, 31 (1954), 11–12 (quote, from the Alterati's minutes).

76. Belloni, *Seicento* (1943), 570; Bigi, in Maccagni, *Saggi*, 3:2 (1972), 526–7; Wlassics, *Galilei* (1974), 38–9, 46, 55, 65; Della Terza, *Forma* (1979), 197–9.

77. Barbi, *Accademico* (1900), 18–19.

78. Vaccalluzzo, *Gal. lett.* (1896), 48, 53–4, 58–62; Ottone, *Aevum*, 46 (1972), 315–16.

79. Galileo, on GL, 1.46 (Chiari, 502). Cf. Vaccalluzzo, *Gal. lett.*, 37. This famous comparison of poems to collections was not altogether fresh with Galileo; Bolzoni, *Galilaeana*, 4 (2007), 160–2.

80. Galileo, "Considerazioni," 6:27, in Chiari, 573–4.

81. GL, 12.52–66, 19.104–14. Cf. Del Lungo, *Patria* (1909), 1, 376–7; Colapietra, *Belfagor*, 11 (1956), 569.

82. GL, 16.16–30; Galileo, on GL, 16.18 (Chiari, 625).

83. Galileo, on OF, 5.51, 7.29, 11.69 (Chiari, 390, 401, 408). Cf. Varanini, *Galileo critico* (1967), 20–2.

84. Galileo, on GL, 12.27–28 (Chiari, 411); Vaccalluzzo, *Gal. lett.* (1896), 49, 74q, 90.

85. OF, 27.103–12, 125–6q, ed. Waldman, 337; Ottone, *Aevum*, 46 (1972), 324; Colapietro, *Belfagor*, 11 (1956), 569.

86. Galileo, on *GL*, 16.1 (Chiari, 617–18). Among other criticisms of want of verisimilitude: on *GL*, 1.33, 5.92, 7.8–9, 14.31, 37 (Chiari, 499, 568–9, 589–90, 603–5).

87. Maggini, *Convivium*, 3:6 (1949), 848q, 853–5. Cf. Ottone, *Aevum*, 46 (1972), 318–20; Colapietro, *Belfagor*, 11 (1956), 558, 561; Vaccalluzzo, *Gal. lett.* (1896), 49, and De Sanctis, in ibid., 55–6, 72–4.

88. Vaccalluzzo, *Gal. lett.* (1896), 409, 50, 52, 70–1.

89. Battistini, *Intro.* (1989), 9.

90. Galileo, on *GL*, 19.7 (Chiari, 633): "Truly a wonderful, noble, and most generous reply, perhaps without an equal in the rest of the book." Cf. Del Lungo, *Patria* (1909), 1, 380–2, and Bigi, in Maccagni, *Saggi*, 3:2 (1972), 533–4.

91. Galileo, on *GL*, 1.1 and 19.26 (Chiari, 493–4, 635).

92. BH, *Dial.*, 362; "istorici o dottori di memoria" (*Dial.*, 7:139).

93. Varanini, *Galileo critico* (1967), 33q. Cf. Del Longo, *Patria* (1909), 1, 384; Olschki, in McMullin, *Galileo* (1967), 147–8.

94. Belloni, *Seicento* (1943), 535–6.

95. Wlassics, *Galilei* (1974), 116.

96. Viviani, in Op. 19:627. Cf. Vaccalluzzo, *Gal. lett.* (1896), 83–6; Del Lungo, *Patria* (1909), 1, 365; Spongano, *Prosa* (1949), 103.

97. Mestica, *Scritti* (1889), xvii.

98. OF, 25.62. Cf. Aquila, in Giandomenico and Guaragnella, *Prosa* (2006), 249–50 (Ariosto goes with Copernicus, Tasso with Ptolemy).

99. OF, 34.52–72, ed. Cavetti, 2, 1036–42 (tr. Waldman, 417–19).

100. OF, 34.73–87, ed. Cavetti, 2, 1042–44 (tr. Waldman, 419–20); Fumagalli, *Unità* (n.d.), 119–20.

101. The documents, collected by Favaro, are in Op. 19:42–108. They do not appear to have been exploited.

102. Op. 19:78, 86–8, 94–7, 99, 106–7; Favaro, in Campion and Kollerstrom, *Galileo's astrology* (2003), 10–11, and Rutkin, *Galilaeana*, 2 (2005), 110, on the young Galileo's competence in astrology.

103. Op. 8:591–4; Barra, *LCF*, 101–5. The frequency of 10 is 27/216, of 9, 25/216. Cf. *Assayer*, 318.

104. OF, 23.101–36.

105. Op. 19:54–6 (quotes); Maggi, in Clericuzio and Ernst, *Rinascimento* (2008), 289–93, for Maddalena.

106. GG to Giacomini, 5 Oct. 1589 [1590] (10:41–2).

107. Op. 19:60, 63, 83, 105.

108. Burton, *Anatomy*, Part. 3, 4.1.1 ("Religious melancholy"), ed. Jackson (2001), 3, 311–432.

109. Mancini, in Congresso, xx, *Atti* (1964), 284–5, citing Viviani and Gherardini.

110. Op. 19:45–6; Favaro, *Scampoli* (1992), 1, 256–7.

111. GG to Welser, 3 Feb. 11 (11:41q).

Chapter 2: A Tuscan Archimedes

1. Toussaint, *Enfer* (1997), 61–3, 69–72; cf. Brieger, in Brieger et al., *Illum. manuscr.* (1969), 1, 99.

2. Barbi, *Fortuna* (1890), 142, 216–35; Kleiner, *Mismapping* (1994), 24–5, 33, 149n14; Foà, in Foà and Gentili, *Dante* (2000), 188–9. Engel, *Dantes Inferno* (2006), 77–88, gives an abstract of Benivieni's dialogue and all the pictures.

3. Giambullari, *De 'l sito* (1544), 4; Barbi, *Fortuna* (1890), 137–8; abstract and illustrations in Engel, *Dantes Inferno* (2000), 180–1, 183, 187.

4. Vellutello, résumé with illustrations, in Engel, *Dantes Inferno* (2006), 145–52; Kleiner, *Mismapping* (1994), 30.

5. Giambullari, *De 'l sito* (1544), 6–7, 8–9q, 10–12; Barbi, *Fortuna* (1890), 139–40, 141q.

6. Giambullari, *De 'l sito* (1544), 10q, 34–5.

7. Cf. Kleiner, *Mismapping* (1994), 49.

8. Barbi, *Fortuna* (1890), 144.

9. Op. 9:32–4.

10. Brunner, *Illustrierung* (1999), 108–11, 128–32; Engel, *Dantes Inferno* (2006), 163–6, 169; Alamanni to Strozzi, 7 Aug. 1594 (10:17). Alamanni, a patrician literary type like Strozzi, heard Galileo's lectures to the Fiorentina and a few years later gave similar ones to the Alterati; *DBI*, 1:571.

11. Op. 9:47–50.

12. *De caelo*, 2.4 (esp. 311^b1–20, 34–6).

13. Op. 9:52, 55; Zingarelli, in Benivieni, *Dialogo* (1897), 23–4.

14. See Section 7.2 below; cf. Patterson, *AJP*, 70 (2002), 575–80.

15. Benivieni, *Dialogo* (1897), 99–100, 138.

16. Ibid., 92–3.

17. Cf. Grünbein, *Galilei* (1996), 91–101.

18. Vitruvius, 9:3, intro., §§ 9–12, ed. Morgan, 253–4; Helbing, *Filosofia* (1989), 215–25.

19. "On floating bodies," Bk. 1, props. 5, 7, in Archimedes, *The works* (1912), 257–8; Op. 1:216–17.

20. Archimedes' lever law requires $(W - w)CA = (W)EC$; but $CA - EC = AE$.

21. The weight d at G is at the center of gravity of the constituents W(gold) at E and W(silver) at F, so W(gold):W(silver) = FG:EF.

22. Op. 1:218–20.

23. "On floating bodies," Bk 2, Prop. 2, 4, in *The works* (1912), 265, 268.

24. Commandino, *Liber* (1565), ff. +2, +3r; Torelli, *Archimedis quae supersunt omnia* (1792), as reviewed in *Bibliothèque britannique*, 1 (1796), 59, 710–11.

25. Commandino, *Liber* (1565), f. 41v–45r.

26. *TNS*, 261–3.

27. *TNS*, 272–4.

28. *TNS*, 274–8.

29. *TNS*, 279–80. Balancing moments around B, we have $BX([B] - [A]) = AB[A] = (3/4)(h - k)[A]$, where $h = OQ$, $k = OP$. With $PB = (3h/4 - k)$ and $QB = h/4$, $PX{:}QX = (3b^4 - 4ab^3 + a^4){:}(3a^4 - 4ba^3 + b^4)$, which, after canceling out the common factor $(a - b)^2$, is equivalent to Galileo's expression.

30. GG to Clavius, 8 Jan., and reply, 16 Jan. 1588 (10:22–5). Cf. Drake, *Galileo at work* (1978), 13–14, and Wallace, *Sources* (1984), 226–7, who suggests, implausibly, that Clavius' difficulty with the proof drove Galileo to the study of scholastic logic.

31. Guidobaldo to GG, 16 Jan., 24 Mar., and 28 May 1588 (10:25–6, 33–4); Rose, in *DSB*, 9:487–8; Arrighi, Acc. luchese, *Atti*, 12 (1965), 198–9.

32. Guidobaldo to GG, 5 Mar. 1588 and 8 Dec. 1590 (10: 29–30, 45).

33. Coignet to GG, 31 Mar. 1588 (10:31–2); cf. Favaro, *Amici* (1983), 2, 874, 880–1.

34. *DBI*, 6:300–3 (Bardi); Op. 1:183 (12 Dec. 87); Favaro, *Galileo* (1883), 1, 21–2, and *Nuovi studi* (1891), 9–10.

35. Borro, "Multae sunt nostrarum ignorantium causae," in Schmitt, in Mahoney, *Philosophy* (1976), 469–70, 475; and Borro, *De motu gravium et levium* (1575), 214–15, in Schmitt, *Physis*, 14 (1972), 268–70. Cf. Viviani, *Medici* (1923), 104, 109.

36. Borro, *De motu* (1576), +2v, 186–7q. "My business is not to set forth what others have done well or badly, but what Aristotle taught and delivered"; Borro, *De methodo* (1584), 39, quoted by Schmitt, in Mahoney, *Philosophy* (1976), 467.

37. Quoted in Schmitt, *Physis*, 14 (1972), 267n; although Montaigne did not specify Borro, the attribution is generally accepted.

38. Quoted in Schmitt, *Physis*, 14 (1972), 268n, from Borro, *De peripatetica docendi atque addiscendi metodo* (Florence, 1584), 20; Gilbert, *Concepts* (1960), 186–92.

39. Duhem, *Etudes* (1984), 3, 205–6.

40. Quoted from Naudé, who may not be always reliable, by Spini, in Maccagni, *Saggi*, 3:2 (1972), 427.

41. Cf. Helbing, *Filosofia* (1989), 346–8, and in Laird and Rioux, *Mechanica* (2008), 188.

42. Borro, *De motu* (1576), 3q, 120–1, 126, 212, 217–24.

43. Ibid., 213–16.

44. Ibid., 126–32, 243–5.

45. Ibid., 201–8.

46. Borro, *Del flusso* (1583), a.5; on Giovanna's interest in astronomy, Strozzi, *Orazioni* (1635), 53.

47. Boro, *Del flusso* (1583), 1, 4, 9–11.

48. Ibid., 12–13, 14–15q.

49. Aristotle, *De caelo*, 1:1, 268ª–268ᵇ11.

50. Borro, *Del flusso* (1583), 16–18, 19q, 107ff. (the tides). The answers to the questions are on pp. 113–17.

51. Camerota and Helbing, *Early sci. med.*, 5 (2000), 331–48.

52. De Pace, in *De motu* (1990), 7–8, 15–16, 19–20, 25–32, 38–9, 47–52, 55.

53. Helbing, *Filosofia* (1989), 28–36, 44–7, 352–4, 360–1, 365–9; Favaro, "Libreria" (1886), nos. 51, 259, 260.

54. Helbing, *Filosofia* (1989), 58, and *Physis*, 18 (1976), 49, 52.

55. Other topics were the usual problems *de motu* and the theory of floating bodies. Helbing, *Filosofia* (1989), 70, 350; Op. 4:80; Section 5.3 below.

56. Helbing, *Physis*, 18 (1976), 47.

57. Helbing, *Filosofia* (1989), 201–2, 358, 364–5, 458.

58. Ibid., 196–200; Helbing, in Bucciantini and Torrini, *Diffusione* (1997), 62q.

59. Cesalpino, *Peripateticae quaestiones* (1571), Bk 3, Qu. 5; Helbing, *Filosofia* (1989), 200; Viviani, *Cesalpino* (1922), 108–9.

60. Helbing, *Filosofia* (1989), 46, 62–3, 348–9, 362; Spini, in Maccagni, *Saggi*, 3:2 (1972), 427q.

61. Helbing, *Filosofia* (1989), 48–9, 53, and *Physis*, 18 (1976), 42–4; Weinberg, *History* (1961), 1, 554.

62. Thorndike, *Sphere* (1949), 5–14, 42, 118–42 (the entire English translation); GG to V. Galilei, 15 Sep. 1590 (10:44–5).

63. Purnell, *Physis*, 14 (1972), 273–5; Mazzoni, *De triplici hominum vita* (Caserna, 1576); Vannucci, *Dante* (1965), 63–6.

64. Mazzoni, *Praeludia* (1597); Purnell, *Physis*, 14 (1972), 244–7.

65. Crombie, in Righini Bonelli and Shea, *Reason* (1975), 158, 160–2; Wallace, in Butts and Pitt, *New perspectives* (1978), 98–100, 107–9, 115–17, 121–3; Carugo and Crombie, IMSS, *Ann.*, 8:2 (1983), 27–8.

66. Carugo and Crombie, IMSS, *Ann.*, 8:2 (1983), 35–6.

67. Wallace, in Maurer, *St Thomas* (1974), 2, 297, 299, 320–1, 324.

68. Adapted from Koyré, *Etudes* (1939), 1, 18–41.

69. Drake, *Stud. hist. phil. sci.*, 17 (1986), 430–1, 439–40; Fredette, LCF, 168–75.

70. Favaro, in Op. 1:248. Drake, *Stud. hist. phil. sci.*, 17 (1986), 439, dates the dialogue to 1586–7 mainly because the manuscript has a watermark not otherwise

represented in Galileo's papers and he spent some of that time in Siena. Fredette, *Physis*, 14 (1972), 328, 333, suggests 1590. The latest and most thorough investigation, Giusti, *Nuncius*, 13:2 (1998), 435–42, puts the dialogue and the treatise in the Pisan period.

71. Valerio to GG, 4 Apr. 09, and Cigoli to GG, quoting Valerio, 9 Apr. 09 (10:240–1); Favaro, *Amici* (1983), 1, 6, 379; Napolitani, in Lomonaco and Torrini, *Galileo e Napoli* (1987), 161, 167–70.

72. Serassi, *Vita* (1710), 91q, 107–9.

73. Galileo took copious notes from Jesuit handbooks and commentaries on Aristotle's *Libri naturales*; Carugo, *Hist. techn.*, 4 (1987), 323–5, 328–32, identifies some.

74. On Galileo's borrowings from Borro, De Pace, *De motu* (1990), 56, 63.

75. For efficiency, some of the dialogue is given here in paraphrase though attributed to the relevant speaker. Direct quotations are indicated by quotation marks.

76. Op. 1:367–8. Cf. Galileo, "De motu" (1960), 123–9, and Moody, in Wiener and Noland, *Roots* (1957), 181–96.

77. Op. 1:368.

78. Ibid., 269–72, 377q; cf. "De motu" (1960), 76–84.

79. In the treatise "De motu" (1960, p. 15), Galileo set a higher value on this argument: earth is denser than other elements in order to put more matter in a more restricted space; "if not necessary, it [is] at least useful and reasonable."

80. Op. 1:374–80. Alexander refers to "his" invention of the *bilancetta* when Dominicus mentions the story of the bathtub. Cf. "De motu" (1960), 16–23.

81. Op. 1:388–93. Cf. "De motu" (1960), 94–100.

82. Op. 1:394–401. Cf. Galileo, "De motu" (1960), 41–50, 61–3.

83. Op. 1:402–4.

84. Op. 1:405–8.

85. Mazzoni, *Praeludia* (1597), 187q, 190q, 191, 192q; De Pace and Barozzi, *Matematiche* (1993), 327–33. Cf. Koyré, in Wiener and Noland, *Roots* (1957), 167–8.

86. Mazzoni, *Praeludia* (1597), 193–4.

87. Mazzoni, *Praeludia* (1597), 194; Galileo, "De motu" (1960), 70–1, 97.

88. GG to Mazzoni, 30 May 1597 (2:197). Cf. Purnell, *Physis*, 14 (1972), 293, and De Pace and Barozzi, *Matematiche* (1993), 311n, 325–6.

89. Borro, *De motu* (1576), 124–5, 251–2.

90. Galileo, "De motu" (1960), 79, 116–23; Fredette, *Physis*, 14 (1972), 336, 341, 345–7, and *LCF*, 173.

91. Galileo, "De motu" (1960), 63–5.

92. Galileo, "De motu" (1960), 67–9.

93. Galileo, "De motu" (1960), 72–6. Cf. Helbing, in Bucciantini and Torrini, *Diffusione* (1997), 49–50, 60, on Platonic ideas about persistent circular motion.

94. Cf. Festa and Roux, *Galilaeana, 3* (2006), 140–3, and Büttner, ibid., 5 (2008), 34, 57–60 (flywheels).

95. Galileo, "De motu" (1960), 101, 105, 107q, 109–10.

96. Nagler, *Theater* (1964), 58–61.

97. Palisca, *Camerata* (1989), 6q; Saslow, *Wedding* (1996), 28; Nagler, *Theater* (1964), 75–6.

98. Saslow, *Wedding* (1996), 18, 28, 42–5, 76–8, 95–9, 102–3, 126–7, 195; Nagler, *Theater* (1964), 70–3; Bartelà and Tofani, *Feste* (1969), 240–2.

99. Palisca, *Camerata* (1989), 208, 211, 221 (quote from Strozzi); Beijer, *Visions* (1956), fig. 7, opp. 413; Nagler, *Theater* (1964), 85–7.

100. Saslow, *Wedding* (1996), 50–1, 54, 140, 161–5, 171; Nagler, *Theater* (1964), 91–2.

101. Saslow, *Wedding* (1996), 144–5.

102. Op. 1:58, 72, 80, 99.

103. Saslow, *Wedding* (1996), 59q, 61–3, 177.

104. Op. 1:65, 73.

105. GG to Cappone Capponi, 2 June 1590 (10:43–4); Saslow, *Wedding* (1996), 63 (master tailor at 35 lire/month). In 1590/1 Galileo missed 24 lectures for each of which he forfeited one lira; Caffarelli, *Galileo* (2004), 35, 37.

106. GG to his father, 26 Oct. 90 (10:46); testimony re business (19:63, 81).

107. Segrè, *Stud. hist. phil. sci.,* 20 (1989), 436–8, 442–4, 450; Viviani, as translated by Cooper, *Aristotle* (1935), 26.

108. Viviani, *Vita* (2001), 32–3 (19:608); Lazzarini, *Lampada* (1998), 11, 14–16, 22–3; Caffarrelli, *Galileo* (2004), 22.

109. Ongaro, in Santorio, *Medicina* (2001), 8–10, 25–6; Proverbio, *NCCS,* 67; Santorio, *Methodus vitandorum errorum omnium qui in arte medica contingunt* (1602).

110. Notably Koyré, notably in Wiener and Noland, *Roots* (1957), 151, 171–5.

111. Settle, in McMullin, *Galileo* (1967), 322, 325–6, 331; Camerota and Helbing, *Early sci. med.,* 5 (2000), 345, 364.

112. Chiari, 3–12; Vaccalluzzo, *Galileo letterato* (1896), 114–16, 117q.

113. Reynolds, *Crit. lett.,* 8 (1980), 428.

114. Ibid., 419, 423–31, and *Ren. hum.* (1997), 4, 27, 29; Graf, *Cinquecento* (1926), 139–78 ("I pedanti").

115. The original has "like two chickens in the market." Reynolds, *Nuncius,* 17:1 (2002), 55–62, gives a literal translation of the entire *capitolo* in blank semiverse.

116. Capponi was *prevveditore generale* from Nov. 88 to Oct. 91; Favaro, *Scampoli* (1992), 1, 376.

117. Mercuriale to GG, 3 Mar. 93 (10:54–5); Favaro, Ist. ven., Atti., 72 (1917), 96, = Amici, 2, 1634.

118. G. del Monte to GG, 10 Apr. 90, 21 Feb. 92, and 10 Jan. 93q (10:42–3, 47, 53–4); Viviani, Vita (2001), 38, 90nn57–8.

119. Giovanni Uguccioni (Tuscan agent in Venice) to Belisario Vinta (Tuscan secretary of state), 21 Sep. 92, and to Grand Duke Ferdinando, 26 Sep. 92 (10:49, 50); Benedetto Giorgio Zorzi (Venetian official) to GG, 12 Dec. 92 (10:51).

Chapter 3: Life in the Serenissima

1. Kuhn, Ven. Arist. (1996), 131, 474; Poppi, Ricerche (2001), 168.

2. Cremonini, Orazioni (1998), 13–45; cf. Poppi, Ricerche (2001), 171–6.

3. Cremonini, Orazioni (1998), 46–9, and "Le nubi," quoted by Montanari, in Cesare Cremonini (1990), 136.

4. Favaro, Studio (1878), 29–59, = Studio (1966), 1, 58–75; Kuhn, Ven. Arist. (1996), 99–103.

5. Cremonini, Orazioni (1998), 59–69.

6. Piaia, in Santinello, Galileo (1992), 79–81; Sangalli, in Riondata and Poppi, Cremonini (2000), 1, 207–9.

7. Intus ut libet, foris ut moris est, attributed to Cremonini by Naudé; Bosco, in Cesare Cremonini (1990), 251. Kuhn, in Riondato and Poppi, Cremonini (2000), 153, 162–4, ascribes four masks to Cremonini without unveiling what if anything lay under them.

8. Sangalli, Acc. pat., Atti mem., 90 (1997/8), 249, 338–53.

9. Possevino, Bibl. sel. (1593), 2, 182–201, (1607), 2, 223, 240, and Coltura (1598), 100, 108; Biondi, in Brizzi, Ratio (1981), 44–7; Balsamo, Possevino (2006), 71.

10. Cantimori, in Maccagni, Saggi, 3:2 (1972), 411–12.

11. Possevino, Coltura (1598), 61, and Bibl. sel. (1593), as cited by Biondi, in Brizzi, Ratio (1981), 55.

12. Possevino, Bib. sel. (1593), 2, 61, quoting from De recta philosophandi ratione (1577) by the cardinal bishop of verona, Agostino Valier. Cf. Biondi, in Brizzi, Ratio (1981), 49.

13. Possevino, Bibl. sel. (1593), 2, 107–8, 120–1 (1603), 54q; Biondi, in Brizzi, Ratio (1981), 60.

14. Biondi, in Brizzi, Ratio (1981), 69–71, 73q; Possevino, Coltura (1598), 108 ("imagini di nude done, o anco di cose più vergognose").

15. Donnelly, Rocky Mtn. Med. Ren. Assn., Jl., 3 (1982), 158, 159q, 160.

16. Possevino, Coltura (1598), 86–7; Blum, Nouv. Rep. Lett., 1983:1, 116–26; Judith, chs. 10–13.

17. Possevino, *Bibl. sel.* (1583), 175–7; Tessari, *AHSJ*, 52 (1983), 251–3; Balsamo, *Possevino* (2006), 179–81. Possevino had the expert advice of Clavius in drawing up his section on mathematics.

18. Donnelly, *AHSJ*, 55 (1986), 3–4, 16–23 (Possevino probably was of Jewish descent); Balsamo, *Possevino* (2006), 41–6.

19. Pinelli to GG, 3, 9, 25 Sep. 1592 (10:47–8, 50); Raugei, *Pinelli* (2001), 1, xiv–xviii.

20. Cozzi, *Ambiente* (1997), 218–21.

21. Favaro, *Scampoli* (1992), 1, 106–8, and *Amici* (1983), 2, 938–41, 625–6; Raugei, *Pinelli* (2001), xix.

22. Favaro, *Studio* (1966), 1, 259, 2, 55 (Ghetaldi), 151–4 (Badouère), 63 (Pignoria), and *Amici* (1983), 2, 946–57 (Segeth). Cf. Camerota, 81.

23. Favaro, *Scampoli* (1992), 1, 157–9; Section 5.1 below.

24. Stella, in Santinello, *Galileo* (1992), 308–9, 317–19; Rose, *Physis*, 18 (1976), 117–18, 126; Thoren, *Lord* (1990), 97.

25. Op. 10:51–2.

26. Op. 19:119–20; Favaro, *Studio* (1966), 1, 113; Section 3.2 below.

27. Quoted by Carugo, in *Storia*, 4/2 (1984), 185.

28. Quoted by Stocchi, in *Storia*, 4/2 (1984), 48–9.

29. Bosco, *Cesare Cremonini* (1990), 272, 281–2. Cf. Poppi, *Intro.* (1970), 37–8, and Schmitt, *Cremonini* (1980), 13, 15.

30. Kennedy, *Vivarium*, 18 (1980), 144 ("Here lies everything that was Cesare Cremonini"). Bosco, in *Cesare Cremonini* (1990), 255, 257, 263–4, 267, and Stella, *Acad. pat., Atti mem., 103* (1990/1), 64, for accusations of libertinism.

31. Kennedy, *Vivarium*, 18 (1980), 150–7; Davi, in Riondato and Poppi, *Cremonini* (2000), 1, 125, 127, 129; Kuhn, *Ven. Arist.* (1996), 475.

32. Introduction to *De anima*, and "Proteste" to *Clorinda e Valliero* (1623), quoted by Montanari, in *Cremonini* (1990), 144–5, 128–9, resp.

33. Cf. Schmitt, *Cremonini* (1980), 8–9.

34. Montanari, in *Cesare Cremonini* (1990), 127–33; Kuhn, in Riondato and Poppi, *Cremonini* (2000), 1, 158, 161, and in *Ven. Arist.* (1996), 113, 126, 484–6; Bosco, in *Cesare Cremonioni* (1990), 269.

35. Op. 2:212–19; Wallace, *Sources* (1984), 255–60.

36. Op. 2:220–3.

37. Quoted by Carugo, in *Storia*, 4/2 (1984), 174–5.

38. Op. 2:223–4.

39. Op. 2:225–55.

40. Moletti, in Laird, *Unfinished mechanics* (2000), 87–93, 125 (text of 1576); Laird, ibid., 18–19, 23, 25–32; Helbing, *LCF*, 217–36.

41. Aristotle, *Mech.*, 847^{b}17– 848^{a}15; Vilain, in Laird and Roux, *Mechanics* (2008), 154, 166–7.

42. Moletti, in Laird, *Unfinished mechanics* (2000), 75, 83; Section 4.2 below.

43. Galileo, *Mech.* (1960), 148–9, 153–6, 159.

44. Ibid., 163–76; cf. Camerota, 88–94.

45. Galileo, *Mech.* (1960), 180; *TNS*, Day 5; Section 8.2 below.

46. Del Monte, *Mech.* (1969), quotes on 241, 247, 243, resp.; Mamiami, *Elogi* (1828), 60–3, 79–80.

47. Del Monte, *Mech.* (1969), 259–304.

48. Ibid., 310–17.

49. Galileo, *Mech.* (1960), 167–9. For the odd case (Figure 3.6a), $X = P = W/3$ and the "lever arm" CD is in equilibrium between the moments (taken around C) of $(W–X)/2$ and P; for the even case (Figure 3.6b), $X = 2P$ and CD again balances between the moments $(W–X)/2$ and P.

50. Galileo, *Mech.* (1960), 163–79, 77q; Benedetti, *Speculations* (1969), 191–2.

51. Cf. Galluzzi, *Momento* (1979), 203–26.

52. Lane, *Venice* (1973), 362–3, 373–4; Tenenti, *Piracy* (1967), 110–11, 124–5; Naish, in Singer, *History*, 3 (1957), 471–4.

53. Carugo, in *Storia*, 4/2 (1984), 176; Tafuri, *Venice* (1995), 121, 131–5.

54. Aristotle, *Mech.*, 850^{b}10–27, to explain why rowers amidships are more effective than others (greater leverage as more oar is within the ship there).

55. GG to Contarini, 22 Mar. 1593, and reply (10:55–7, 57–60); Renn and Valleriani, *Nuncius*, 16 (2001), 488; see also Section 2.1 above (the scale model of Hell).

56. Lane, *Venice* (1973), 362, 364 (100 galleys in two months in 1570).

57. Camerota, 86–7; Op. 19:126–7. The Venetian authorities granted the patent in February 1594.

58. Favaro, *Padova* (1968), 85–6, and *Scampoli* (1992), 1, 186. The marriage contract was signed at the villa of Francesco Contarini.

59. Favaro, *Arch. ven.*, 5 (1893), 201–5; Cozzi, *Doge* (1958), 57, 49n, *Sarpi* (1979), 134, 141–2, and in Sarpi, *Pensieri* (1976), xxviii–xxix.

60. Bouwsma, *Venice* (1968), 233–35, 244–8, 251–2; Cozzi, *Doge* (1958), 218–24.

61. Bouwsma, *Venice* (1968), 236–7.

62. Cozzi and Cozzi, in *Storia* 4/2 (1984), 3. "Servite" refers to the order of the Servi di Maria.

63. Favaro, *Studio* (1966), 1, 55–6, and *Amici* (1983), 2, 914–15, 918–21, 929–30.

64. Anon., *Del genio* (1785), 1, 51–3, on Sarpi's interest in Viète.

65. Cozzi, in Sarpi, *Pensieri* (1976), xix (quote, from Fabricius)–xx, xlii–xliv; Della Porta, *Nat. magick* (1658), 190.

66. Cozzi, in Sarpi, *Pensieri* (1976), xiii, xxiii–xxiv (quote), xxxiii, xxxvi–xxxvii; Cozzi and Cozzi, in *Storia*, 4/2 (1984), 6–9. Cf. Bouwsma, *Venice* (1968), 532–3.

67. Cozzi, in Sarpi, *Pensieri* (1976), xviii–xix, xxv, and *Sarpi* (1979), 161.

68. Quoted by Spini, in Maccagni, *Saggi*, 3:2 (1972), 417; Bouwsma, *Venice* (1968), 516–19.

69. Micanzio, quoted in Cozzi, in Sarpi, *Pensieri* (1996), xxxix–xl; Wootton, *Sarpi* (1983), 21–2, 28, 36; Sarpi, *Opere* (1969), ed. Cozzi, 90–3, and *Arte di ben pensar*, in Sarpi, *Scritti* (1951), 140–1.

70. Cozzi, *Sarpi* (1979), 157, 161–2; Favaro, *Studio* (1966), 1, 152, 176.

71. Sagredo to Mark Welser, 4 Apr. 14 (12:45); Favaro, *Studio* (1966), 2, 77, 86.

72. Favaro, *Arch. ven.*, 4 (1902), 328–9 (Jesuit joke); Op. 12:492, 497 (art); ibid., 200 (Sagredo to GG, 17 Oct. 15q), 349 (20 Oct. 17), 416 (13 Oct. 18), amours.

73. Sagredo to GG, 23 Nov. 06 (10:164); Wilding, *Galilaeana*, 3 (2006), 237–45.

74. A bon mot adopted from Dante, *Purgatorio*, 25. 76–8; Favaro, *Scampoli* (1992), 2, 746–7. Cf. GG to anon., 27 Oct. 06 (10:163), suggesting payment for an instrument in good wine.

75. Chemello, in *Corte* (1980), 2, 127–8; Bassanese, *Texas stud.*, 30:3 (1988), 296–9.

76. Graf, *Attraverso* (1926), 188, 191; Rosenthal, *Hon. court.* (1992), 65, 72–3.

77. Rosenthal, *Hon. court.* (1992), 102–11; *Dangerous beauty*, Regency Enterprises and Warner Brothers, 1998.

78. Rosenthal, *Hon. court.* (1992), 15, 66, 83–5, 89; Zannini, *Motivi* (1982), 22–5, 36–8; Chemello, in Fonte, *Merito* (1988), x–xii; Ruggiero, *Boundaries* (1985), 33–48.

79. Chojnacka, *Working women* (2001), 55.

80. Rosenthal, *Hon. court.* (1992), 76–8, 82, 161–2. The fee of 2 scudi comes from a guidebook of 1576 reprinted in Barzaghi, *Donne* (1980), 155–67.

81. Rosenthal, *Hon. court.* (1992), 21–3, 268n44; Fortini Brown, *Private lives* (2004), 161–3.

82. Fortini Brown, *Private lives* (2004), 167; Bassanese, *Texas stud.*, 30:3 (1988), 297.

83. Bassanese, *Texas stud.*, 30:3 (1988), 302–6, 310; Richter, *Forum italicum*, 3:1 (1969), 26.

84. Rosenthal, *Hon. court.* (1992), 89–98.

85. Vianello, *Stud. fil. ital.*,14 (1956), 217–18, 229, 239; cf. Daniele, in Riondato and Poppi, *Cremonini* (2000), 27.

86. Battistini, *Intro.* (1989), 10; Vianello, *Stud. fil. ital.*, 14 (1956), 244–9, and *Lett. ital.*, 6 (1954), 283.

87. Vianello, *Stud. fil. ital.*, 14 (1956), 233–5, 242–3; Galileo, in Chiari, "Postille al Petrarca," at 105. 72, 64. 13, 262. 5–6. Cf. 40. 12–14, 43. 14.

88. Rosenthal, *Hon. court.* (1992), 80.

89. Favaro, *Arch. ven.*, 4 (1902), 318 (12:139).

90. Chiari, 19–20, 22.

91. Chiari, 13–14.

92. Pictures in Fortini Brown, *Private lives* (2004).

93. Della Porta, *Nat. Magick* (1658), 233–4.

94. Chiari, 30–2.

95. Chiari, 33–44.

96. Vaccalluzzo, *Galilo lett.* (1896), 100–1, and Del Lungo, *Patria italiana* (1909), 1, 352, respectively.

97. Firpo, *Giorn. crit. fil. ital.*, 35 (1956), 544–6; Aquilecchia, in Torrini, *Della Porta* (1990), 219; Clubb, *Della Porta* (1965), ch. 7. The evidence for the meeting of 1593 is a lost letter from Campanella dated 1636 and known only from a copyist's summary. The original went astray after its theft by a greater sinner than any of Campanella's quartet, Guillaume Libri, who used his position as overseer of the archives of France in the 1840s to steal whatever caught his eye.

98. Firpo, *Giorn. crit. fil. ital.*, 20 (1939), 7–14, 21, and *DBI*, 17:876–7.

99. Clubb, *Della Porta* (1965), 33–4.

100. Favaro, *Studio* (1966), 2, 41–3; Agostino da Mula to GG, "at the Santo," 3 July 1599 (10:74), is the evidence that GG by then occupied the property in Via Vignoli.

101. GG to Girolamo Quaratesi, in Florence, 24 Aug. 07 (10:178–9). Galileo started to take paying guests in 1602 according to Favaro, *Studio* (1966), 2, 79.

102. Valleriani, *LCF*, 285–6, 290. During the time Galileo had 28 students he sold 23 compasses.

103. Bellinati, in Santinello, *Galileo* (1992), 258 (rent); Lazzarini, in ibid., 169–70; Bedini, in *Galileo* (1967), 90–3 (Mazzoleni); letters to GG from Antonio Quirini, 24 Aug., and Sagredo, 1 Sep. 1599 (10:76–8).

104. Camerota, 113–14; Op. 19:147–9.

105. Estimated on the assumption that Galileo had half a dozen or more people in service (instrument maker, copyist, factotum, housekeeper, maids, gardener) and that a family of four needed a minimum of 25 or 30 scudi a year. Cf. Kuhn, *Arist. Ven.* (1996), 76, 78.

106. Letters from Livia, 1 May, and from Giulia (GG's mother), 29 May 1599 (10:60, 61).

107. Op. 19:571–3.

108. GG to his mother, 25 Aug. 1600 (10:81–2); Girolamo Mercuriale to GG, 29 May 1593 (10:74–5).

109. GG to his brother, 20 Nov. 1601 (10:84–5); Bellinati, in Santinello, *Galileo* (1992), 261 (Livia's wedding); Camerota, 111–12; Op. 10:88–9, 103–4 (two advances, May 1602 and Feb. 1603).

110. Sagredo to GG, 12 Apr. 04, 26 Apr. 08 (10:105, 203). Galileo should have received his third six-year contract in 1604, but the Serenissima pleaded shortage of funds.

111. Favaro, *Scampoli* (1992), 1, 186–90; Michelangelo Galilei to GG, 4 Mar. 08 (10:192–3). Galileo had the Venetian suit quashed by friends who managed to switch the jurisdiction to Florence.

112. Martini, Inst. ven. sci., *Atti*, 145 (1986/7), 304, 313–15, 322–5; Cowan, in Kittell and Madden, *Venice* (1999), 283–6, 289–90; Waterworth, *Canons* (1848), 202–3, 270–1.

113. M. Galilei to GG, 4 Nov. 08 (10:193); cf. Spini, in Maccagni, *Saggi*, 3:2 (1972), 423.

114. Favaro, *Studio* (1966), 1, 147, 2, 151–3, and *Mente e cuore*, 1881, 6.

115. Favaro, *Mente e cuore*, 1881, 4–6; letters to GG from Sagredo, 18 Oct. 02, and Gualterotti, 29 Mar. 08 (10:96, 198); Landolfi, *Stud. seic.*, 29 (1988), 131–2.

116. Cf. Campion and Kollerstrom, in *Galileo's astrology* (2003), 97.

117. Cf. Brady, in *Galileo's astrology* (2003), 116–24, 132–5.

118. Edwards, in *Galileo's astrology* (2003), 145–6; Swerdlow, *JHA*, 35 (2004), 136–40, for the charts.

119. Chiari, 15.

120. Favaro, *Studio* (1966), 2, 158.

121. Ptolemy, *Tetrabiblos* (1964), 333, 347, 357, 359.

122. Moletti, as quoted in Carugo, in *Storia*, 4/2 (1984), 172.

123. Francesco Rasi to GG, 28 Jan. 13 (11:472–3q); letters to GG from Franciotto Orsini, 24 Aug. 13, and from Cardinal Alessandro d'Este, 2 Mar. 18 (11:556, 12:375), asking for astrological advice.

124. Ernst, NCCS, 265–6; Sarpi to Groslot de l'Isle, 7 July 09, in Sarpi, *Pensieri* (1996), 381q.

125. Raugei, *Pinelli* (2001), xxvii–xxx; Favaro, *Studio* (1966), 2, 57–9.

126. Gualdo to Querenghi, 6 Nov. 1599, in Motta, *Querenghi* (1997), 151, 155q.

127. Benetti, in Billanovich, *Viridarium* (1984), 320, 326, 329.

128. Motta, *Querenghi* (1997), 192–3, 196, 199–200, 204–10; Grafton, *Cardano's cosmos* (1999), 134–55.

129. Gamba and Rossetti, *Giornale* (1999), 3q, 4.

130. Ibid., 14–15, 35.

131. Ibid., 12, 19; Maggioli, *Soci* (1983), t.p., 136; Motta, *Querenghi* (1997), 164.

132. Boethius, *Consolation*, 3:10bis. 6, tr. "I.T.," 277.

133. *Odyssey*, 13.110–12, tr. Butcher and Lang, 198.

134. Motta, *Querenghi* (1997), 159–60; Riondato, in Riondato and Poppi, *Cremonini* (2000), 15, 17.

135. Benzoni, in Riondato, *Accademie* (2001), 23, 45, 55.

136. Gilbert, *De magnete* (1600), 5.5, tr. Thompson, 195.

137. Sarpi, *Pensieri* (1996), 170, 210, 510, 578–9; Cozzi, in Sarpi, *Pensieri* (1976), xxxviii.

138. Sarpi to GG, 2 Sep. 02 (10:91–2); Gilbert, *De magnete* (1600), 5.6–8, tr. Thompson, 200–2.

139. Sagredo to GG, 30 Dec. 02 (10:100–1).

140. Sagredo to GG, 8 Aug. and 28 Sep., 18 Oct. 02 (10:89, 96).

141. Correspondence between GG and officials in Florence (Curzio Picchena, Belisario Vinta), 16 Nov. 07, 4 and 17 Jan., 8 Feb., 14 and 22 Mar. 08 (10:184–90, 194–8).

142. Gilbert, *De magnete* (1600), 2.17–20, 25, tr. Thompson, 86–8, 92–4. Galileo had the magnet for four or five months. Cf. Loria, in Maccagni, *Saggi*, 3:2 (1972), 225.

143. GG to Vinta, 4 Apr. (1st quote) and 3 May 08 (2nd and 3rd) (10:200, 205–9).

144. Stella, Acc. patr., *Atti mem.*, 95 (1982/3), 176–9, and in Santinello, *Galileo* (1992), 314–15.

145. Maschietto, in Santinello, *Galileo* (1992), 434–8; Masetti Zannini, *Vita* (1961), 12, 18; Section 4.1 below.

146. Galileo, *Operations* (1978), 92.

147. Ibid., 43–51; Bedini, in McMullin, *Galileo* (1967), 262–8; Schneider, *Proportionalzirkel* (1970), 5–12.

148. Hulsius, *Grund. Bericht* (1604), 6–11; Rose, *Physis*, 10 (1968), 62–4; Schneider, *Proportionalzirkel* (1970), 45–8.

149. Drake, in Galileo, *Operations* (1978), 9–22.

150. Galileo, *Operations* (1978), 52–9.

151. Ibid., 60–4.

152. Ibid., 65–72.

153. Ibid., 73–91.

154. Drake, in Galileo, *Operations* (1978), 23–4.

155. Camerota, 124–30.

156. Drake, in Galileo, *Operations* (1978), 24–5; Milani, in *Galileo* (1992), 183–4. Cf. Cosmo Pinelli (Pinelli's nephew) to GG, 3 Apr. 99 (10:73).

157. Galileo, *Difesa* (1607), in Op. 2:515–601.

158. Quaranta, in Santinello, *Galileo* (1992), 218–21; Cozzi, *Sarpi* (1979), 173–4; Schneider, *Proporzionalzirkel* (1970), 21–2; Silvio Piccolomini to GG, 8 Oct. 07, and Duodo to GG, 29 June 09 (10:181, 247).

159. Poppi, *Cremonini* (1993), 9–11, 51–4 (Pagnoni's testimony); Black, *Inquisition* (2009), 62, 74–5 (confession and inquisition).

160. Poppi, *Cremonini* (1993), 54.

161. Poppi, *Cremonini* (1973), 55 (document of 5 May 04).

162. Spini, *Ricerca* (1983), 161–6, 389–91, and *Galileo* (1996), 23–30; Cozzi, *Sarpi* (1979), 153–5; Op. 7:641 ("animalaccio").

163. Sagredo to GG, 7 Feb. 15 (12:139); Spini, *Ricerca* (1983), 155–7; Spruit, in Riondato and Poppi, *Cremonini* (2000), 1, 95.

164. Poppi, *Cremonini* (1993), 56. Cf. Black, *Inquisition* (2009), 72, 117.

165. Spruit, in Riondato and Poppi, *Cremonini* (2000), 1, 196–7; Kuhn, *Ven. Arist.* (1996), 481.

166. Poppi, *Cremonini* (1993), 9–15, 57–79, esp. 71; Favaro, *Padova* (1968), 24.

167. Bouwsma, *Venice* (1968), 39–54, 370–7.

168. GG to his brother, 11 May 06 (10:157–8).

169. Bouwsma, *Venice* (1968), 417.

170. Cremonini, *Orazioni* (1998), 109–47 (10 May 1606).

171. Bouwsma, *Venice* (1968), 420, 427, 454, 456–7, 461–3, 466, 522–3, 544–6.

172. Bouwsma, *Venice* (1968), 306–12, 469–70, 471q, 524–5.

173. Bouwsma, *Venice* (1968), 483, 488, 511–12.

174. Sarpi to "Rossi," 18 Aug. 09, in *Scelte lett.* (1848), 119–20, 140–1; cf. Ibid., 119–20.

175. Bouwsma, *Venice* (1968), 492–4, 505–11; Anon., *Del genio* (1785), 1, 166–75, 185–6.

176. Bouwsma, *Venice* (1968), 513, 515q. "Le locuste ed…loro capitano": Sarpi to Foscarini, 26 May 09, in Sarpi, *Scelte lett.* (1848), 127.

177. Sarpi to "Rossi," 18 Aug. 09, in Sarpi, *Scelte lett.* (1848), 141–2, and Sarpi, *Lettere* (1863), 1, 287 .

Chapter 4: Galilean Science

1. Shakespeare, *Taming of the shrew*, 1.1.

2. Op. 2:198.

3. Mazzoni, *Praeludia* (1597), 196–7; Aristotle, *Meteorol.*, 350^{a}28–34; Chiaramonti, *Opuscola* (1653), 297–9, 316.

4. Mazzoni, *Praeludia* (1597), 131–2.

5. Ibid., 201–2.

6. Segonds, in Kepler, *Secret* (1984), xxxvi–xxxvii; Kepler to Michael Mästlin, Oct. 1597, ibid., lv, n. 51, and Op. 10:69 (surprise at the name); Drake, *Galileo studies* (1970), 123–7; Biancarelli Martinelli, *Galilaeana*, 1 (2004), 174–9 (identification of Homberg).

7. GG to Kepler, 4 Aug. 1597 (10:68).

8. Kepler to Galileo, 13 Oct. 1597 (10:69–71); Bucciantini, *Galileo e Keplero* (2003), 49–68.

9. Bucciantini, *Galileo e Keplero* (2003), 74, 77–80; Kepler, *Harm. mundi* (1619), 5: proem., ed. Aiton et al., 391q.

10. Kepler, *Secret* (1984), 21–2; Imerti, in Bruno, *Expulsion* (2004), 45, 52–3.

11. Kepler, *Secret* (1984), 22–5, 26q.

12. GG to Kepler, 4 Aug. 97 (10:68); Herwart von Hohenburg to Kepler, 12 Mar., and reply, 26 Mar. 98, in Segonds, in Kepler, *Secret* (1984), xxix, and Op. 10:72, resp.

13. Sosio, in *Galileo e la cultura* (1995), 307; Sarpi, *Pensieri* (1996), 423–7, and Sosio, ibid., cli–clviii.

14. Drake, *Galileo studies* (1970), 201–4, gives the texts and strong reasons for assigning the invention to Galileo; Sosio, in *Galileo* (1995), 308, gives equally plausible reasons for thinking Sarpi the originator.

15. Drake, *Essays* (1999), 1, 355–6, 362.

16. Kepler to Herwart von Hohenburg, 26 Mar. 98 (10:72).

17. Kepler, *Mysterium*, chs. 20–1, = *Secret* (1948), 135–50.

18. Clavelin, *Philosophie* (1968), 181–5; Bucciantini, *Galileo e Keplero* (2003), xxii, 65–7, 94, and Camerota, 102–5, summarize the vexed question of the relation of Galileo's mechanics and astronomy during his Paduan time. They agree with Clavelin, *Philosophie* (1968), 156.

19. Circumference $C = v\tau$; $v/2 = b/\tau$; thus $C = 2b$, where b is the drop.

20. Cf. Barcaro, *NCCS*, 117–18. The documents in the case, which contain only one word, abbreviated, in a flurry of calculations, were discovered and first analyzed by Drake, *JHA*, 4 (1973), 174–91, who recognized their connection with Kepler but erred in reconstructing Galileo's arguments. Meyer, *Isis*, 80 (1989), 456–68, began the reassessment now elegantly reworked by Büttner, *LCF*, 391–401. Cf. Drake, *Essays* (1999), 1, 366, 375–7, and Bucciantini, *Galileo e Keplero* (2003), 105–15.

21. Bruce to Kepler, 15 Aug. 02 and 21 Aug. 03 (10:90, 104); Bucciantini, *Galileo e Keplero* (2003), 105–15.

22. Segonds, in Kepler, *Secret* (1984), xxxi–xxxiv.

23. Caspar, *Kepler* (1959), 96–102, 121–2.

24. Bucciantini, *Galileo e Keplero* (2003), 37, 42–8, 84–92.

25. Tycho to Pinelli, 3 Jan. 1600 (10:78–9). Cf. Drake, *Gal. studies* (1970), 130–1.

26. Tycho to GG, 4 May 1600 (10:79–80).

27. Bucciantini, *Galilo e Keplero* (2003), 85–7; Drake, *Galileo at work* (1978), 50.

28. Op. 2:278–9 (text of 1604).

29. Tengnagel to Magini, 1603 (10:104–5); Bucciantini, *Galileo e Keplero* (2003), 118; Drake, *Galileo studies* (1970), 126–31. Sarpi sought to defeat some of Tycho's arguments against Copernicus, not as a Copernican, but as a sceptic; Sosio, in Sarpi, *Pensieri* (1996), xcv, xcvi, clx, clxxi–clxxiii.

30. Op. 2:278.

31. Bucciantini, *Galileo e Keplero* (2003), 124–5, 129–38; Op. 2:281q, 282–84.

32. Drake, *Galileo studies* (1970), 125, and *Pioneer scientist* (1990), 131–2, and Wallace, *Galileo and his sources* (1984), 259–60, suggest the doubt; Bucciantini, *Galileo e Keplero* (2003), 119–27, rejects it.

33. Meli, *Nuncius*, 7:1 (1992), 27–8.

34. Del Monte to P.M. Giordano, 6 and 31 Dec. 04, in Arrighi, Acc. Lucchesi, *Atti*, 12 (1965), 194–5.

35. Same to same, 20 Jan. 05, ibid., 189; Meli, *Nuncius*, 7:1 (1992), 29–30.

36. Altobello to GG, 3 Nov. 04 (10:116–17); Drake, *Galileo against the philosophers* (1976), 8.

37. Letters to GG from, resp., Leonardo Tedeschi, 22 Dec, and Ottavio Brenzoni, 15 Jan. 05 (10:122–32, 138–41).

38. Kepler, *De stella nova* (1606), in Ricci, *Tre studi* (1994), 22.

39. Altobello to GG, 3 Nov. 04 (10:116–17q). Drake, *Galileo against the philosophers* (1976), 9–15, and Tomba, in *Cesare Cremonini* (1990), 87–8q, give reasons for assigning parts of Lorenzini's tract to Cremonini.

40. Newton, *Principia* (1687), Bk 3, Rule 3, ed. Cohen, 2, 795.

41. Kuhn, *Ven. Arist.* (1966), 437, 450–3, 457, 465, 468.

42. Capra, *Considerazione astronomica circa la nuova e portentosa stella* (1605); Lorenzini, *Discorso intorno alla nuova stella* (1605); Bucciantini, *Galileo e Keplero* (2003), 142; Camerota, 117–21.

43. Cf. Altobello to GG, late Nov. 04 (10:120), and Motta, *Querenghi* (1997), 127, who spies behind Cecco's satire "the initial symptoms of a deep and distinct cultural transformation."

44. Cremonini, *De Caelo* (1613), quoted by Kuhn, *Ven. Arist.* (1996), 453.

45. Lovarini, *Studi* (1965), 379–80, 385–6, 406, 409.

46. Ibid., 409; Camerota, 123–4; Favaro, *Galileo* (1881), 6; Milani, in Santinello, *Galileo* (1992), 189–91; Motta, *Aevum*, 67 (1993), 597–600; Pignoria to GG, 27 Dec. 19 (12:502), confirms Spinelli's authorship. Galileo needed a collaborator who spoke Padovan fluently.

47. Favaro, *Galileo* (1881), 20–8, 43–6; the excerpts quoted are from Drake's complete translation in *Galileo against the philosophers* (1976), 36–51.

48. Drake, *Galileo against philosophers* (1976), 27–31.

49. Quoted in Mauri, *Considerazioni* (1606), in Drake, *Galileo against the philosophers* (1976), 110–19, 122, 129.

50. Drake, ibid., 58, and Mauri, in ibid., 82, 85, 95. There is also the typically Galilean paradox that Aristotelians should be delighted with equants and eccentrics and other mathematical constructs because they allow preservation of the physical principle that celestial motions must be circular (102, 124).

51. Mauri, *Considerazioni* (1976), ibid, 76. There are also Mauri's preference for the generation of novae by conjunctions and his supposition that the moon has a dull light of its own (108, 106).

52. Delle Colombe to GG, 24 June 07 (10:176–7). Gualtieri has a much better claim than Galileo to have been the main target of delle Colombe's *Discorso*; Jacoli, *Bulletino*, 7 (1874), 381. Cf. Lovarini, *Studi* (1965), 408.

53. Drake, *Essays*, 1 (1999), 174–80; Renn et al., in Renn, *Galileo* (2001), 99–100.

54. Settle, *Science*, 133 (1961), 19–38; Naylor, *BJHS*, 7 (1974), 103–34, *Isis*, 66 (1975), 394–6, *Isis*, 71 (1980), 550–70, and *Ann. sci.*, 37 (1980), 371–7; MacLachlan, *Isis*, 66 (1975), 402–3.

55. E.g., Drake, *Galileo against the philosophers* (1976), 136–44, and in Coelho, *Music* (1992), 15.

56. E.g., Naylor, *Ann. sci.*, 34 (1977), 287–92, and many others.

57. E.g., Wallace, *Sources* (1984), 338–47.

58. Laird, *Unfinished mechanics* (2000), 101, 121, 125–7, 149.

59. This is the strategy of Renn et al., in Renn, *Galileo* (2001), 56–61, 80–5.

60. Sarpi, *Pensieri* (1996), nos. 535–8, 541–2, 547; Sarpi's editors claim these discoveries for him (ibid., 400). Cf. Renn et al., in Renn, *Galileo* (2001), 82–3; Sosio, in *Galileo* (1995), 302.

61. Wisan and Galluzzi also accept that the documents do not support unambiguous reconstruction of Galileo's discoveries in mechanics; Galluzzi, *Momento* (1979), 273–80.

62. See above, Section 2.3. Cf. the good résumé in Naylor, *Ann. sci.*, 37 (1980), 364–71.

63. Cf. Galluzzi, *Momento* (1979), xi–xii.

64. Cf. Settle, in McMullin, *Galileo* (1967), 324, 331–3.

65. Del Monte, note made between 1587 and 1592, inclusive. Renn et al., in Renn, *Galileo* (2001), 45, 56–60, make it plausible that Galileo and Guidobaldo did or discussed the roof experiment together. A slip in their translation (p. 45) and in Damerow et al., *Exploring* (1992), 337, which follows Naylor, *Isis*, 71 (1980), 551, makes hash of the last two sentences of Guidobaldo's text: they render *a fatto* as "at all" rather than "completely."

66. Naylor, *Physis*, 16 (1974), 327–33, 337.

67. Palladino, *Giorn. crit. fil. ital.*, 63 (1984), 378–80, and in Lomonaco and Torrini, *Galileo e Napoli* (1987), 388–9; Festa and Roux, *Galilaeana*, 3 (2006), 134–7; Boyer, in McMullin, *Galileo* (1967), 236–42.

68. Cf. Wisan, *NCCS*, 42.

69. *Elements*, 3:37. If you have forgotten your Euclid, draw CD and CG the bisector of AD. Then $\Delta DCG \sim \Delta AFC$ and AD:2AC = AC:AF.

70. Drake, *Galileo* (1976), 78–81; see Section 8.2 below. Neither proposition is true.

71. GG to Guidobaldo, 29 Nov. 1602 (10:97–100).

72. Cf. Naylor, in Gooday et al., *Uses* (1989), 118.

73. The mistake was to neglect rolling, which occurs at lower inclinations and slows the motion. Naylor, *Isis*, 67 (1967), 409–19.

74. Drake, *Galileo* (1978), 92–4. Drake, *Pioneer* (1990), 12–27, suggested, improbably, that Galileo discovered the t^2–square rule by experimental comparison of

lengths of times required to drop through h and $h/2$ in connection with determining the ratio of the period of a pendulum to the time of free fall through its length.

75. Cf. Sosio, in *Galileo* (1995), 284–7, 296–7; Renn et al., in Renn, *Galileo* (2001), 60–1.

76. GG to Sarpi, 16 Oct. 1604 (10:115–16), tr. after Drake, *Galileo* (1978), 103; Galluzzi, *Momento* (1979), 270–88.

77. Wallace, *Sources* (1984), 264–76.

78. Drake, *Galileo* (1978), 90, *Pioneer* (1990), 9–11, and in Coelho, *Music* (1992), 11–15; Naylor, *Ann. sci.*, 34 (1977), 373–5, *Ann. sci.*, 37 (1980), 373–7, and *Isis*, 71 (1980), 554–5.

79. Cf. Drake, *Pioneer* (1990), 102, 105–6, and *Galileo* (1978), 99–103.

80. GG to Guidobaldo, 29 Nov. 1602 (10:97–100); Renn et al., in Renn, *Galileo* (2001), 133–5; Sosio, in *Galileo* (1995), 288, 297.

81. Sarpi to GG, 9 Oct. 1604, and reply, 16 Oct. (10:114, 115–16).

82. For details of the following reconstruction, Naylor, *Ann. sci.*, 33 (1976), 154, 157–8, 166, and *Ann. sci.*, 34 (1977), 383–92.

83. Naylor, *Isis*, 67 (1976), 408–10, and *Isis*, 71 (1980), 553–7.

84. A body that descends an inclined plane of length k in time t will cover $2k$ in the same time if allowed to continue its motion along the horizontal.

85. Cf. Drake, *Pioneer* (1990), 110, 120–8.

86. GG to Baliani, 7 Jan. 39 (18:11–12), and to Carcavy, 5 June 37 (17: 89–90); Archimedes, *Works*, ed. Heath (1897), 154–5; Cesalpino, quoted by Helbing, in Laird and Roux, *Mechanics* (2008), 190.

87. We have again (see above, Section 4.1) $h_1 = \pi r_1$; instead of $\tau_1 : \tau_2 = \pi r_2 : h_2$, we have $\tau_1 : \tau_2 = \pi r_2 : \sqrt{(h_1 h_2)}$.

Chapter 5: Calculated Risks

1. *Galileo's astrology* (2003), 59–63, 147–9; *SN*, 31.

2. Mercuriale to GG, 29 Mar. 01 (10:83–4).

3. Letters to GG from Alessandro Sertini and Vincenzo Giugni, 16 Apr. and 4 June 05 (10: 142–4).

4. Giovanni del Maestro to GG, 15 Aug. 05 (10:146).

5. Asdruble Barbolani to Ferdinando, 10 June, to Vinta, 12 Aug., and to Vincenzo Giugni, 23 Sep.; Cipriano Saracinelli to GG, 26 May, 30 Sep., all 1606 (10:158–62).

6. Barbolani to Ferdinando, 29 Oct.; Vincenzo Giugni to GG, 5 Nov.; GG to Christina, 11 Nov.; Cipriano Saracinelli to GG, 5 Dec., all 1605 (10:147–51).

7. GG to Cosimo, 29 Dec. 05, and reply, 9 Jan. 06 (10:153–5).

8. GG to Christina, 8 Dec. 06 (10:164–6).

9. GG to Vinta, 4 Apr., 30 May, and 20 June 08 (10:199–200, 210–13, 215); Vinta to GG, 12 Apr. and 11 June 08 (10:200–1, 214–15).

10. Anon., in Targioni-Tozzetti, *Notizie* (1780), 2:1, 103, 107–8; *DBI*, 25:148; Costanzo, *Critica* (1969), 1, 148–50.

11. Cf. Biagioli, *Courtier* (1993), 106–12.

12. Gealasso Calderara, *Amazzone* (1985), 35–6, 57.

13. Ibid., 43–50, 54–5, 63 (quoting della Porta, *Physiognomonia coelestis* (Naples, 1603)).

14. GG to Christina, 19 Dec. 08 and 11 Feb. 09 (10:225, 227–8).

15. GG to Cosimo, 26 Feb., and to "Sig. Vesp.," ? Feb. 09 (10:230–3).

16. Letters to GG from Antonio de' Medici, 6 Mar., Cosimo, 7 Mar., Giovancosimo Geraldini, 12 Mar., Enea Piccolomini Aragona, 27 June, all 1609 (10:235–6, 239, 246–7).

17. Letters to GG from Duodo, 10 Mar., Cigoli, 22 May, Valerio, 4 Apr. and 30 May; GG to Riformatori, 9 Mar. and 4 Nov., all in 1609 (10:238, 239, 241, 244–6). On Sarrocchi, an orphan protected by a cardinal and raised in a nunnery, see Gabrieli, *Contributi* (1989), 1, 843q, 844, and Favaro, *Amici* (1983), 1, 8–14.

18. GG to A. de' Medici, 11 Feb. 09, and Valerio to GG, 4 Apr. and 18 July 09 (10:229–30q, 239, 248–9).

19. Sarpi to to F. Castrino, 21 July 09, in Sarpi, *Lettere* (1863), 1, 279, and *Lettere prot.* (1931), 2, 45.

20. Sarpi to Jérôme Groslot de l'Isle, 6 Jan. 09, in *Lettere* (1863), 1, 181, and *Lett. prot.* (1931), 1, 58.

21. Paolo Pozzobonelli, Padua, to GG, 12 Sep. 02 (10:93); Giulia Ammannati to Alessandro Piersanti, 9 Jan. 09, in Favaro, *Studio* (1966), 2, 228–9.

22. Reeves, *Glassworks* (2008), 34–46, 74–9, 165–6.

23. Montano, *Elucidationes in quatuor evangelia* (Antwerp: Plantin, 1575), p. 192, note e. "Hoc potuit effici prospectivae sive opticae artis vi, quam diabolus non ignorat; ut eadem arte à nobis conficiuntur inspicilla, quae longissimè distantes res oculis exactissimè subiicunt."

24. Pignoria to Gualdo (then in Rome), 1 Aug. 09, and Bartoli to Vinta, 22 and 29 Aug. 09 (10:250, 255).

25. Chevalley, *NCSS*, 172; Dupré, *Galilaeana*, 2 (2005), 146–8, 152–3; Sagredo to GG, 2, 6, 30 June, 7 July 12 (11:315, 331, 349–50, 356).

26. Dupré, *Galilaeana*, 2 (2005), 171–9, for fitting; Zik, *Nuncius*, 14:1 (1999), 41–51, 60–7, and 17:2 (2002), 459–63 (cut-and-try inspired by perspective theory).

27. Sarpi to Jacques Lechassier, 16 Mar. 10, in Anon., *Del genio* (1785), 1, 254; Sosio, in Sarpi, *Pensieri* (1996), clxvi–clxvii. Sosio, ibid., clxix–clxxv, gives further indirect evidence of Sarpi's involvement.

28. GG to Donà, 24 Aug. 09 (10:250–1). Notes for the letter to Donà, perhaps written during a conversation with Sarpi, survive; Drake, *Essays* (1999), 1, 389. Galileo, *Assayer*, 212, says he made the gift at the suggestion of a patron.

29. Sarpi to Groslot, 10 May 10, in Sarpi, *Lettere prot.* (1931), 1, 122: "The wonders that are discovered with this device belong to the profession of perspective, which includes the way vision is performed and the theory of spectacles for weak and short sight."

30. GG to Landucci, 29 Aug. 09 (10:253–4). Favaro regarded this letter, which is not in Galileo's hand, as a fake, but the information it contains agrees perfectly with all the other known facts—except Galileo's continuation of his quest for Cosimo's patronage.

31. Bartoli to Vinta, 29 Aug. 09 (10:255).

32. Piccolomini Aragona to GG, 29 Aug., and Strozzi to GG, 19 Sep. 09 (10:254–5, 259).

33. Bartoli to Vinta, 24 Sep,, 3 and 24 Oct., 31 Oct., and 7 Nov. (quote), all 1609 (10:260–1, 264, 267).

34. Van Helden, IMSS, *Ann.*, 2:1 (1976), 25–6; North, in Shirley, *Harriot* (1974), 146–7, 160.

35. Van Helden, in *SN*, 9–11. Galileo says that he began to view the moon when it was five days old (letter to A. de' Medici, 30 Jan. 10 (10:273)). The balance of probability identifies the lunation as that begun on 27 Nov.: Righini, *Contributo* (1978), 26–35, and in Righini-Bonelli and Shea, *Reason* (1975), 67–76; Drake, *Essays* (1999), 1, 381–2; Whitaker, *JHA*, 9 (1978), 155–69.

36. Gingerich and Van Helden, *JHA*, 34 (2003), 254, 260–2.

37. Egerton, *Heritage* (1991), 225, 244–5 (drawing); Dupré, *Nuncius*, 5:2 (2000), 557, 587–8 (commonality of perspective); Bertola, in Santinello, *Galileo* (1992), 267–75 (keenness of sight).

38. Bloom, *JHA*, 9 (1978), 117–18; Shirley, *Harriot* (1983), 249–67, 397–8; *DSB*, 6:124–9; Alexander, *Stud. hist. phil. sci.*, 29 (1988), 346–51, 358–64.

39. Bredekamp, *Galilei* (2007), 92–4, 212–13.

40. Bredekamp, *Galilei* (2007), 189–210.

41. Cf. Gingerich, in Righini-Bonelli and Shea, *Reason* (1975), 86–7.

42. Montgomery, *Scientific voice* (1996), 226–9; Dupré, *Nuncius*, 15:2 (2000), 553–5; *SN*, 47.

43. *SN* (1989), 51–3, and (1993), 110–11. Galileo must have meant that in proportion to their radii, his moon mountain was four times as high as the highest earth mountain.

44. Aristotle, *Meteorol.*, $345^{b}32$–$346^{b}15$.

45. Sarpi to Jacques Lechassier, 16 Mar. 10, in Sarpi, *Lettere* (1863), 2, 41–2, and to Groslot, in ibid., 71, and *Lett. prot.* (1931), 1, 122.

46. GG to Vinta, 30 Jan. 10 (10:280).

47. Letter of 21 Nov. 09 (10:268–9), complaining of no news for several weeks; Favaro, *Studio* (1966), 2, 225–6.

48. Ammannati to Piersanti, 9 Jan. 10 (10:279); Favaro, *Studio* (1966), 2, 228–9. Acquapendente's pills were a compound of aloe and rose water baked in the sun (19:201).

49. GG to Vinta, 30 Oct. and 20 Nov. 09, and to Christina, 16 Jan. and 11 Feb. 09 (10:263, 268, 226–8). Christina delivered; Vinta traced the defaulters on the loan to Poland; the evidence that Piersanti turned over the letters is their survival in Galileo's papers.

50. *SN*, 65 (fate). Around "opposition," when closest to earth, Jupiter, Mars, and Saturn (the "superior planets") appear to "retrograde," moving from east to west in the direction of the diurnal motion of the stars, whereas most of the time they move "directly," from west to east, in the direction of the sun's apparent path along the ecliptic. On the Copernican theory, their retrogradation is a consequence of the placement of the earth's orbit within that of the superior planets; on the Ptolemaic theory, the phenomenon is just a brute fact. See Section 7.2 below.

51. Cozzi and Cozzi, in *Storia*, 4:2 (1984), 23; G. Cozzi, *Paolo Sarpi* (1979), 79–87.

52. Glorioso to Terrentius, 29 May 10 (10:363–4); Favaro, *Amici* (1983), 1, 329–33, and *Studio* (1966), 2, 90–1, 256–8.

53. Drake, *Essays* (1999), 1, 387–90; *SN*, 64–9.

54. Dupré, in Folkerts and Kühne, *Astronomy* (2006), 361. Galileo also believed before 1610 that the stars reflect sunlight if indeed he was "Alimberto Mauro" (see Section 4.1 above); Drake, *Galileo against the philosophers* (1976), 90–1, 97, 120.

55. Reeves, *Painting* (1997), 23–38, and Egerton, *Heritage* (1991), 248–50, suggest that Galileo's training in perspective, which included rendering secondary light, assisted his recognition of the cause of moonglow.

56. Kepler to GG, 19 Apr. 10 (10:322, 327, 331–2), and in *Disc.* (1993), 20–2; Pantin, in ibid., lxviii–lxix.

57. Sarpi, *Pensieri* (1996), 33–8, esp. 38, and Sosio, in ibid., clxxii–clxxiv. Jacoli, *Bulletino*, 7 (1874), 386, 390, points out that Gualterotti explained moonglow as a consequence of earthshine in 1605 in a published discourse that Galileo might well have known.

58. Drake, *Essays* (1999), 1, 391–2; *SN*, 84.

59. Op. 10:280.

60. Ramsay and Licht, *Comet* (1997), 62–4.

61. Vinta to GG, 6 and 20 Feb., and GG to Vinta, 6 Feb. 10 (10:281–5).

62. *SN*, 31–2 (dedication); cf. Biagioli, *Courtier* (1993), 127–30.

63. Sarpi to Lechassier, 5 Apr. 10, in *Lettere* (1863), 2, 61–5; Sosio, in Sarpi, *Pensieri* (1996), clxx–clxxiii; Pignoria to Gualdo, 15 Oct. 09, 19 and 26 Sep. 10, 19 Jan. 11 (10:260, 434, 436, 11:28), and in Benetti, in Billanovich, *Viridarium* (1984), 323–4.

64. GG to Vinta, 13 Mar. 10 (10:289), and to Buonarroti, 4 Dec. 09 (10:271).

65. Vinta to GG, 19 and 27 Mar. 10 (10:302–3, 307–8).

66. Enea Piccolomini to GG, 27 Mar. 10 (10:305).

67. GG to Vinta, 19 Mar. 10 (unsent draft) (10:297–9); GG to Vinta, 19 Mar. 10 (10:299–302).

68. Letter to Vinta, in Drake, *Discoveries* (1957), 61; Pantin, in Kepler, *Disc.* (1993), xviii; GG to Matteo Carosio, 24 May 10 (10:357–8).

69. Drake, *Discoveries* (1957), 62–4.

70. Vinta to GG, 22 May and 5 June 10 (10:355–6, 369); Favaro, *Nuovi studi* (1891), 375–6.

71. GG to Vinta, 18 June, and to V. Giugnio, 25 June 10 (10:373–4, 379–82).

72. An example in point, an obstreperous critic in Mantua, refrained from attacking Galileo in the presence of the duke and the cardinal, but not otherwise. Biancani to Grienberger, 11 June 11 (11:126); cf. Op. 3:301–7.

73. Andrea Labia to GG, 29 May 10 (10:361), on behalf of Borghese; Cigoli to GG, 18 Mar. 10 (10:290–1), *re* del Monte; Giuliano de' Medici to GG, 6 Sep. 10 (10:427–8), *re* the emperor; Pantin, in Kepler, *Disc.* (1993), xvi–xvii, xxx–xxxiv. The queen received hers in September, acknowledged in Andrea Cioli to Vinta, 13 Sep. 10 (10:430).

74. Christina to V. Giugni, 18 Apr., and letters to GG from del Monte, 28 Apr., Tommaso Mermanni, 12 May, Berlinghiero Gessi *re* Scipione Borghese, 30 June, M. Galilei, 14 Apr., all 1610 (10:318, 343–4, 354, 385, 312–14).

75. M. Galilei to GG, 14 Apr. 10 (10:312).

76. Letter of 27 Mar. 10 (10:306).

77. Watson, *Hist. ophth.*, 54 (2009), 630–40; Pantin, in Kepler, *Disc.* (1993), lxx.

78. Letters to Kepler from Horky, 16 Apr., and Magini, 20 Apr., and Hasdale to GG, 31 May 10 (10:316, 341, 365–7); Pantin, in Kepler, *Disc.* (1993), xix, xxv–xxvi, lix–lxii; Hasdale to GG, 28 Apr. and 7 June 10 (10:344–6, 370).

79. Pantin, in Kepler, *Disc.* (1993), xl–li, lxiv–lxv.

80. Horky to Kepler, 27 Apr. 10 (10:342–3).

81. Mancini, Congresso, xx, *Atti* (1964), 285–94, gives a convenient though incomplete list of Galileo's physical problems drawn from his correspondence; GG to ?, 27 Oct. 06 (10:162–3q); GG to Christina, 8 Dec. 06, and to Vinta, 25 and 30 May 08 (10:164–5, 209, 212).

82. Kepler, *Disc.* (1993), 9–10.

83. Ibid., 9.

84. Letters to Kepler from Horky, 27 Apr., and Magini, 26 May, and Hasdale to GG, 31 May 10 (10:342–3, 359, 365).

85. Kepler to Magini, 10 May 10 (10:353), and *Disc.* (1993), 8–9.

86. Chevally, *NCCS*, 169–74; Rothman, *JHA*, 40 (2209), 409–11, on Kepler's strategic promotion of Galileo in the Copernican cause.

87. GG to Kepler, 19 Aug. 10 (10:422–3); Kepler, *Disc.* (1993), 40–1; Pantin, in ibid., lxxi–lxxvi; Van Helden, *Osiris*, 9 (1994), 12–13.

88. The news was out by the Feast of San Giovanni, 24 June. Antonio Santini to GG, 24 June, and Barbolani to Vinta, 26 June 10 (10:377–8, 384); cf. Del Lungo, *Patria* (1912), 2, 243.

89. GG to Clavius, 17 Sep. 10 (10:431–2). Cf. Tafuri, *Venice* (1995), 194–5, Quaranta, in Santinello, *Galileo* (1992), 226–8, and Cozzi, *Sarpi* (1979), 187.

90. Daniello Antonini to GG, 9 Apr. 11 (11:84).

91. Cf. Favaro, *Studio* (1966), 2, 47, 50.

92. Alessandro Sertini to GG, 27 Mar. 10 (10:306); Fortunio Liceti to GG, 22 Oct. and 31 Dec. 10 (10:449–50, 505–6), on transmitting money to Marina.

93. GG to Liceti, 23 June 40 (18:209q); cf. Biagioli, *Courtier* (1993), 120–38 (Galileo's understanding of court dynamics).

94. GG to Vinta, 20 Aug. 10 (10:424).

95. Horky to Kepler, 31 Mar. 10 (10:308); *GL*, 13.30 ("magic tricks, perhaps, or great prodigies of nature").

96. Antonio Santini to GG, 25 Dec. 10 (10:495).

97. Manso to Beni, ? Mar. and to GG, 18 Mar. 10 (10:292–4, 296); Battistini, *Ann. Ital.*, 10 (1992), 117.

98. Letters to GG from Giuliano de' Medici, 19 Apr., and Orazio del Monte, 16 June 10 (10:318–19); Segeth, quoted in Vaccalluzzo, *Galileo* (1910), xxxv, and letter to GG, 24 Oct. 10 (10:454–5).

99. Resp., Pignoria to GG, 4 Mar. 11 (11:66), and G.B. Marino, "A Galileo Galilei" (Venice, 1618), quoted in Vacculluzzo, *Galileo* (1910), xxxiv, 79–80.

100. Letters to GG from Altobello, 17 Apr. 10 (10:317), and Castelli, 3 Apr. 11 (11:82); Kepler, *Disc.* (1993), 9; *Acts of the apostles*, 1:11, 2:7. Castelli did not believe by faith alone. He had a 9x telescope with which he could make out lunar features.

101. Iezzi, in Lomonaco and Torrini, *Galileo e Napoli* (1987), 148–9, 153–5.

102. Cf. Vaccalluzzo, *Galileo* (1910), xxvii–xxx; GG to Vinta, 21 May 10 (10:447), passing along Girolamo Maganati, *Meditazione poetica sopra i pianeti Medicei*; GG to Buonarroti, 16 Oct. 10 (10:446–7); GG to Alessandro Sertini, 27 Mar. 10 (10:305–6).

103. Rossi, *Galilaeana*, 4 (2007), 190–1. The original intent appears from the title under which the *Nuncius* was submitted to the Vernetian censors: *denunciatio*, "announcement."

104. Cf. Swerdlow, in Machamer, *Companion* (1998), 255–9, and Van Helden, *Osiris*, 9 (1993), 15–16.

105. GG to Giuliano de' Medici, 11 Dec. 10 (10:483), and to Vinta, 1 Apr. 11 (11:80q); Santini to GG, 20 July 11 (11:155, finding the periods will make you immortal).

106. Magini favored a semi-Tychonic system compatible, he thought, with Aristotle. Peruzzi, in Bucciantini and Torrini, *Diffusione* (1997), 84–6.

107. GG (still in Padua) to Vinta, 30 July, and to Giuliano de' Medici (quotes), 13 Nov. (10:410, 474); Hasdale to GG, 19 Dec. 10 (10:491). "I saw the most distant planet as three-bodied." Kepler did not find the observation much easier than the anagram. Kepler to GG, 9 Jan. 11 (11:15–16).

108. Maclean, *Rev. hist. math.*, 11 (2005), 66–7.

109. Hasdale to GG, 24 Aug. and 19 Dec. 10 (10:426–7); Magini to GG, 28 Sep. 10 (a mirror for 500 scudi that could kindle fire at 2.5 feet), 2 Oct. 10 (please arrange to sell it to the grand duke), 20 Nov. 10 (will stay in Bologna with increased salary) (10:437–8, 443, 476); GG to G. de' Medici, 1 Oct. 10 (glad if Kepler gets Paduan job) (10:339–40).

110. Copernicus, *De rev.* (1543), 10.1 (ed. Rosen, 18).

111. Palmieri, *JHA*, 32 (2001), 112–15, and in Montesinos and Solís, *LCF*, 435–9. If Venus stayed above the sun, she would not have crescent phases, and if below, she could not be gibbous.

112. GG to Clavius, 17 Sep. and 30 Dec., and to Castelli, 30 Dec. 10 (10:431–2, 499–505); cf. GG to ?, 25 Feb. 11 (11:53–4).

113. Castelli to GG, 5 Dec. 10 (10:482); Westfall, *Isis*, 76 (1985), 25–9; Drake, *Essays* (1999), 1, 406.

114. Cf. Favaro, *Studio* (1966), 196–201, and *Scampoli* (1992), 2, 720–3; Sosio, in Sarpi, *Pensieri* (1996), clxix, clxix.

115. As is well argued by Westfall, *Isis*, 76 (1985), 24–9, to whom Palmieri, *JHA*, 32 (2001), 109–20, gives a convincing answer.

116. Kepler to GG, 28 Mar. 11 (11:77–8); Drake, *Essays* (1999), 1, 403–4.

117. GG to G. de' Medici, 1 Jan. 11 (11:11–12).

118. G.L. Romponi (Bologna) to GG, 23 July 11 (11:160).

119. Cf. Federico Cesi to Francesco Stelluti, 30 Apr. 11 (11:99).

120. Kepler to GG, 9 Jan. and 28 Mar. 11 (11:15–16, 77–8).

121. Gualdo to GG, 4 and 10 Feb. 11 (11:28, 41–4).

122. Magini to GG, 11 Jan. 11 (11:19–20); Welser to Gualdo, enclosed in Gualdo to GG, 10 Feb. 11 (11:44); Welser to GG, 29 Oct., and reply, 8 Nov. 10 (10:460, 465–6); Gabrieli, *Contributi* (1989), 2, 990.

123. GG to Clavius, 17 Sep., and Cigoli to GG, 1 Oct. 10 (10:431–2, 441–2).

124. Santini to GG, 25 Sep., 9 Oct., 4 Dec. 10 (10:435–6, 444–5, 479–80).

125. GG to G. de' Medici, and to Gualdo, both 17 Dec. 10 (10:483–4); Welser to Clavius, 7 Jan., and Grienberger to GG, 22 Jan. 11 (11:14, 31–4). Cf. Lattis, *Clavius* (1994), 180–7.

126. Clavius to GG, 17 Dec. 10 (10:484–5), confirming Jupiter's moons and Saturn's appearance. "Truly Your Lordship deserves great praise."

127. GG to Welser, ? Feb. 11 (11:41).

128. GG to Vinta, 15 Jan. and 19 Mar., and replies, 20 Jan. and 19 Mar., and to Clavius, 5 Mar. 11; Cosimo to del Monte, 27 Feb., and Niccolini (Florentine ambassador to Rome) to Cosimo, 30 Mar. 11; in Op. 11:27, 71, 28–9, 72, 67, 60–1, 78–9.

129. Dante, *Paradiso*, 19.79–81: "Or tu che se', che vuo' a scranna / per giudicar di lungi mille miglia / con la veduta corte d'una spanna?" The question is the celestial reply to Dante's query why heathens who knew not Christ could not be saved. Cf. Section 2.1 above.

130. Letters to GG from Santini, 9 Mar., and Kepler, 28 Mar. 11 (11:69, 77); GG to Salviati, 22 Apr. 11 (11:91); Drake, *Essays* (1999), 1, 442–4; Redondi, *NCCS*, 182–4.

131. Giovanni Antonini, Brussels, to GG, 9 and 29 Apr. 11 (11:84, 98); Welser to Clavius and to Gualdo, both 7 Jan., and to Clavius, 11 Feb. 11 (quote) (11:14, 15, 45).

132. Bellarmine to the mathematicians of the Roman College, 19 Apr., and response, 24 Apr. 11 (11:87, 93). Bellarmine had already seen the Galileian sights on the moon and Venus.

133. Dini to Cosimo Sassetti, 7 May 11 (11:101–2); Cigoli to GG, 11 Aug. 11 (11:168).

134. Favaro, *Scampoli* (1992), 1, 227–30; Pantin, *Galilaeana*, 2 (2005) 26–42; GG to Gallanzone Gallanzoni, 26 Jan. 11, and to Grienberger, 1 Sep. 11 (11:131–2, 182–5, 191–3); Kepler to GG, 19 Apr. 10 (10:329), citing Lucianus on the moon as cheese.

135. Dollo, in Bucciantini and Torrini, *Diffusione* (1997), 158–60, 162q.

136. Letters to Clavius from Welser, 7 Jan., and delle Colombe, 27 May 11; Gallanzoni to GG, 26 June, and reply, 16 July 11 (11:14, 118, 131–2, 141–9).

137. Cf. Baroncini, *Forme* (1992), 68–77, 86–95.

138. Welser to GG, 17 June 11, and Grégoire de Saint Vincent, SJ, to Jacques van der Stratten, 23 July 11 (11:127, 162–3); Landis, *Clavius* (1994), 193–5.

139. Cigoli to GG, 13 Nov. 10 (10:475), *re* Buonarroti's ability to show the Jovian stars convincingly; Barberini to Buonarroti and to A. de' Medici, both 2 Apr. 11 (11:80–1).

140. Hammond, in Coelho, *Music* (1992), 68; Cole, *Muse* (2007), 13–15, 180q ("ohimè, ohimè, che se 'l divinar mio / lui se fosse un dì Papa immaginato / sarei ito dormir su una ponda o 'n su l'estremo tegol d'una groma."

141. Barberini, "Epigrammata LXI," in Wiendlocha, *Jugendgedichte* (2005), 190–1q.

142. Rietbergen, *Power* (2006), 98–100; Pastor, *Popes, 28* (1938), 26–36; Kelley, *Popes* (1988), 280.

143. GG to Salviati, 27 Apr. 11 (11:89–90); Camillo Borsarchi to GG, 3 July 11 (11:137).

144. Valerio to GG, 23 Oct. 10 (10:451–2); Favaro, *Amici* (1983), 1, 14–31; Gabrieli, *Contributi* (1989), 1, 841; Cesi to GG, 14 Dec. 12 (11:446).

145. Biagioli, *Courtier* (1993), 292–4, quoting a contemporary MS; Bellini, *Umanisti* (1997), 23–7.

146. Olmi, in Lomonaco and Torrini, *Galileo e Napoli* (1987), 24, 29, 31–4q; Freedberg, *Eye* (2002), 92.

147. Ibid., 36, 38; Rosen, *Naming* (1947), 87n, 107; Gabrieli, *Contributi* (1989), 2, 913, 916.

148. Rosen, *Naming* (1947), 30–1, 35, 53–4, 57, 61. Previously Galileo had used *perspicilum*, *occhiale*, and *istrumento*.

149. The lincei conceded della Porta a role in the invention on the strength of Kepler's interpretation of the relevant part of *Natural magick*; Rosen, *Naming* (1947), 16–20, 23–4.

150. Cesi to GG, 30 Nov. 12 (11:439); Kepler, in Chevalley, *NCCS*, 173.

151. Eck, *De nova stella* (1605); Ricci, *Tre studi* (1994), 8, 10–11, 15–17, 24–5n.

152. Correspondence of June/July 1612, in Ricci, *Tre studi* (1994), 26n23, 27n24; Cesi to GG, 20 June and 21 July 12, and reply, 30 June 12 (11:332–3, 344–5, 366–7); Gabrieli, *Contributi* (1989), 1, 348–51.

153. Freedberg, *Eye* (2003), 67–73, 79–80.

154. Cf. Ferrone, *NCCS*, 246–7.

155. Del Monte to Cosimo, 31 May, and Grienberger to GG, 24 June 11 (11:119, 131). GG left Rome on 4 June; Guicciardini to Vinta, 4 June 11 (11:121).

156. For Galileo's illnesses, 1611–13, Op. 11:46, 247, 266–7, 291, 293, 295, 297, 299, 302, 465.

157. Gabrieli, *Contributi* (1989), 1, 970–1, 975, 985–6, expanded and corrected by Biagioli, *Nuncius*, 7:2 (1992), 82–7, 92–5. Cesi was to marry into the influential Roman branch of the Salviatis.

158. Drake, *Galileo at work* (1981), 169–70; Biagioli, *Courtier* (1993), 160, 170–7.

159. Ibid., 171–4; *OF*, 11.46–51. The virgin that Orlando saved from the orc was not his girlfriend Angelica, whom Ruggiero had rescued from a similar fate (see Section 1.3 above).

160. Biagioli, *Courtier* (1993), 161–5, 169; Vinta to GG, 12 Jan. 11, and reply, 15 Jan. 11, and Pappazoni to GG, 1 Mar. 11 (11:63).

161. Drake, *Galileo at work* (1978), 174–5, and *Cause* (1981), 31–2.

162. Drake, *Cause* (1981), 41–6, 26q. The business is one of pressure: that exerted at S by the wood is $(abc)\delta/ab$, and that by the water prism IIS, $bOS \cdot QS/bOS$; since the pressures must be equal, $c\delta = QS$. Cf. ibid., 48–50.

163. Ibid., 34, 39–40, 47.

164. Shea, *Intellectual revolution* (1977), 21–3, and, in general, 14–43.

165. Drake, *Cause* (1981), 89, 94; 70, 83, 115, 118; 85–6, 94–102, 122–3.

166. Ibid., 125–6, 135–6, 140–3, 148–9.

167. Drake, *Galileo at work* (1981), 187–8, 200.

168. GG to Baliani, 7 Jan. 39 (18:12, "liquidi, materia alla mia mente molto oscura e piena di difficoltà"); Biagioli, *Courtier* (1993), 191–4, 205–7qq. The quoted opinions

come from Pisan academics, the first from Vincenzo di Grazia, the second from an anonymous colleague.

169. Shea, *Intellectual revolution* (1977), 31.

170. Favaro, *Nuovi studi* (1891), 375–6; Biagioli, *Courtier* (1993), 159.

171. Campanella to GG, 8 Mar. 14 (12:32).

172. Cf. Biagioli, *Courtier* (1993), 68–71, 76–8.

173. Scheiner, *Tres epistolae* (1612), in Op. 5:25–33; Scheiner to Welser, 12 Nov. 11 (first letter), in OS, 58–65; Shea, *Isis*, 61 (1970), 499–500; Biagioli, *Credit* (2006), 164–72.

174. Daxecker, *Hauptwerke* (1996), 5, and *Physicist* (2004), 19–21, 95–153; Wade, *Galilaeana*, 4 (2007), 285–9; OS, 36–45.

175. Scheiner to Welser, 19 Dec. (on Venus) and 26 Dec. 11 (on starlets), in OS, 65–73; Bredekamp, *Galilei* (2007), 218–222, 366–82; Van Helden, APS, *Proc.*, 140 (1996), 373. Cf. Reeves, *Galilaeana*, 4 (2007), 62–72.

176. GG to Welser, 4 May 12 (first letter), in OS, 89–105, 95q.

177. Jacoli, *Bulletino*, 7 (1874), 389; Van Helden, APS, *Proc.*, 140 (1995), 373; Reeves, *Painting* (1997), 78–81; Op. 2:14–16; OS, 18–22 (Gualterotti and Kepler), 26–9 (Harriot).

178. Johannes Fabricius, *De maculis in sole observatis*, in OS, 30–4.

179. Micanzio to GG, 22 Sep. 31 (14:299); Bredekamp, *Galilei* (2007), 274 (5:137–8); Biagioli, *Credit* (2006), 201–2; OS, 74–7.

180. Cf. Biagioli, *Credit* (2006), 16.

181. Letter of 4 May 12, in OS, 89q, 91–2, 94–9.

182. Galileo, in OS, 93, 98, 99q.

183. Ibid., 101–4, 102q; Biagioli, *Credit* (2006), 205–6.

184. Cf. Swerdlow, in Machamer, *Companion* (1998), 255.

185. Drake, *Cause* (1981), 18–20.

186. Bredekamp, *Galilei* (2007), 230–43, 248–50, 251q, 254, 258–60, 425–78 (Galileo's and Cigoli's drawings, May–Aug. 1612, in large scale (solar disk of 5 inches)); GG to Welser, 14 Aug. 12, in OS, 129q; Castelli to GG, 8 May 12 (11:294).

187. GG to Welser, 14 Aug. 12, in OS, 108–68, on 113–16. $XY^2 = PQ^2 + (QY − PX)^2$; QY and PX may be found by the Pythagorean theorem and the measured values OQ, OP. For the orthographic projection, which Scheiner did not use in his analysis, see Camerota, *Galilaeana*, 1 (2004), 151–61, and 4 (2007), 83, 85, 88, 92.

188. Cf. Shea, *Isis*, 61 (1970), 503–6, and OS, 357–9.

189. SN (1989), 84, (1993), 172–3.

190. Galileo, in OS, 124–5; cf. Camerota, 250–2.

191. See Sections 2.3 and 3.4 above. Cf. Festa and Roux, *Galilaeana*, 3 (2006), 125.

192. Galileo, in OS, 124.

193. Galileo, in OS, 128–9; GG to Cesi, 12 May 12 (11:295–7); Grant, *Planets and stars* (1994), 324–70.

194. Scheiner, *Accuratior disquisitio* (1612), in OS, 205, 210–11, 225q, 226–9; Reeves and Van Helden, in OS, 175–81.

195. Letters to GG from Cesi and from Cigoli, 3 and 12 Nov., and GG to Cesi, 4 Nov. and reply, 10 Nov. 1612 (11:423, 425–6, 429).

196. Welser to GG, 5 Oct. 12, in OS, 252–3, and reply, 1 Dec., ibid., 253–304, 266q.

197. Van Helden, APS, *Proc.*, 140 (1995), 378–81.

198. Camerota, 259; Bucciantini, *Galileo e Keplero* (2003), 217–18. Cf. Clavelin, *LCF*, 20–3, 26–7.

199. Cigoli to GG, 1 Oct. 10 (10:441–2); GG to Gualdo, 16 June 12 (11:327), in Drake, *Galileo at work* (1976), 286, 287q.

200. De Filiis, in Galileo, *Istoria*, "Praefatio," in OS, 374–8.

201. Letters to GG from Cesi, 22 Feb., and from Cigoli, 26 Feb. (11:483–4, 489–90).

202. Gabrieli, *Contributi* (1989), 2, 991, 995; Ferrone, NCCS, 250–1.

203. GG to Welser, 1 Dec. 12, in OS, 295, 296q; Van Helden, JHA, 5 (1974), 105–11; Deiss and Nebel, JHA, 29 (1998), 215–20; cf. GG to ?, 15 Jan. 39, and to Castelli, 28 Aug. 40 (18:19, 288).

204. OF, 7.27, 8.1.

205. Camerota, 189; F.M. del Monte to GG, 18 Nov. and 16 Dec. 11 (11:234–5, 245); Maschietto, in Santinello, *Galileo* (1992), 438.

206. Lodovica Vinta to GG, 2 July 14 (12:80–1); Rietbergen, *Power* (2006), 104; Sobel, *Galileo's daughter* (1999), 379. The fact that the sister of Galileo's friend Vinta was abbess of San Matteo no doubt made the place more palatable to Galileo.

207. Sarrocchi to GG, 29 June 11 (11:163–4); Zannini, *Motivi* (1982), 40–1; OF, 20.1–2.

208. Sobel, *Letters* (2001), comprises Maria Celeste's literary remains.

209. V. Landucci to GG, 21 Dec. 39, in Favaro, *Scampoli* (1992), 1, 130.

210. Bellinati, in Santinello, *Galileo* (1992), 261–2; Favaro, *Scampoli* (1992), 1, 125–9, and *Arch. ven.*, 4 (1902), 302–4, 404.

211. Favaro, *Scampoli* (1992), 1, 131–4; Benetti, in Billanovich, *Viridarium* (1984), 324; Section 8.3 below.

212. Letter of 27 Apr. 11 (11:95–7).

213. Galileo to Clavius, 30 Dec. 10 and 5 Mar. 11, and to Sarpi, 12 Feb. 11; letters to GG from Cigoli, 26 Nov. 10, and Gualdo, 12 July 11 (10:478, 499, 11:46–7, 67, 139).

214. Cigoli to GG, 1 July 11 (11:132); Maschietto, in Santinello, *Galileo* (1992), 439–40.

215. Contarini to Senate, 27 Feb. 16; Poppi, Acc. pat., Cl. sci. mor., *Atti*, 105 (1992/3), 42.

216. Castelli to GG, 1 Apr. 07 (10:169, 170q); to D. Ermagora (Padua), 24 Oct. 10 (10:183–4); and to GG, 27 Sep. and 24 Dec. 10 (10:436, 493–4).

217. Favaro, *Archeion*, 1 (1919), 29–6; *DSB*, s.v., "Cavalieri;" Maschietto, in Santinello, *Galileo* (1992), 436–8; Masetti Zannini, *Vita* (1961), 67.

218. Masetti Zannini, *Vita* (1961), 28q, 37; *DBI*, 21:688.

219. Gualdo to GG, 6 May, 12 and 29 July 11 (11:100, 139, 165); Kuhn, *Ven. Arist.* (1996), 394–402, 408–10; Poppi, *Ricerche* (2001), 255–9; Riondato and Poppi, *Cremonini* (2000), 36–7. On Gualdo's joking, Roncone, Acc. pat., Cl. sci. mor., *Atti*, 105 (1992/3), 108–9, 116.

220. Kuhn, *Ven. Arist.* (1996), 482, 489; Stella, Acc. pat., Cl. sci. mor., *Atti*, 103 (1990/1), 66–9.

221. Campanella to GG, 8 Mar. 14 (12:32–3); Ernst, *Religione* (1991), 241.

222. Berti, in *Galileo a Padova* (1983), 535–6; Spruit, in *Cremonini* (2000), 1, 197–203.

223. Decree of 17 May 11, in Genovesi, *Processi* (1966), 15–16.

224. Cremonini, as quoted in Gualdo to GG, 29 July 11 (11:165).

225. Sosio, in Sarpi, *Pensieri* (1996), cii, cxxv; Sarpi, ibid., 238 (document dated between 1585 and 1587).

226. Bucciantini, *Galileo e Keplero* (2003), 224–9.

227. GG to Dini, 23 Mar. 15 (5:302); Bucciantini, *Galileo e Keplero* (2003), 226, and *Nuncius*, 9:1 (1994), 29–32; Applebaum and Baldasso, *LCF*, 385–8.

228. GG to Sarpi, 12 Feb. 11 (11:47–50).

229. Micanzio to GG, 26 Feb. 11 (11:57–8); cf. Cozzi, *Sarpi* (1979), 165, and Sosio, in Sarpi, *Pensieri* (1996), clxxvi, clxxi.

230. Gualdo to GG, 6 May 11 (11:100–1); Sagredo to GG, 7 Feb. 15 (12:139); Ferrone, *NCCS*, 242–3.

231. Liceti to GG, 31 Dec. 10 (10:505–6), and Pignoria to Gualdo, 10 Jan. 11 (11:28).

232. Cozzi, *Sarpi* (1979), 218–19, 321; Sosio, in *Galileo e la cultura* (1995), 310; Cozzi and Cozzi, in *Storia*, 4:2 (1984), 33–4; Norwich, *Venice* (1989), 519–23.

233. Stella, in Santinello, *Galileo* (1992), 321; Sarpi to Tommaso Contarini, 7 May 16, in Lerner, *Primi lincei* (2005), 391–2.

234. Sagredo to GG, 13 Aug. 11 (11:171).

235. Favaro, *Arch. ven.*, 4 (1902), 347–53, 359–61, 378–80.

236. Favaro, *Studio* (1966), 2, 9; Sagredo to GG, 18 Aug. and 22 Sep. 12 (11:379, 398, resp.).

237. GG to Giuliano de' Medici, 1 Oct. 10 (10:439–41); Varetti, Acc. Naz. Lincei, Cl. sci. mor., stor., fil., *Rend.*, 3–4 (1939), 219–21.

238. Favaro, *Archivio veneto*, 4 (1902), 350–1, 353 (12:404–5); Bedini, in McMullin, *Galileo* (1967), 93–5, 98–103; Sagredo to GG, 23 Apr. 16 (12:257–8, on 1% yield), 107–10.

239. Gabrieli, *Contributi* (1989), 1, 970–2; Biagioli, *Nuncius*, 7:2 (1992), 87–90.

Chapter 6: Miscalculated Risks

1. Session IV (8 Apr. 1546), tr. Waterworth, *Canons* (1848), 20, and Blackwell, *Galileo* (1991), 183.

2. Gualdo to GG, 6 and 27 May, the last forwarding Welser to Gualdo, 27 May 11 (11:100–1, 117).

3. Conti to GG, 7 July 12 (11:354–5); *DBI*:28, 376–8.

4. GG to Gallanzoni, 16 July, and Antonini to GG, 24 June 11 (11:152–3, 129–30); Delle Colombe, *Contro il moto della terra* (3:253–90); Favaro, *Padova* (1966), 1, 151–2.

5. Shea and Artigas, *Galileo* (2003), 49–50.

6. Redondi, *Galilaeana*, 1 (2004), 118–19, 122–3; Cesi to GG, 30 Nov. 12 (11:438–9); Stabile, *Nuncius*, 9:1 (1994), 44–6.

7. Castelli to GG, 6 Nov. 13 (11:590); Favaro, *Studio* (1966), 127–30.

8. Solerti, *Musica* (1905), 72; the duke was Francesco Maria II della Rovere.

9. Castelli to GG, 14 Dec. 13 (11:605, FA, 47–8); Stabile, *Nuncius*, 9:1 (1994), 49. Antonio was born before his parents married.

10. Blackwell, *Galileo* (1991), 83–5; Leonardi, in *Galileo in Padova* (1983), 136–46, and in Santinello, *Galileo* (1992), 421–3.

11. GG to Castelli, 21 Dec. 13 (FA, 49–52).

12. FA, 51q, 53–4. Cf. GG to Gallanzoni, 16 July 11 (11:147–8).

13. Leonardi, in *Galileo a Padova* (1983), 154–7, and in Santinello, *Galileo* (1992), 375–83; "Raffiniert ist der Herr Gott, aber boshaft ist er nicht"; Einstein, in Pais, *Subtle is the Lord* (1982), [vi].

14. Galileo, *Letter to Christina*, 1615 (FA, 93).

15. Cf. Poppi, *Ricerche* (2001), 198, 207–9; Pesce, *Ann. stor. eseg.*, 4 (1987), 248, 252–61.

16. Guerrini, *Galileo* (2009), 35–8, 55q, 61–3, 71, 77. Guerrini, ibid., 64, 70, places the undated attack on Hell in 1611/12.

17. Ibid., 55–6: "The world also is stablished that it cannot be moved" (Ps. 93:1); "With the bread of understanding shall she [the law] feed him, and give him the water of wisdom to drink" (Ecclesiasticus 15:3).

18. Guerrini, *Galileo* (2009), 29–30, 60 (text of 1615); "he looks for a spot on the sun."

19. Ibid., 19–20, 23, 27; Cigoli to GG, 16 Dec. 11 (11:241–2).

20. Ricci Riccardi, *Galilei* (1902), xiv–xv, 15–16, 31–2, 35–7, 40, 61–7.

21. Attavanti, deposition of 14 Nov. 15 (FA, 143–5).

22. Ximenes, deposition of 13 Nov. 15 (FA, 141–3). Ximenes said that Caccini's sermons on Joshua provided the occasion for discussing Copernicus with Attavanti (FA, 136).

23. M. Caccini to T. Caccini, 2 Jan. 15, and to Alessandro Cassini (a third brother), 2 and 9 Jan. 15, in Ricci Riccardi, *Galilei* (1902), 69–71, 80; Guerrini, *Galileo* (2009), 34–5, 67–9.

24. *FA*, 135–7, 139.

25. *FA*, 138–41.

26. Ibid.; Shea and Artigas, *Galileo* (2003), 52–3.

27. Lorini to Sfondrati, *circa* 7 Feb. 15 (*FA*, 134–5).

28. Consultor's report, n.d. (*FA*, 135–6).

29. Drake, *Galileo at work* (1978), 238–9. Cf. Bucciantini, *Contro Galileo* (1995), 34–6.

30. GG to Dini, 16 Feb. 15 (*FA*, 55–8).

31. Dini to GG, 7 Mar. 15 (*FA*, 58–9); Baldini, *NCCS*, 293–305; McMullin, in McMullin, *Church* (2005), 174–82.

32. GG to Dini, 23 Mar. 15 (*FA*, 60–2).

33. Ibid., 62–7 (5:297, 305); Poppi, *Ricerche* (2001), 201.

34. *Letter to Christina*, 1615 (*FA*, 101q); Stabile, *Nuncius*, 9:1 (1994), 56, 62; Pesce, in Prodi, *Disciplina* (1994), 164.

35. Caroti, in Lomonaco and Torrini, *Galileo e Napoli* (1987), 86, 90–1, 99–104; Basile, *Riv. lett. ital.*, 1 (1983), 74–5, 86–8, 93–4.

36. Blackwell, *Galileo* (1991), 87–97; Cesi to GG, 7 Mar. and 20 June 15 (12:150–1, 189–90); Ciampoli to GG, 21 Mar. 15 (12:160–1), urging caution.

37. Castelli supplied some of the theological references, a Barnabite father others (Castelli to GG, 6 Jan. 15 (13:126–7)), and Foscarini most; Leonardi, in Santinelli, *Galileo* (1992), 406–7; McMullin, in McMullin, *Church* (2005), 99–112.

38. Letter to Christina, FA, 87–90, 98q, 113q.

39. *FA*, 91, 98.

40. *FA*, 111, 112; Rosen, *Isis*, 49 (1958), 329–30. Even St Thomas has passages helpful to Galileo; Fabris, *Orientamenti* (1986), 37n, gives citations to the *Summa theologica*.

41. *FA*, 92–6, 107–9.

42. *FA*, 91, 100, 116.

43. *FA*, 94, 104–5.

44. *FA*, 94, 96, 103–5, 114, 116–18; Blackwell, *Bellarmine* (1991), 75–82; Fabris, *Orientamenti* (1986), 43–4.

45. Quoted by Blackwell, *Galileo* (1991), 253–4.

46. Quoted ibid., 255–61.

47. Bellarmine to Foscarini, 12 Apr. 15, in Blackwell, *Galileo* (1991), 265–6 (*FA*, 67–8). Bellarmine had taught this literalism in his anti–Protestant apologetics, *Disputationes de controversiis christianae fidei* (1586, 1596); Fabris, *Orientamenti* (1986), 34–6.

48. Blackwell, *Galileo* (1991), 266–7 (*FA*, 68–9).

49. Castelli to GG, 12 Mar. 15, as translated in Shea and Artigas, *Galileo* (2003), 63–4.

50. Cesi to GG, 12 Jan. 15 (12:128–9); Barberini's opinion as reported and glossed in Ciampoli to GG, 27 Feb. 15 (12:146).

51. As reported in Dini to GG, 7 Mar. 15 (12:151).

52. Ciampoli to GG, 21 Mar. 15 (12:160), and GG to Dini, ? May 15 (12:183–5), in Shea and Artigas, *Galileo* (2003), 71.

53. Cosimo to Piero Guicciardini (ambassador), and Picchena to Primi, both 28 Nov. 15 (12:203, 205), first and second quotes, resp.

54. Guicciardini to Picchena, 5 Dec. 15 (12:207).

55. Guicciardini to Cosimo, 4 Mar. 16 (12:241–3).

56. Cosimo to del Monte, 28 Nov., and to Borghese, 2 Dec. 11 (12:203–4, 205–6); Bucciantini, *Contro Galileo* (1995), 70–2.

57. GG to Picchena, 12 and 26 Dec. (12:208–9, 211–12); Section 5.1 above (remarks of the ambassador).

58. GG to Picchena, 1 and 8 Jan., and 6 Feb. 16 (12:220, 222–3, 230–2).

59. GG to Picchena, 8 Jan. and 6 Feb. 16 (quote) (12:222–3, 230).

60. Quotes from GG to Picchena, 23 Jan. and 20 Feb. 16 (12:228, 238, resp.); Querenghi to Alessandro d'Este, 30 Dec. 15 and 20 Jan. 16 (12:212, 226–7).

61. Querenghi to Alessandro d'Este, 5 Mar. 16, in Motta, *Aevum*, *67* (1993), 614.

62. Sarpi, *Pensieri* (1996), 423–7. Although the Paduan tidal theory does not hold water, partisans have tried to assign it to their favorite; for Galileo, Drake, *Galileo studies* (1970), 200–13, and for Sarpi, Sosio, in Sarpi, *Pensieri* (1996), cl–clix.

63. *FA*, 119–22, 124, 126, 128q, 130q, 133.

64. *FA*, 126. The tidal generating force is the difference between the pull of the moon (and the sun) on the earth and on the oceans.

65. *FA*, 133.

66. Guicciardini to Cosimo, 4 Mar. 16 (12:242–3); Fantoli, *Galileo* (1996), 232–3, 267–9, notes 97–8.

67. *FA*, 146–7.

68. Cf. McMullin, in McMullin, *Church* (2005), 173, 183–4.

69. *FA*, 147–8; Pagano, 45–6 (doc. 21, 16 Feb. 16). Historians have rendered *successive et incontinenti* as "immediately after," "later on," "in connection with the foregoing," according to whether they think Seghizzi or Galileo precipitated the second step. Mayer, *Nuncius*, 24:1 (2009), 81–4.

70. Beretta and Lerner, *Galilaeana*, 3 (2006), 207–16; Job, 9:8; Leonardi, in Santinello, *Galileo* (1992), 377–8.

71. *FA*, 148–9; Ciampoli to GG, 28 Feb. 15 (12:146).

72. Pagano, 46–7 (doc. 22, 5 Mar. 16).

73. Lerner, in *Primi lincei* (2005), 327–35, 346–56. Lerner observes (pp. 342–4) that Ingoli missed many passages in *De revolutionibus* that on his terms deserved cancellation.

74. Gingerich, *Eye* (1993), 274–84; Pizzamiglia, *Inseg. mat.*, 27:2 (2004), 167, 169; Bucciantini, *Contro Galileo* (1995), 44–8, 103; Shea and Artigas, *Galileo* (2003), 85–7; Lerner, in *Primi lincei* (2005), 382–3.

75. Baldini, in *Inquisizione* (2000), 331–8, 362.

76. Ibid., 352–4. Cf. Spini, *Galileo* (1996), 36–7, and in Maccagni, *Saggi*, 3:2 (1972), 432, 436.

77. Letter of 4 Mar. 16 (12:242–3), in Shea and Artigas, *Galileo* (2003), 80–1.

78. GG to Picchena, 6 Mar. 16 (FA, 150–1).

79. The unflattering portrait of Paul is Sarpi's; Benrath, *Neue Briefe* (1909), 59 (letter of 28 Oct. 11).

80. GG to Picchena, 12 Mar. 16 (FA, 151–3).

81. Certificate, 26 May 16 (FA, 153); Spini, *Galileo* (1996), 35; letters to GG from Castelli, 20 Apr., and Sagredo, 23 Apr. 16 (12:254, 257–8).

82. Bucciantini, *Contro Galileo* (2005), 23, 39–43, 53–9.

83. Beretta, *Galilée* (1998), 267–78; Bucciantini, *Contro Galileo* (1995), 48, 52.

84. Castelli to GG, 16 Mar. 30 (14:87), reporting Campanella's account of a conversation with Barberini, then Pope. Campanella: some Germans decided not to convert to Rome because of the decree of 1616. The Pope: "that was never our intention, etc." Cf. Bolzoni, Scuola Norm., Pisa, *Ann.*, 19 (1989), 306.

85. Binachi, *LCF*, 578–82; Job, 38:4–6.

86. Milton, *Paradise Lost*, 8.72–8.

87. Galileo, *Cause* (1981), 25q, 151.

88. *OS*, 121. Cf., Carugo and Crombie, IMSS, *Ann.*, 8:2 (1983), 46–51.

89. Buonamici, quoted in Helbing, *Physis*, 18 (1976), 53; Waterworth, *Canons* (1848), Session 13, canons 1–2, p. 82; Camerota, 408–10.

90. Speller, *Galileo's inquisition* (2008), 392–5, argues that since such fiddling would be deceptive, God's absolute goodness would preclude it. But since our ideas of reason and justice do not restrain His actions, it is unlikely that our ideas of propriety will.

91. Oreggi, *De Deo uno tractatus primus* (1629), in Favaro, *Scampoli* (1991), 2, 614–17. The exchange goes beyond the letter of Oreggi's report from "I do not mean" to "our senses," but does not depart from its spirit.

92. Bacon, *Essays* (1625), in *Major works*, ed. Vickers (2002), 351.

93. Guerrini, *Galileo* (2009), 90, 95.

94. Guicciardini to Cosimo, 4 Mar., and to Picchena, 13 May 16 (12:241–2, 259).

95. Guicciardini to Picchena, 13 May 16 ("strano e scandoloso lavoro," 12:259); Shea and Artigas, *Galileo* (2003), 92.

96. Rietbergen, *Power* (2006), 111–15.

97. Spini, *Galileo* (1996), 41–4, on 44; Vaccalluzzo, *Galileo* (1910), 129–31.

98. Pieralisi, *Urbano VIII* (1875), 15–18.

99. Spini, *Galileo* (1996), 43.

100. Naudé, quoted in Pieralisi, *Urbano VIII* (1875), 20; Bolzoni, Scuola Norm., Pisa, *Ann.*, 19 (1989), 289–90, 292; Costanzo, *Critica*, 2 (1970), 16–17.

101. Costanzo, *Critica*, 1 (1969), 71, 148–50.

102. Guglielminetti and Masoero, *Stud. seic.*, 1 (1978), 135–8; Ciampoli to Barberini, 10 Nov. 10, and to Borromeo, 7 Nov. 11 and 7 Apr. 13, in ibid., 155–9; Gualdo to GG, 16 Dec. 11 and 8 Jun 12 (11:243, 320).

103. Ciampoli, *Poemetto* (1615), in Costanzo, *Critica*, 1 (1969), 102–4 (quote, 103–4); Ciampoli to Borromeo, 24 Oct. 16, in Guglielminetti and Masoero, *Stud. seic.*, 19 (1978), 160–1.

104. *DBI*, 24:198–200; Redondi, *Heretic* (1987), 89–92.

105. Bellini, *Umanisti* (1997), 2, 12–13, 16, 27–8, 31–4, 41, 52; Gabrieli, *Contributi* (1989), 1, 767–8, 771; Cesarini to GG, 31 Dec. 16 and 1 Oct. 18 (12:299, 413–15).

106. Guglielminetti and Masoero, *Stud. seic.*, 19 (1978), 139, 142–44, 145q; Ciampoli to Barberini, 18 Mar. 19, 6 Feb. and 18 Nov. 20, in ibid., 169, 170, 174.

107. Letters of 24 Mar. 19 and 6 Feb. 20, in Gabrieli, *Contributi* (1989), 1, 774–5, 800–2.

108. Rietberger, *Power* (2006), 101–2, 105–9.

109. Flora, *Storia* (1967), 3, 374q, 323q; Costanzo, *Storia*, 1 (1969), 73–6, and 2 (1970), 19–22, 40–2.

110. *DBI*, 25:149; Ciampoli to Borromeo, 3 July 21, and to GG, same date, in Guglielminetti and Masoero, *Stud. seic.*, 19 (1978), 174, and Op. 13:69, resp.

111. Bellini, *Umanisti* (1992), 47–9, 54–9; Section 6.3 below.

112. See Section 1.2 above.

113. Flora, *Storia* (1967), 3, 458–81.

114. Marino, *L'Adone* (1623), 10.39–41, ed. Pozzi (1976), 1, 528–9, 2, 440–2.

115. Marino, *L'Adone* (1623), 10.45, ed. Pozzi (1976), 1, 530. Cf. Battistini, *Ann. ital.*, 10 (1992), 126, and Flora, *Storia* (1967), 3, 274.

116. Strozzi, *Venetia edificata* (Venice, 1623), 7.52–7, quoted in Vaccalluzzo, *Galileo* (1910), lx–lxi. Merlin appears in *OF*, 3.10–19.

117. Sagredo to GG, 15 Mar. 15 (12:156); Vaccalluzzo, *Galileo* (1910), xiv–xvii; Luigi Maraffi to GG, 12 Dec. 15 (12:210), quoting from "Stanze sopra le nuove stelle e macchie solari" dedicated to Cardinal Aldobrandini.

118. Vaccalluzzo, *Galileo* (1910), xii (cf. Op. 11:86, 262, 324; 12:413, 13:254, 15:96); Armour, in Aquilecchia et al., *Essays* (1971), 151–2, 160; Gualdo to GG, 5 July and 13 Dec. 14 (12:81–2, 118–19).

119. Favaro, *Scampoli* (1992), 1, 231–3. Illness prevented Galileo from serving.

120. Campanella to GG, 5 Aug. 32 (14:366); *TCWS*, 164 ("una commedia in commedia," Op. 7:190). It is possible to see something of Sarpi in the stage Sagredo; Cozzi and Cozzi, in *Storia*, 4:2 (1984), 13n37.

121. Cf. Battistini, *Lett. ital.*, 30 (1978), 320–3. The falsity of isochronism is easily shown: two pendulums of the same length, one set swinging at an amplitude of 3°, the other at 80°, are 180° out of phase in four full swings. Giulini, in Renn, *Galileo* (2001), 139–40.

122. "Fantasia ingengniosa," "capriccio matematico," "poema," "favole;" Op. 7:188, TCWS, 162; Battistini, *Lett. ital.*, 30 (1978), 182, 217; Jardine, in Kelly and Popkin, *Shapes* (1991), 104–15.

123. Cf. Rossi, *Galilaeana*, 4 (2007), 199–200; Reynolds, *Nuncius*, 17:1 (2002), 51.

124. Armour, in Aquilecchia, *Essays* (1971), 150–1.

125. Letter of 6 Apr. 16, in Gabrieli, *Contributi* (1989), 1, 847.

126. Gabrieli, *Contributi* (1989), 1, 839, 847–51, 854–5; Napolitani, in Lomonaco and Torrini, *Galileo e Napoli* (1987), 166; Dini to GG, 7 Mar. 15 (12:151–2); Bucciantini, *Contro Galileo* (1995), 117–47; Pastor, *Popes*, 25 (1937), 294–5.

127. Blackwell, *Galileo* (1991), 137–42, 140q.

128. Baldini, *Legem* (1992), 195–281, esp. 217–50, and Grienberger's comments, pp. 232–6.

129. Biancani, *Sphaera* (1620), "Praefatio," 250, 281, 130 (quotes, resp.); cf. Baldini, *Legem* (1992), 231, 235–6.

130. Biancani, *Sphaera* (1620), 197.

131. Ibid., 292q, 299, 305–7.

132. Baldini, *Legem* (1992), 255–6; the philosopher's lecture is printed, ibid., 257–71. Aristotle, *Meteorol.*, 344ᵃ5–345ᵃ10, at 344ᵃ5–7: "we consider a satisfactory explanation of phenomena to be given when our account of them is free from impossibilities."

133. These epithets come from Galileo's marginalia in one of Grassi's books; Battistini, *Lett. ital.*, 30 (1978), 311. "Fat pig" is a crude play on *grasso* ("fat").

134. Bösel, *Grassi* (2004), 24–5, 28–34.

135. Tassinari, in Bösel, *Grassi* (2004), 222–7; Dorian Recordings 93243, with introduction and translation by T. F. Kennedy. The composer, the Venetian Johannes Hieronymous Kapsburger, was an excellent lutanist in the style of Vincenzo Galilei.

136. Grassi, *Astr. disp.* (1619), in Drake and O'Malley, *Controversy* (1960), 5q, 17, and BH, *Disc.*, 255, 283, 447.

137. Grassi, in Drake and O'Malley, *Controversy* (1960), 11–17, and in BH, *Disc.*, 259–83, 404–5.

138. BH, *Disc.*, (2003), 442–7, emphasize the iconography and the projection, and reproduce the drawings (after p. 252).

139. Galileo would return to the shrine, for the same purpose and to the same effect, in 1628; Michelangelo Galilei to GG, 5 Apr. 28 (13:408); Spini, in Maccagni, *Saggi*, 3:2 (1972), 416; Favaro, *Scampoli* (1992), 1, 205–6; Gabrieli, *Contributi* (1989), 1,

196. Some may not know that angels brought Mary's house from the Holy Land to the Apennines in 1294.

140. Bedini, *Pulse* (1991), 12–17; Castelli to GG, 14 Feb. 18 (12:373–4); GG to Archduke Leopold, 23 May 18 (12:389–92).

141. Drake, *Galileo at work* (1978), 262–4; BH, *Disc.*, 26–34, 37–9.

142. Viviani, *Vita* (2001), 55–6.

143. Guiducci, in Drake and O'Malley, *Controversy* (1960), 22, 24, 40–1, 48.

144. Ibid., 59, 41, resp.

145. Ibid., 45–7, 53–7.

146. Ibid., 53, 50, 63, 31, resp.

147. Van Helden, IMSS, *Ann.*, 7:2 (1982), 71, 73, 82; Magini's textbook of 1599 assigns to first magnitude stars a disk of 10′; to second, 50″; to sixth, 1″ (p. 74).

148. Van Helden, IMSS, *Ann.*, 7:2 (1982), 75–6, 83; GG to Sarpi, 12 Feb. 11 (11:49).

149. Kepler, *Astronomia* (1604), in BH, *Sagg.*, 35, 586–7; Favaro, "Libreria," (1886), #115.

150. Despite the heroic efforts in Drake, *Galileo at work* (1978), 270–3, to save it, Galileo's comet theory is just wrong. Cf. BH, *Disc.*, 421–30.

151. Kepler, *De cometis* (1619), in BH, *Sagg.*, 11–13; Camerota, 371–5; Casanovas, *NCCS*, 311–12.

152. Ciampoli to GG, 12 July and 6 Dec. 19 (12:465–6, 499).

153. Grienberger to Ricardo de Burgo, 1619, in Baldini, *Legum* (1993), 195; Biancani, *Sphaera* (1670), 281 ("acutissimus Galilaeus").

154. [Grassi], *Astro. balance* (1619), in Drake and O'Malley, *Controversy* (1960), 71, 70 (resp.).

155. Ibid., 71. Cf. Clavelin, *LCF*, 34–6, and Shea, *Intell. rev.* (1977), 85–7.

156. Biagioli, *Courtier* (1993), 286; Camerota, 375–7.

157. Op. 6:116.

158. Cesi, in Altieri Biagi and Basile, *Scienziati* (n.d.), 49, 53, 54, 61, 69 (*nullius in verba iurare*, a line from Horace later taken by the Royal Society of London as its motto); Aristotle, *Metaphysics*, 980^{b}22.

159. Biagioli, *Courtier* (1993), 268q, 276–80, 286–8.

160. BH, *Disc.*, 322, 450; Op. 6:386; Guiducci to Cesi, 19 June 20 (13:41).

161. Favaro, *Nuovi studi* (1891), 205.

162. Grassi, *Balance* (1619), in Drake and O'Malley, *Controversy* (1960), 81q, 75q, 77–9, 85.

163. Ibid., 97, 98q.

164. Ibid., 105q, 106–11; BH, *Sagg.*, 591–2.

165. *Assayer*, 117, 118q, 120q, 121; BH, *Sagg.*, 612–13. The phenomenon is a consequence of the decomposition of bodies buried without coffins.

166. Ibid., 127, 129–30, 109q.

167. Daniel, 3:25.

168. Guiducci, *Letter* (1620), in Drake and O'Malley, *Controversy* (1960), 142q, 138, 139q, 140.

169. Ibid., 145, 147q.

170. Ibid., 149–50.

171. Cesi to Bellarmine, 4 Aug. 18, and reply, 25 Aug., in Altieri Biagi and Basile, *Scienziati* (n.d.), 9–37, 38q.

172. Ciampoli to GG, 12 July and 6 Dec. 19 (12:465–6, 498–9), and 2 Aug. 20 (13:46–7); Camerota, 378.

173. Biagioli, *Courtier* (1997), 290–1; BH, *Sagg.*, 49–52; Bellini, *Umanisti* (1992), 47–9, 54–6, 66–9.

174. *Assayer*, 331. "Sarsiad" is Ciampoli's quip (letter to GG, 15 Jan. 22, 13:84).

175. *Assayer*, 170–1, 176–7, 186q, 195.

176. OF, 23.81, tr. Waldman, 276.

177. *Assayer*, 163–9, 179.

178. Ibid., 152, 154–5, 162 (the last translation slightly altered for scansion); Eszer, *Angelicum*, 60 (1983), 429.

179. As this description might be controversial, a few examples of each may be in order. Quibbles and bombast: *Assayer*, 173–5, 192, 199, 211, 216, 218, 221, 253–4, 256, 261, 263, 264. Bamboozlement: Galileo's role in Guiducci's essay, 169, 176–7, and in the invention of the telescope, 211–13. Misrepresentations: almost everything concerning Tycho, 180–3, 191, 258, 271; and, of course, Sarsi, who did not sting, unprovoked, like a scorpion (172). High moral tone: 188, 255, 260, 270, 272, 274.

180. BH, *Sagg.*, 36–7, 40, 457–69; Galileo, *Difesa contro alle calumnie ed imposture di Baldassare Capra* (1607) (2:519–99).

181. Galileo, *Difesa* (1607) (2:517–18).

182. *Assayer*, 183–4.

183. Plato, *Timaeus*, 53C–55C, in Cornford, *Plato's cosmology* (n.d.), 210–19; Koyré, in Wiener and Noland, *Roots* (1957), 171–5; BH, *Sagg.*, 487–8.

184. *Assayer*, 181–4. Galileo charges Sarsi with ineptitude in mathematics on 249, 252, 263.

185. Kepler, *Appendix* (1625), in Drake and O'Malley, *Controversy* (1960), 339–40; BH, *Sagg.*, 480–3, 488–91.

186. *Assayer*, 235–6; BH, *Sagg.*, 539–42, suggesting Buonamici and Aristotle as possible sources for the story. Varanini, *Galileo critico* (1967), 89, takes the passages about bird and insect songs as exemplary of Galileo's integration of "writer and scientist."

187. *Assayer*, 189; BH, *Sagg.*, 575.

188. BH, *Sagg.*, 500–1.

189. *Assayer*, 201–10, 216, 220, 223.

190. Ibid., 205–6; cf. 255.

191. Cf. Ronchi, *Cannochiale* (1964), 104–10, and GG to Dini, 21 May 11 (11:115).

192. *Assayer*, 227–30; BH, *Sagg.*, 493.

193. *Assayer*, 184; BH, *Sagg.*, 492–4, 634–5; Kepler, *Appendix*, in Drake and O'Malley, *Controversy* (1960), 344–5; Van Helden, IMSS, *Ann.*, 7:2 (1982), 79–80. Cf. *TCWS*, 360–1.

194. *Assayer*, 180, 185; Kepler, *Appendix* (1960), 342–4.

195. *OF*, 11.2–7; Section 1.3 above; *Assayer*, 268.

196. BH, *Sagg.*, 42; *Assayer*, 192, 194, referring to Kepler, *De cometis* (1619).

197. *Assayer*, 262, 269q; Galileo ranks his cometary theory as a mere supposition or possibility on 227, 231, 233, 260, 275.

198. BH, *Sagg.*, 44–5, 562–3, 569; Redondi, *Heretic* (1989), 142–3, 168–72. Galileo quickly dropped his comet theory except for a passing mention in the *Dialogue*.

199. *OF*, 30.49, tr. Waldman, 365; *Assayer*, 295.

200. *Assayer*, 301.

201. *Assayer*, 293–5.

202. Ibid., 308–9.

203. *Assayer*, 309q, 310–11, 311–12q, 313. Cf. BH, *Sagg.*, 618–19, 624–5, and Gómez, *Galilaeana*, 5 (2008), 219, 222–3, 226–9.

204. *Assayer*, 278, 288, 290 (puerilities); 291, 314, and elsewhere (misrepresentations); 326 ("your repeated error").

205. *Assayer*, 321–5, 327q.

206. Cesi to GG, 9 Sep. 28 (13:448); Biagioli, *Courtier* (1993), 309.

Chapter 7: Vainglory

1. Ciampoli to Buonarroti, 22 July and 19 Aug. 23q, in Guglielminetti and Masoero, *Stud. seic.*, 19 (1978), 182, 183, and to GG, 22 July 23 (13:119); Pastor, *Popes*, 28 (1937), 25.

2. Pastor, *Popes*, 28 (1937), 11–15, 21–5; Scott, *Images* (1991), 70, 75–7, and plates 136, 140, 141.

3. Pastor, *Popes*, 28 (1937), 40–8, makes the total take of Francesco 63 million and of the pope's breeding nephew Taddeo 42 million; Hammond, *Music* (1994), 4–6, 26–30, 71–2.

4. Freedberg, *Eye* (2002); 55–8, 74, 76; Hammond, *Music* (1994), 19; Guerrini, *Galilaeana*, 5 (2008), 268.

5. Ciampoli to Giambattista Strozzi, 6 May 24, in Guglielminetti and Masoero, *Stud. seic.*, 19, (1978), 185; Barbi, *Accademico* (1900), 61.

6. Ciampoli to Strozzi, [1624], and to Giuseppe Gualdo, 4 July 26, in Guglielmi-netti and Masoero, *Stud. seic.*, 19 (1978), 190–3, 196–7; *DBI*, 25:149.

7. Quoted by Cole, *Muse* (2007), 179, who is not responsible for the translation.

8. Rietberger, *Power* (2005), 133–6; Wiendlocha, *Jugendgedichte* (2005), 303–5; see Section 6.3 above.

9. Cole, *Muse* (2007), 178–9, and *Ren. quart.*, 60 (2007), 771–2.

10. Belloni, in *Primi lincei* (2005), 68–9 (Mascardi's text of 1625); Fumaroli, *Poussin* (1989), 55–7, 62; Pozzi, in Marino, *L'Adone* (1988), 2, 137–40.

11. Scott, *Images* (1991), 3, 16, 18, 127, 130, 133, 135, 139.

12. Young, *Medici* (1926), 2, 388, 391–4, 400–1, 411–13. Christina died in 1626, Maria Maddalena in 1631, en route to visit her brother, another Ferdinando II, the Holy Roman Emperor.

13. Ibid., 394–6, 403–4; Pastor, *Popes*, 28 (1937), 59–61, 97–9.

14. Ranke, *Die Päpste* (1854), 2, 532q, 533–8; Pastor, *Popes*, 28 (1937), 50.

15. Bracciolini, *L'Elettione* (1628), 23. 77, p. 483; Hammond, *Music* (1994), 63.

16. Favaro, *Studio* (1878), 65.

17. Pastor, *Popes*, 29 (1937), 41; Rietbergen, *Power* (2006), 386; Shea and Artigas, *Galileo in Rome* (2003), 105–9.

18. *Assayer*, 153q; Rinuccini to GG, 3 Nov. 23 (13:145–6).

19. Bellini, *Umanisti* (1997), 73, 81–2; Ciampoli to Borromeo, 18 May 24, in Guglielminetti and Masoero, *Stud. seic.*, 19 (1978), 187; Gabrieli, *Contributi* (1989), 1, 199, 203–5.

20. Ciampoli to Borromeo, 18 May 24, in Gabrieli, *Contributi* (1989), 1, 808; Motta, *Querenghi* (1997), 303–8, 319–28.

21. Cesarini to GG, 12 Jan. 23 (13:105–7); Freedberg, in Barnes and Wheelock, *Van Dyck* (1994), 158–62, 165; Bellini, in *Primi lincei* (2005), 72, 75, 77; Cesi, in Altieri Biagi and Basile, *Scienziati* (n.d.), 51q, 52–3.

22. Bellini, *Umoristi* (1997), 51, 56–84, and in *Primi lincei* (2005), 61–3.

23. GG to Cesi, 8 June 24 (13:182–3); Camerota, 416–17.

24. GG to Cesi, 9 Oct. 23 (13:134–5).

25. GG to Cesi, 15 May and 8 June 24 (13:179, 182).

26. Cesarini to GG, 3 Feb. 23 (13:109); Faber to Cesi, 1 June 24 (quote), and GG to Cesi, 8 June 24, on angels (13:181, 183).

27. Faber to Cesi, 11 May, and GG to Cesi, 23 Sep. 24 (13:177, 208–9); Op. 19:609–10; Freedberg, *Eye* (2002), 151–3; Drake, *Galileo at work* (1978), 286.

28. Cesi, *Apiarium* (1626), ed. Guerrini (2005); Freedberg, *Eye* (2002), 146, 154–72.

29. TCWS, 446–7. The final text of 1632, quoted here, differs little from the ver-sion completed in 1625; Drake, *Galileo at work* (1978), 295–6.

30. *OF*, 23.103–4, tr. Waldman, 278.

31. *TCWS*, 447.

32. *TCWS*, 452–3q.

33. *TCWS*, 454q–7.

34. *TCWS*, 454.

35. *TCWS*, 462.

36. Ancient astronomy saved the inequalities in the sun's apparent motion, assumed constant in its orbit, by displacing the center of the orbit from the center of the earth. Kepler showed that, in addition, there was an inequality arising from inconstant motion along the orbit. This produced the famous and instructive battle between the whole- and the half-eccentrics; Heilbron, *Sun* (1999), 102–9.

37. *TCWS*, 424, 430–1; 447; 444; 436–41; 457–60.

38. *TCWS*, 460, 445q.

39. Bucciantini, *Contro Galileo* (1995), 85, 88–97, 107–9, and 159–74 (a transcript of Ingoli's manuscript response to Kepler); Lerner, in *Primi lincei* (2005), 367–72.

40. *FA*, 155–6.

41. *FA*, 157–65.

42. *FA*, 177–80, 182q; cf. Bucciantini, *Contro Galileo* (1995), 159–74.

43. *FA*, 182–8.

44. *FA*, 188–91, 190q.

45. *FA*, 192q, 184–5q, 195.

46. Redondi, *Heretic* (1987), 130–4; Guiducci to GG, 15 Oct. 24 (13:216).

47. Redondi, *Heretic* (1987), 107–18; De Dominis, *Euripus, sive de fluxu et refluxu maris sententia* (Oct. 1623).

48. Redondi, *Heretic* (1987), 179–80; Camerota, 382; Stellutti to GG, 4 Nov. 23, and Tommaso Rinuccini to GG, 3 Nov. 23 (13:147, 145).

49. Redondi, *Heretic* (1987), 181–9; letters to GG from Rinuccini, 2 Dec. 23, Guiducci, 21 June 24, and Bartolomeo Imperiali, 27 Feb. 26 (13:153–4, 186–7, 307); Favaro, *Nuovi studi* (1891), 207–8, 215–16.

50. *Ratio ponderum librae et simbellae* (1626).

51. Session 13, chs. 1–8, and canons i–xi, tr. Waterworth, *Canons* (1848), 82.

52. Redondi, *Heretic* (1987), 68, 136, 193–5.

53. Sobel, *Letters* (2001), 87 (prison), 101.

54. Ibid., 35q, 37q, 127; 11, 13, 43, 135, 163, 171 (cold and hunger), 25, 45, 103 (poor health), 49, 53 (toothache).

55. Ibid., 93q, 73, 75q, 85, 87, 95, 97, 113.

56. Ibid., 139, 155, 375.

57. Ibid., 51, 53, 67, 83, 129, 145, 147, 149.

58. Ibid., 55, 69, 71 (garden).

59. M. Galilei to GG, 14 July 27 and 26 Feb. 28 (13:365–7, 394–5).

60. Same to same, 27 Apr. and ? June 28 (13:414–18, 438–9); Sobel, *Letters* (2001), 41, 47, 155, 159.

61. Sobel, *Letters* (2001), 75, 77, 109, 131, 169.

62. Ibid., 183, 187, 225.

63. Ibid., 69, 79, 91, 99, 109, 113 (family get-togethers). Galileo had been stingy with his visits when living in Bellosguardo; ibid., 17, 39, 45.

64. Maria Celeste to GG, 10 Aug. 23 and 26 Apr. 24, in ibid., 3q, 31q; ibid., 19, 117.

65. Ibid., 19, 125, 179, 181, 189, 191, 219, 221.

66. Ibid., 197, 203.

67. Maria Celeste to GG, 22 Mar. 28 and 4 June 31, ibid., 81q, 179q, 199.

68. Ibid., 15, 83, 103, 107, 117, 123, 125q.

69. Ibid., 9, 11, 51, 53, 111, 117.

70. Ibid., 129; BH, *Dial.*, 166, 195; TCWS, 22–7.

71. TCWS, 9–15, 18, 23–8; BH, *Dial.*, 149–52. Cf. Clavelin, NCCS, 18–21.

72. TCWS, 33q, 34–7q, 41, 46.

73. BH, *Dial.*, 138–40, 279.

74. TCWS, 56q, 57q.

75. Ibid., 51–56, 54q.

76. TCWS, 61, 67–8, 84–5, 96, 100–1; BH, *Dial.*, 259, 273–4, 278–9, 293–4, 309–10.

77. Ciampoli to GG, 28 Feb. 15 (12:146), and Barberini's advice (Section 6.1 above); BH, *Sagg.*, 259.

78. TCWS, 91–9; BH, *Dial.*, 174, 244, 305–6, 312–13.

79. BH, *Dial.*, 239–43, 251–3, 282–3, 300, 303, 328.

80. TCWS, 53, 103, 163q, 105q; BH, *Dial.*, 327, 333–6. The dichotomy intrinsic/extrinsic, and many other epistemological distinctions and principles Galileo uses ("Natura nihil frustra facit," "Eadem est ratio totius et partium"), are found in Buonamici; BH, *Dial.*, 154, 170, 171, 181, 184, 194, 199, 202, 258.

81. TCWS, 211q, 256q, 131q, 162q; cf. 118, 122, 254, 167.

82. TCWS, 112q, 113q.

83. TCWS, 115–19.

84. TCWS, 126–7, 141–5, 171–3, 186–7; BH, *Dial.*, 369–70, 396–8, 401, 409, 431, 488–91.

85. TCWS, 148–54, 155q; BH, *Dial.*, 447–8.

86. TCWS, 165–6; Heilbron, *Sun* (1999), 178–80, 344n13.

87. TCWS, 163q, 222q, 223–4, making acceleration under gravity about ¾ its correct value. There are further glimpses into the unpublished *oeuvre* on pp. 226–30. BH, *Dial.*, 544, 554; Baliani to Castelli, 20 Feb. 27 (13:348).

88. *TCWS*, 194–9, 200q, 201, 203q, 207 ("filosofo geometra"), 216–17. The recent effort to salvage something from this argument by Palmieri, *JHA*, 39 (2008), 431–51, is subjected to detailed analysis by Finocchiaro, *Defending Copernicus and Galileo* (2010), 97–120.

89. Cf. Koyré, *Galileo studies* (1978), 191–5 (the figure here, p. 191, and in BH, *Dial.*, is faulty); Shea, *Intell. rev.* (1977), 140–2; Clavelin, *Philosophie* (1968), 247–50.

90. Sosio, in Sarpi, *Pensieri* (1996), cxlviii–cl; Sarpi, ibid., 400–1.

91. *TCWS*, 190–1, 192q, 196, 185q.

92. *TCWS*, 112q, 130q, 131, 272. Consider the contradictory propositions, "You have stopped [not stopped] beating your wife."

93. *TCWS*, 231–40, 243, 245; Koyré, APS, *Trans.*, 45 (1955), 331–3; BH, *Dial.*, 534–8. Locher and Scheiner exposed themselves to unnecessary ridicule by what appears to be a typo. They say that at least six days are needed for fall from the moon at constant velocity u and also that $u = u_A$, u_A being the velocity of the point A on the lunar sphere, which is nonsense (the journey would take under 4 hours). It appears from their numbers for u_A and for the corresponding velocity u_Y for a point Y above Ingolstadt ($u_Y = 2u_A/3$) that they took for the *daily* travel of the balls along the vertical, the *hourly* travel of the point Y. In the latter case, $u = (2/3)(u_A/24)$ and the time of the journey is $R/u = 36R/u_A = 36/2\pi \approx 6$ hours, which Scheiner–Locher miswrote as 6 days.

94. *TCWS*, 246, 243, 247; BH, *Dial.*, 579–82; Gapaillard, *Et pourtant, elle tourne* (1993), 204–14, and in *Sciences tech. persp.*, 2 (1990/1), 1–10.

95. *TCWS*, 248, 253–5, 256q; 257–60; 261–6.

96. *TCWS*, 268–70, 271q.

97. *TCWS*, 274, 275q.

98. *TCWS*, 276, 277q, 278, 281.

99. *TCWS*, 225–8, 289q, 290–2; BH, *Dial.*, 633, 675.

100. *TCWS*, 293–310, 309q; BH, *Dial.*, 649, 655–6, 663.

101. *TCWS*, 311–15.

102. *TCWS*, 319–26, 325q.

103. *TCWS*, 328, 407, the first regarding Copernicus, the second Gilbert.

104. *TCWS*, 328–9, 335q, 336–42.

105. *TCWS*, 345q; Scheiner, *Rosa* (1630), lib. 4, pt. 2, cap. 1; Daxecker, *Hauptwerk* (1996), 53; BH, *Dial.*, 723.

106. BH, *Dial.*, 724–7, 730; Daxecker, *Hauptwerk* (1996), 57.

107. *TCWS*, 346q, 347q; BH, *Dial.*, 733, 737. That Galileo did not discover the tilt is demonstrated in Drake, *Galileo studies* (1970), 183–5. Galileo may have known about it before 1630; if so, he did not recognize its importance to Copernican cosmology; OS, 316–27.

108. BH, *Dial.*, 721–2; GG to Marsili, 21 Apr. 29 (14:36), and to Micanzio, 9 Feb. 36 (16:391); Castelli to GG, 26 Sep. 31 (14:296–8). Scheiner published a preliminary version of *Rosa* in 1629; Camerota, *Galilaeana*, 2 (2005), 50n.

109. Cf. Cavalieri to GG, 18 Nov. 31 (14:308–9); Van Helden, IMSS, *Ann.*, 2:1 (1976), 27–8, 34–5; Bedini, in McMullin, *Galileo* (1967), 282.

110. Scheiner to Gassendi, 23 Feb. 33 (15:47), and GG to Micanzio, 9 Feb. 36 (16:391).

111. Scheiner, *Rosa* (1630), lib. 2.2, caps. 1–2, 11; Daxecker, *Hauptwerk* (1996), 53–4, 57; *TCWS*, 352–5.

112. *TCWS*, 536.

113. *TCWS*, 398q, 399; BH, *Dial.*, 754–8, 803, 806.

114. Smith, *Isis*, 76 (1985), 543–51; Drake, *Galileo studies* (1970), 191–6. The third motion cannot be eliminated because without a fixed axis of solar spin the annual changes in sunspot trajectories depicted in figure 6.7 would be even harder to save. Cf. Hutchinson, *Isis*, 81 (1990), 70, and Topper, *Isis*, 90 (1999), 763–7.

115. Vickers, IMSS, *Ann.*, 8:2 (1983), 72–6, 83–6; *TCWS*, 107, 96, 205qq; ibid., 153–4, 262, 271 (opposite is true).

116. Hammond, *Music* (1994), 110, 124, 200; Rietbergen, *Power* (2006), 108–9.

117. Guglielminetti and Masoero, *Stud. seic.*, 19 (1978), 229–36, 235q, 229q. Cf. Costanzo, *Critica, I* (1969), 76–7.

118. Cf. Fumaroli, *Age* (1980), 203–4.

119. Motta, *Querenghi* (1997), 251–7; Formichetti, in Formichetti, *Mago* (1985), 199; Campanella to Paul V, 1606, in ibid., 208q.

120. Ernst, in Campanella, *Opuscoli* (2003), 9–10, 13, 16, and in Galluzzi, NCCS, 256–8; Bolzoni, Scuola Norm., Pisa, *Ann.*, 19 (1989), 310; Campanella to GG, 8 Mar. 14 (12:32–3).

121. Spini, *Galileo* (1996), 48, 53; Blackwell, in Campanella, *Defense* (1994), 23–4, 39; Cesarini to GG, 12 Jan. 23 (13:105–7).

122. Campanella, *Defense* (1994), 67–8, 88–91, 93.

123. Ibid., 62q, 81q, 63, 64q, 84.

124. Ibid., 71q, 72, 74, 76, 55 (re heresy).

125. Ibid., 80, 109–10, 120, 121q, 122. The story of Pythagoras' Jewish antecedents, of which the earliest surviving notice dates from 300 years after his death, was recorded in Josephus and repeated by Origen and other Fathers. Joost-Gaugier, *Measuring heaven* (2006), 22, 265n62.

126. Campanella, *Defense* (1994), 78–9, 97, 113q.

127. Corsano, *Giorn. crit. fil. ital.*, 39 (1960), 357–8, 361–4, 374.

128. Spini, *Galileo* (1996), 55, 49q, 50; Formichetti, *Campanella* (1983), 20, 25, 26–7q, 15q; Rietbergen, *Power* (2006), 121–2, 124–5.

129. Pieralisi, *Urbano VIII* (1875), 25–6, 27q.

130. Spini, *Galileo* (1996), 53–4; Formichetti, *Campanella* (1983), 16.

131. Ernst, in Campanella, *Opuscoli* (2003), 17–21.

132. Ibid., 22–3.

133. Ibid., 25–6; Walker, *Magic* (1975), 205–23; Ernst, *Religione* (1991), 273–9; Formichetti, *Mago* (1985), 208 (wrestling); Rietbergen, *Power* (2006), 344–5.

134. Grillo, *Questioni* (1961), 8–9, 13, 18–19, 21–2. Cf. Firpo, *Rev. fil.*, 30 (1939), 203–11, 213–15.

135. Grillo, *Questioni* (1961), 10; Ernst, in Campanella, *Opuscoli* (2003), 22–3, 32–3; Formichetti, *Mago* (1985), 200–3. Most of the evidence for this plot comes from Campanella.

136. Ernst, in Campanella, *Opuscoli* (2003), 34q, 35.

137. Guerrini, *Brun. Camp.*, 7 (2000), 239–444; Ernst, in Canone, *Bibliothecae* (1993), 222–9.

138. Ernst, in Canone, *Bibliothecae* (1993), 229–31, 237–44.

139. Buonarroti to GG, 3 June 30 (14:111).

140. Ernst, in Campanella, *Opuscoli* (2003), 42–4, and *Religione* (1991), 275–8; Formichetti, in Formichetti, *Mago* (1985), 206–7.

141. Quoted in Ernst, *Religione* (1991), 279.

142. Daxecker, *Physicist* (2004), 95–103, on Scheiner's *Tractatus de tubo optico* (1615); Pantin, *Galilaeana*, 2 (2005), 40; Scheiner to Faber, 9 Nov. 12 (11:428).

143. Scheiner-Locher, *Disquisitiones* (1614), 178, and Scheiner to GG, 6 Feb. 15 (12:137–8), both from *OS*, 308–9.

144. Scheiner, *Rosa* (1630), 1:1 (ed. Daxecker, 11). Cf. BH, *Sagg.*, 460.

145. Scheiner, *Sol* (1615), 6q, 10, 13–17, 28–30 (ed. Daxecker, 18, 21–8, 43–5); Scheiner to GG, 11 Apr. 15 (12:170–1).

146. *Assayer*, 316; BH, *Sagg*, 630–1. Scheiner calculated a good value for the difference between the axes of the solar disk at sunset, some 5'; Scheiner, *Reactiones coelestes* (1615), in Daxecker, *Physicist* (2004), 118.

147. Scheiner, *Sol* (1615), 21–2, 24 (ed. Daxecker, 34–8); Heilbron, *Sun* (1999), 127–34.

148. Daxecker, *Physicist* (2004), 9–10, 21, 60–4, 88–90, 93–4.

149. Daxecker, *Physicist* (2004), 73, 111–15; Acquaviva to Scheiner, 13 Dec. 14, ibid., 112; Scheiner to GG, 2 Feb. 15 (12:137); Section 6.3 above; BH, *Dial.*, 174, 305f, 313, 543f, 579ff, 688, 716.

150. Vitelleschi to Scheiner, 15 July 17, 27 Feb, 7 and 21 Aug, 4 Sep, 27 Nov., 11 Dec., all 1621, in Daxecker, *Physicist* (2004), 74–76q.

151. Cf. Spini, in Maccagni, *Saggi* (1972), 3:2, 438–9; Op. 19:410; Paschini, *Vita* (1965), 2, 587–90.

152. Daxecker, *Physicist* (2004), 76q.

153. Daxecker, *Physicist* (2004), 21–7, 67–9.

154. Ibid., 18, 28, 31–2, 67; GG to Leopold, 23 May 18 (12:389–92).

155. Guiducci to GG, 18 Apr. 25 (13:265–6).

156. Scheiner to Leopold, 10 Jan. and 19 Dec. 26, in Daxecker, *Physicist* (2004), 65, 67; Scheiner, *Rosa* (1629), 6–66 (ed. Daxecker, 12–16); see also Section 6.1 above.

157. Bellarmine to Cesi, 25 Aug. 18, in *Rosa* (1630), 783.

158. Daxecker, *Physicist* (2004), 42–3; Castelli to GG, 26 Sep. 31 and 20 Feb. 32 (14:296–8, 330).

159. Scheiner to Leopold, 29 Nov. 30; Vitelleschi to Scheiner, 8 July and 16 Sep. 34, 10 Mar. and 4 Aug. 35, 12 Apr. and 25 Oct. 36, 17 Jan. 37, 6 Feb. and 4 Dec. 38; in Daxecker, *Physicist* (2004), 70–1, 82–3, 85–8.

160. Caraffa to Scheiner, 9 and 16 Jan. 1649, in Daxecker, *Physicist* (2004), 90–1.

161. Costantini, *Baliani* (1969), 73; Camerota, 378.

162. Redondi, *Heretic* (1987), 120–6.

163. Redondi, *Heretic* (1987), 192–4, 335–6q.

164. Ibid., 198–9; Camerota, 389–91; Castelli to GG, 22 Jan. 28, and Cesi to GG, 9 Sep. 28 (13:388–9, 448–9).

165. Cerbu, *LCF*, 588–91.

166. Quoted ibid., 588.

167. Ibid., 591–2; Blackwell, *Behind the scenes* (2006), 31–2.

168. Ciampoli to GG, 10 July 27 (13:365), and 5 Jan. 30 (14:64q); BH, *Dial.*, 25; GG to George Fortescue, ? Feb. 30 (14:85); Maddison, in *Atti* (1967), 72.

169. Castelli to GG, 9 Feb. and 6 Apr. 30 (14:77–80, 89–90); Buonarroti to GG, 3 June 30 (14:111–13).

170. GG to Elia Diodati, 29 Oct. 29 (14:49).

171. GG to Cesi, 23 Sep. 24 (13:208), and to Marsili, 7 Dc 24 (13:235–6).

172. Orso d'Elci to GG, 3 June 30 (14:113); BH, *Dial.*, 29; Shea and Artigas, *Galileo* (2003), 139, quoting the *Avvisi*.

173. Pagano, xcix–ciii; Ciampoli to GG, 28 Dec. 25 (13:295).

174. Cf. Favino, *LCF*, 872, and Castelli to GG, 21 Nov. 30 (14:150).

175. Pagano, cix–cxiii, cxxi, following Camerota, 408–35; BH, *Dial.*, 29; Morandi to GG, 24 May 30 (14:107).

176. Riccardi to Egidi, 24 May 31 (Pagano, 53, doc. 25A).

177. Pagano, 51, 52 ("medicine of the end"), doc. 25; ibid., 56q (doc. 25C), 57q (doc. 25D); these last are the draft letter and the covering letter dated 19 July 31, resp.

178. Visconti to GG, 16 June 32 (14:120); Castelli to GG, 21 Sep. 30 (14:150–1); Pagano, cxiv; Camerota, 426–7; Niccolini to Cioli, 29 June 30 (14:121q); Ciampoli to GG, 13 July 30 (14:123).

179. GG to Andrea Cioli (Florentine Secretary of State), 3 May 31 (14:259); Spini, in Maccagni, *Saggi* (1972), 3:2, 437; Favaro, *Nuovi studi* (1891), 342–51. Galileo had to mobilize Castelli and Micanzio to collect the 40 scudi; Castelli to GG, 29 Mar. and 26 Apr. 31 (14:235–6, 255–6); Favaro, *Nuovi studi* (1891), 366–71.

180. Cebru and Lerner, *Nuncius*, 15:2 (2000), 595–7, 603, 607–8; Allacci, *Apes* (1633), 118–19; Garcia, *Diodati* (2004), 65.

181. Daxecker, *Physicist* (2004), 53, 70; Allacci, *Apes* (1633), 68–71 (Scheiner), 118–19 (Galileo), 136–7 (Grassi), 156–7 (Ciampoli), 190–2 (Inchofer), 199–202 (Riccardi).

182. Maria Celeste to GG, 21 July 30, in Sobel, *Letters* (2001), 119; Section 6.2 above; Rietbergen, *Power* (2006), 346–9.

183. Scheiner, *Rosa* (1630), 604; Daxecker, *Hauptwerk* (1996), 57; *TCWS*, 109–10 (Day 2); BH, *Dial.*, 40.

184. Freedberg, *Eye* (2002) 65–6, 76–7.

185. Maria Celeste to GG, 26 Oct. 30, in Sobel, *Letters* (2001), 125; Guiducci to Castelli, 26 Oct. 30, *re* the death of Galileo's "contadino," in Castelli, *Cart.* (1988), 115–16; Nussdorfer, *Politics* (1992), 145–61.

186. BH, *Dial.*, 32–4; Castelli to GG, 24 Aug. and 30 Nov. 30 (14:135, 169).

187. Caterina Niccolini to GG, 19 Oct. and 12 Nov. 30 (14:157, 167); Castelli to GG, 21 Sep. 30 (14:150); BH, *Dial.*, 36–7.

188. Niccoloni to Cioli, 19 and 17 May 31 (14:251, 261); Riccardi to Niccolini, 25 Apr. 31 (14:254), and to Egidi, 24 May 31, in BH, *Dial.*, 39q.

189. BH, *Dial.*, 39–42.

190. GG to Diodati, 16 Aug. 31 (14:289); Maria Luisa Altieri Biagi, in BH, *Dial.*, 828.

191. *TCWS*, 5–6.

192. *TCWS*, 464–5.

193. BH, *Dial.*, 888–902, 901q, quoting Oreggio, *De deo uno* (1629), 193–5. Cf. Battistini, *Lett. ital.*, 30 (1978), 326, and Section 6.1 above.

194. In presenting a conjecture on the nature of comet tails in the *Assayer*, 247, Galileo had remarked that he "could offer others, while, perhaps, there are still others unthinkable to us." He could not have made a similar statement at the end of the *Dialogue*. He was unsure about comet tails but not about the cause of the tides.

195. Galileo's persistence in claiming innocence appears from his letters to Fabri de Peiresc, 21 Feb. 35, and to Diodati, 7 Mar. 34; Speller, *Inquisition* (2008), 346–7.

196. Micanzio to GG, 15 May 32 (14:350), almost quoting 1 Kings 10:20, " there was not the like made in any kingdom." Cf. Battistini, *Lett. ital.*, 30 (1978), 318–27, and Reynolds, *Crit. lett.*, 8 (1980), 421.

197. Castelli to GG, 29 May 32 (14:357–8); Shea and Artigas, *Galileo* (2003), 161.

198. Riccardi to Inquisitor of Florence, 25 July 32 (20:571–2).

199. Naudé, as quoted in Rietbergen, *Power* (2006), 372; Naudé to Gassendi, ? Apr. 33 (15:88–9); Camerota, *Galilaeana*, 2 (2005), 47–8.

200. Redondi, *Heretic* (1987), 229–32; Section 6.2 above; Ciampoli to Strozzi, [1624], in Guglielminetti and Masoero, *Stud. seic.*, 19 (1978), 190; Favino, *LCF*, 865–9, 874–7.

201. Biagioli, *Courtier* (1993), 333–6; Shea and Artigas, *Galileo* (2003), 164–5.

202. Cf. Ciampoli to "mio figlio" (the castrato Loreto Vittorio, a favorite of Urban's) 26 Sep. 28, in Guglielminetti and Masoero, *Stud. seic.*, 19 (1978), 203–5, and Redondi, *Heretic* (1987), 266; Hammond, *Music* (1994), 65.

203. Castelli to GG, 23 and 30 Oct., 20 and 27 Nov. 32 (14:416, 420, 430, 433); Guglielminetti and Masoero, *Stud. seic.*, 19 (1978), 207, 219–20.

204. Magalotti to Guiducci, 7 Aug. 32 (14:370), and GG to Diodati, 15 Jan. 33 (15:25–6); Rietbergen, *Power* (2006), 267, quoting Lucas Holste, Francesco Barberini's librarian.

205. Redondi, *Heretic* (1987), 227–32.

206. Grassi to Girolamo Bardi, 22 Sep. 33, in Favaro, *Nuovi studi* (1891), 218–20.

207. BH, *Dial.*, 34–5; Shea and Artigas, *Galileo* (2003), 165–6, 168, 172–3; Niccolini to Cioli, 5 Sep. 32 (*FA*, 229).

208. Eszer, *Angelicum*, 60 (1983), 434–43; *DBI*, 25:150.

209. Niccolini to Cioli, 5 and 11 Sep. 32 (*FA*, 230, 236 [last two quotes]).

210. Ibid., *FA*, 229–30.

211. Niccolini to Cioli, 11 and 18 Sep. 32 (*FA*, 233q, 234–5); Special Commission report, Sep. 32 (*FA*, 219). Cf. Pieralisi, *Urbano VIII* (1875), 341–56.

212. Beretta, *LCF*, 560–5; Speller, *Inquisition* (2008), 219–26.

213. Niccolini to Cioli, 11 Sep. 32 (*FA*, 233); Section 6.2 above.

214. Cerbu, *LCP*, 592–8; Artigas et al., in McMullin, *Galileo* (2005), 220–7; Heilbron, in ibid., 284–8; Camerota, 393–6.

215. McMullin, *JHA*, 40 (2009), 205–6.

216. Speller, *Inquisition* (2008), 397–9.

217. In this way, Redondi's insight in *Galileo heretic* (1989), that consideration of the eucharist played an important role in Galileo's trial, receives confirmation without requiring his special argument, to which he no longer holds, that the trial of 1633 was a concession, allowing Galileo to plead guilty to a lesser offense than undercutting the eucharistic miracle. Eucharistic considerations did play a major part later in the condemnation of Descartes' writings; Armogathe and Carraud, *Nouv. Rep. Lettres*, 2001:2, 104–10.

218. *FA*, 218–22; BH, *Dial.*, 334, on God, Galileo and geometry.

219. GG to Francesco Barberini, 13 Oct. 32 (*FA*, 357n11; *Op.* 14:408–10).

220. Niccolini to Cioli, 24 Oct., 6 and 13 Nov. 32 (*FA*, 236–9); Garcia, *Diodati* (2004), 339; Magalotti to Guiducci, 4 Sep. 32 (14:380).

221. Niccolini to Cioli, 13 Nov. 32 (*FA*, 239–40).

222. Same to same, 26 Dec. 32 (*FA*, 240).

223. Buonarroti to F. Barberini, 12 Oct. 32 (19:332–3).

224. "Fede medica," 17 Dec. 32q, order to Egidi, 30 Dec. 32, and Egidi to Cardinal Antonio Barberini, 22 Jan. 33, in Pagano, 64–6 (docs. 34–36); Niccolini to Cioli, 15 Jan. 33 (*FA*, 241).

225. Nussdorfer, *Politics* (1992), 156q, 157–8.

226. Niccolini to Cioli, 14 and 19 Feb. 33 (*FA*, 242, 243q, 244q).

227. Niccolini to Cioli, 27 Feb. 33 (*FA*, 244–5, 246q).

228. Niccolini to Cioli, 13 Mar. 33 (*FA*, 247–8).

229. *FA*, 262 (Orezzi's report), 271–6, 275q (Pasqualigo's).

230. *FA*, 262–70, 262q, 267q, 265q.

231. Cf. Speller, *Inquisition* (2008), 249–54.

232. Naudé to Gassendi, ? Apr. 33 (15:87–8); Camerota, *Galilaeana*, 2 (2005), 47–8.

233. Scheiner to Gassendi, 23 Feb. and 7 July 33, in Daxecker, *Physicist* (2004), 57–8 (15:47, 183q); Grassi to Girolamo Bardi, 22 Sep. 33 (15:273); Shea and Artigas, *Galileo* (2003), 171–2: "Galileo caused his own ruin by thinking too highly of himself and despising others."

234. Birely, *Jesuits* (2003), chapt. 5.

235. GG to Diodati, 25 July 34, in Blackwell, *Behind the scenes* (2006), 29 (16:116–17).

236. Niccolini to Cioli, 16 and 23 Apr., 1 May 33 (*FA*, 250–4, 252q).

237. Niccolini to Cioli, 3, 16, 22 May 33 (*FA*, 253–4).

238. Niccolini to Cioli, 27 Feb., 16 Apr., and 19 June 33 (*FA*, 245, 250, 254).

239. Beretta, *Nuncius*, 16 (2001), 633.

240. Blackwell, *Behind the scenes* (2006), 8–10; *FA*, 256–9.

241. *FA*, 259–62.

242. *FA*, 262q

243. Blackwell, *Behind the scenes* (2006), 13–16; Maculano to Francesco Barberini, 28 Apr. 33 (*FA*, 276–7); Fantoli, *Galileo* (1996), 314–16.

244. Maculano to Francesco Barberini, 28 Apr. 33 (*FA*, 276–7).

245. *FA*, 277–9.

246. GG to Maria Celeste, *circa* 1 May 33, as inferred from her letters to him of 7 and 14 May, and Maria Celeste to GG, 21 May 33q, in Sobel, *Letters* (2001), 229, 231, 237, 241q.

247. *FA*, 280–1.

248. Blackwell, *Behind the scenes* (2006), 13–21; *FA*, 281–6.

249. Blackwell, *Behind the scenes* (2006), 22.

250. Heilbron, in McMullin, *Church* (2005), 281–3. Cf. McMullin, *JHA*, 40 (2009), 205–6.

251. *FA*, 287–91. In his *Memorie* (1648), 124, Bentivoglio gave as grounds for Galileo's condemnation "having wished to publish his new opinions about the motion of the earth against the true consensus of the faith." Cf. Castelli to GG, 12 Sep. 26 (13:341).

252. *FA*, 291–2; Scheiner to Athanasius Kircher, 16 July 33 (15:184).

Chapter 8: End Games

1. Inchofer, *Tractatus* (1633), in Blackwell, *Behind the scenes* (2006), 142, 146, 150, 153–6.

2. Ibid., 107, 129–30, 156, 204; Eccl. 11:3.

3. Inchofer, *Tractatus* (1633), in Blackwell, *Behind the scenes* (2006), 157q, 159–63, 164q.

4. Ibid., 199q, 108, 167, 202q.

5. Gorman, *Persp. sci.*, 4 (1996), 295, 314–16. Grassi also rejected the assertion of heresy; ibid., 296 (6:485–90).

6. Inchofer, *Tractatus* (1633), in Blackwell, *Behind the scenes* (2006), 106q; Vitelleschi to Scheiner, 12 Aug. and 21 Oct. 34, in Daxecker, *Physicist* (2004), 31, 82, 83.

7. Daxecker, *Physicist* (2004), 47q; Pieroni to GG, 4 Jan. and 11 Aug. 35 (16:189, 301); GG to Micanzio, 9 Feb. 36 (16:391) ("bamboccerie," "fantoccerie di questo animalaccio").

8. Daxecker, *Physicist* (2004), 152–3; Camerota, *Galilaeana*, 2 (2005), 43–7; Scheiner, *Prodromus pro sole mobile et terra stabili contra…Galilaeum* (1651).

9. Bösel, *Grassi* (2004), 39–41.

10. Redondi, *Heretic* (1987), 262; Grassi to Baliani, ? Aug. 52, in Costantini, *Baliani* (1969), 108q, 95–109; Dollo, in Bucciantini and Torrini, *Diffusione* (1997), 144–5, 157–61, details the relevant views of the Jesuit generals.

11. Riccardi, *Historiae synopsis* (n.d.); Eszer, *Angelicum*, 60 (1983), 444–50.

12. Bolzoni, Scuol. Norm., Pisa, *Ann.*, 19 (1989), 309, quoting Campanella to A. Barberini, 1 Feb. 35; Ernst, in Campanella, *Opuscoli* (2003), 45–9, 54.

13. Eszer, *Angelicum*, 60 (1983), 449–61; Castelli to GG, 7 June 39 (18:57–8).

14. Inchofer, *Oratio* (1639), v, viiq.

15. Blackwell, *Behind the scenes* (2006), 43–4; Gorman, *Persp. sci.*, 4 (1996), 303–4.

16. Ciampoli to GG, 5 Apr. and 14 June 33 (15:80, 154).

17. Redondi, *Heretic* (1987), 266–71; Bellini, *Umanisti* (1997), 45; Ciampoli, *Frammenti* (1654); Torrini, *NCCS*, 269, 271–5.

18. Ragazzini, in *Convivium*, 27 (1959), 52–3, 55; Bellini, *Primi lincei* (2005), 90–1; Masetti Zannini, *Vita* (1961), 72, 92.

19. Speller, *Inquisition* (2008), 164–70, 266–8, 297–8, 304–5, 309–12, 318–20, 333, 375; Niccolini to Cioli, 19 Feb., and Guiducci to GG, 19 Mar. 33 (15:45, 71).

Freedberg, in Barnes and Wheelock, *Van Dyck* (1995), 156–7, 171n, demonstrates Bentivoglio's connections with the lynxes.

20. Pagano, civ–cvi, 106–139; Guiducci to GG, 27 Aug. 33 (15:241); Genovesi, *Processi* (1966), 283, 331–2.

21. Camerota, 514–21; Lerner, *LCF*, 516–18; Beaulieu, *NCCS*, 374–6.

22. The consensus is that the proceedings of 1616 did not make Copernican-ism a heresy. Cf. Speller, *Inquisition* (2008), 149; Lerner, in *Primi lincei* (2005), 322; Beretta, *LCF*, 567–8.

23. Heilbron, in McMullin, *Church* (2005), 281–4, 315n13; Beretta, *Galilée* (1998), 98–9, 104–13, 270–1, 276–8, and in *Primi lincei* (2005), 282, 287–8; Speller, *Inquisition* (2008), 344; emphasis added to the quotation.

24. Garzend, *Inquisition* (1912), 6–7, 14–23, 56–7, 106–12, 429–73, 501–16; cf. Beretta, *Gregorianum*, 84 (2003), 163–92, and Speller, *Inquisition* (2008), 22–3, 27.

25. Pagano, 110, and Genovesi, *Processi* (1966), 292 (6 Aug. 33).

26. Descartes to Mersenne, Apr. 1634, in Descartes, *Corresp.* (2003), 262, 264, 266q.

27. Dandelet, *Spanish Rome* (2001), 188–201.

28. Garcia, *Diodati* (2004), 321; Fabri de Peiresc to Francesco Barberini, 5 Dec. 34, 2 and 31 Jan. 35 (16:169–71, 187, 202).

29. Firpo, *Giorn. crit. fil. ital.*, 20 (1939), 21–4, 33–6; Black, *Inquisition* (2009), 90–2.

30. Maschietto, in Santinello, *Galileo* (1992), 438–9 (tobacco), 440; Masetti Zan-nini, *Vita* (1961), 44, 52–4, 75, 83, 134–5; *DBI*, 21:689; Castelli to GG, summer 1637, in Castelli, *Cart.* (1988), 168–80 (Castelli's appointment as abbot of S.M. di Praglia near Padua).

31. Maffioli, *Galilaeana*, 5 (2008), 190–4, 199–202. The episode took place in 1630/1.

32. Cf. Masetti Zannini, *Vita* (1961), 73; Castelli to GG, 8 and 29 Jan. 39 (18:14–15, 23–4); GG to Castelli, 8 Aug. 39 (18:81–21).

33. Camerota, 522–3, 526.

34. Bucciantini and Camerota, *Galilaeana*, 2 (2005), 231.

35. Caspar, *Kepler* (1959), 338–50; Wedgwood, *Thirty Years War* (1961), 314–17; Schiller, *Piccolomini* (1800).

36. Rutkin, *Galilaeana*, 2 (2005), 124–5, 138–42; Ernst, *Religione* (1991), 247–8, 251–2; Section 3.2 above.

37. GG to Diodati, 15 Jan. 33 (15:24).

38. Pagano, 146–7 (doc. 101). Nothing seems to have come of the charge, pre-ferred in January 1634.

39. Favaro, *Scampoli* (1992), 2, 673–8. The painting, made around 1650, was in pri-vate hands when Favaro found it.

40. GG to Cioli, 23 July, and Niccolini to same, 7 Aug. 33 (15:187–8q, 217).

41. Maria Celeste to GG, Autumn 1633, in Sobel, *Letters* (2001), 209, 255, 275, 277, 315.

42. Ibid., 197, 235, 241q (21 May 33), 247, 271, 279q (28 Jul), 291, 293, 297.

43. Ibid., 311, 327, 331, 337, 371, 377q.

44. Ibid., 261 (2 July), 265 (13 July).

45. Ibid. 295, 299, 317, 331, 343, 375.

46. Ibid., 267 (13 July 33); Aggiunti to GG, 27 Dec. 33 (15:364–5).

47. Sobel, *Letters* (2001), 325 (prison); 273, 285, 287, 289 (the enlargement, the adjacent house, bought jointly by Galileo and Bocchineri); 275, 279 (Landucci); 313 (Giulia's family); 343, 347, 367 (other philanthropy).

48. Ibid., 351q, 359, 363. Maria Celeste died on 2 Apr. 34.

49. GG to Diodati, 25 July 34 (16:116).

50. Ibid.; Sobel, *Letters* (2001), 355, 373.

51. Pagano, 192–3 (docs. 138–9, 16 June 33); *FA*, 292; Aggiunti to GG, 27 Dec. 33 (15:364–5); GG to Micanzio, 19 Nov., and to Diodati, 21 Dec. 34 (16:162–3, 177); Giovanni Peroni to GG, 4 Jan. 35 (16:188–90).

52. Micanzio to GG, 10 Feb. 35 (16:209).

53. Micanzio to GG, 27 Jan, 3 Feb. 35, and 11 Nov. 34 (16:200–1, 203, 154–5); Favaro, *Arch. ven.*, 13 (1907), 51, 57–9.

54. Bernegger to Wilhelm Schickard, 25 Feb. 34 (16:54); Garcia, *Diodati* (2004), 286–90, 327–8, 334–6; Garcia, in McMullin, *Church* (2005), 266–72, 273q.

55. GG to Fabri de Peiresc, 21 Feb. 35 (16:215–16); Castelli to GG, 9 Aug. 36 (16:273, 17:126).

56. GG to Diodati, 9 June 35 (16:273).

57. Pieroni to GG, 4 Jan., 11 and 18 Aug. (quote), 21 Oct., 15 Dec. 35, 9 Feb., 1 Mar., 19 Apr. 36 (16:188–90, 300–2, 304q, 359–60, 393–4, 398, 419–20); GG to Diodati, 18 Dec. 35 (16:361q).

58. Garcia, *Diodati* (2004), 300–3; Pagano, 204 (doc. 154); Drake, in *TNS*, xi, xv.

59. GG to Diodati, ? Aug. 1638 (17:370, 373).

60. Noailles to GG, 20 July 38 (17:357). Cf. Garcia, *Diodati* (2004), 257–60, 277–8.

61. *TNS*, ix–xi, 6–7q.

62. *TNS*, 11–14. Cf. Section 3.1 above.

63. *TNS*, 17–27; Baliani to GG, 27 July 30, and reply, 6 Aug. 30 (14:124–5, 127–9), on the limited rise of water in a siphon; Nenci, *Galilaeana*, 5 (2008), 81–6. Sarpi had considered the question of the cohesive force of vacuum (*Pensieri* (1996), 372–3), and Galileo had built a pump to exploit it during their early acquaintance (Section 3.1 above).

64. *TNS*, 29–39, 34q, 51, 54q, 57. Cf. Frajese, *Galileo matematico* (1964), 53–65, 201–3.

65. *TNS*, 55–7.

66. Sarpi, *Pensieri* (1996), 273–4 (Pen. 345); cf. ibid., 130–6 (Pen. 111–16).

67. GG to Micanzio, 19 Nov. 34 (16:163); *TNS*, 55–7.

68. *TNS*, 66–72, 87–8, 97–9, 104q; 106–7; Section 1.2 above.

69. *TNS*, 37, 57, 83; 15, 17, 51, 108.

70. *TNS*, 115, 118–9. $Ra \propto a^3$ and $V = a^2b$; hence for equal lengths, $R_1:R_2 = [V_1:V_2]^{3/2}$.

71. *TNS*, 120

72. *TNS*, 121–29.

73. *TNS*, 135–41, 141q.

74. *TNS*, 144–5.

75. *TNS*, 147–60, 161q. If $v \propto s$ and $t = s/v$, t = constant.

76. *TNS*, 164–70.

77. *TNS*, 171–8, 172q.

78. *TNS*, 170–81, 182q; the vertical circle is Prop.VI, p.178.

79. *TNS*, 186.

80. *TNS*, 202.

81. *TNS*, 215; Descartes to Mersenne, 11 Oct. 38, in Descartes, *Lettere* (2005), 880.

82. *TNS*, 217q.

83. *TNS*, 218–22.

84. *TNS*, 223–7, 225q.

85. *TNS*, 251.

86. *TNS*, 243–8.

87. *TNS*, 233–4.

88. The angles in question are $\alpha_0 \pm \tan^{-1}(2x/R)$.

89. *TNS*, 256–7; cf. Naylor, *Physis*, 16 (1974), 337–9.

90. *TNS*, 38, 141–2, 260, 281.

91. GG to Peiresc, 22 Feb. 35 (16:215–16).

92. GG to Peiresc, 16 Mar. 35 (16:234–5).

93. GG to Micanzio, 26 July 36 (16:455). Margolis, *Stud. hist. phil. sci.*, 22 (1991), 260n, rightly sees that anger at the supposed insult is an insufficient explanation of Urban's behavior, but substitutes the less probable thesis that the pope was incensed at Galileo's treatment of Tycho (272–5), whose system the *Dialogue* mentions only once. BH, *Dial.*, 717.

94. Examples of denials, 1636/7, 1641, in Pagano, 201–3, 205, 212 (docs. 151, 153, 155, 164).

95. Micanzio to GG, 21 June 36, and reply, 28 June 36 (16:443, 445q); Black, *Inquisition* (2009), 162.

96. GG to Carcavy, 5 June 37 (price of books), to Bernegger, 15 July 36, and to Peiresc, 16 Mar. 35 (nothing in bookstores) (17:89, 16:451–2, 235).

97. *OF*, 37.106, quoted, almost verbatim, in GG to Diodati, 24 Apr. 37 (17:63); GG to Micanzio, 30 Jan. 38 (269–71).

98. The translator, Marco Ambrogetti, stayed at Arcetri from 1 June 37 to 25 Jan. 39 (20:369); the 100 copies, in GG to Micanzio, 28 June 36 (16:445); Elzevier's interest, Micanzio to GG, 5 July 36 (16:446–7).

99. GG to Micanzio, 28 June (quote) and 26 July 36, and to Diodati, 14 Aug. 1638 (16:445q, 455, 17:373).

100. Varetti, Acc. naz. Lincei, Cl. sci. mor., stor. fil., *Rend.*, 15:3–4 (1939), 236–42.

101. Bedini, in McMullin, *Galileo* (1967), 280–3.

102. GG to Hortensius, 1636 (16:536); *DSB*, 6:520; Section 6.9 above; Garcia, *Diodati* (2004), 305–12.

103. Lorenzo Realio (Admiral Lorenzo Reael) to GG, 3 Mar., GG to Reael, ? June, Hortensius to Diodati, 22 June, and States General to GG, 25 Apr. 37 (17:40–1, 96–105, 119–20, 66).

104. Reael to GG, 22 June, GG to Diodati, 16 July (quote), and GG to Reael, 22 Aug. 37 (quote) (17:116–19, 136–7q, 175q); Bedini, in McMullin, *Galileo* (1967), 280–3 (Jovilabium); Varetti, Acc. naz. Linei, *Rend.*, 15:3–4 (1939), 248–51.

105. Muzzarelli to Barberini, 26 June and 25 July 38 (19:396–8); Micanzio to GG, 17 Sep., 8 and 22 Oct. 39 (18:104–5, 112, 115).

106. GG to Diodati, 14 Aug. 1638 (17:372).

107. GG to Diodati, 30 Dec. 39, and to Hugo Grotius, 15 Jan. 40; Micanzio to GG, 4 Jan. 42 (17:132–3, 140–1). Cf. Bedini, *Pulse* (1991), 18–21.

108. GG to States General of the Netherlands, 15 Aug. 36 (16:463–9); Bedini, *Pulse* (1991), 1–5, 20, 23–41, Plate 5.

109. Huygens, *Horologium* (1673), tr. Pighetti (1963), 31–45.

110. GG to Micanzio, 30 Jan. 38 ("il mio inquieto cervello") (17:271).

111. GG to Micanzio, 7 Nov. 37 (17:214).

112. Day 1, *TCWS*, 65–6. Galileo described the parallactic libration at length in a letter to Alfonso Antonini, 20 Feb. 38 (17:291–7); cf. BH, *Dial.*, 268–72.

113. Righini, *Contributo* (1978), 35–40.

114. GG to Diodati, 2 Jan. 38 (17:247).

115. Galileo to Baliani, 1 Aug. and 1 Sep. 39 (18:76–8, 93); Drake, *Galileo at work* (1978), 364, 398–400, 403, 419–20.

116. GG, ["Astronomical operations"] (8:453–66). Cf. the reports on Francesco Fontana's lenses, in Antonio Sontini to Castelli, 6 Oct., and reply, 18 Oct. 38, and Castelli to Cavalieri, 2 Oct. 38, in Castelli, *Cart.* (1988), 190, 192, 188.

117. Antonio Sontini to Castelli, 8 Oct. 38, in Castelli, *Cart.* (1988), 190–1.

118. Aproino to GG, 3 Mar. 35 (16:218–20); Gabrieli, *Contributi* (1989), 1, 980–1.

119. *TNS*, 242, 281–303; Drake, *Galileo at work* (1978), 383–4, 402; Dugas, *Mécanique* (1954), 72: "sur la percussion, Galilée s'est montré très confus."

120. Torricelli, *Opere* (1919), 2, 5–14, 5q; Moscovici, in McMullin, *Galileo* (1967), 434–8, 441–5. The scare quotes indicate that the words are not to be taken in the sense in which modern physics employs them.

121. GG to Liceti, 25 Aug. 40 (18:232–6); Camerota, 563–4; Drake, *Galileo at work* (1987), 405–8q; Galileo's lengthy reply, in the form of a letter to Leopoldo de' Medici, to Liceti's arguments about moonglow is in Op. 8:487–542.

122. GG to Liceti, ? 15 Sep. 40 (18:247–51); Drake, *Galileo at work* (1978), 408–10.

123. GG to Liceti, Jan. 41 (18:293–5); Drake, *Galileo at work* (1978), 411–12; Cavalieri to GG, 5 June 1641 (18:201), a book a week.

124. *TNS*, 149–50; Drake, *Galileo at work* (1978), 422–3q; Euclid, 6: Def. 5, ed. Heath, 2, 114q, 120–6. Expressed in algebra, $a:b = c:d$ if, given any numbers x, y, when $xa >$, =, or $< yc$, then $xb >$, =, or $< yd$.

125. Drake, *Galileo at work* (1978), 424–30.

126. *DBI*, 25:150 ("Ciampoli"); Godoli and Paoli, *Dimora* (1979), 8–9; Gabrieli, *Contributi* (1989), 1, 812–15; Castelli to GG, 1 Oct. 39 (18:109–10), first quote; Torricelli, *Opere* (1919), 2, 59q.

127. GG to Micanzio, 21 June and 16 Aug. 36 (16:441, 475q); Alberto Cesare Galilei to GG, 1 Aug. 36 (16:459–60).

128. Harris, IMSS, *Ann.*, 10:2 (1985), 5–9; Campbell and Corns, *Milton* (2008), 112–13; quotes from, resp., Byron, *Childe Harold's pilgrimage*, 4.54, and Milton, *Aereopagitica* (1644), in *Major works* (2003), 258–9.

129. Pagano, 208 (doc. 159, 13 July 38); GG to Baliani, 7 Jan. 39 (18:11); to Castelli, 8 Aug. 39 and 16 Apr. 40 (18:81, 178–9); Giuseppe Calasanzio (founder of the Piarists) to G.D. Romani (supervisor of the order in Florence), 16 Apr. 39 (18:41). Among other visitors: Cavalieri ("un ingegnio mirabile"), GG to Micanzio, 28 June and 16 Aug. 36 (16:444–5, 475q); Famiano Michelini to Castelli, 2 Sep. 34, in Castelli, *Cart.* (1988), 135.

130. GG to Buonarroti, ? Jan. 37 (17:24).

131. Pagano, 209 (doc. 161, 25 Nov. 38); Op. 19:38 (1638).

132. GG to Rinuccini, 5 Nov. 39 (18:120–1); Mestica, *Scritti* (1889), 199. The parallels (T/A): flight of Erminia (7.3)/flight of Angelica (1.33); duel of Argante and Tancredi (6.20)/duels of Rinaldo and Sacripante (2.5), Ruggiero and Mandricardo (30.45), and Ruggiero and Rodomonte (46.103); Rinaldo's leaving Armida (14.57)/Ruggiero's leaving Alcina (6.16); discord in Godfrey's camp (8.57)/discord in Agramonte's (27.40); Rinaldo in Jerusalem (19.30)/Rodomonte in Paris (16.20, 17.6, 18.8).

133. Rinuccini to GG, 23 Mar., and reply, 29 Mar. 41 (18:311, 314–16); Drake, *Galileo at work* (1978), 417–18q.

NOTES TO PAGES 355–361

134. Watson, *Survey ophth.*, 54 (2009), 638–9.

135. Grondona, in *Atti* (1967), 142, 149; GG to Micanzio, 7 Nov. 37 (17:211–12).

136. Castelli to GG, 13 Jun 37 (17:111–12), and to Dini, 25 July 37, in Castelli, *Cart.* (1988), 167q, 184q, quoting from Ps 126:5–6, rendered by King James' translators as "They that sow in tears shall reap in joy/He that goeth forth and weepeth, bearing precious seed, shall doubtless come again with rejoicing."

137. Francesco Barberini to Giovanni Muzzarelli, 6 Mar. 38 (17:310–11); Decrees of the Holy Office, Feb.–Mar. 38 (19:286–8) = Pagano, 206–7 (docs. 156–8, 4 and 25 Feb., 29 Mar. 38).

138. Castelli to GG, 12 Dec. 37 and 9 Jan. 38 (17:234, 254–5); Francesco Barberini to Giovanni Muzzarelli, 6 Mar. and 3 Apr., and Muzzarelli to Barberini, 13 Feb. (quote) and 10 Mar. 38 (quote) (17:310–11, 324, 290, 312–13).

139. GG (in Florence) to Diodati, Aug. 1638 (17:370).

140. GG to Castelli, 3 Sep. 39 and 24 Feb. 40 (18:97, 153–4).

141. Rinuccini to Leopoldo de' Medici, 15 Nov. 41 (18:368).

142. Niccolini to G.B. Gondi, 25 Jan. 42, and F. Barberini to Muzzarelli, same date (18:378–80); Galluzzi, in Machamer, *Companion* (1998), 417–20q, and in Lunardi and Sabbatini, *Il rimembrar* (2009), 203–6.

143. Vasari, *Lives*, ed. Burroughs (1946), 296q; Galluzzi, in Machamer, *Companion* (1998), 421–31, and in Lunardi and Sabbatini, *Il rimembrar* (2009), 207–20; Viviani to Lorenzo Magalotti, 24 July 73, in Garcia, *Diodati* (2004), 238–9.

144. Lunardi and Sabbatini, *Il rimembrar* (2009), 23.

145. Rendered from the transcription in Lunardi and Sabbatini, *Il rimembrar* (2009), 23–39.

146. D'Addio, *Considerazioni* (1985), 117–19; Brandmüller and Greipl, *Copernico* (1992), 183–484; Maffei, *Giuseppe Settele* (1987), and *RG*, 193–218, reprint documents in the case.

147. Calmet, in Galileo, *Opere* (1744), 4, 1–2.

148. For details of the story to follow, Heilbron, in McMullin, *Church* (2005), 279–322.

149. Riccioli, *Alm. nov.* (1651), xix, 478.

150. Riccioli, *Apologia* (1669), 4.

151. Riccioli, *Alm. nov.* (1651), 52.

152. Ibid., 304, 309.

153. Divini [i.e. Fabri], *Pro sua annotatione* (1661), 46–9.

154. Mayaud, *Condamnation* (1997), 110–11.

155. Torrini, in Lomonaco and Torrini, *Galileo e Napoli* (1987), 359, 364–73; Osbat, *Inquisitione* (1974), 15–25, 43–53, 63–4, 101–22, 172–6, 247–52.

156. Heilbron, in Kockel and Sölch, *Bianchini* (2005), 71–80.

157. Ceva, *Philosophia* (1704), 9, 28–30, 48, 68, 88.

158. Russell, *Ann. sci.*, 46 (1989), 388–9.

159. Marchetti, *Breve introduzione* (1738), "A chi legge," and 95, 101. Galleo's remote successor at Pisa was Alessandro Marchetti.

160. Frisi, *De motu* (1756), "Praefatio."

161. Heilbron, in McMullin, *Church* (2005), 309–10, 313–14.

162. *RG*, 263–6.

163. The quote is from an eyewitness: Blaschke, *Galileo und Kepler* (1943), 11.

164. Heilbron, *Sun* (1999), 210–11; *RG*, 318–37.

165. *RG*, 338–57; Brandmüller, *Galileo e la chiesa* (1992). Cf. Galli, *Angelicum*, 60 (1983), 420–7.

166. *RG*, 340–1.

167. Philippe de la Trinité, *Divinitas*, 25 (1959), 36, quoted by Poupard, *Galileo* (1987), xv.

168. GG to Picchena, 6 Mar. 16 (*FA*, 151)(12:244, "ogni diabolica suggestione").

169. Viviani, *Vita* (2001), 52q, 84q; Nelli, *Vita* (1793), x, 3q, 4q, 8q.

170. Fantoli, *Galileo* (1996), 511.

171. Belloni, *Scrittori* (1933), 6.

172. Favaro, *Scampoli* (1992), 2, 680, 683–7; Gori, *Reliquie* (n.d.), 17–37. The notice of the reopening of Florence's Museum of the History of Science as the Museo Galileo in *The New York Times*, 23 Jul. 2010, p. A1, makes a feature of these body parts.

173. Giuseppe Bianchini, "Sonnetto…composto dopo aver veduto il dito di gran Galileo," cited in Cochrane, *Academies* (1961), 114.

WORKS CITED

Abetti, Giorgio. *Amici e nemici di Galileo*. Milan: Bompiani, 1945.

Addio, Mario d'. *Considerazioni sui processi a Galilei*. Rome: Herder, 1985.

Adorno, Francesco and Luigi Zangheri. *Gli statuti dell'Accademia del Disegno*. Florence: Olschki, 1998.

Agnelli, Giovanni. *Topo-cronografia del viaggio dantesco*. Milan: Hoepli, 1891.

Alexander, Amir. "Lunar maps and coastal lines: Thomas Hariot's mapping of the moon." *Studies in the history and philosophy of science*, 29 (1998), 345–68.

Allacci, Leone. *Apes urbanae, sive De viris illustribus, qui ab anno MDCXXX per totum MDCXXXII Romae adfuerunt, ac typis aliquid evulgarunt*. Rome: L. Grignano, 1633.

Altieri Biagi, Maria Luisa and Bruno Basile, eds. *Scienziati del seicento*, vol. 2. Milan: Riccardi, 1980.

Anonymous. *Del genio di F. Paolo Sarpi in ogni facoltà scientifica*. 2 vols. Venice: L. Bassagia, 1785.

——. "Prefazione." In Galileo, *Considerazioni* (1793), iii–x.

Applebaum, Wilbur and Renzo Baldasso. "Galileo and Kepler on the sun as planetary power." *LCF*, 381–90.

Aquila, Giulia dell'. "Galileo tra Ariosto e Tasso." In Giandomenico and Guaragnella, *Prosa* (2006), 239–64.

Aquilecchia, Giovanni. "'In facie prudentis relucet sapientia.' Appunti sulla letteratura metoposcopica tra cinque e seicento." In Torrrini, *Della Porta* (1990), 199–228.

Archimedes. *De iis quae vehuntur in aqua libri duo*. Ed. Federico Commandino. Bologna: A. Benari, 1565.

——. *The works*. Ed. T.L. Heath. Cambridge: Cambridge University Press, 1912.

Ariosto, Ludovico. *Orlando furioso*. Ed. Lanfranco Caretti. 2 vols. Milan: Einaudi, 1992.

——. *Orlando furioso*. Tr. Guido Waldman. Oxford: Oxford University Press, 1998.

Aristotle. *The works*. Ed. Willam David Ross. Oxford: Oxford University Press, 1908–52.

Armogathe, Jean Robert and Vincent Carraud. "La première condamnation des *Oeuvres* de Descartes d'après des documents inédits aux Archives du Saint-Office." *Nouvelles de la République des Lettres*, 2001:2, 103–37.

Armour, Peter. "Galileo and the crisis in Italian literature of the early seicento." In Giovanni Aquilecchia et al., eds. *Collected essays on Italian language & literature presented to Kathleen Speight*. Manchester: Manchester University Press, 1971, 143–69.

Arrighi, Gino. "Un grande scienziato italiano: Guidobaldo del Monte in alcune carte inedite della Biblioteca Oliveriana di Pesaro." Accademia lucchese di scienze, lettere ed arti, *Atti*, 12 (1965), 181–99.

Artigas, Mariano, Rafael Martínez, and William R. Shea. "New light on the Galileo affair?" In McMullin, *Church* (2005), 213–33.

Atti del Symposium internazionale di storia, metodologia, logica e filosofia della scienza "Galileo nella storia e nella filosofia della scienza." Florence: G. Barbèra, 1967.

Bacon, Francis. *The essays, or counsels, civil and moral* [1625]. In Bacon. *The major works*. Ed. Brian Vickers. Oxford: Oxford University Press, 2002.

Baffetti, Giovanni. "Federico Borromeo e i Lincei: La spiritualità della nuova scienza." In Andrea Battistini, ed. *Mappe e letture*. Bologna: Il Mulino, 1994, 85–102.

Baldini, Ugo. "L'astronomia del Cardinale Bellarmino." NCCS, 293–305.

——. "La 'nova' del 1604 e i matematici e i filosofi del Collegio Romano." IMSS, *Ann.*, 6:2 (1981), 63–98. (Also in Baldini, *Legem impone*, 155–82.)

——. *"Legem impone subactis." Studi su filosofia e scienza dei Gesuiti in Italia, 1540–1632.* Rome: Bulzoni, 1992.

——. "Le congregazioni romane dell'Inquisizione e dell'Indice e le scienze, dal 1542 al 1615." In *L'Inquisizione e gli storici: Un cantiere aperto*. Rome: Accademia Nazionale dei Lincei, 2000, 329–64.

Balsamo, Luigi. *Antonio Possevino, S.J. Bibliografo della controriforma*. Florence: Olschki, 2006.

Banfi, Antonio. *Vita di Galileo Galilei*. Milan: La Cultura, 1930.

Barbero, Giliola, Massimo Bucciantini, and Michele Camerota. "Uno scritto inedio di Federico Borromeo." *Galilaeana*, 4 (2007), 309–41.

Barbi, Adrasto Silvio. *Un accademico mecenate e poeta, Giovan Battista Strozzi il giovane (1551–1634)*. Florence: Sansoni, 1900.

Barbi, Michele. *Della fortuna di Dante nel secolo xvi*. Pisa: T. Nistri, 1890.

Barcaro, Umberto. "Riflessioni sul mito platonico del *Dialogo*." NCCS, 117–25.

Baroncini, Gabriele. *Forme di esperienze e rivoluzione scientifica*. Florence: Olschki, 1992.

Barra, Mario. "Galileo Galilei e la probabilità." LCF, 101–18.

Barzaghi, Antonio. *Donne o cortigiane? La prostituzione a Venezia: Documenti di costume dal xvi al xviii secolo*. Verona: Bertani, 1980.

Barzman, Karen—edis. *The Florentine Academy and the early modern state*. Cambridge: Cambridge University Press, 2000.

Basile, Bruno. "Galileo e il teologo 'Copernicano' Paolo Antonio Foscarini." *Rivista di letteratura italiana*, 1 (1983), 63–96.

——. *Il tempo e la memoria. Studi di critica testuale*. Modena: Mucchi, 1996.

Bassanese, Fiora A. "Private lives and public lies. Texts by courtesans of the Italian Renaissance." *Texas studies in literature and language*, 30:3 (1988), 295–319.

Battistini, Andrea. "Gli 'aculei' ironici della lingua di Galileo." *Lettere italiane*, 30 (1978), 289–332.

——. *Gli scrittori. Introduzione a Galilei*. Rome: Laterza, 1989.

——. "'Cedat Columbus' e 'Vicisti, Galilaee!' Due esploratori a confronto nell'immaginario barocco." *Annali d'italianistica*, 10 (1992), 116–32.

——. "'Girondole' verbali e 'severità di geometriche dimonstazioni.' Battaglie linguistiche nel Saggiatore." *Galilaeana*, 2 (2005), 87–106.

Beaulieu, Armand. "Les réactions des savants français au début du xviie siècle devant l'héliocentrisme de Galilée." *NCCS*, 373–81.

——. "L'influence de Della Porta sur la physique en France au xviie siècle." In Torrini, *Della Porta* (1990), 291–309.

Bedini, Silvio. "The instruments of Galileo Galilei." In McMullin, *Galileo* (1967), 256–92.

——. "The makers of Galileo's scientific instruments." In *Atti* (1967), 89–114.

——. *The pulse of time. Galileo Galilei, the determination of longitude, and the pendulum clock*. Florence: Olschki, 1991.

Beijer, Agne. "Visions célestes et infernales dans le théâtre du moyen âge et de la renaissance." In Jacquot, *Fêtes* (1956), 1, 405–17.

Bellinati, Claudio. "Integrazioni e correzzioni alle pubblicazioni di A. Favaro." In Santinello, *Galileo* (1992), 257–65.

——. "Galileo e il sodalizio con ecclesiastici padovani." In Santinello, *Galileo* (1992), 327–57.

Bellini, Eraldo. *Umanisti e lincei: letteratura e scienza a Roma nell'età di Galileo*. Padua: Antenore, 1997.

——. *Agostino Mascardi tra "ars poetica" e "ars historica."* Milan: Vita e pensiero, 2002.

——. "'Il papato dei virtuosi.' I lincei e i Barberini." In *Primi lincei* (2005), 45–97.

Belloni, Antonio. *Scrittori italiani con notizie storiche e analisi estetiche. Galilei*. Turin: Paravia, 1933.

——et al., eds. *Storia letteraria d'Italia. Il seicento*. 3rd edn. Milan: Villardi, 1943.

Belloni, Luigi. "Il microscopio applicato alla biologia da Galileo e della sua scuola (1610–1661)." In Maccagni, *Saggi*, 3:2 (1972), 689–730.

Benedetti, Giovanni Battista. *Diversarum speculationum mathematicarum liber* [1685]. Tr. I. E. Drabkin. In Drake and Drabkin, *Mechanics* (1969), 166–237.

Benetti, Francesca Zen. "Per la biografia di Lorenzo Pignoria, erudito padovano." In Maria Chiara Billanovich et al. eds. *Viridarium floridum*. Padua: Antenore, 1984, 317–36.

Benivieni, Hieronymo. *Dialogo di Antonio Manetti cittadino fiorentino circa il sito, forma e misura dello Inferno, di Dante Alighieri* [1506]. Ed. Nicola Zingarelli. Castello: S. Lapi, 1897.

Benrath, D. Karl. *Neue Briefe von Paolo Sarpi (1608–1616)*. Leipzig: R. Haupt, 1909.

Bentivoglio, Guido. *Memorie...con le quali descrive la sua vita*. Venice: Baglioni, 1648.

Benzoni, Gino. "La storiografia e l'erudizione storico-antquaria. Gli storici municipali." In *Storia*, 4/2 (1984), 67–93.

——. "I Ricovrati nel '600." In Riondato, *Accademia* (2001), 11–57.

Beretta, Francesco. *Galilée devant le Tribunal de l'Inquisition*. Fribourg: Université de Fribourg, 1998.

——. "Un nuovo documento sul processo di Galileo Galilei." *Nuncius*, 16 (2001), 629–41.

——. "Urbain VIII Barberini protagoniste de la condamnation de Galilée." *LCF*, 549–74.

——. "L'Affaire Galilée et l'impasse apologétique." *Gregorianum*, 84 (2003), 163–92.

——. "Rilettura di un documento celebre: Redazione e diffusione della sentenza e abiura di Galileo." In *Primi lincei* (2005), 273–319.

——. "Les dominicains, le procès de Galilée et les historiens." *Histoire dominicaine*, 20 (2006), 55–66.

—— and M.P. Lerner. "Un édit inédit. Autour du placard de mise à l'Index de Copernic par le maître du Sacré Palais Giacinto Petroni." *Galilaeana*, 3 (2006), 199–216.

Bernabeo, Raffaele. "La mancata venuta di Galileo Galilei allo studio di Bologna." In Congresso, xx, *Atti* (1964), 196–205.

Bertacchi, Cosimo. *Dante geometra. Note di geografia medioevale*. Turin: Istituto Fornaris-Marocco, 1887.

Berti, Enrico. "Galileo e l'Aristotelismo patavino del suo tempo." In *Galileo a Padova* (1983), 527–45.

Bertola, Francesco. "Le osservazioni di Galilei del pianeta Nettuno." In Santinello, *Galileo* (1992), 267–75.

Bertolà, Giovanna and Annamaria Petrioli Tofani, eds. *Feste e apparati medici da Cosimo I a Cosimo II. Mostra di disegni e incisioni*. Florence: Olschki, 1969.

Bertolotti, A. "Giornalisti, astrologi e negromanti in Roma nel secolo xvii." *Rivista europea*, 5 (1878), 466–514.

Besomi, Ottavio, ed. *Ariosto, Boiardo, Marino, Pulci, Tasso*...Hildesheim: Olms, 1994.

Biagioli, Mario. "Filippo Salviati: A baroque virtuoso." *Nuncius*, 7:2 (1992), 81–96.

——. *Galileo courtier. The practice of science in the age of absolutism*. Chicago: University of Chicago Press, 1993.

——. *Galileo's instruments of credit. Telescopes, images, secrecy*. Chicago: University of Chicago Press, 2006.

Biancani, Giuseppe. *Sphaera mundi, seu Cosmografia demonstrativa*. Bologna: S. Bonomi, 1620.

Bianchi, Luca. "Agostino Oreggi, qualificatore del *Dialogo*, e i limiti della conoscenza scientifica." *LCF*, 575–84.

Bigi, Emilio. "Galileo lettore." In Maccagni, *Saggi*, 3:2 (1972), 519–40.

Binacarelli Martinelli, Roberto. "Paul Hamberger: Il primo intermedio tra Galileo e Keplero." *Galilaeana*, 1 (2004), 171–81.

Biondi, Albano. "La *Bibliotheca selecta* di Antonio Possevino. Un progetto di egemonia culturale." In Gian Paolo Brizzi, ed. *La "Ratio studiorum." Modelli culturali e practiche educative dei Gesuiti in Italia tra cinque e secento*. Rome: Bulzone, 1981, 43–75.

Birley, Robert. *The Jesuits and the Thirty Years War*. Cambridge: Cambridge University Press, 2003.

Black, Christopher F. *The Italian inquisition*. New Haven: Yale University Press, 2009.

Blackwell, Richard J. *Galileo, Bellarmine, and the bible*. Notre Dame: Notre Dame University Press, 1991.

——. "Introduction." In Campanella, *Defense* (1994), 1–34.

——. *Behind the scenes at Galileo's trial. Including the first English translation of Melchior Inchofer's* Tractatus syllepticus. Notre Dame: Notre Dame University Press, 2006.

Blaschke, Wilhelm. *Galileo und Kepler*. Leipzig: Teubner, 1943.

Bloom, Terrie. "Borrowed perceptions: Harriot's maps of the moon." *JHA*, 9 (1978), 117–22.

Blum, P.R. "Die geschmückte Judith. Die Finalisierung der Wissenschaften bei Antonio Possevino, S.J." *Nouvelles de la République des Lettres*, 1983:1, 113–26.

Boethius, A.M.S. *Consolation of philosophy*. Tr. "I.T." (1609), revised by H.F. Stewart. Cambridge: Harvard University Press, 1962.

Bolzoni, Lina. "Un modo da commentare alla fine dell'umanesimo: I 'Commen-taria' del Campanella ai 'Poemata' di Urbano VIII." Scuola normale superiore, Pisa, Classe di lettere e filosofia, *Annali*, 19 (1989), 289–311.

——. "Giochi di prospettiva sui testi: Galileo lettore di poesia." *Galilaeana*, 4 (2007), 157–75.

Bonelli, Maria Luisa. "Note about Galileo's instruments." In *Atti* (1967), 125–7.

Bonifazi, Neuro. *Jacopo Mazzoni: La Defesa di Dante. Passi scelti*. Urbino: Angalia, 1982.

Booth, Sara Elizabeth and Albert van Helden. "The virgin and the telescope. The moons of Cigoli and Galileo." In Renn, *Galileo in context* (2001), 193–216.

Borro, Girolamo. *De motu gravium, & levium*. Florence: G. Marescotti, 1576.

——. *Del flusso e riflusso del mare, & Dell'inondatione del Nilo*. 3rd edn, Florence: G. Marescotti, 1583.

Bösel, Richard. *Orazio Grassi. Architetto e matematico gesuita*. Rome: Argos, 2004.

Borsetto, Luciana and Bianca Moria da Rif. *Formazione e fortuna del Tasso nella cultura della Serenissima*. Venice: Istituto veneto di scienze, 1997.

Bosco, Domenico. "Cremonini e le origini del libertinismo." In *Cesare Cremonini* (1990), 249–99.

Bossi, Albertazzi. "Galileo tecnico." In *Atti* (1967), 79–87.

Bouwsma, William J. *Venice and the defense of republican liberty*. Berkeley: University of California Press, 1968.

Boyer, Carl B. "Galileo's place in the history of mathematics." In McMullin, *Galileo* (1967), 232–55.

Bracciolini, Giuliano. *L'Elettione di Urbano VIII*. Rome, n.p., 1628.

Brady, Bernadotte. "Four Galilean horoscopes: An analysis of Galileo's astrological techniques." In *Galileo's astrology* (2003), 113–44.

Branca, Vittore. "Galileo fra Petrarca e l'umanesimo veneziano." In *Galileo Galilei* (1995), 339–50.

Brand, Charles Peter. *Torquato Tasso*. Cambridge: Cambridge University Press 1965.

——. *Ariosto. A preface to the "Orlando furioso."* Edinburgh: Edinburgh University Press, 1974.

Brandmüller, Walter. *Galileo e la chiesa, ossia il diritto di errore*. Vatican City: Libreria vaticana, 1992.

——and Egon Johannes Geipl, eds. *Copernico, Galileo e la chiesa: Fine della controver-sia (1820)*. Florence: Olschki, 1992.

Bredekamp, Horst. "Zwei frei Skizzenblätter Galileo Galileis." In *Ars naturam adi-uvans. Festschrift für Matthias Winner*. Ed. Victoria von Flemming and Sebastian Schütze. Mainz: Zabern, 1996, 477–84.

——. *Galileo der Künstler: der Mond, die Sonne, die Hand*. Berlin: Akademie Verlag, 2007.

Brieger, Peter. "Pictorial commentaries to the Commedia." In *Illuminated manuscripts of the "Divine Comedy."* Ed. Peter Brieger et al. 2 vols. Princeton: Princeton University Press, 1969. Vol. 1, pp. 83–113.

Brooks, Julian, ed. *Graceful and true. Drawing in Florence c. 1600*. Oxford: Ashmolean Museum, 2003.

Brown, Howard Meyer. "Vincenzo Galilei in Rome: His first book of lute music." In Coelho, *Music* (1992), 153–84.

Brown, P.H. "In defense of Ariosto: Giovanni de' Bardi and Lionardo Salviati." *Studi seicenteschi*, 12 (1971), 3–27.

Brunner, Michael. *Die Illustrierung von Dantes Divina commedia in der Zeit der Dante–Debatte (1570–1600)*. Munich: Deutscher Kunstverlag, 1999.

Bruno, Giordano. *The expulsion of the triumphant beast* [1584]. Tr. A.D. Imerti. Lincoln: University of Nebraska Press, 2004.

Bucciantini, Massimo. "Dopo il 'Sidereus nuncius': Il copernicanismo in Italia tra Galileo e Keplero." *Nuncius*, 9:1 (1994), 15–35.

——. *Contro Galileo. Alle origini dell'Affaire*. Florence: Olschki, 1995.

——. *Galileo e Keplero. Filosofia, cosmologia e teologia nell'età della Controriforma*. Turin: Einaudi, 2003.

——. "Reazione alla condanna di Copernico: Nuovi documenti e nuove ipotesi di ricerca." *Galilaeana*, 1 (2004), 3–19, reprinted in *Primi lincei* (2005), 301–19.

—— and Michele Camerota. "Once more about Galileo and astrology." *Galilaeana*, 2 (2005), 229–32.

—— and Maurizio Torrini, eds. *Geometria e atomismo nella scuola galileiana*. Florence: Olschki, 1992.

——. *La diffusione del Copernicanismo in Italia 1543–1610*. Florence: Olschki, 1997.

Buonamici, Francesco. *De motu libri x*. Florence: B. Sernatelli, 1591.

Burke, Peter. "The great unmasker: Paolo Sarpi." *History today*, 15:6 (1965), 436–42.

Burton, Robert. *The anatomy of melancholy* [1621]. Ed. Holbook Jackson. New York: New York Review of Books, 2001.

Büttner, Jochen. "Galileo's cosmography." *LCF*, 391–401.

——. "Big wheel keep on turning." *Galilaeana*, 5 (2008), 33–62.

Butts, R.E. and J.C. Pitt, eds. *New perspectives on Galileo*. Dordrecht: Kluwer, 1978.

Caffarelli, Roberto Vergare. *Galileo e Pisa*. Pisa: Felici, 2004.

Cagnoli, Antonio. *Notizie astronomiche, adatte all'uso commune*. 2 vols. Reggio: Fiaccadori, 1827.

Calmet, Agostino. "Dissertatione sopra il sistema del mondo degli antichi ebrei." In Galileo Galilei. *Opere*. 4 vols. Padua: Stamperia del Seminario, 1744. Vol. 4, pp. 1–20.

Camerota, Filippo. "Il contributo di Galileo alle matematizzazione delle arti." *Galilaeana*, 4 (2007), 79–103.

Camerota, Michele. *Galileo Galilei e la cultura scientifica nell'età della Controriforma*. Rome: Salerno, 2004. Abbreviated "Camerota."

——. "Aristotelismo e nuova scienza nell'opera di Christoph Scheiner: Intorno ad un commenatrio *De caelo*." *Galilaeana*, 2 (2005), 43–85.

——and Mario O. Helbing. "Galileo and Pisan Aristotelianism: Galileo's 'De motu antiquiora' and the 'Quaestiones de motu elementorum' of the Pisan professors." *Early science and medicine*, 5 (2000), 319–65.

Campanella, Tommaso. *A defence of Galileo the mathematician from Florence* [1622]. Tr. Richard J. Blackwell. Notre Dame: University of Notre Dame Press, 1994.

——. *Opuscoli astrologici*. Ed. Germana Ernst. Milan: Rizzoli, 2003.

Campbell, Gordon and Thomas N. Corns. *John Milton. Life, work, and thought*. New York: Oxford University Press, 2008.

Campinotti, Veronica. "Galileo contro Aristotele nello Studio di Pisa: Resoconto di una *Disputatio circularis* di Alessandro Marchetti sulla natura delle comete." *Galilaeana*, 3 (2006), 217–28.

Canone, Eugenio, ed. *Bibliothecae selectae. De Cusano a Leopardi*. Florence: Olschki, 1993.

Cantimori, Delio. "Galileo e la crisi della Controriforma." In Maccagni, *Saggi*, 3:2 (1972), 401–15.

Capra, Baldassar. *Usus et fabrica circini cuiusdam proportionis*. Padua: P. Tozzi, 1607.

Carella, Candida. "Antonio Possevino e la biblioteca 'selecta' del principe cristiano." In Canone, *Bibliothecae* (1993), 507–16.

Caroti, Stefano. "Un sostenitore Napoletano della mobilità della terra: il padre Paolo Antonio Foscarini." In Lomonaco and Torrini, *Galileo e Napoli* (1987), 81–121.

Carugo, Adriano. "L'insegnamento della matematica all'Università di Padova prima e dopo Galileo." In *Storia*, 4/2 (1984), 151–99.

——. "Les Jésuites et la philosophie naturelle di Galileo." *History and technology*, 4 (1987), 321–33.

——and Alistair C. Crombie. "The Jesuits and Galileo's ideas of science and nature." IMSS, *Ann.*, 8:2 (1983), 3–67.

Casanovas, Juan. "Il P. Orazio Grassi e le comete dell'anno 1618." *NCCS*, 307–13.

Caspar, Max. *Kepler*. London: Abelard-Schuman, 1959.

Castellani, C., ed. *Lettere inedite di Fra Paolo Sarpi a Simone Contarini ambasciatore veneto in Rome, 1915.* Venice: Visentini, 1892. (Deputazione veneta di storia patria, *Miscellanea*, 12:1.)

Castelli, Benedetto. *Carteggio.* Ed. Massimo Bucciantini. Florence: Olschki, 1988.

Castelloni, Giuseppe. "La vocazione alla Compagnia di Gesù del p. Antonio Possevino da una relazione inedita del medesimo." *Archivum historicum Societatis Jesu,* 14 (1945), 102–24.

Ceccarelli, Ibaldo. "Brani di lettere da alcuni medici a Galileo." In Congresso, xx, *Atti* (1964), 371–81.

Cerbu, Thomas. "Melchior Inchofer, 'un homme fin & rusé.'" *LCF,* 587–611.

——and Michel-Pierre Lerner. "La disgrâce de Galilée dans les *Apes urbanae*: Sur la fabrique du texte de Leone Allacci." *Nuncius,* 15:2 (2000), 589–610.

Cesare Cremonini (1550–1631). Il suo pensiero e il suo tempo. Convegno di studi. Cento: Centro Studi "Girolamo Baruffaldi," 1990.

Cesi, Federico. "Del natural desiderio di sapere [1616]." In Altieri Biagi and Basile, *Scienziati* (1980), 39–70.

——. "De caeli unitate…ex sacris litteris epistola [1618]." In Altieri Biagi and Basile, *Scienziati* (1980), 9–35.

——. *Apiarium* [1626]. Ed. Luigi Guerrini, tr. Marco Guardo. Rome: Accademia Nazionale dei Lincei, 2005 (http://brunelleschi.imss.fi.it/apiarium/index.asp).

Ceva, Tommaso. *Philosophia antiqua-nova* [1704]. Milan: Bellagatta, 1718.

Chemello, Adriana. "Donna di palazzo, moglie, cortigiana: Ruoli e funzioni sociali delle donne in alcuni trattati del cinquecento." In *La corte e il "cortegiano."* 2 vols. Rome: Bulzoni, 1980. Vol. 2, pp. 113–32.

——. "Introduzione." In Fonte, *Merito* (1988), ix–lxvi.

Chevalley, Catherine. "Kepler et Galilée dans la bataille du *Sidereus nuncius.*" *NCCS,* 167–75.

Chiaramonti, Scipione. *Opuscola varia mathematica.* Bologna: C. Zeneti, 1653.

Chiari, Alberto. "Galileo Galilei. Due lezioni all'Accademia fiorentina circa la figura, sito e grandezza dell'Inferno di Dante." In Chiari. *Scritti letterari.* Florence: F. Le Monnier, 1943.

Chojnacka, Monica. *Working women of early modern Venice.* Baltimore: Johns Hopkins University Press, 2001.

Ciardi, Roberto Paolo. "I Vallombrosiani e le arti figurative." In Ciardi, *Vallombrosa* (1999), 27–107.

——, ed. *Vallombrosa. Santo e meraviglioso luogo.* Pisa: Pacini, 1999.

Clavelin, Maurice. *La philosophie naturelle de Galilée.* Paris: A. Colin, 1968.

———. "Le 'Dialogue' ou la conversation rationnelle. A propos de la première journée." *NCCS*, 17–29.

———. "Galilée astronome philosophe." *LCF*, 19–29.

Clericuzio, Antonio and Germana Ernst, eds. *Il Rinascimento italiano e l'Europa*. Vol. 5. *Le scienze*. Treviso: Angela Colla, 2008.

Clubb, Louise George. *Giambattista della Porta dramatist*. Princeton: Princeton University Press, 1965.

Clucas, Stephen. "Galileo, Bruno and the rhetoric of dialogue in seventeenth-century natural philosophy." *History of science*, 46 (2008), 405–29.

Cochrane, Eric. *Tradition and enlightenment in the Tuscan academies, 1690–1800*. Chicago: University of Chicago Press, 1961.

———. *Florence in the forgotten centuries 1527–1800*. Chicago: University of Chicago Press, 1973.

Coelho, Victor, ed. *Music and science in the age of Galileo*. Dordrecht: Kluwer, 1992.

Cohen, I. Bernard. "Galileo and Newton." In Maccagni, *Saggi*, 3:2 (1972), 376–400.

Colapietra, Raffaele. "Il pensiero estetico galileiano." *Belfagor*, 11 (1956), 557–69.

Cole, Jamie. "Cultural clientelism and brokerage networks in early modern Florence and Rome: New correspondence between the Barberini and Michelangelo Buonarroti the Younger." *Renaissance quarterly*, 60 (2007), 729–88.

———. *A muse of music in early baroque Florence: The poetry of Michelangelo Buonarroti il giovane*. Florence: Olschki, 2007.

Commandino, Federico. *Liber de centro gravitatis solidorum*. Bologna: A. Benaci, 1565.

Congresso nazionale di storia della medicina, xx. *Atti*. Rome: E. Cossidente, 1964.

Cooper, Lane. *Aristotle, Galileo, and the Tower of Pisa*. Ithaca: Cornell University Press, 1935.

Copernicus, Nicholas. *De revolutionibus orbium coelestium* [1543]. Tr. Edward Rosen. Baltimore: Johns Hopkins University Press, 1992.

Cornford, Francis M. *Plato's cosmology. The Timaeus of Plato*. Indianapolis: Bobbs-Merrill, n.d.

Corsano, Antonio. "La poetica del Campanella." *Giornale critico della filosofia italiana*, 39 (1960), 357–72.

———. "Campanella e Galileo." *Giornale critico della filosofia italiana*, 44 (1965), 313–32.

Costantini, Claudio. *Baliani e i Gesuiti. Annotazioni in margine alla correspondenza del Baliani con Gio. Luigi Confalonieri e Orazio Grassi*. Florence: Giunti, 1969.

Costanzo, Mario. *Critica e poetica del primo seicento*. Vol. 1. *Inediti di Giovanni Ciampoli (1590–1643)*. Rome: Bulzoni, 1969.

——. *Critica e poetica del primo seicento*. Vol. 2. *Maffeo e Francesco Barberini, Cesarini, Pallavicino*. Rome: Bulzoni, 1970.

Cowan, Alexander. "Patricians and partners in early modern Venice." In Ellen E. Kittell and Thomas F. Madden, eds. *Medieval and Renaissance Venice*. Urbana: University of Illinois Press, 1999, 276–93.

Cozzi, Gaetano. *Il doge Nicolò Contarini. Ricerche sul patriziato veneziano agli inizi del seicento*. Venice: Istituto per la collaborazione culturale, 1958.

——. "Cultura politica e religione nella 'pubblica storiografia' veneziana del '500." Istituto di storia della società e dello stato veneziano, *Bolletino*, 5–6 (1963–64), 215–94.

——. "Padri, filii e matrimoni clandestini (metà sec. xvi–metà sec. xviii)." *La cultura*, 14 (1976), 169–213.

——. *Paolo Sarpi tra Venezia e l'Europa*. Turin: Einaudi, 1979.

——. "Note introduttive." In Sarpi, *Pensieri* (1996), xi–cxxxix.

——. *Ambiente veneziano, ambiente veneto. Saggi su politica, società, cultura nella Repubblica di Venezia in età moderna*. Venice: Marsilio, 1997.

—— and Luisa Cozzi. "Paolo Sarpi." In *Storia* 4/2 (1984), 1–56.

Crapulli, Giovanni. *Mathesis universalis: Genesi di una idea nel xvi secolo*. Rome: Ateneo, 1969.

Crasso, Lorenzo. *Elogii degli huomini letterati*. 2 vols. Venice: Combi & La Noù, 1666.

Cremonini, Cesare. *Le orazione*. Ed. Antonino Poppi. Padua: Antenore, 1998.

Crombie, Alistair C. "Sources of Galileo's early natural philosophy." In Righini Bonelli and Shea, *Reason* (1975), 157–95.

Damerow, Peter and Jürgen Renn. "Galileo at work. His complete notes on motion in an electronic representation." *Nuncius*, 13 (1998), 781–9.

—— et al. *Exploring the limits of preclassical mechanics*. New York: Springer, 1992.

Dandelet, Thomas James. *Spanish Rome 1500–1700*. New Haven: Yale University Press, 2001.

Daniele, Antonio. "'Un pura disputa di cose poetiche, senza rancore di sorte alcuna.' Alessandro Tassoni, Cesare Cremonini e Giuseppe degli Aromatori." In Riondato and Poppi, *Cremonini* (2000), 19–41.

Davi, Mariarosa. "L'immortalità dell'anima nei manoscritti cremoniniani della Bibliothèque nationale de Paris." In Riondato and Poppi, *Cremonini* (2000), 1, 125–9.

David, F.N. *Games, gods, and gambling*. New York: Hafner, 1962.

Daxecker, Franz. *Das Hauptwerk des Astronomen P. Christoph Scheiner S.J. Rosa ursina sive sol. Eine Zusammenfassung*. Innsbruck: Universitätsverlag Wagner, 1996. (Naturwissenschaftlich–Medizinische Verein, Innsbruck. *Berichte*, suppl. 13.)

——. *The physicist and astronomer Christopher Scheiner*. Innsbruck: University of Innsbruck, 2004.

Deiss, Bruno M. and Volker Nebel. "On a pretended observation of Saturn by Galileo." *JHA*, 29 (1998), 215–20.

Delius, Walter. *Antonio Possevino S.J. und Ivan Groznyj. Ein Beitrag zur Geschichte der kirchlichen Union und der Gegenreformation des 16. Jahrhunderts.* Stuttgart: Evangelisches Verlagswerk, 1962. (*Kirche im Osten, 3:* Beiheft.)

Della Porta, Giovambattista. *Natural Magick…in twenty books* [1589]. London: T. Young and S. Speed, 1658.

Della Terza, Dante. "Galileo letterato: 'Considerazioni al Tasso.'" In Della Terza. *Forma e memoria*. Rome: Bulzoni, 1979, 197–221.

Del Lungo, Isidorio. *Patria italiana*. 2 vols. Bologna: Zanichelli, 1909.

——. *Del carteggio e dei documenti. Pagine di vita di Galileo*. Florence: Sansoni, 1968.

——and Antonio Favaro. *La prosa di Galileo per saggi criticamente disposti ad uso scolastico e di cultura*. Florence: Sansoni, 1911, 1957.

Del Monte, Guidobaldo. *Le mechaniche* [1581]. Tr. Stillman Drake. In Drake and Drabkin, *Mechanics* (1969), 239–328.

Del Torre, Maria Assunta. *Studi su Cesare Cremonini. Cosmologia e logica nel tardo aristotelismo padovano*. Padua: Antenore, 1968.

——. "Gli aspetti complessivi dell'opera di Cesare Cremonini. In *Cesare Cremonini* (1990), 15–28.

De Pace, Anna. "Galileo lettore di Girolamo Borrri nel 'De motu'." In *De motu. Studi di storia del pensiero su Galileo, Hegel, Huygens e Gilbert*. Milan: Cisalpino Istituto editoriale universitario, 1990, 3–69.

——and Francesco Barozzi. *Le matematiche e il mondo. Ricerche su un dibattito in Italia nella seconda metà del cinquecento*. Milan: FrancoAngeli, 1993.

Descartes, René. *Tutte le lettere, 1619–1650. Testo francese, latino e olandese*. Ed. Giulia Belgioso. Milan: Bompiani, 2005.

Diffley, P.B. *Paolo Berni. A biographical and critical study*. Oxford: Oxford University Press, 1988.

Divini, Eustachio [Fabri, Honoré]. *Septempedanus pro sua annotatione in systema saturnium*. Rome: Dragondelli, 1661.

Dollo, Corrado. "L'egemonia dell'Archimedismo in Galilei." In Dollo, *Archimede* (1992), 199–223.

——, ed. *Archimede. Mito tradizione scienza*. Florence: Olschki, 1992.

——. "La ragione del geocentrismo nel Collegio romano (1572–1612)." In Bucciantini and Torrini, *Diffusione* (1997), 99–167.

Donnelley, John Patrick. "Antonio Possevino, S.J., as a Counter-Reformation critic of the arts." Rocky Mountain Medieval and Renaissance Association, *Journal*, 3 (1982), 153–64.

——. "Antonio Possevino and Jesuits of Jewish origin." *Archivum historicum Societatis Jesu*, 55 (1986), 3–31.

Drabkin, I.E. "A note on Galileo's *De motu*." *Isis*, 51 (1960), 271–7.

Drabkin, I.E. and Stillman Drake. *Galileo Galilei on motion and on mechanics*. Madison: University of Wisconsin Press, 1960.

Drake, Stillman. Stillman Drake. *Discoveries and opinions of Galileo*. New York: Anchor, 1957.

——. "Galileo gleanings, V. The earliest version of Galileo's Mechanics." *Osiris*, 13 (1958), 262–90.

——. "Galileo gleanings, VII. An unrecorded manuscript copy of Galileo's cosmology." *Physis*, 1 (1959), 294–306.

——. "Galileo gleanings, XII. An unpublished letter of Galileo to Peiresc." *Isis*, 53 (1962), 201–11. (*Essays*, 3, 45–56.)

——. *Galileo studies. Personality, tradition, and revolution*. Ann Arbor: University of Michigan Press, 1970.

——. "Galileo gleanings, XXI. On the probable order of Galileo's notes on motion." *Physis*, 14 (1970), 55–68. (*Essays*, 1, 171–84.)

——. "Galileo's 'Platonic' cosmogony and Kepler's Prodromus." *JHA*, 4 (1973), 174–91.

——. "Galileo's language: Mathematics and poetry in a new science." *Yale French studies*, 49 (1973), 13–27. (*Essays*, 1, 50–62.)

——. "Copernicanism in Bruno, Kepler, and Galileo." *Vistas in astronomy*, 17 (1975), 177–90. (*Essays*, 1, 325–39.)

——. "Kepler and Galileo." *Vistas in astronomy*, 18 (1975), 237–47. (*Essays*, 1, 340–50.)

——. *Galileo against the philosophers in his Dialogue of Cecco di Ronchitti (1605) and Considerations of Alimberto Mauri (1606) in English translations*. Los Angeles: Zeitlin & Ver Brugge, 1976.

——. "Galileo gleanings, XXIV. The evolution of De Motu." *Isis*, 67 (1976), 239–50. (*Essays*, 1, 201–14.)

——. *Galileo at work. His scientific biography*. Chicago: University of Chicago Press, 1978.

——. "Introduction." in Galilei, *Operations* (1978), 9–35.

——. *Cause, experiment, and science: A Galilean dialogue, incorporating a new English translation of Galileo's "Bodies that stay atop water, or move in it."* Chicago: University of Chicago Press, 1981.

——. "Galileo's pre–Paduan writings: Years, sources, motivations." *Studies in the history and philosophy of science*, 17 (1986), 429–48. (*Essays*, 1, 215–35.)

——. "Galileo's steps to full Copernicanism and back." *Studies in the history and philosophy of science*, 18 (1987), 93–105. (*Essays*, 1, 351–63.)

——. *Galileo: Pioneer scientist*. Toronto: University of Toronto Press, 1990.

——. "Music and philosophy in early modern science." In Coelho, *Music* (1992), 3–16.

——. *Essays on Galileo and the history and philosophy of science*. Ed. N.M. Swerdlow and T.H. Levere. 3 vols. Toronto: University of Toronto Press, 1999.

——and C.D. O'Malley. *The Controversy on the comets of 1618*. Philadelphia: University of Pennsylvania Press, 1960.

——and I.E. Drabkin. *Mechanics in sixteenth-century Italy. Selections from Tartaglia, Benedetti, Guido Ubaldo, & Galileo*. Madison: University of Wisconsin Press, 1969.

Dugas, René. *La mécanique au xviie siècle*. Neuchâtel-Suisse: Editions du Griffon, 1954.

Duhem, Pierre. *Etudes sur Léonard da Vinci* [1906–13]. 3 vols. Paris: Archives contemporains, 1984.

Dupré, Sven. "Mathematical instruments and the 'Theory of the concave spherical mirror.' Galileo's optics beyond art and science." *Nuncius*, 15:2 (2000), 551–88.

——. "Ausonio's mirrors and Galileo's lenses: The telescope and sixteenth–century practical optical knowledge." *Galilaeana*, 2 (2005), 145–80.

——. "Optica est ars bene videndi: From Gemma's radius to Galileo's telescope." In Menso Folkerts and Andreas Kühne, eds. *Astronomy as a model for the sciences in early modern times*. Augsburg: E. Rauner, 2006, 355–68.

Edwards, Michael. "A sonnet by Galileo Galilei." *Culture and cosmos*, 7:1 (2003), 145–6.

Egerton, Samuel Y. *The heritage of Giotto's geometry: Art and science on the eve of the scientific revolution*. Ithaca: Cornell University Press, 1991.

Engel, Henrik. *Dantes Inferno: Zur Geschichte der Höllenvermessung und des Höllentrichtermotivs*. Munich: Deutscher Kunstverlag, 2006.

Ercolani, M. "Galileo Galilei novizio vallombrosiano." *Rivista storica benedettina*, 2 (1907), 569–80.

Ernst, Germana. "Aspetti dell'astrolgia e della profezia in Galileo e Campanella." *NCCS*, 255–66.

——. "Della Bolla 'Coeli et terrae' all' 'Inscrutabilis.' L'astrologia tra religione, natura e politica nell'età della Controriforma." In Ernst, *Religione* (1991), 255–79.

——. *Religione, ragione e natura. Ricerche su Tommaso Campanella e il tardo Rinascimento*. Milan: FrancoAngeli, 1991.

——. "Scienza, astrologia e politica nella Roma barocca. La biblioteca di don Orazio Morandi." In Canone, *Bibliothecae selectae* (1993), 217–52.

——. "Introduzione." In Campanella, *Opuscoli* (2003), 5–60.

Eszer, Ambrogio. "Niccolò Riccardi, O.P. 'Padre Mostro' (1585–1639)." *Angelicum*, 60 (1983), 428–61.

Euclid. *Elements*. Tr. T.L. Heath. 3 vols. New York: Dover, 1956.

Fabris, Rinaldo. *Galileo Galilei e gli orientamenti esegetici del suo tempo*. Vatican City: Pontificia academia scientiarum, 1986.

Fantoli, Annibale. *Galileo for Copernicus and for the Church*. Tr. G.V. Coyne. 2nd edn. Vatican City: Vatican Observatory, 1996.

——. "The disputed injunction and its role in Galileo's trial." In McMullin, *Church* (2005), 117–49.

Favaro, Antonio. *Lo studio di Padova e la compagnia di Gesù, sul finire del secolo decimosesto*. Venice: Antonelli. 1878.

——. "Galileo astrologo." *Mente e cuore*, 1881, 1–10.

——. *Galileo Galilei ed il "Dialogo di Cecco di Ronchitti di Bruzene de la stella nova."* Venice: Antonelli, 1881.

——. "La libreria di Galileo Galilei." *Bulletino di biografia e di storia delle scienze matematiche e fisiche*, 19 (1886), 219–93.

——. "Galileo Galilei e il padre Orazio Grassi." Istituto lombardo di scienze, lettere ed arti, *Memorie*, 24 (1891), 203–20.

——. *Nuovi studi galileiani*. Venice: Antonelli, 1891.

——, ed. *Omaggi. Galileo Galilei per il terzo centenario della inaugurazione del suo insegnamento nel Bò*. Padua, G.B. Randi, 1892.

——. "Un ridotto scientifico in Venezia al tempo di Galileo Galilei." *Nuovo archivio veneto*, 5:1 (1893), 196–209.

——. "Giovanfranceso Sagredo e la vita scientifica in Venezia al principio del xvii secolo." *Nuovo archivio veneto*, 4 (1902), 313–442. (*Amici*, 1, 193–322.)

——. "Fulgenzio Micanzio e Galileo Galilei." *Nuovo archivio veneto*, 13 (1907), 3–36. (*Amici*, 2, 703–36.)

——. "Galileo e Guidobaldo del Monte." Accademia di scienze, lettere ed arti, Padua, *Atti*, 30 (1914), 54–61.

——. "Sulla vericità del 'Racconto istorico della vita di Galileo' dettato di Vincenzio Viviani." *Archivio storico italiano*, 1915, 323–80.

——. "Giuseppe Moletti." Istituto veneto di scienze, lettere ed arti, *Atti*, 72:2 (1917/8), 47–118. (*Amici*, 2, 1585–1656.)

——. "Galileo Galilei, Benedetto Castelli e la scoperta delle fasi di Venere." *Archeion*, 1 (1919), 283–96.

——. *Galileo Galilei e lo Studio di Padova* [1883]. 2 vols. Padua: Antenori, 1966.

——. *Gaileo Galilei a Padova*. Padua: Antenori, 1966.

——. *Amici e corrispondenti di Galileo*. 3 vols. Florence: Salimbeni, 1983.

——. *Scampoli galileiani*. Ed. Lucia Rossetti and Laura Soppelsa. 2 vols. Trieste: Lint, 1992.

Favino, Federica. "'Quel petardo di mia fortuna.' Riconsiderando la 'caduta' di Giovan Battista Ciampoli." *LCF*, 863–82.

Ferguson, Ronnie. *The theatre of Angelo Beolco (Ruzante). Text, context and performance*. Ravenna: Longo, 2000.

Ferrone, Vincenzo. "Galileo tra Paolo Sarpi e Federico Cesi. Premessa per una ricerca." *NCCS*, 239–53.

Festa, Egidio and Sophia Roux. "La moindre petite force peut mouvoir un corps sur un plan horizontal. L'émergence d'un principe mécanique et son devenir cosmologique." *Galilaeana*, 3 (2006), 123–47.

Finocchiaro, Maurice A. *The Galileo affair*. Berkeley: University of California Press, 1989. Abbreviated *FA*.

——. *Retrying Galileo, 1633–1992*. Berkeley: University of California Press, 2005. Abbreviated *RG*.

——. *Defending Copernicus and Galileo. Critical reasoning in the two affairs*. Dordrecht: Springer, 2010.

Firpo, Luigi. "I primi processi campanelliani in una ricostruzione unitaria." *Giornale critico della filosofia italiana*, 20 (1939), 5–43.

——. "Il Campanella e i suoi persecutori romani." *Rivista di filosofia*, 30 (1939), 200–15.

——. "Appunti campanelliani, XXV. Storia di un furto." *Giornale critico della filosofia italiana*, 10 (1956), 541–9.

Flora, Francesco. *Storia della letteratura italiana*. 5 vols. Verona: Mondadori, 1967.

Foà, Simona. "Il *Dialogo sul sito, forma e misura dell'inferno* di Girolamo Benivieni e un particolare aspetto dell'esegesi dantesca tra xv e xvi secolo." In Foà and Sonia Gentili. *Dante e il locus inferni*. Rome: Bulzoni, 2000, 179–90.

Fonte, Moderata. *Il merito delle donne* [1600]. Ed. Adriana Chemello. Venice: Eidos, 1988.

Formazione e fortuna del Tasso nella cultura della Serenissima. Venice: Istituto veneto di scienze, lettere ed arti, 1997.

Formichetti, Gianfranco. "Campanella a Roma. I 'Commentaria' ai 'Poemata' di Urbano VIII." *Studi romani*, 30 (1982), 325–39.

——. *Campanella critico letterario. I 'Commentaria' ai 'Poemata' di Urbano VIII*. Rome: Bulzoni, 1983.

——. Il 'De sideriali fato vitando' di Tomasso Campanella." In Formichetti, *Mago* (1985), 199–217.

——, ed. *Il mago, il cosmo, il teatro degli astri*. Rome: Bulzoni, 1985.

——. *Tommaso Campanella. Eretico e mago alla corte dei papi*. Casale Monferrato: Edizione Piemme, 1999.

Fortini Brown, Patricia. *Private lives in Renaissance Venice*. New Haven: Yale University Press, 2004.

Frajese, Attilio. *Galileo matematico*. Rome: Studiorum, 1964.

Fredette, Raymond. "Galileo's 'De motu antiquiora.'" *Physis*, 14 (1972), 321–48.

——. "Galileo's 'De motu antiquiora.' Notes for a reappraisal." *LCF*, 165–81.

Freedberg, David. "Van Dyck and Virginio Cesarini: A contribution to the study of Van Dyck's Roman sojourn." In Susan J. Barnes and Arthur K. Wheelock, eds. *Van Dyck 350*. Washington: National Gallery of Art, 1994, 152–74.

——. *The eye of the lynx. Galileo, his friends, and the beginnings of modern natural history*. Chicago: University of Chicago Press, 2002.

Frisi, Paolo. *De motu diurno terrae dissertatio*. Pisa: Giovanelli, 1756.

Fumagalli, Giuseppina. *Unità fantastica dell'Orlando Furioso*. Messina: G. Principato, n.d.

Fumaroli, Marc. *L'âge de l'éloquence. Rhétorique et res letteraria de la Renaissance au seuil de l'époque moderne*. Geneva: Droz, 1980.

——. *L'inspiration du poète de Poussin*. Paris: Réunion des musées nationaux, 1989.

Furfaro, Domenico. "Consulto medico di Giovanni Trullio sulla cecità di Galileo." In Congresso, xx *Atti* (1964), 387–93.

Gabrieli, Giuseppe. *Contributi alla storia della Accademia dei Lincei*. 2 vols. Rome: Accademia Nazionale dei Lincei, 1989.

Galilei, Galileo. "De motu antiquiora [ca. 1590]." In Drabkin and Drake, *Galilei* (1960), 3–131.

——. "Le mecchaniche [1600]." In Drabkin and Drake, *Galilei* (1960), 147–86.

[——]. *Dialogo in perpuosito de la stella nuova* [1605]. See Ronchitti, Cecco di.

——. *Le operazioni del compasso geometrico et militare* [1606]. In Stillman Drake. *Operations of the geometric and military compass*. Washington, D.C.: Smithsonian Institution, 1978, 37–92.

——. *Difesa contro alle calumnie ed imposture di Baldassar Capra* [1607]. In Op, 2: 515–99.

——. *Sidereus nuncius* [1610]. *The sideral messenger*. Tr. Albert van Helden. Chicago: University of Chicago Press, 1989. Abbreviated *SN*.

——. *Sidereus nuncius* [1610]. Ed. Andrea Battistini, tr. Maria Timpanaro Cardini. Venice: Marsiglio, 1993.

——. *Discorso intorno alle cose che stanno in su l'acque o che in quella si muovono* [1612]. In Drake, *Cause* (1981).

——. *Istoria e dimostrazioni intorno alle macchie solari* [1613]. OS, 89–105, 108–68, 253–304.

[——]. *Discorso delle comete* [1619]. See Guiducci, Mario.

——. *Il saggiatore* [1623]. *Edizione critica e commento*. Ed. Ottavio Besomi and Mario Helbing. Padua: Antenore, 2005. Abbreviated BH, *Sagg*.

——. *Il saggiatore* [1623]. *The assayer*. In Drake and O'Malley, *Controversy* (1960), 151–336.

——. *Dialogo sopra i due massime sistemi del mondo tolemaico e copernicano* [1632]. *Edizione critica e commento*. Ed. Ottavio Besomi and Mario Helbing. 2 vols. Padua: Antenore, 1998. Abbreviated BH, *Dial*.

——. *Dialogo* [1632]. *Dialogue concerning the two chief world systems*. Tr. Stillman Drake. Berkeley: University of California Press, 1953. Abbreviated *TCWS*.

——. *Discorsi e dimostrazioni matematiche, intorno a due nuove scienze attenenti alla mecanica & i movimenti locali* [1638]. *Discourses and mathematical demonstrations concerning two new sciences pertaining to mechanics and local motions*. Tr. Stillman Drake. Madison: University of Wisconsin Press, 1974. Abbreviated *TNS*.

——. *Considerazioni al Tasso*. Rome: Pagliarini, 1793.

——. *Opere*. Ed. Antonio Favaro. 20 vols. Florence: Giunti-Barbera, 1890–1909.

——. *Scritti letterari*. Ed. Alberto Chiari. Florence: F. Le Monnier, 1970. Abbreviated Chiari.

Galilei, Vincenzo. *Fronimo. Dialogo…sopra l'arte del bene intavolare et rettamente sonare la musica negli strumenti artificiali…& in particolare nel liuto* [2nd edn, 1584]. Bologna: Forni, 1969.

——. *Discorso di Vincentio Galilei intorno all'uso delle disonanze*. Ed. Annibale Gianuario. Florence: Fondazione Centro studi rinascimento musicale, 2002.

——. *Dialogo della musica antica e della moderna* [1581] *Dialogue on ancient and modern music*. Tr. C.V. Palisca. New Haven: Yale University Press, 2003.

Galileo Galilei a Padova. Libertà di indagine e principio di autorità. Special issue of *Studia patavina*, 29 (1983), 497–777.

Galileo Galilei e la cultura veneziana. Venice: Istituto veneto di scienze, lettere ed arti, 1995.

Galileo's astrology. Ed. Nicolas Campion and Nick Kollerstrom. *Culture and cosmos*, 7:1 (2003), 179p.

Galli, Mario. "L'argomentazione di Galileo in favore del sistema copernicano dedotta del fenomeno delle maree." *Angelicum*, 60 (1983), 386–427.

Galluzzi, Paolo. *Momento. Studi galileiani.* Rome: Ateneo & Bizzarri, 1979.

——, ed. *Novità celesti e crisi del sapere.* Florence: IMSS, 1983. Abbreviated *NCCS.*

——. "The sepulchres of Galileo: The 'living' remains of a hero of science." In Machamer, *Companion* (1998), 417–47; also in Lunardi and Sabbatini, *Il rimembrar* (2009), 203–55.

Gamba, Antonio and Lucia Rossetti, eds. *Giornale della gloriossisima Accademia ricovrata. A. Verbali delle adunanze accademiche dal 1599 al 1624.* Trieste: Lint, 1999.

Gamba, Enrico. *Galileo Galilei e gli scienziati del ducato di Urbino.* Urbino: Quattroventi, 1989.

Gapaillard, Jacques. "Galilée et l'expérience de Locher." *Sciences et techniques en perspective,* 2 (1990/1), 1–10.

——. *Et pourtant, elle tourne!* Paris: Seuil, 1993.

Garcia, Stéphane. *Elie Diodati et Galilée. Naissance d'un réseau scientifique dans l'Europe du xviie siècle.* Florence: Olschki, 2004.

——. "Galileo's relapse. On the publication of the letter to the Grand Duchess Christina (1636)." In McMullin, *Church* (2005), 265–78.

Gardner, Edmund Garratt. *The king of court poets. A study of the work, life, and times of Ludovico Ariosto.* New York: Haskell, 1968.

Garin, Eugenio. *Scienza e vita civile nel Rinascimento italiano.* Bari: Laterza, 1965.

Garzend, Léon. *L'Inquisition et l'hérésie: Distinction de l'hérésie théologique de l'hérésie inquisitoriale.* Paris: Desclée, 1912.

Gelasso Calderara, Estella. *La granduchessa Maria Maddalena d'Austria. Un amazzone tedesca nella Firenze medicea del '600.* Genoa: Sagep, 1985.

Genovesi, Enrico. *Processi contro Galileo.* Milan: Casa editrice ceschina, 1966.

Giacobbe, Giulio Cesare. "Il 'Commentarium de certitudine mathematicarum disciplinarum' di Alessandro Piccolomini." *Physis,* 14 (1972), 162–93.

——. "Franceso Barozzi e la 'Quaestio de certitudine mathematicarum.' *Physis,* 14 (1972), 357–74.

——. "La riflessione metamatematica di Pietro Catena." *Physis,* 15 (1973), 178–96.

——. "Epigoni nel seicento della 'Quaestio de certitudine mathematicarum': Giuseppe Bianconi." *Physis,* 18 (1976), 5–40.

——. "Un gesuita progessista nella 'Quaestio de certitudine mathematicarum' rinascimentale: Benito Peyreira." *Physis,* 19 (1977), 51–86.

Giacomini Tebalducci Malespina, Lorenzo. "Del furor poetico discorso." In Giacomini, *Orazioni e discorsi.* Florence: Sermartelli, 1597, 53–73.

Giambulari, Pierfrancesco. *De 'l sito, forma e misura dello Inferno di Dante.* Florence: N. Dortelata, 1544.

Giandomenico, Mauro di and Pasquale Guaragnella, eds. *La prosa di Galileo. La lingua, la retorica, la storia.* Lecce: Argo, 2006.

Gianuario, Annibale. "Vincenzo Galilei, la disonanza e la seconda pratica." In V. Galilei, *Discorso* (2002), 19–36.

Gigli, Ottavio, ed. *Studi sulla Divina commedia, di Galileo Galilei, Vincenzo Borghini ed altri.* Florence: Le Monnier, 1855.

Gilbert, Neal Ward. *Renaissance concepts of method.* New York: Columbia University Press, 1960.

Gilbert, William. *De magnete* [1600]. *On the magnet, magnetick bodies also, and on the great magnet the earth.* Tr. Silvanus P. Thompson. London: Chiswick Press, 1900.

Gilson, Simon A. "Medieval science in Dante's *Commedia*." *Reading medieval studies,* 27 (2001), 39–77.

——. *Dante and Renaissance Florence.* Cambridge: Cambridge University Press, 2005.

Gingerich, Owen. "Dissertatio cum professore Righini et siderio nuncio." In Righini-Bonelli and Shea, *Reason* (1975), 77–88.

——. *The eye of heaven. Ptolemy, Copernicus, Kepler.* New York: American Institute of Physics, 1993.

——and Albert van Helden. "From occhiale to printed page: The making of Galileo's *Sidereus nuncius.*" *JHA, 34* (2003), 251–67.

Girardi, Maria Teresa. "Tasso, Speroni e la cultura padovana." In Borsetto and Da Rif, *Formazione* (1997), 63–77.

Giulini, Domenico. "Galileo's claims from the perspective of modern physics." In Renn, *Galileo* (2001), 138–45.

Giusti, Enrico. "Elements for the relative chronology of Galileo's De motu antiquiora." *Nuncius, 13* (1998), 427–60.

Godoli, Antonio and Paolo Paoli, *L'ultima dimora di Galileo: la Villa 'il Gioiello' ad Arcetri.* Florence: Giunti-Barbèra, 1979. 16p.

Gómez, Susana. "The mechanization of light in Galilean science." *Galilaeana, 5* (2008), 207–44.

Gori, Pietro. *La preziossisime reliquie di Galileo Galilei. Reintegrazione storica.* Florence: Galletti and Cocci, n.d.

Gorman, Michael John. "A matter of faith? Christopher Scheiner, Jesuit censorship, and the trial of Galileo." *Perspectives on science, 4:3* (1996), 283–320.

Gradenigo, Pietro. "La malattia che determinò la cecità di Galileo." Istituto veneto di scienze, lettere ed arti, *Atti, 56* (1897/8), 421–30.

Grant, Edward. *Planets, stars, and orbs: The medieval cosmos, 1200–1687.* New York: Cambridge University Press, 1994.

Grassi, Orazio. *De tribus cometis disputatio* [1619]. In Drake and O'Malley, *Controversy*, (1960), 3–19.

———, *Libra astronomica ac philosophica in qua Galilaei Galilaei opiniones de cometis...examinantur* [1619]. In Drake and O'Malley, *Controversy* (1960), 67–132.

Graf, Arturo. *Attraverso il cinquecento*. Turin: G. Chiantore, 1926.

Grafton, Anthony. *Cardano's cosmos. The worlds and works of a Renaissance astrologer*. Cambridge: Harvard University Press, 1999.

Grillo, Francesco. *Questioni campanelliane. La stampa fraudolenta e clandestina degli "Astrologicorum libri."* Cosenza: n.p., 1961.

Grmek, D.M. "L'enigme des relations entre Galilée et Santorio." In *Atti* (1967), 155–61.

Grondona, F. "In tema di eziogenesi della cecità di Galileo." In *Atti* (1967), 141–54.

Grossi, Pisana. "La magia rinascimentale e il 'Furioso.'" In Formichetti, *Mago* (1985), 115–34.

Grünbein, Durs. *Galilei vermisst Dantes Hölle und bleibt an den Massen hängen*. Frankfurt/M.: Suhrkampf, 1996.

Guerrini, Luigi. "'Luz pequeña.' Galileo fra gli astrologi." *Bruniana e campanelliana*, 7 (2000), 233–44.

———. "'Ogni speculazione del suo sovrano ingegno.' Niccolò Aggiunti a Galileo in un inedito frammento di carteggio del 1634." *LCF*, 895–901.

———. "Galileo, Alessandro Padovani e la prima diffusione del *Dialogo del flusso e riflusso dal mare* (1616)." *Galilaeana*, 4 (2007), 183–203.

———. "Due sconosciute testimonianze lincee." *Galilaeana*, 5 (2008), 267–72.

———. *Galileo e la polemica anticopernicana a Firenze*. Florence: Polistampa, 2009.

Guglielminetti, Marziano and Mariarosa Masoero. "Lettere e prose inedite (o parzialamente edite) di Giovanni Ciampoli." *Studi seicenteschi*, 19 (1978), 131–237.

Guiducci, Mario. *Discorso delle comete* [1619]. *Discourse on the comets*. In Drake and O'Malley, *Controversy*, (1960), 21–65.

———. *Lettera a Tarquinio Galluzzi* [1620]. *Letter to Tarquinio Galluzzi*. In Drake and O'Malley, *Controversy* (1960), 133–50.

Gullino, Giuseppe. "I Corner e l'Accademia." In Riondato, *Accademia* (2001), 59–73.

Gunter, Edmond. *The description and use of the sector...*London: W. Jones, 1624, 1636.

Hall, A. Rupert. *Ballistics in the seventeenth century*. Cambridge: Cambridge University Press, 1952.

Hammond, Frederick. "The artistic patronage of the Barberini and the Galileo affair." In Coelho, *Music* (1992), 67–89.

——. *Music and spectacle in baroque Rome. Barberini patronage under Urban VIII.* New Haven: Yale University Press, 1994.

Harris, Neil. "Galileo as symbol. The 'Tuscan artist' in *Paradise Lost.*" *IMSS, Ann.*, 10:2 (1985), 3–29.

Hartner, Willy. "Galileo's contribution to astronomy." In McMullin, *Galileo* (1967), 178–94.

Heilbron, John L. *The sun in the church.* Cambridge: Harvard University Press, 1999.

——. "Bianchini as an astronomer." In Valentin Kockel and Brigitte Sölch, eds. *Francesco Bianchini (1662–1729) und die europäische gelehrte Welt um 1700.* Berlin: Akademie Verlag, 2005, 57–82.

——. "Censorship of astronomy in Italy after Galileo." In McMullin, *Church* (2005), 279–322.

——. "Coming to terms with the Scientific Revolution." *European review*, 15 (2007), 473–89.

Helbing, Mario Otto. "Un capitolo del *De motu* di Francesco Buonamici e alcune informazioni sull'autore e sulle sue opere." *Physis*, 18 (1976), 41–63.

——. *La filosofia di Francesco Buonamici, professore di Galileo a Pisa.* Pisa: Nistri-Lischi, 1989.

——. "Galileo e le *Questioni meccaniche* attribuite ad Aristotele. Alcune indicazioni." *LCF*, 217–36.

——. "Mobilità della terra e referimento a Copernico nelle opere dei professori dello Studio di Pisa." In Bucciantini and Torrini, *Diffusione* (1997), 57–66.

——. "Mechanics and natural philosophy in late 16th century Pisa: Cesalpino and Buonamici, humanist masters of the faculty of arts." In Laird and Roux, *Mechanics* (2008), 185–93.

——. "Spigolature attinenti alla tecnologia nel Dialogo." *Galilaeana*, 5 (2008), 17–32.

Hempfer, Klaus W. *Diskrepante Lektüren: Die Orlando-Furioso-Rezeption im Cinquecento.* Stuttgart: F. Steiner, 1987.

Hulsius, Levinus. *Gründlicher / Augenscheinlicher Bericht dess neuen geometrischen gruntreissenden Instruments / Planimetra genandt.* Frankfurt: at the author's, 1604.

Hutchinson, Keith. "Sunspots, Galileo, and the orbit of the earth." *Isis*, 81 (1990), 68–74.

Huygens, Christiaan. *Horologium oscillatorium e Traité de la lumière.* Tr. Clelia Pighetti. Florence: Barbèra, 1963.

Iezzi, Benito. "Un gesuita estimatore napoletano di Galilei: p. Costanzo Pulcarelli." In Lomonaco and Torrini, *Galileo a Napoli* (1987), 141–57.

Imeriti, Arthur D. "Introduction." In Bruno, *Expulsion* (2004), 3–65.

Inchofer, Melchior. *Tractatus syllepticus* [1633]. In Blackwell, *Behind the scenes* (2006), 105–206.

——. *Oratio funebris qua reverendissimo patri F. Nicolao Riccardio parentabat M. Inchofer.* Rome: L. Gignani, 1639.

Ingegno, Alfonso. "Galileo, Bruno, Campanella." In Lomonaco and Torrini, *Galileo a Napoli* (1987), 123–39.

I primi lincei e il Sant'Uffizio. Questioni di scienza e di fede. Rome: Bardi, 2005.

Jacoli, Ferdinando. "Intorno a due scritti di Raffaele Gualterotti fiorentino relativi all' apparizione di una nuova stella avvenuta nell'anno 1604." *Bulletino di bibliografia e di storia delle scienze mathematiche e fisiche*, 7 (1874), 377–415.

Jacquot, Jean, ed. *Les fêtes de la Renaissance.* 3 vols. Paris: CNRS, 1956–75.

Jaffe, David. "The Barberini circle. Some exchanges between Peiresc, Rubens and their contemporaries." *Journal of the history of collections*, 1 (1989), 119–47.

Jardine, Nicholas. "Demonstration, dialectic and rhetoric in Galileo's *Dialogue*." In D.R. Kelley and Richard Popkin, eds. *The shapes of knowledge from the Renaissance to the Enlightenment.* Dordrecht: Kluwer, 1991, 101–22.

Joost-Gaugier, Christiane L. *Measuring heaven. Pythagoras and his influence on thought and art in Antiquity and the Middle Ages.* Ithaca: Cornell University Press, 2006.

Kelly, J.N.D. *The Oxford dictionary of popes.* Oxford: Oxford University Press, 1986.

Kennedy, Leonard A. "Cesare Cremonini and the immortality of the soul." *Vivarium*, 18 (1980), 143–58.

Kepler, Johannes. *Mysterium cosmographicum* [1596]. *Le secret du monde.* Tr. Alain Segonds. Paris: Belles Lettres, 1984.

——. *Dissertatio cum nuncio siderio* [1610] and *Narratio de observatis Jovis satellitibus* [1611]. *Discussion avec le messager celeste; Rapport sur l'observation des satellites de Jupiter.* Tr. Isabelle Pantin. Paris: Belles Lettres, 1993.

——. *Harmonice mundi* [1619]. *The harmony of the world.* Tr. E.J. Aiton et al. Philadelphia: American Philosophical Society, 1997. (APS, *Memoirs*, vol. 209.)

Kleiner, John. *Mismapping the underworld. Daring and error in Dante's "Comedy."* Stanford: Stanford University Press, 1994.

Kollerstrom, Nick. "Galileo's astrology." *LCF*, 421–31.

Koyré, Alexandre. "Galileo and Plato." *Journal of the history of ideas*, 4 (1943), 400–28; reprinted in Wiener and Noland, *Roots* (1957), 146–75, and in Koyré, *Metaphysics* (1968), 16–43.

——. "A documentary history of the problem of fall from Kepler to Newton." American Philosophical Society, *Transactions*, 45 (1955), 329–95.

"Giambattista Benedetti, critic of Aristotle." In McMullin, *Galileo* (1967), 98–117.

——. *Metaphysics and measurement*. London: Chapman & Hall, 1968.

——. *Galileo studies*. Hassocks, Sussex: Harvester, 1978.

Kuhn, Heinrich C. *Venetischer Aristotelismus am Ende der aristotelischen Welt. Aspekte der Welt und des Denkens des Cesare Cremonini (1550–1631)*. Frankfurt/M.: Peter Long, 1996.

——. "Cesare Cremonini volti e maschere di un filosofo scomodo per tre secoli e mezzo." In Riondato and Poppi, *Cremonini* (2000), 153–67.

Laird, Walter Roy. *The unfinished mechanics of Giuseppe Moletti. An edition and English translation of his Dialogue on mechanics, 1576*. Toronto: University of Toronto Press, 2000.

——and Sophie Roux, eds. *Mechanics and natural philosophy before the Scientific Revolution*. Berlin: Springer, 2008.

Landolfi, Domenica. "Don Giovanni de' Medici 'principe intendissimo in varie scienze.'" *Studi seicenteschi, 29* (1988), 125–62.

Lane, Frederic C. *Venice: A maritime republic*. Baltimore: Johns Hopkins University Press, 1973.

Lattis, James M. *Between Copernicus and Galileo. Christopher Clavius and the collapse of Ptolemaic cosmology*. Chicago: University of Chicago Press, 1994.

Lazzarini, Lino. "Galileo, Padova e l'Accademia dei Ricovrati." In Santinello, *Galileo* (1992), 165–78.

Lazzarini, Maria Teresa. *La "lampada di Galileo" nel Duomo di Pisa*. Pisa: Opera della primaziale pisana, 1998.

Leigh, Marcella Diberti. *Veronica Franco. Donna, poetessa e cortigiana del Rinascimento*. Turin: Priuli and Verlucca, 1988.

Leman, Auguste. *Urbain VIII et la rivalité de la France et de la maison d'Autriche de 1631 à 1635*. Lille: Giard; Paris, Champion, 1920.

Leonardi, Giovanni. "Verità e libertà di ricerca nell'ermeneutica biblica cattolica nell'epoca galilaeana e attuale." In *Galileo in Padova* (1983), 109–47.

——. "Galileo: originalità e fonti dei suoi documenti teologici." In Santinello, *Galileo* (1992), 373–430.

Lerner, Michel–Pierre. "La réception de la condamnation de Galilée en France au xviie siècle." *LCF*, 513–45.

——. "The heliocentric 'heresy' from suspicion to condemnation." In McMullin, *Church* (2005), 11–37.

——. "Copernic suspendu et corrigé: Sur deux décrets de la Congrégation Romaine de l'Index (1616–1620)." In *Primi lincei* (2005), 321–403.

Litchfield, R. Burr. *Emergence of a bureaucracy. The Florentine patricians 1530–1790*. Princeton: Princeton University Press, 1986.

Lodrini, Antonio. "Relazione sull'origine ed i primi anni di Benedetto Castelli." In Masetti Zannini, *Vita* (1961), 131–40.

Lomonaco, Fabrizio and Maurizio Torrini, eds. *Galileo e Napoli*. Naples: Guida, 1987.

Lupi, F. Walter. "Galilée criminalisé. Eroismi e 'crisi' della cultura italiana nel giudizio degli oltremontani." *NCCS*, 399–409.

Loria, Mario. "William Gilbert e Galileo Galilei. La terella e la calamita del Granduca." In Maccagni, *Saggi*, 3:2 (1972), 208–47.

Lovarini, Emilio. *Studi sul Ruzzante e la letteratura pavana*. Ed. Gianfrancesco Folena. Padua: Antenore, 1965.

Lunardi, Roberto and Oretta Sabbatini. *Il rimebrar delle passate cose. Una casa per memoria: Galileo e Vincenzo Viviani*. Florence: Edizioni polistampa, 2009.

Maccagni, Carlo, ed. *Saggi su Galileo Galilei*. 3 vols. Florence: Barbèra, 1972–74.

Machamer, Peter. "Galileo and the causes." In Butts and Pitt, *Perspectives* (1978), 161–80.

——, ed. *The Cambridge companion to Galileo*. Cambridge: Cambridge University Press, 1998.

MacLachlan, James. "Reply to the Shea-Wolf critique." *Isis*, 66 (1975), 402–3.

Maclean, Ian. "Thomas Harriot on combinations." *Revue d'histoire des mathématiques*, 11 (2005), 57–88.

Maddison, R.E.W. "Spread of the knowledge of Galileo's work in England in the first half of the seventeenth centurty." *Atti* (1967), 67–77.

Maffei, Paolo. *Giuseppe Settele, il suo diario e la questione galileiana*. Foligno: dell'Aquarto, 1987.

Maffioli, Cesare S. "Galieo Galilei and the engineer Bartolotti on the Bisenzio River." *Galilaeana*, 5 (2008), 179–2006.

Maggi, Armando. "La melancolia tra medicina e demonologia." In Clericuzio and Ernst, *Rinascimento* (2008), 281–95.

Maggini, Francesco. "Galileo studioso di letteratura." *Convivium*, 3:6 (1949), 847–61.

Maggioli, Attilo. *I soci dell'Accademia patavina dalla sua fondazione (1549)*. Padua: Accademia patavina, 1983.

Magrini, Silvio. "Il 'De magnete' del Gilbert e i primordi della magnetologia in Italia in rapporto alla lotta intorno ai massimi sistemi." *Archeion*, 8 (1927), 17–39.

Mamiani della Rovere, Giuseppe. *Elogi storici di Federico Commandino, G. Ubaldo del Monte, Giulio Carlo Fagani*. Pesaro: Nobili, 1828.

Mancini, Clodomiro. "Le malattie di Galileo." In Congresso, xx, *Atti* (1964), 284–94.

Manetti, Antonio di Tuccio. *The life of Brunelleschi*. Ed. Howard Saalman. University Park: Pennsylvania State University Press, 1968.

Marchetti, Angelo. *Breve introduzione alla cosmografia*. 2nd edn. Pistoia: Bracali, 1738.

Margolis, Howard. "Tycho's system and Galileo's Dialogue." *Studies in the history and philosophy of science*, 22 (1991), 259–75.

Marini, Marino. *Galileo e l'Inquisizione*. Rome: Propaganda Fidei, 1850.

Marino, Giovan Battista. *L'Adone* [1623]. Ed. Giovanni Pozzi. 2 vols. Milan: Mondadori, 1976.

Martin, Craig. "Rethinking Renaissance Averroism." *Intellectual history review*, 17 (2007), 3–19.

Martini, Gabriele. "La donna veneziana del '600 tra sessualità leggitima e illegittima." Istituto veneto di scienze, lettere ed arti, *Atti*, 145 (1986/7), 301–39.

Maschietto, Ludovico. "Giralomo Spinelli e Benedetto Castelli, benedittini di Santa Giustina, discepoli e amici di Galileo Galilei." In Santinello, *Galileo* (1992), 431–44.

Masetti Zannini, Gian Ludovico. *La vita di Benedetto Castelli*. Brescia: Camera di Commercio, 1961.

——. *Motivi storici della educazione femminile*. Naples: M. D'Auria, 1982.

Mauri, Dalimberto. *Considerazioni sopra alcuni luoghi del Discorso di Ludovico delle Colombe intorno alla stella apparita 1604* [1606]. In Drake, *Galileo against the philosophers* (1976), 74–130.

Mayaud, Pierre–Nöel. *La condamnation de livres coperniciens et sa révocation à la lumière des documents inédits des Congrégations de l'Index et de l'Inquisition*. Rome: Università Gregoriana, 1997.

Mayer, T.F. "The status of the Inquisition's precept to Galileo (1616) in historical perspective." *Nuncius*, 24:1 (2009), 61–95.

Mazzoni, Jacopo. *Della difesa della Comedia di Dante*. Cesena: np, 1587.

——. *In universam Platonis et Aristotelis philosophiam praeludia, sive De comparatione Platonis et Aristotelis*. Venice: Guerilio, 1597.

——. *On the defence of the Comedy of Dante. Introduction and summary*. Tr. Robert L. Montgomery. Tallahassee: Florida State University Press, 1983.

McMullin, Ernan, ed. *Galileo, man of science*. New York: Basic Books, 1967.

——. "The conception of science in Galileo's work." In Butts and Pitt, *Perspectives* (1978), 209–57.

——. "Galileo's theological venture." In McMullin, *Church* (2005), 88–116.

——. "The church's ban on Copernicanism." In McMullin, *Church* (2005), 150–90.

——. "The Galileo affair: Two decisions." *JHA*, 40 (2009), 191–212.

——, ed. *The Church and Galileo*. Notre Dame: Notre Dame University Press, 2005.

Meli, Domenico Bartoloni. "Guidobaldo del Monte and the Archimedean revival." *Nuncius*, 7:1 (1992), 3–32.

Mestica, Enrico. *Scritti di critica letteraria di Galileo Galilei, raccolti ed annotati per uso delle scuole*. Turin: Loescher, 1889.

Meyer, Eric R. "Galileo's cosmological calculations." *Isis, 80* (1989), 456–68.

Michelangeli, Luigi Alessandro. *Sul disegno dell'Inferno dantesco*. Bologna: Zanichelli, 1886.

——. "Il disegno dell'Inferno dantesco." *Giornale dantesco, 9* (1902), 225–36.

Middleton, W.E. Knowles. *The history of the barometer*. Baltimore: Johns Hopkins University Press, 1964.

——. *The experimeters. A study of the Accademia del Cimento*. Baltimore: Johns Hopkins University Press, 1971.

Milani, Marisa. "Galileo Galilei e la letteratura pavana." In Santinello, *Galileo* (1992), 179–202.

Milton, John. *The major works*. Oxford: Oxford University Press, 2003.

Moletti, Giuseppe. [*Dialogue on mechanics*] In Laird, *Unfinished mechanics* (2000).

Montanari, Ugo. "L'opera letteraria di Cesare Cremonini." In *Cesare Cremonini* (1990), 125–247.

Montano, Benito Arias. *Elucidationes in quattuor evangelia*. Antwerp: Plantin, 1575.

Montesinos, José and Carlos Solís, eds. *Largo campo de filosofare. Eurosymposium Galileo 2001*. La Orotava: Fundación canaria Orotava de historia de la ciencia, 2001. Abbreviated *LCF*.

Montgomery, Robert L. "Preface." In Mazzoni, *Defense* (1983), 1–27.

Montgomery, Scott L. *The scientific voice*. New York: Guilford, 1996.

——. *The moon and the Western imagination*. Tucson: University of Arizona Press, 1999.

Moody, Ernest A. "Galileo and Avempace. The dynamics of the leaning tower experiment." *Journal of the history of ideas*, 12 (1951), 163–93, 375–422. Partly reprinted in Wiener and Noland, *Roots* (1967), 176–206.

Morganti, Camillo. "La documentata fama del medico verolano Giovanni Trulli...ed una sua consulenza medica sulle infermità di Galileo Galilei." *Pagine di storia della medicina*, 3:4 (1959), 40–9.

Moscovici, Serge. "Torricelli's *Lezioni accademiche* and Galileo's theory of percussion." In McMullin, *Galileo* (1967), 432–48.

Motta, Uberto. "Querenghi e Galileo. L'ipotesi copernicana nelle imagini di un umanista." *Aevum, 67* (1993), 595–616.

——. *Antonio Querenghi (1546–1633): Un letterato padovano nella Roma del tardo Rinascimento*. Milan: Vita e pensiero, 1997.

Musacchio, Enrico and Gigino Pellegrini, eds. *Jacopo Mazzoni: Introduzione alla difesa della "Commedia" di Dante*. Bologna: Cappelli, 1982.

Muscetta, Carlo. "Simplicio e la 'commedia filosofica' dei Massimi sistemi." In Muscetta, *Realismo neorealismo controrealismo*. Milan: Garzante, 1976.

Nagler, A.M. *Theater festivals of the Medici, 1539–1637*. New Haven: Yale University Press, 1964.

Naish, G.P.B. "Ships and shipbuilding." In Charles Singer et al., eds. *A history of technology*. Vol. 3. *From the Renaissance to the Industrial Revolution*. Oxford: Oxford University Press, 1957, 471–500.

Napolitani, Pier Daniele. "Galileo e due matematici napolitani: Luca Valerio e Giovanni Camillo Glorioso." In Lomonaco and Torrini, *Galileo a Napoli* (1987), 159–85.

——. "La matematica nell'opera di Giovan Battista Della Porta." In Torrini, *Della Porta* (1990), 113–66.

Naylor, Ronald H. "Galileo and the problem of free fall." *British journal for the history of science*, 7 (1974), 105–34.

——. "The evolution of an experiment: Guidobaldo del Monte and Galileo's 'Discorsi' demonstration of the parabolic trajectory." *Physis*, 16 (1974), 323–46.

——. "An aspect of Galileo's study of the parabolic trajectory." *Isis*, 66 (1975), 394–6.

——. "Galileo: Real experiment and didactic demonstration." *Isis*, 67 (1976), 398–419.

——. "Galileo: The search for the parabolic trajectory." *Annals of science*, 33 (1976), 153–72.

——. "Galileo's theory of motion: Processes of conceptual change in the period 1604–1610." *Annals of science*, 34 (1977), 365–92.

——. "The role of experiment in Galileo's early work on the law of fall." *Annals of science*, 37 (1980), 363–78.

——. "Galileo's theory of projectile motion." *Isis*, 71 (1980), 550–70.

——. "Galileo's physics for a rotating earth." *LCF*, 337–55.

——. "Galileo's experimental discourse." In David Gooding et al., eds. *The uses of experiment: Studies in the natural sciences*. Cambridge: Cambridge University Press, 1989, 113–34.

Nel terzo centenario della morte di Galileo Galilei. Saggi e conferenze. Rome: Università del S. Cuore, 1942.

Nelli, Giovanni Battista Clemente. *Vita e commercio letterario di Galileo Galilei*. 2 vols. Lausanne: n.p., 1793.

Nenci, Elio. "Galileo and the 'Boboli fontanieri': The problem of the hydraulic pumps between philosophers and practioners." *Galilaeana*, 5 (2008), 63–87.

Newman, William R. and Anthony Grafton, eds. *Secrets of nature: Astronomy and alchemy in early modern Europe*. Cambridge, MA: MIT Press, 2002.

Newton, Isaac. *Philosophiae naturalis principia mathematica* [1687]. Ed. I.B. Cohen et al. 2 vols. Cambridge, MA: Harvard University Press, 1972.

North, John. "Thomas Harriot and the first telescopic observations of sunspots." In Shirley, *Harriot* (1974), 129–65.

Norwich, John Julius. *A history of Venice*. New York: Vintage, 1989.

Nussdorfer, Laurie. *Civic politics in the Rome of Urban VIII*. Princeton: Princeton University Press, 1992.

Olivieri, Achille. "I Ricovrati e le trasformazioni dell'idea di prudenza: 'L'antro e la parola' (1599–1601)." In Riondato, *Accademia* (2001), 361–4.

Olmi, Giuseppe. "La colonia lincea a Napoli." In Lomonaco and Torrini, *Galileo a Napoli* (1987), 23–58.

Olschki, Leonardo. "Galileo's literary formation." In McMullin, *Galileo* (1967), 140–59.

Ongaro, Giuseppe. "L'opera medica di Fortunio Liceti." In Congresso, xx, *Atti* (1964), 235–44.

——. "Introduzione." In Santorio, *Medicina* (2001), 5–47.

Osbat, Luciano. *L'Inquisizione a Napoli: Il processo agli ateistici, 1688–1697*. Rome: Storia e letteratura, 1974.

Ottone, Giuseppe. "'Postille' e 'Considerazioni' galileiane." *Aevum*, 46 (1972), 312–24.

Pagano, Sergio, ed. *I documenti vaticani del processo di Galileo Galilei (1611–1741)*. 2nd edn. Vatican City: Archivio segreto vaticano, 2009. Abbreviated Pagano.

Pais, Abraham. *"Subtle is the Lord…" The science and the life of Albert Einstein*. Oxford: Oxford University Press, 1982.

Palisca, Claude V. *Girolamo Mei (1519–1594). Letters on ancient and modern music to Vincenzo Galilei and Giovanni Bardi. A study with annotated texts*. n.l.: American Institute of Musicology, 1960.

——. "Scientific empiricism in musical thought." In H.H. Rhys, ed. *Seventeenth century science and the arts*. Princeton: Princeton University Press, 1961, 91–137.

——. "The Alterati of Florence." In *New looks at Italian opera. Essays in honor of Donald G. Grout*. Ithaca: Cornell University Press, 1968, 9–38.

——. *The Florentine camerata*. New Haven: Yale University Press, 1989.

——. "Was Galileo's father an experimental scientist?" In Coelho, *Music* (1992), 143–51.

——. "Intoduction." In Galilei, *Dialogue* (2003), xvii–lxix.

Palladino, Franco. "Origine e diffusione della filosofia del calcolo differenziale in Italia." *Giornale critico della filosofia italiana*, 63 (1984), 377–405.

——. "La geometria di Galileo e l'introduzione del calcolo a Napoli." In Lomonaco and Torrini, *Galileo a Napoli* (1987), 385–98.

Palmieri, Paolo. "Galileo did not steal the discovery of Venus' phases. A counter-
argument to Westfall." *LCF*, 433–44.

——. "Galileo and the discovery of the phases of Venus." *JHA*, 32 (2001), 109–29.

——. "'Galileus deceptus, non minime decepit.' A reappraisal of a counter-
argument in [the] *Dialogo* to the extrusion effect of a rotating earth." *JHA*, 39
(2008), 425–52.

Panofsky, Erwin. *Galileo as a critic of the arts*. Nijhoff: The Hague, 1954.

——. "More on Galileo and the arts." *Isis*, 47 (1956), 183–5.

Panteleoni, Marina and Raffaele Barnabeo. "L'opera scientifica di fra Paolo Sarpi
ed i suoi rapporti con le le concezioni galileiani." In Congresso, xx, *Atti* (1964),
269–83.

Pantin, Isabelle. "La Dissertatio et la Narratio dans leur contexte." In Kepler, *Dissertatio* (1993), ix–cxxvi.

——. "Galilée, la lune et les jésuites: à propos du *Sidereus nuncius Collegii Romani*
et du 'Problème de Mantoue.'" *Galilaeana*, 2 (2005), 19–42.

Paschini, Pio. *Vita e opere di Galileo Galilei*. 2 vols. Rome: Herder, 1965.

Passerini, Luigi. *Genealogia e storia delle famiglie Ricasoli*. Florence: M. Cellini, 1861.

Pastor, Ludwig. *The history of the popes from the close of the middle ages*. Tr. Ernst Graf.
Vols. 25–9. London: Kegan Paul, Trench, Teubner, 1937–38.

Patterson, Mark A. "Galileo's discovery of scaling laws." *American journal of physics*,
70 (2002), 575–80.

Pedote, Vittorio. "La misura dell'accelerazione del polso da Galileo Galilei a
Santorio Santorio." In Congresso, xx, *Atti* (1964), 357–70.

*Per il terzo centenario della inaugurazione dell'insegnamento di Galileo Galilei nello Studio
di Padova*. Florence: G. Barbèra, 1892.

Perazzi, Francesco and Anna Maria Perazzi. "Considerazioni sull'applicazione
del metodo sperimentale in medicina da parte di Galileo Galilei." In Congresso,
xx, *Atti* (1964), 655–71.

Peruzzi, Enrico. "Critica e rielaborazione del sistema copernico in Giovanni
Antonio Magini. In Bucciantini e Torrini, *Diffusione* (1997), 83–98.

Pesce, Mauro. "L'interpretazione della bibbia nella lettera di Galileo a Cristina e la
sua ricezione." *Annali di storia dell'esegesi*, 4 (1987), 239–84.

——. "L'indisciplinabilità del metodo e la necessità politica della simulazione
e dissimulazione in Galileo dal 1609 al 1642." In Paolo Prodi, ed. *Disciplina
dell'anima, disciplina del corpo, disciplina della società tra medioevo ed età moderna*.
Bologna: Il Mulino, 1993, 161–84.

Piaia, Gregorio. "Chiesa e stato nei trattatisti padovani al tempo di Galileo." In Santinello, *Galilei* (1992), 79–85.

Pieralisi, Sante. *Urbano VIII e Galileo Galilei. Memorie storiche.* Rome: Propaganda fide, 1875.

Pino, Guido di. "Letteratura e scienza in Italia nella prima metà del seicento." In *Literature and science.* Oxford: Blackwell, 1955, 138–44.

Piranesi, Giorgio. *Di un pubblico festeggiamento tenuto in Firenze il iii di novembre mdcviii in occasione delle nozze di Cosimo II de' Medici con Maria Maddalena d'Austria.* Florence: Aldino, 1905. 23p.

Pitt, Joseph. "The heavens and the earth: Bellarmine and Galileo." In Peter Barker and Roger Ariew, eds. *Revolution and continuity.* Washington: Catholic University of America Press, 1991, 131–42.

Pizzamiglia, Pierluigi. "L'opera astronomica di Nicolò Copernico (1473–1543) e la censura ecclesiastica." *L'insegnamento della matematica e delle scienze integrate,* 27:2 (2004), 154–73.

Pizzorusso, Claudio. "Galileo in giardino." *Galilaeana,* 4 (2007), 211–24.

Poppi, Antonino. *Introduzione all'aristotelismo padavano.* Padua: Antenore, 1970.

———. "Lo studio di Aristotele nella scuola di Padova." *Scienza e cultura,* 2 (1980), 137–58.

———. "L'eco dei processi romani contro Galileo." Accademia patavina di scienze lettere ed arti, Classe di scienze morale, lettere ed arti, *Atti e memorie, 105* (1992/3), 35–45.

———. *Cremonini, Galilei e gli inquisitori del Santo a Padova.* Padua: Centro studi antoniani, 1993.

———. *Ricerche sulla teologia e la scienza nella scuola padovana del cinque e seicento.* Catanzaro: Rubbettino, 2001.

Possevino, Antonio. *Bibliotheca selecta rationum studiorum.* Rome: Typ. Apostolica Vaticana, 1593; 2nd edn., 2 vols., Cologne: J. Gymnicus, 1602, 1609.

———. *Coltura degl'ingegni.* Vicenza: G. Greco, 1598.

Poupard, Paul. *Galileo Galilei: Toward a solution of 350 years of debate.* Pittsburgh: Duquesne University Press, 1987.

Pozzi, Giovanni. "Guida alla lettura." In Marino, *L'Adone* (1976), 2, 9–140.

[I] primi lincei e il Sant'Uffizio: Questioni di scienza e di fede. Rome: BARDI, 2005.

Proverbio, Edoardo. "Galileo e il problema della misura del tempo." *NCCS,* 63–72.

Ptolemy, Claudius. *Tetrabiblos.* Tr. F.E. Robbins. Cambridge, MA: Harvard University Press, 1964.

Purnell, Frederick, Jr. "Jacopo Mazzoni and Galileo." *Physis,* 14 (1972), 273–94.

Quaranta, Mario. "Galileo Galilei e l'Accademia Delia di Padova." In Santinello, *Galileo* (1992), 203–31.

Ragazzini, Vittorio. "Evangelista Torricelli e Giovanni Ciampoli." *Convivium*, 27 (1959), 51–5.

Ramat, Raffaello. *La critica ariostesca dal secolo xvi ad oggi.* Florence: Nuova Italia, 1954.

Ramelli, Agostino. *The various and ingenious machines* [1588]. Tr. Martha Teach Gnudi and Eugene S. Ferguson. New York: Dover, 1976.

Ramsay, John T. and A. Lewis Licht. *The comet of 44 B.C. and Caesar's funeral games.* Atlanta: Scholar's Press. 1997.

Rapp, Rolf. "Premessa." In Galilei, *Fronimo* (1969), [i–ii].

Ranke, Leopold. *Die römischen Päpste, ihre Kirche und ihr Staat im sechzehnten und siebzehnten Jahrhundert.* 3 vols. Berlin: Duncker and Humblot, 1854–57.

Raugei, Anna Maria. "Introduction." In Raugei, *Pinelli* (2001), 1, ix–cxxvi.

——. *Gian Vincenzo Pinelli et Claude Dupuy. Une correspondance entre deux humanistes.* 2 vols. Florence: Olschki, 2001.

Redondi, Pietro. "La luce 'messagio celeste.'" NCCS, 177–86.

——. *Galileo heretic.* Tr. R. Rosenthal. Princeton: Princeton University Press, 1987.

——. "Fede lincea e teologia tridentina." *Galilaeana*, 1 (2004), 117–41.

Reeves, Eileen. *Painting the heavens. Art and science in the age of Galileo.* Princeton: Princeton University Press, 1997.

——. "Mere projections. Sunspots and the *Camera obscura*." *Galilaeana*, 4 (2007), 47–77.

——. *Galileo's glassworks. The telescope and the mirror.* Cambridge, MA: Harvard University Press, 2008.

——and Albert Van Helden, tr. *On sunspots: Galileo Galilei and Christoph Scheiner.* Chicago: University of Chicago Press, 2010. Abbreviated OS.

Renn, Jürgen, ed. *Galileo in context.* Cambridge: Cambridge University Press, 2001.

——, Peter Damerow, and Simone Rieger. "Hunting the white elephant: When and how did Galileo discover the law of fall?" In Renn, *Galileo* (2001), 29–149.

—— and Matteo Valleriani. "Galileo and the challenge of the Arsenal." *Nuncius*, 16 (2001), 481–503.

Reynolds, Anne. "Il 'Capitolo contra il portar la toga' di Galileo Galileio." *Critica letteraria*, 8 (1980), 419–35.

——. "Galileo Galilei's poem 'Against wearing the toga.'" *Italica*, 59:4 (1982), 330–41.

——. *Renaissance humanism at the court of Clement VII. Francesco Berni's Dialogues against poets in context.* New York: Garland, 1997.

———. "Galileo Galilei and the satirical poem 'Contro il portar la toga': The literary foundations of science." *Nuncius*, 17:1 (2002), 45–62.

Ricci, Saverio. "Federico Cesi e la nova dal 1604. La teoria della fluidità del cielo e un opuscolo dimenticato di Joannes van Heeck." Accademia nazionale dei Lincei, Classe di scienze morali, storiche e filologiche, *Rendiconti, 385* (1988), 111–33. (Ricci, *Tre studi*, 7–31.)

———. *"Una filosofia milizia." Tre studi sull'Accademia dei Lincei.* Udine: Campanotto, 1994.

———. "Nicola Antonio Stigiola, enciclopedista e linceo." Accademia nazionale dei Lincei, Classe di scienze morali, storiche e filologiche, *Memorie, 8* (1996), 149p.

———. *Il sommo inquisitore. Giulio Antonio Santori tra autobiografia e storia (1532–1602).* Rome: Salerno, 2002.

Ricci Riccardi, Antonio. *Galileo Galilei e fra Tommaso Caccini. Il processo del Galilei nel 1616 e l'abiura segreta rivelata delle carte caccini.* Florence: Le Monnier, 1902.

Riccioli, Giovambattista. *Almagestum novum.* Frankfurt: Bayer, 1651.

———. *Apologia.* Venice: Salerni and Cagnolini, 1669.

Richter, B.L.O. "Petrarchism and anti-Petrachism among the Venetians." *Forum italicum, 3:1* (1969), 20–42.

Rietbergen, Peter. *Power and religion in baroque Rome. Barberini cultural politics.* Leyden: Brill, 2006.

Righini, Guglielmo. "L'oroscopio galileiano di Cosimo II de' Medici." IMSS, *Annali, 1:1* (1976), 28–36.

———. *Contributo alla interpretazione scientifica dell'opera astronomica di Galileo.* Florence: Giunti, 1978. (IMSS, *Annali*, 3:2, Suppl.)

Righini-Bonelli, Maria Luisa and William R. Shea, eds. *Reason, experiment, and mysticism in the Scientific Revolution.* New York: Science History 1975.

——— and Albert van Helden. *Divini and Campani. A forgotten chapter in the history of the Accademia del Cimento.* Florence: IMSS, 1981. (IMSS, *Annali*, 6:1, Suppl.)

Riondato, Ezio. "Cremonini e l'Accademia dei Ricovrati." In Riondato and Poppi, *Cremonini* (2000), 9–18.

———, ed. *Dell'Accademia dei Ricovrati all'Accademia galilaeana.* Padua: La Garangola, 2001.

——— and Antonino Poppi, eds. *Cesare Cremonini. Aspetti del pensiero e scritti.* 2 vols. Padua: Accademia galilaeana di scienze lettere ed arti in Padova, 2000.

Ronchi, Vasco. *Galileo e il cannochiale.* Turin: Boringhieri, 1964.

Ronchitti, Cecco di [Girolamo Spinelli et al.]. *Dialogo in perpuosito de la stella nova.* Padua: Tiozzi, 1605. In Favaro, *Galileo* (1881), 49–76, and Drake, *Galileo against the philosophers* (1976), 34–53.

Roncone, Giorgio. "Paolo Gualdo, Antonio Querenghi e le accademie." Accademia patavina di scienze, lettere ed arti, Classe di scienze morali, lettere ed arti, *Atti e memorie*, 105 (1992/3), 101–19.

——. "Paolo Gualdo e Galileo." In Santinello, *Galileo* (1992), 359–71.

Rose, Paul L. "The origins of the proportional compass from Mordente to Galileo." *Physis*, 10 (1968), 53–69.

——. "Jacopo Contarini (1536–1595), a Venetian patron and collector of mathematical instruments and books." *Physis*, 18 (1976), 117–30.

Rosen, Edward. *The naming of the telescope*. New York: H. Schuman, 1947.

——. "The correspondence between J. Lipsius and T. Seget." *Latomus*, 8 (1949), 63–7.

——. "Galileo's misstatements about Copernicus." *Isis*, 49 (1958), 319–30.

Rosenthal, Margaret. *The honest courtesan: Veronica Franco, citizen and writer in sixteenth-century Venice*. Chicago: University of Chicago Press, 1992.

Rospogliosi, Giulio. "Discorso…sopra l'Elettione di Urbano VIII. Poema del Sig. Francesco Bracciolini Dell'Api." In Bracciolini, *Elettione* (1628), 484–93.

Rossi, Massimiliano. "La crusca nell'occhio. L'Empoli tra Galileo e Michelangelo il giovane." *Galilaeana*, 4 (2007), 189–209.

Rothman, Aviva. "Forms of persuasion: Kepler, Galileo and the dissemination of Copernicanism." *JHA*, 40 (2009), 403–19.

Rotondò, Antonio. "Cultura e difficoltà di censori. Censura ecclesiastica e discussioni cinquecentesche sul Platonismo." In *La Pouvoir et la plume*. Paris: Nouvelle Sorbonne, 1982, 16–30.

Ruggiero, Guido. *The boundaries of eros, sex, crime and sexuality in Renaissance Venice*. New York: Oxford University Press, 1985.

Russell, John L. "Catholic astronomers and the Copernican system after the condemnation of Galileo." *Annals of science*, 46 (1989), 365–86.

Rutkin, H. Darrel. "Galileo astrologer." *Galilaeana*, 2 (2005), 107–43.

Sangalli, Maurizio. "Apologia dei padri Gesuiti contro Cesare Cremonini, 1592." Accademia patavina di scienze, lettere ed arti, Classe di scienze morali, lettere ed arti, *Atti e memorie*, 90 (1997/8), 241–355.

——. "Cesare Cremonini, la Compagnia di Gesù e la Repubblica di Venezia: Eterodossia e protezione politica." In Riondato and Poppi, *Cremonini* (2000), 207–18.

——. "Paolo Beni: Da Gesuita a Ricovrato." In Riondato, *Accademia* (2001), 491–503.

Santinello, Giovanni. "Galileo: Autorità della storia e libertà della ricerca." In *Galileo Galilei a Padova* (1983), 149–57.

——, ed. *Galileo e la cultura padovana. Convegno…1992*. Padua: CEDAM, 1992.

Sarpi, Paolo. *Scelte lettere inedite*. Lugano: Tip. svizzera italiana, 1848.

——. *Lettere*. Ed. F.L. Polidoro. 2 vols. Florence: Barbèra, 1863.

——. *Lettere ai protestanti*. Ed. Manlio Duilio Busnelli. 2 vols. Bari: Laterza, 1931.

——. *Scritti filosofici e teologici editi ed inediti*. Ed. Romano Amerio. Bari: Laterza, 1951.

——. *Opere*. Ed. Gaetano and Luisa Cozzi. Milan: Riccardi, 1969.

——. *Pensieri naturali, metafisici e matematici*. Ed. Luisa Cozzi and Libero Sosio. Riccardi: Milan, 1996.

Saslow, James. *The Medici wedding of 1589*. New Haven: Yale University Press, 1996.

Sari, Buonardo [Urbano d'Aviso, SJ]. *Trattato della sfera di Galileo Galilei*. Rome: Tionazzi, 1656.

Scheiner, Christoph. *Tres epistolae de maculis solaribus* [Jan. 1612]. In OS, 59–73.

——. *De maculis solaribus et stellis circa Jovem errantibus, accuratior disquisitio*. [Sep. 1612]. In OS, 189–230.

——. *Sol ellipticus* [1615]. Ed. Franz Daxecker and Lav Subaric. Innsbruck: University of Innsbruck, 1998.

——. *Rosa ursina, sive Sol*. Bracciano: Typ. Ducale, 1630.

Schiavo, Armando. "Notizie riguardanti la Badia di Passignano." *Benedictina*, 9 (1955), 31–92.

Schiller, Friedrich. *The Piccolomini, or the first act of Wallenstein, a drama in five acts*. Tr. S.T. Coleridge. London: Chapman and Hall, 1800.

Schmitt, Charles B. "The faculty of arts in Pisa at the time of Galileo." *Physis*, 14 (1972), 243–72.

——. "Girolamo Borro, 'Multae sunt nostrarum ingnorantiarum causae.'" In Edward P. Mahoney, ed. *Philosophy and humanism: Renaissance essays in honour of Oskar Kristeller*. Leiden: Brill, 1976, 462–76.

——. *Cesare Cremonini, un aristotelico al tempo di Galilei*. Venice: Centro tedesco di studi veneziani, 1980. 21p.

Schneider, Ivo. *Der Proportionalzirkel, ein universelles Analogrecheninstrument der Vergangenheit*. Munich: Oldenburg, 1970. (Deutches Museum, *Abhandlungen und Berichte, 38* (1970:2), 1–96.)

Schrade, Leo. "Les fêtes du mariage de Francesco dei Medici et de Bianca Capello." In Jacquot, *Fêtes* (1956), 1, 107–31.

Scolari, Antonio. "L'*Orlando Furioso* e la critica del secolo xvi." In Scolari, *Scritti di varia letteratura e di critica*. Bologna: Zanichelli, 1937, 41–64.

Scott, John Bolton. *Images of nepotism. The painted ceilings of Palazzo Barberini*. Princeton: Princeton University Press, 1991.

Segni, Pier. "Orazione…di M. Jacopo Mazzoni." In *Prose fiorentine*, vol. 1. Venice: D. Occhi, 1735, 109–24. (Vol. 1, part 1, orazione 8.)

Segonds, Alain. "Introduction." In Kepler, *Secret* (1984), ix–lviii.

Segrè, Michael. "Galileo, Viviani and the Tower of Pisa." *Studies in the history and philosophy of science,* 20 (1989), 435–51.

Serassi, Pier Antonio. *La vita di Jacopo Mazzoni.* Rome: Pagliarini, 1790.

Settle, Thomas B. "An experiment in the history of science." *Science,* 133 (1961), 19–23.

——. "Galileo's use of experiment as a tool of investigation." In McMullin, *Galileo* (1967), 315–37.

——. "Antonio Santucci, his 'New tractatus on comets,' and Galileo." *NCCS,* 229–38.

Shea, William R. "Galileo, Scheiner, and the interpretation of sunspots." *Isis,* 61 (1970), 488–519.

——. *Galileo's intellectual revolution. Middle period, 1610–1632.* New York: Science History, 1977.

——. "Melchior Inchofer's 'Tractatus syllepticus': A consultor of the Holy Office answers Galileo." *NCCS,* 283–92.

—— and Mariano Artigas. *Galileo in Rome. The rise and fall of a troublesome genius.* New York: Oxford University Press, 2003.

Sherberg, Michael. "The Accademia fiorentina and the question of the language. The politics of theory in ducal Florence." *Renaissance quarterly,* 56 (2003), 26–55.

Shirley, John W. *Thomas Harriot. A biography.* Oxford: Oxford University Press, 1983.

——, ed. *Thomas Harriot, Renaissance scientist.* Oxford: Oxford University Press, 1974.

Smith, A.M. "Galileo's proof for the earth's motion from the movement of sunspots." *Isis,* 76 (1985), 543–51.

Sobel, Dava. *Galileo's daughter.* London: Fourth Estate, 1999.

——. *Letters to father. Suor Maria Celeste to Galileo 1623–1633.* New York: Walker, 2001 (bilingual edition).

Solerti, Angelo. *Musica, ballo e drammatica alla corte medicea dal 1600 al 1637.* Florence: Bemporad, 1905.

Soppelsa, Maria Laura. "Un dimenticato scolaro galileiano: il padre Girolamo Spinelli." Museo civico, Padua, *Bolletino,* 60:2 (1971), 97–114.

——. *Genesi del metodo galileiano e tramonto dell'aristotelismo nella scuola di Padova.* Padua: Antenore, 1974.

Sosio, Libero. "I pensieri di Paolo Sarpi sul moto." *Studi veneziani,* 13 (1971), 315–92.

——. "Galileo Galilei e Paolo Sarpi." *Galileo e la cultura veneziana* (1995), 269–311.

——. "Fra Paolo Sarpi e la cosmologia." In Sarpi, *Pensieri* (1996), lxxxix–cxciv.

Speller, Jules. *Galileo's inquisition trial revisited.* Frankfurt: Peter Lang, 2008.

Spini, Giorgio. "La religiosità di Galileo." In Maccagni, *Saggi*, 3:2 (1972), 416–40. (Spini, *Galileo, Campanella*, 23–40.)

——. *Ricerca dei libertini. La teoria dell'impostura delle religioni nel seicento italiano.* Rome: Nuova Italia, 1983.

——. *Galileo, Campanella e il "divinus poeta."* Bologna: Il Mulino, 1996.

Spongano, Raffaele. *La prosa di Galileo e altri scritti*. Messina: G. D'Anna, 1949.

Spruit, Leen. "Cremonini nelle carte del Sant'Uffizio romano." In Riondato and Poppi, *Cremonini* (2000), 193–204.

Stabile, Giorgio. "Il primo oppositore al *Dialogo*: Claude Bérigard." *NCCS*, 277–82.

——. "Linguaggio della natura e linguaggio della Scrittura in Galilei. Dalla 'Istoria' sulle macchie solari alle lettere copernicane." *Nuncius*, 9:1 (1994), 37–64.

Stella, Aldo. "Galileo e i 'padovani polacchi.'" Accademia patavina di scienze, lettere ed arti, Classe di scienze morali, lettere ed arti, *Atti e memorie*, 95 (1982/3), 175–81.

——. "L'Università di Padova al tempo del Cremonini." In *Cesare Cremonini* (1990), 69–82.

——. "Cesare Cremonini 1550–1631. Il suo pensiero e il suo tempo." Accademia patavina di scienze, lettere ed arti, Classe di scienze morali, lettere ed arti, *Atti e memorie*, 103 (1990/1), 63–74.

——. "Galileo, il circolo di Gian Vincenzo Pinelli e la 'Patavina libertas.'" In Santinello, *Galileo* (1992), 307–25.

Stochi, Manilio Pastore. "Il periodo veneto di Galileo Galilei." In *Storia*, 4/2 (1984), 37–66.

Storia della cultura veneta. Il seicento. Vol. 4/2. Venice: Neri Pozza, 1984.

Strozzi, Giovanni Battista, il giovane. *Orazioni et altre prose*. Rome: L. Grignani, 1635.

Swerdlow, Noel. "Galileo's discoveries with the telescope and their evidence for the Copernican theory." In Machamer, *Galileo* (1998), 244–70.

——. "Galileo's horoscopes." *JHA*, 35:2 (2004), 135–41.

Tafuri, Manfredo. *Venice and the Renaissance.* Cambridge, MA: MIT Press, 1995.

Tannery, Paul. "Galileo and the principles of dynamics." In McMullin, *Galileo* (1967), 163–77.

Targioni-Tozzetti, Giovanni. *Notizie degli aggrandimenti delle scienze fisiche accaduti in Toscana nel corso di anni lx. del secolo xvii.* 3 vols. in 4. Florence: G. Bouchard, 1780.

Tassinari, Magda. "Scenografia." In Bösel, *Grassi* (2004), 222–7.

Tasso, Torquato. *Gerusalemme liberata*. Ed. Lanfranco Caretti. Turin: Einaudi, 1993.

———. *The liberation of Jerusalem*. Tr. Max Wickert. Oxford: Oxford University Press, 2009.

Tenenti, Alberto. *Piracy and the decline of Venice, 1580–1615*. Berkeley: University of California Press, 1967.

Tessari, Antonio Secondo. "Antonio Possevino e l'archittetura." *Archivum historicum Societatis Jesu*, 52 (1983), 247–61.

Thoren, Victor. *The Lord of Uraniborg. A biography of Tycho Brahe*. Cambridge: Cambridge University Press, 1990.

Thorndike, Lynn. *The Sphere of Sacrobosco and its commentators*. Chicago: University of Chicago Press, 1949.

Tofani, Annamaria Petrioli. "L'illustrazione teatrale e il significato dei documenti figurativi per la storia dello spettacolo." In Elizabeth Cropper et al., eds. *Documentary culture. Florence and Rome from Grand-Duke Ferdinando I to Pope Alexander VII*. Bologna: Nuova Alfa, 1992, 49–62.

Tomba, Tullio. "L'osservazione della stella nova del 1604 nell'ambito filosofico scientifico padovano." In *Cesare Cremonini* (1990), 83–95.

Tomlinson, Gary. *Music in Renaissance magic. Toward a historiography of others*. Chicago: University of Chicago Press, 1993.

Tongiorgio Tomasi, Lucia. "La conquista del visibile. Rimeditando Panofsky, rileggendo Galilei." *Galilaeana*, 4 (2007), 5–46.

Topper, David. "Galileo, sunspots, and the motions of the earth." *Isis*, 90 (1999), 757–67.

Torelli, Giuseppe. *Archimedis quae supersunt omnia*. Oxford: Oxford University Press, 1792.

Torricelli, Evangelista. *Lezioni accademiche*. In Torricelli, *Opere* (1919), 2, 1–99.

———. *Opere*. 4 vols. in 5. Ed. Gino Loria and Giuseppe Vassura. Faenza: G. Montanai, 1919 (vols. 1–4); Faenza, F. Laga, 1944 (vol. 5).

Torrini, Maurizio. "Giovanni Ciampoli filosofo." *NCCS*, 267–75.

———. "La discussione sullo statuto delle scienze tra la fine del' 600 e l'inizio del' 700." In Lomonaco and Torrini, *Galileo a Napoli* (1987), 357–83.

———, ed. *Giovan Battista della Porta nell'Europa del suo tempo*. Naples: Guida, 1990.

Tosi, Alessandro. "'Vallis ego memor umbrosa.' Artisti, poeti e viaggiatori nella Vallombrosa." In Ciardi, *Vallombrosa* (1999), 255–323.

Toussaint, Stéphane. *De l'enfer à la coupole. Dante, Brunelleschi et Ficin*. Rome: L'Erma di Bretschneider, 1997.

Vaccalluzzo, Nunzio. *Galileo letterato e poeta; le rime inedite di Vincenzo Galilei*. Catania: N. Giornatta, 1896.

———. *Galileo Galilei nella poesia del suo secolo. Raccolta di poesie edite ed inedite scritte da' contemporanei in lode di Galileo.* Milan: Sandron, 1910.

Vaccheri, Giulio Giuseppe and Cosimo Bertacchi. *Cosmografia della Divina commedia. La visione di Dante Alighieri considerata nello spazio e nel tempo.* Turin: G. Candeletti, 1981.

Valeri, Diego. *L'Accademia dei Ricovrati.* Padua: Accademia dei Ricovrati, 1987.

Valleriani, Matteo. "A view on Galileo's *Ricordi autografi.* Galileo practitioner in Padua." *LCF,* 281–91.

Van Helden, Albert. "The telescope in the seventeenth century." *Isis,* 65 (1974), 38–58.

———. "Saturn and his anses." *JHA,* 5 (1974), 105–21.

———. "The 'astronomical telescope' 1611–1650." IMSS, *Annali,* 2:1 (1976), 13–35.

———. "Galileo on the sizes and distances of the planets." IMSS, *Annali,* 7:2 (1982), 65–86.

———. "Telescopes and authority from Galileo to Cassini." *Osiris,* 9 (1994), 9–29.

———. "Galileo and Scheiner on sunspots. A case study in the visual language of astronomy." American Philosophical Society. *Proceedings,* 140 (1995), 358–96.

———. "Introduction." *SN,* 1–24.

Vannucci, Marcello. *Dante nella Firenze del '500. Studi danteschi della Accademia fiorentina.* Florence: Istituto "Leonardo da Vinci," 1965.

Varanini, Giorgio. *Galileo critico e prosatore. Note e ricerche.* Verona: Fiorini and Ghidini, 1967.

Varetti, Carlo Vittorio. "L'artifice di Galileo. Ippolito Francini detto Tordo." Accademia nazionale dei Lincei. Classe di scienze morali, storiche e filologiche, *Rendiconti,* 15: 3–4 (1939), 204–97.

Varignon, Pierre. *Conjectures sur la pesanteur.* Paris: J. Boudot, 1690.

Vasari, Giorgio. *Lives of the artists* [1568]. Ed. Betty Burroughs. New York: Scribners, 1946.

Vianello, Nereo. "Le inedite 'Postille al Petrarca.'" *Lettere italiane,* 6 (1954), 283–6.

———. "Le postille al Petrarca di Galileo Galilei." *Studi di filologia italiana,* 14 (1956), 211–433.

Vickers, Brian. "Epideictic rhetoric in Galileo's Dialogue." IMSS, *Annali,* 8:2 (1983), 69–102.

Vilain, Christiane. "Circular and rectilinear motion in the *Mechanica* and in the 16th and 17th century." In Laird and Roux, *Mechanica* (2008), 149–72.

Vinaty, Bernard T., OP. "La formation du système solaire d'après Galilée." *Angelicum,* 60 (1983), 333–85.

Vitruvius. *The ten books on architecture*. Tr. Morris Hicky Morgan. Cambridge: Harvard University Press, 1914.

Viviani, Ugo. *Vita ed opere di Andrea Cesalpino*. 2nd edn. Arezzo: Viviani, 1922.

——. *Medici, fisici e cerusici della Provincia aretina vessuti dal v al xvii secolo*. Arezzo: Viviani, 1923.

Viviani, Vincenzo. *Vita di Galileo* [1654]. Ed. Bruno Basile. Rome: Salerno, 2001.

Wade, Nicholas J. "Galileo and the senses: Vision and the art of deception." *Galilaeana*, 4 (2007), 297–307.

Walker, D.P. *Spiritual and demonic magic from Ficino to Campanella*. Notre Dame: Notre Dame University Press, 1975.

Wallace, William A. "Galileo and the Thomists." In Armand A. Maurer, ed. *St Thomas Aquinas commemorative studies 1274–1974*. 2 vols. Toronto: Pontifical Institute of Medieval Studies, 1974. Vol. 2, pp. 293–330.

——. "Galileo Galilei and the doctores parisienses." In Butts and Pitt, *New perspectives* (1978), 87–138.

——. "Galileo's early arguments for geocentrism and his later rejection of them." *NCCS*, 31–40.

——. *Galileo and his sources. The heritage of the Collegio Romano in Galileo's science*. Princeton: Princeton University Press, 1984.

——. "Galileo's Pisan studies in science and philosophy." In Machamer, *Companion* (1998), 27–52.

Waterworth, J. *The canons and decrees of the Sacred and Oecumenical Council of Trent...to which are prefixed essays on the external and internal history of the council*. London: Burns and Oates, [1848].

Watson, Peter G. "The enigma of Galileo's eyesight: Some novel observations on Galileo Galilei's vision and his progression to blindness." *History of ophthalmology*, 54 (2009), 630–40.

Waźbiński, Zygmunt. *L'Accademia medicea del disegno a Firenze nel cinquecento*. 2 vols. Florence: Olschki, 1987.

Wedgwood, C.V. *The Thirty Years War*. Harmondsworth: Penguin, 1961.

Weinberg, Bernard. "Argomenti di discussione letteraria nell'Accademia degli Alterati (1570–1600)." *Giornale storico della letteratura italiana*, 131 (1954), 175–94.

——. "The Accademia degli Alterati and literary taste from 1570 to 1600." *Italica*, 31 (1954), 207–14.

——. *History of literary criticism in the Italian Renaissance*. 2 vols. Chicago: University of Chicago Press, 1961.

Westfall, Richard S. "Science and partronage. Galileo and the telescope." *Isis*, 76 (1985), 11–30.

Westman, Robert. "The reception of Galileo's 'Dialogue.' A partial world census of extant copies." *NCCS*, 329–71.

Whitaker, Ewen. "Galileo's lunar observations and the dating of the composition of *Sidereus nuncius*." *JHA*, 9 (1978), 155–69.

Wiendlocha, Jolanta. *Die Jugendgedichte Papst Urbans VIII*. Heidelberg: Winter, 2005.

Wiener, Philip P. and Aron Noland, eds. *Roots of scientific thought. A cultural perspective*. New York: Basic Books, 1967.

Wilkinson, L.P. *Horace and his lyric poetry*. Cambridge: Cambridge University Press, 1968.

Wilding, Nick. "Galileo's idol: Gianfrancesco Sagredo revealed." *Galilaeana*, 3 (2006), 229–45.

Winkler, Mary G. and Albert van Helden. "Representing the heavens: Galileo and visual astronomy." *Isis*, 83 (1992), 195–217.

Wisan, Winifred L. "The new science of motion: A study of Galileo's 'De motu locali.'" *Archive for history of exact science*, 13 (1974), 103–306.

——. "Galileo's scientific method: A re-examination." In Butts and Pitt, *New perspectives* (1978), 1–58.

——. "Galileo's 'De systemate mundi' and the new mechanics." *NCSS*, 41–7.

——. "On the chronology of Galileo's writings." IMSS, *Annali*, 9:2 (1984), 85–8.

Wlassics, Tibor. *Galilei critico letterario*. Ravenna: Longo, 1974.

Wohlwill, Emil. "The discovery of the parabolic shape of the projectile trajectory [1899]." In Renn, *Galileo* (2001), 375–410.

Wootton, David. *Paolo Sarpi. Between Renaissance and Enlightenment*. Cambridge: Cambridge University Press, 1983.

Young, G.F. *The Medici*. 3rd edn, 2 vols. New York: Dutton, 1926.

Zeccaria, Vittorio. "Le accademie padane cinquecentesche e il Tasso." In Borsetto and Da Rif, *Formazione* (1997), 35–61.

Zangheri, Luigi. *Gli accademici del Disegno. Elenco alphabetico*. Florence: Olschki, 2000.

Zik, Yaakov. "Galileo and the telescope." *Nuncius*, 14:1 (1999), 31–67.

——. "Galileo and optical aberrations." *Nuncius*, 17:2 (2002), 455–65.

Zingarelli, Nicola. "Introduzione." In Benivieni, *Dialogo* (1897), 7–30.

INDEX

For abbreviations, see headnote to the Notes. In addition, CT signifies "Copernican theory" or "Copernican system," and an asterisk distinguishes a name that appears in the Glossary.

Voltaire 17
voluntarism 47, 222, 221–3, 321, 428n90
  and SEND 248, 321; *see also*
    hypothesis

Wallace, William 397n30
*Wallenstein, Albrecht von 90, 326
water 181–2
*Welser, Mark 170, 177, 183, 185, 186,
  190, 201

and GG 172, 182, 191–2
Wilde, Oscar 21

*Xavier, Francis, Saint 227, 234
*Ximenes, Ferdinando 206, 214

*Zarlino, Gioseffè 2, 9–10
*Zieckmesser, Jan Eutel 103
Zollern *see* Eitel
*Zuccari, Federico 3, 568